D1761547

GAUGE THEORY OF THERMODYNAMICS

RICHARD A. WEISS

U. S. Army Engineer Waterways Experiment Station

Vicksburg, Mississippi 39180

K & W PUBLICATIONS

Vicksburg, Mississippi

1989

TO BETH

Published by K & W Publications
A subsidiary of K & W Research Company
212 Buena Vista Drive, Vicksburg, Mississippi 39180

Library of Congress Catalog Card Number: 89-63492

ISBN 0-9624789-0-3

Printed in the United States of America

PREFACE

This book is a collection of thirteen revised and corrected papers that were originally presented at a series of U. S. Army Conferences on Applied Mathematics and Computing held during the years 1985 through 1989. The papers represent several years of research in the areas of gauge theory and broken symmetry and their application to thermodynamics, condensed matter physics, real gas theory, electromagnetism and gravity. Many books treat these subjects individually but not collectively in a unified fashion. Therefore I decided to combine the papers on these subjects together with an Introduction into a book which will give the reader a sense of the gestalt that comes from the central threads of gauge theory and broken symmetry that bind these subjects together.

I thank Elizabeth K. Klein for typing and editing the individual papers and this book. Neither could have been produced without her kind help.

Richard A. Weiss

Vicksburg, Mississippi
August, 1989

CONTENTS

4. LAGRANGIAN FORMULATION OF RELATIVISTIC THERMODYNAMICS

5. THERMODYNAMIC GAUGE THEORY OF SOLIDS AND QUANTUM LIQUIDS WITH INTERNAL PHASE

6. RELATIVISTIC THERMODYNAMICS OF REAL GASES WITH BROKEN INTERNAL SYMMETRY

7. MAXWELL'S EQUATIONS WITH BROKEN INTERNAL SYMMETRIES

8. GAUGE THEORY OF ATOMIC PROCESSES

9. THE BROKEN SYMMETRY OF SPACE AND TIME IN BULK MATTER AND THE VACUUM

10. NEWTONIAN GRAVITY IN MATTER WITH BROKEN INTERNAL SYMMETRY

11. THE INTERNAL PHASE STRUCTURE OF ATOMS

12. WAVE PROPAGATION IN ASYMMETRIC MEDIA

13. ULTRAFAST THERMODYNAMIC PROCESSES

INTRODUCTION

Gauge theories have unified the forces of nature because it is thought that gravitation, electromagnetism and the weak and strong nuclear forces are gauge fields.[1] The electroweak interactions are described by the gauge group $SU(2)_L \times U(1)_Y$ where $SU(2)_L$ is the local symmetry group of 2 x 2 matrices for weak isospin. The gauge group of quantum chromodynamics is $SU(3)_C \times SU(2)_L \times U(1)_Y$ where $SU(3)_C$ is the local group of 3 x 3 matrices for colored quarks. Even higher dimensional local gauge groups exist that allow gravity to be included into a common unified description of the four forces of nature.[2] There is an enormous power in the gauge principle that dictates the form of the interaction by requiring local phase invariance of a wave function that describes elementary particles.[1,2] Each gauge group requires the introduction of a new field (the gauge field) in order for the theory to be gauge invariant. For instance, for electromagnetism the gauge field is the magnetic vector potential.

Are there similar unifying gauge groups for macroscopic physics? For example, can thermodynamics be unified with mechanics by means of a gauge theory? Another way of asking these questions is to inquire whether the zero temperature ground state of a thermodynamic system, such as a $T = 0$ solid or Fermi liquid, can be described completely in terms of the zero temperature values of the pressure and internal energy without any reference to the concept of $T \neq 0$ states, or whether the $T = 0$ ground state internal energy must be calculated in conjunction with other quantities which require the existence of $T \neq 0$. In order to answer this question consider the standard way of calculating ground state energy. The conventional way of calculating the $T = 0$ ground states of nuclear matter and liquid helium is to treat them as quantum mechanical systems whose internal energy is given by the following standard self consistent field expression[3]

$$U_o^a = N_o^a + 1/2 \sum_{ij} < ij|V_o^a|ij,ji > \qquad (1)$$

where U_o^a = conventional internal energy for a $T = 0$ system, N_o^a = conventional kinetic energy of a $T = 0$ system, V_o^a = conventional two-particle interaction potential and where the sum is over all momentum and spin states. The exchange terms in equation (1) appear only for fermion systems. Equation (1) and its modern sophisticated variations which include many-body effects is a standard way of calculating the ground state energy of a thermodynamic system.[4] There is no indication from equation (1) that requires a $T \neq 0$ state to exist. Is equation (1) correct or must additional quantities (gauge parameters) be introduced which indicate that the mechanical equation (1) describes the $T = 0$ ground state of a more general $T \neq 0$ thermodynamic system?

An affirmative answer to this question was given several years ago when it was suggested that two functions, the Grüneisen function γ and a new thermodynamic function (the Ω function of reference 5) must enter the calculation of the internal energy for the general case of a $T \neq 0$ thermodynamic system, and that for a $T = 0$ system the zero-temperature value of the Grüneisen parameter γ_o enters the ground state energy calculation.[5] Thus γ and Ω are the gauge parameters for a $T \neq 0$ system and γ_o is the gauge parameter for a $T = 0$ system. The gauge parameters γ and Ω enter through the derivative terms that appear in a relativistic trace equation[5]

$$U + T(dU/dT)_{PV} - 3Vd/dV(PV)_U = U^a + T(dU^a/dT)_{P^aV} \tag{2}$$

where U and P = renormalized internal energy and pressure respectively and U^a and P^a = unrenormalized internal energy and pressure respectively. Equation (2) includes special relativity by incorporating the Minkowski metric (1,1,1,-1). The zero-temperature form of equation (2) is[5]

$$E_o - 3[(1 + \gamma_o)P_o - K_o] = E_o^a \tag{3}$$

where E_o , γ_o , P_o and K_o = zero-temperature values of the renormalized energy density, Grüneisen parameter, pressure and bulk modulus respectively, and $E_o^a = U_o^a/V$ = unrenormalized energy density. Thus γ_o appears as a new gauge parameter in equation (3) for calculating E_o . A low temperature power series expansion of equation (2) yields two simultaneous equations for E_o and γ_o for solids and quantum liquids.[5] These simultaneous equations yield a relativistic state equation for Fermi or Bose gases (such as may be found in neutron star matter) which is much softer at high densities than is predicted by conventional calculations.[5] Recent data from pulsars confirms the necessity of this soft equation of state.[6,7] For the real gases the solution of equation (2) is obtained from a virial expansion of the pressure and is given in the form of a relativistic value of the third virial coefficient that includes spacetime vacuum effects.[5] A perturbation calculation applied to equation (2) gives the relativistic equations for waves in a thermodynamic system.[5,8,9] Equation (2) also has vacuum solutions.[5]

The derivative terms in equation (2) are needed in order to make the component terms of the left hand side of equation (2) locally gauge invariant.[8] In this way the Grüneisen function γ and the Ω function (of reference 5) or the b function (of reference 8) assume the roles of gauge parameters needed to form covariant derivatives. The assumption of gauge invariance yields an expression for the pressure dependence of the Grüneisen parameter.[8,9] The gauge group of relativistic thermodynamics is U(1) and represents the invariance of the component parts of the trace equation under local scale transformations of the form $\exp[\pm\phi(P,T)]$ and under local phase rotations of the form $\exp[\pm j\phi(P,T)]$. These transformations are local in the sense that $\phi = \phi(P,T)$ is dependent on the state of the thermodynamic system. A set of scalar renormalization group equations for the ground state and for excited states of a thermodynamic system follows from equation (2).[8,9] The excited states of real gases can be described by a perturbation analysis applied to the trace equation (2) combined with the virial equation of state, and applications to gaseous gravitational wave detectors have been considered.[10] A Lagrangian formulation of relativistic thermnodynamics can be developed and applied to fractal solids, quantum liquids and real gases. In this way the natural void ratios of real gases, liquids and solids can be calculated.[11]

The fact that U(1) is the gauge group for the component parts of the relativistic trace equation (2) suggests that the entire trace equation can describe a broken symmetry of relativistic thermodynamics, and that the trace equation should in fact be written as[12]

$$\bar{U} + T(d\bar{U}/dT)_{\bar{P}V} - 3Vd/dV(\bar{P}V)_{\bar{U}} = U^a + T(dU^a/dT)_{P^aV} \qquad (4)$$

where $\bar{U}$ and $\bar{P}$ = broken symmetry complex number values of the internal energy and pressure respectively. The internal energy and pressure are associated with internal phase angles and are skewed in internal space. This represents a fundamental broken symmetry of the state functions of relativistic thermodynamics. A set of broken symmetry complex number renormalization group equations for both the ground state and excited states of a thermodynamic system can be derived from equation (4), and represent a set of differential equations for the magnitudes and internal phase angles of the pressure and Grüneisen parameter.[12] Equation (4) can be applied to the state equations of solids, Fermi liquids and the real gases.[12,13] For example, the internal phase angles of the pressure, internal energy and entropy can be calculated for the real gases.[13] These phase angles are important for the calculation of the equilibrium configuration of stars. Thermodynamic processes are expected to involve changes in both the magnitudes and internal phase angles of the state functions, but processes involving only changes of the internal phase angles of the internal energy and entropy are possible for matter with broken internal symmetries.[12]

The broken symmetry of the pressure field in matter and the vacuum suggests that space and time coordinates have broken internal symmetries. This follows from Euler's equations of fluid dynamics in a broken symmetry pressure field.[14-20] The broken symmetry of the space and time coordinates affects the description of electromagnetism, atomic processes, atomic structure, dynamics of particles, gravitation and gravity, wave propagation and thermodynamic processes.[14-20] Because the basic kinematical and dynamical parameters of particles have broken internal symmetries the microscopic laws of physics must be written as complex number equations.[14-20] The internal phase angles of the coordinates over large distances are determined mainly by gravity.

Electromagnetism is affected by the broken symmetry of space and time because Maxwell's equations involve the space and time derivatives of the electric and magnetic field vectors, and both coordinates and field vectors are represented by complex numbers having internal phase angles.[14] The Lorentz transformations must also be written as complex number equations with the relative speed of two inertial frames being written as a complex number.[14] The light speed in vacuum, however, is a real number because the photon gas in vacuum is invariant under the action of the trace equation (4), and the internal energy and pressure of a photon gas in vacuum are represented by real numbers. The internal phase angle of the velocity of a body such as earth in its orbit about the sun cannot be obtained from an analysis of the Michelson and Morley experiment.[14] Because over long distances it is gravity that determines the internal phase angles of the space coordinates it follows that electromagnetic effects in matter are affected by the earth's gravity field.

The description of atomic processes such as scattering that occur in broken symmetry spacetime must also include the internal phase angles of space and time coordinates.[15] For example, the microscopic formalism of the laws of conservation of momentum and energy must be written as complex number equations in internal space in order to include the effects of the internal phase angles of geometrical angle, velocity, momentum, energy and frequency.[15] These internal phase angles must be included in the description of such basic atomic processes as the photoelectric effect, Compton scattering, black body radiation in matter, and the various nuclear scattering processes such as Rutherford scattering. Therefore these basic processes must be affected by the earth's gravitational field because over long distances the internal phase angle of the coordinates θ_r , θ_ϕ and θ_ψ are determined mainly by gravity. For instance, Planck's law for the asymmetric photon gas in matter must be written in terms of a broken symmetry complex number frequency whose internal phase angle $\theta_\nu = - \theta_t$.[15] The internal phase angles θ_r , θ_ν and θ_t vary with location within or on the surface of the earth.

Because the coordinates and momenta of particles in bulk matter have internal phase angles it follows that the basic equations of quantum mechanics - the Schrödinger and Dirac equations - must also have imaginary components in internal space, and that the wave function must also have a broken symmetry.[15] It then follows naturally that the magnetic, azimuthal and radial quantum numbers must be represented by complex numbers in internal space, and the energy eigenvalues, eigenfunctions and atomic radii must have an internal phase structure that should be observable in the spectra of one-electron atoms.[16] In particular, the energy eigenvalues of hydrogen will have pressure dependent broken azimuthal symmetry terms that depend on θ_ϕ the internal phase angle of the azimuthal angle. The spectra of hydrogen will therefore be slightly affected by the earth's gravity field and its variations.

The theory of the gravity field of the stars and planets is a fertile area for exploring the effects of the broken symmetry of the coordinates and the thermodynamic functions.[17-18] The gravitational equilibrium of stars and planets residing in space and time with broken internal symmetries can be described by a set of coupled differential equations that determines the internal phase angle of the radial coordinate and the magnitude and internal phase angle of the pressure as functions of the magnitude of the radial coordinate.[17] The broken symmetry of space also enters the gravitational equilibrium equations through a broken symmetry expression for the local mass density.[17-18] The broken symmetry of the radial coordinate as described by the internal phase angle $\theta_r(r,\phi,\psi)$ implies that the value of the gravitational constant G_r depends on the location of the point of measurement and in particular depends on the radial distance from the center of a star or planet.[17-18] In addition, the broken symmetries of the azimuthal angle [described by $\theta_\phi(r,\phi,\psi)$] and the zenith angle [described by $\theta_\psi(r,\phi,\psi)$] suggest that the gravity field of the earth must also include an azimuthal value G_ϕ of the gravitational constant and a zenith value G_ψ of the gravitational constant.[18] The different values of G_r, G_ϕ and G_ψ explain the measured differences of the gravitational constant in mine shafts, boreholes and on towers, and their differences with the results obtained from Eötvös experiments.[18] Therefore ordinary Newtonian gravitation combined with the broken symmetry of space and time is sufficient to acount for the apparent non-Newtonian effects in the earth's gravity field.[18] The values of $\theta_r(r,\phi,\psi)$ can be determined for the earth by performing the Pound-Rebka-Snider photon gravitational red shift experiment at various depths and locations within the earth and on its surface and using the following simple expression[18]

$$\sin^2 \theta_r \sim \frac{z_c - z_m}{z_c} \tag{5}$$

where z_c and z_m = conventionally predicted and measured values respectively of the gravitational red shift of the photon. Thus the Pound-Rebka-Snider experiment can be used to measure the spatial dependence of the apparent non-Newtonian component of the earth's gravity.[18]

Waves and vibrations should also be a good testing ground for the observable effects of the broken symmetry of space and time. The fact that waves and vibrations are observed to have the possibility of being periodic in measured space and time suggests that the phase of the waves must be a real number in internal space.[19] This means that the measured frequency and measured period are not simply given by a reciprocal relation but are connected through the internal phase angle of the frequency.[19] A similar relationship between the measured wave number and the measured spatial wavelength includes terms involving the internal phase angle of the wave number.[19] In general the solution to vibration and wave propagation problems will include the effects of the broken symmetry of coordinates and propagation constants. Therefore wave propagation and vibrations are influenced by gravity through the induced spatial dependence of the internal phase angles of the coordinates and propagation constants.

The fact that thermodynamic functions such as pressure and internal energy have broken symmetries that are described by internal phase angles allows the possibility of thermodynamic preocesses that include rotations in internal space as well as changes in the magnitudes of the thermodynamic functions.[12,20] In fact it is possible to imagine ultrafast thermodynamic processes which involve only rotations of the thermodynamic functions in internal space.[12,20] These processes may have practical applications to the rapid processes that occur in supernova processes, chemical and nuclear explosions and within exotic thermodynamic engines.

The broken symmetry of space implies a further change in the form of the basic trace equation.[20] This is due to the broken symmetry form of the volume element that appears in the first law of thermodynamics and implies that the pressure associated with the volume magnitude V is given by $\bar{P} \sec \beta_{V,V}$ where $\beta_{V,V}$ is related to the internal phase angle of the volume as follows[20]

$$\tan \beta_{V,V} = V \partial \theta_V / \partial V \tag{6}$$

The trace equation (4) becomes[20]

$$\bar{U} + T(d\bar{U}/dT)_{\bar{P}V \sec \beta_{V,V}} - 3V d/dV(\bar{P}V \sec \beta_{V,V})_{\bar{U}} = U^a + T(dU^a/dT)_{P^aV} \tag{7}$$

For the case when θ_V and $\beta_{V,V}$ are independent of temperature it follows that in all thermodynamic relations the substitution $\bar{P} \rightarrow \bar{P}$ sec $\beta_{V,V}$ gives the corresponding relativistic thermodynamic equations for broken symmetry space. For example, the Gibbs-Helmholtz-Maxwell equation for broken symmetry space becomes[20]

$$\partial \bar{U}/\partial V = T\partial/\partial T(\bar{P} \sec \beta_{V,V}) - \bar{P} \sec \beta_{V,V} \tag{8}$$

or

$$\cos \beta_{V,V}\, \partial \bar{U}/\partial V = T\partial \bar{P}/\partial T - \bar{P} \tag{9}$$

because $\beta_{V,V}$ is assumed to be independent of T . Because of the presence of the internal phase angle of the volume (density), thermodynamic processes and equilibrium states are influenced by the earth's gravity field and the pressure in which the thermodynamic system is situated. The internal phase angle θ_V varies with the coordinates r, ϕ and ψ that locate the thermodynamic system within the earth's gravity field, and accordingly the solution of equation (9) is influenced by the coordinate location of the thermodynamic system. Therefore the same broken symmetry of space and time which produces apparent non-Newtonian gravity effects in the earth will also induce anomalous effects in thermodynamic systems.

REFERENCES

1. Dyson, F. J., editor, Symmetry Groups in Nuclear and Particle Physics, Benjamin, New York, 1966.

2. Zee, A., Unity of Forces in the Universe, Vols. 1 and 2, World Scientific, New York, 1982.

3. Bethe, H. A. and Jackiw, R., Intermediate Quantum Mechanics, Benjamin, New York, 1968.

4. Breuckner, K. A., "Properties of Liquid Helium-Three", article in Many-Body Theory, edited Kubo, R., Benjamin, New York, 1966.

5. Weiss, R. A., Relativistic Thermodynamics, Vols. 1 and 2, Exposition Press, New York, 1976.

6. Prakash, M., Ainsworth, T. L. and Lattimer, J. M., "Equation of State and the Maximum Mass of Neutron Stars", Phys. Rev. Lett., 61, 2518, 1988.

7. ter Haar, B. and Malfliet, R., "Equation of State of Dense Asymmetric Nuclear Matter", Phys. Rev. Lett., 59, 1652, 1987.

8. Weiss, R. A., "Relativistic Wave Equations for Solids and Low Temperature Quantum Systems", Third Army Conference on Applied Mathematics and Computing, Georgia Institute of Technology, ARO 86-1, May 13-16, 1985, p. 717.

9. Weiss, R. A., "Scale Invariant Equations for Relativistic Waves", Fourth Army Conference on Applied Mathematics and Computing, Cornell University, ARO 87-1, May 27-30, 1986, p. 307.

10. Weiss, R. A., "Relativistic Wave Equations for Real Gases", Fourth Army Conference on Applied Mathematics and Computing, Cornell University, ARO 87-1, May 27-30, 1986, p. 342.

11. Weiss, R. A., "Lagrangian Formulation of Relativistic Thermodynamics", Fifth Army Conference on Applied Mathematics and Computing, West Point, ARO 88-1, June 15-18, 1987, p. 697.

12. Weiss, R. A., "Thermodynamic Gauge Theory of Solids and Quantum Liquids with Internal Phase", Fifth Army Conference on Applied Mathematics and Computing, West Point, ARO 88-1, June 15-18, 1987, p. 649.

13. Weiss, R. A., "Relativistic Thermodynamics of Real Gases with Broken Internal Symmetry", Sixth Army Conference on Applied Mathematics and Computing, Boulder, ARO 89-1, 31 May - 3 June 1988, p. 203.

14. Weiss, R. A., "Maxwell's Equations with Broken Internal Symmetries", Sixth Army Conference on Applied Mathematics and Computing, Boulder, ARO 89-1, 31 May - 3 June 1988, p. 271.

15. Weiss, R. A., "Gauge Theory of Atomic Processes", Sixth Army Conference on Applied Mathematics and Computing, Boulder, ARO 89-1, 31 May - 3 June 1988, p. 223.

16. Weiss, R. A., "The Internal Phase Structure of Atoms", Seventh Army Conference on Applied Mathematics and Computing, West Point, ARO 90-1, June 6-9, 1989, p. 609.

17. Weiss, R. A., "The Broken Symmetry of Space and Time in Bulk Matter and the Vacuum", Sixth Army Conference on Applied Mathematics and Computing, Boulder, ARO 89-1, 31 May - 3 June, 1988, p. 317.

18. Weiss, R. A., "Newtonian Gravity in Matter with Broken Internal Symmetry", Seventh Army Conference on Applied Mathematics and Computing, West Point, ARO 90-1, June 6-9, 1989, p. 637.

19. Weiss, R. A., "Wave Propagation in Asymmetric Media", Seventh Army Conference on Applied Mathematics and Computing, West Point, ARO 90-1, June 6-9, 1989, p. 691.

20. Weiss, R. A., "Ultrafast Thermodynamic Processes", Seventh Army Conference on Applied Mathematics and Computing, West Point, ARO 90-1, June 6-9, 1989, p. 599.

A

RELATIVISTIC WAVE EQUATIONS FOR SOLIDS AND LOW TEMPERATURE QUANTUM SYSTEMS

Richard A. Weiss
Environmental Systems Division
U. S. Army Engineer Waterways Experiment Station
Vicksburg, Mississippi 39180-0631

ABSTRACT. The local scale invariance of relativistic thermodynamics is established and a set of coupled relativistic wave equations is developed that describe the propagation of small amplitude waves in solids and low temperature interacting Fermi and Bose systems. A first order expansion calculation is done in order to obtain the wave equations from a basic set of material equations. The wave equations determine the relativistic energy density and Grüneisen parameter for small amplitude mechanical or electromagnetic waves. In turn, the wave amplitude and phase velocity are obtained from the energy density and Grüneisen parameter of the waves. At low pressures the wave amplitudes are determined to be not greatly different from those predicted by a nonrelativistic calculation, but at high pressures, such as those expected to occur under nuclear blast loading or in stellar compact objects, the calculated relativistic wave amplitudes can be considerably larger than the corresponding nonrelativistic predictions.

1. INTRODUCTION. Local gauge invariance has become a powerful tool in modern physics.[1] It has unified the electromagnetic force with the weak nuclear force, and possibly also with the strong nuclear force, into a single entity described by local non-Abelian gauge invariance of the fields.[2] The requirement of local gauge invariance necessitates the introduction of new fields which obey sets of coupled differential equations which lead to unification of the fields in a natural way. The gauge transformations correspond to generalized rotations of the fields in the manner $e^{-i\phi(x)}$ where $\phi(x)$ depends on space and time for local gauge invariance.

A theory of relativistic thermodynamics has been developed for solids and quantum liquids which is based on a set of coupled differential equations for the zero temperature values of the Grüneisen parameter and pressure.[3] These coupled equations are derived from a relativistic trace equation that has been gauged (scaled) by the introduction of the Grüneisen parameter. The fact that an additional field (the Grüneisen parameter) has to be considered in order to calculate the relativistic pressure suggests an invariance of the system similar to local gauge invariance. Electromagnetism and the electroweak interactions are examples of fields that are described by local gauge theories in which the gauging is accomplished by the introduction of additional fields. The invariance of relativistic thermodynamics refers to different values of pressure and energy density; and, in analogy to a local gauge transformation, is manifested through a real exponential $e^{-\phi(V,T)}$ corresponding to changes of the local scale of the pressure and energy density of a system.[4] This paper shows that relativistic thermodynamics is scale invariant, and that the local symmetry group for relativistic thermodynamics is the unimodular group of scale transformations $e^{-\phi(V,T)}$.

The trace equation of relativistic thermodynamics is written as[3]

$$U + T\left(\frac{dU}{dT}\right)_{PV} - 3V\frac{d}{dV}(PV)_U = U^a + T\left(\frac{dU^a}{dT}\right)_{P^aV} \tag{1}$$

where U = relativistic internal energy, P = relativistic pressure, T = absolute temperature, V = volume per mole of substance, and U^a and P^a = corresponding nonrelativistic internal energy and pressure. Throughout this paper the index "a" will refer to nonrelativistic calculations. It has been shown that for a physical system described by $U = U_o + U_jT^j + \cdots$ and $P = P_o + P_jT^j + \cdots$ (such as the high temperature Mie-Gruneisen state equation with j = 1 and the Debye state equation with j = 4) the trace equation (1) is equivalent to the following set of coupled differential equations[3]

$$3V^2\frac{d^2P_o}{dV^2} + 3(3+\gamma_o)V\frac{dP_o}{dV} + [3(\gamma_o + V\frac{d\gamma_o}{dV}) + 4]P_o = P_o^a \tag{2}$$

$$U_j\left(1 + j + \frac{j\gamma_oP_o}{P_o - K_o} - 3V\frac{d\gamma_o}{dV}\right) = U_j^a\left(1 + j + \frac{j\gamma_o^aP_o^a}{P_o^a - K_o^a}\right) \tag{3}$$

where the internal energy coefficients are given by[5]

$$\frac{U_j^a}{U_j} = \exp\left[(j-1)\int^V (\gamma_o^a - \gamma_o)\frac{dV}{V}\right] \tag{4}$$

and where P_o, K_o and γ_o = zero temperature values of the relativistic pressure, incompressibility ($-VdP_o/dV$) and Grüneisen parameter respectively, and P_o^a, K_o^a and γ_o^a are the corresponding nonrelativistic values of these quantities. Eqs. (2) and (3) are a set of coupled nonlinear differential equations for P_o and γ_o. The zero temperature value of the Grüneisen parameter is the T = 0 limit of the Grüneisen parameter defined as[5]

$$\gamma = \frac{V}{C_V}\left(\frac{\partial P}{\partial T}\right)_V \tag{5}$$

which for a temperature dependent Debye temperature becomes,[6,7]

$$\gamma = \frac{-\dfrac{V}{\theta_D}\left(\dfrac{\partial\theta_D}{\partial V}\right)_T}{\left[1 - \dfrac{T}{\theta_D}\left(\dfrac{\partial\theta_D}{\partial T}\right)_V\right]} \tag{6}$$

where C_V = heat capacity and θ_D = Debye temperature. The Debye temperature is generally a function of V and T. Equation (6) gives the general nonrelativistic expression for the Grüneisen parameter.

Wave propagation in matter is complicated by the fact that matter is a thermodynamic system which is described by physical quantities such as the heat capacity and Grüneisen parameter.[5,8] Further complications arise because matter is often prestressed as for instance by gravitation in a star or planet.[9] Various assumptions are used to calculate the phase velocity and wave amplitude in terms of the thermodynamic state equation of a material medium, but no completely general procedure exists for calculating these quantities. This paper develops a relativistic calculation of the phase velocity and amplitude for mechanical waves in thermodynamic media.

Specifically, this paper develops a set of coupled first order relativistic equations that govern small amplitude wave propagation in thermodynamic systems such as solids and low temperature Fermi and Bose liquids. These equations determine the relativistic energy density and Grüneisen parameter for mechanical or electromagnetic waves. The phase velocity and wave amplitude can be obtained from the energy density and Grüneisen parameter for the waves, and are expressed in terms of the material parameters of the thermodynamic ground state of the material system.

The wave equations are developed from a small amplitude perturbation expansion of a set of relativistic material equations. This insures that the derived wave equations are intimately connected to the material parameters of the thermodynamic medium. A complicated set of nonlinear wave equations are derived whose complete solution requires numerical computer techniques, however, an order of magnitude approximate solution is given for the case of elastic waves in solids. The procedure followed in this paper is to first review local scale invariance, then the relativistic ground state calculation, and then the relativistic wave equations that govern excitations in a thermal medium.

2. LOCAL SCALE INVARIANCE. The fact that an additional field $\gamma_0(V)$ must be introduced in order to calculate the zero temperature pressure $P_0(V)$ as in Eqs. (2) and (3) suggests that the relativistic trace equation (1) may be invariant under a local scale transformation of the form $\exp[-\phi(V,T)]$. The scale invariance of Eq. (1) will now be demonstrated. To do this the following elementary thermodynamic relationships are used[3]

$$\left(\frac{dU}{dT}\right)_{PV} = \left(\frac{\partial U}{\partial T}\right)_V - \frac{V}{(P - K_T)}\left(\frac{\partial U}{\partial V}\right)_T\left(\frac{\partial P}{\partial T}\right)_V \tag{7}$$

$$\frac{d}{dV}(PV)_U = P - K_T - \gamma[T\left(\frac{\partial P}{\partial T}\right)_V - P] \tag{8}$$

where $K_T = -V(\partial P/\partial V)_T$ = isothermal incompressibility. Using Eqs. (7) and (8) allows Eq. (1) to be written as

$$\left(1 + T\frac{\partial}{\partial T} - bV\frac{\partial}{\partial V}\right)U - 3V\left(1 + \gamma + V\frac{\partial}{\partial V} - \gamma T\frac{\partial}{\partial T}\right)P \tag{9}$$

$$= \left(1 + T\frac{\partial}{\partial T} - b^a V\frac{\partial}{\partial V}\right)U^a$$

where

$$b = \frac{T(\partial P/\partial T)_V}{(P - K_T)} \tag{10}$$

$$b^a = \frac{T(\partial P^a/\partial T)_V}{(P^a - K_T^a)} \tag{11}$$

Note that the Grüneisen parameter γ defined in equation (5) is not independent of the quantity b defined by equation (10). Eq. (9) can be rewritten in a more symmetrical form by writing $U = \mathcal{E}V$, where $\mathcal{E}$ = energy density, with the result

$$\left(1 - b + T\frac{\partial}{\partial T} - bV\frac{\partial}{\partial V}\right)\mathcal{E} - 3\left(1 + \gamma + V\frac{\partial}{\partial V} - \gamma T\frac{\partial}{\partial T}\right)P \tag{12}$$

$$= \left(1 - b^a + T\frac{\partial}{\partial T} - b^a V\frac{\partial}{\partial V}\right)\mathcal{E}^a$$

Eq. (12) can be written as

$$(H_1 - W_1)\mathcal{E} - 3(H_2 - W_2)P = (H_1^a - W_1^a)\mathcal{E}^a \tag{13}$$

where

$$H_1 = T\frac{\partial}{\partial T} - bV\frac{\partial}{\partial V} \tag{14}$$

$$H_2 = V\frac{\partial}{\partial V} - \gamma T\frac{\partial}{\partial T} \tag{15}$$

$$W_1 = b - 1 \tag{16}$$

$$W_2 = -\gamma - 1 \tag{17}$$

Eq. (13) is in a form suitable for demonstrating local scale invariance.

Introduce the following scale transformations $E \to E' = Ee^{-\psi}$ and $P \to P' = Pe^{-\phi}$ where ψ and ϕ are functions of V and T. Also for the nonrelativistic energy density introduce the scale transformation $E^a \to E^{a'} = E^a e^{-\psi^a}$ where ψ^a is also a function of V and T. Under these scale transformations γ and b assume new values $\gamma \to \gamma'$ and $b \to b'$, and equations (14) through (17) become

$$H_1' = T\frac{\partial}{\partial T} - b'V\frac{\partial}{\partial V} \tag{18}$$

$$H_2' = V\frac{\partial}{\partial V} - \gamma' T\frac{\partial}{\partial T} \tag{19}$$

$$W_1' = b' - 1 \tag{20}$$

$$W_2' = -\gamma' - 1 \tag{21}$$

where b' and γ' are to be determined by the local scale invariance conditions. The local scale invariant conditions for the operators in equation (13) are written in a manner similar to local gauge invariance as[2]

$$(H_1' - W_1')Ee^{-\psi} = e^{-\psi}(H_1 - W_1)E \tag{22}$$

$$(H_2' - W_2')Pe^{-\phi} = e^{-\phi}(H_2 - W_2)P \tag{23}$$

$$(H_1^{a'} - W_1^{a'})E^a\, e^{-\psi^a} = e^{-\psi^a}(H_1^a - W_1^a)E^a \tag{24}$$

Equations (22) through (24) allows the trace equation (13) to be written as

$$e^{\psi}(H_1' - W_1')e^{-\psi}E - 3e^{\phi}(H_2' - W_2')e^{-\phi}P = e^{\psi^a}(H_1^{a'} - W_1^{a'})e^{-\psi^a}E^a \tag{25}$$

so that the following operator equations hold

$$H_1 - W_1 = e^{\psi}(H_1' - W_1')e^{-\psi} \tag{26}$$

$$H_2 - W_2 = e^{\phi}(H_2' - W_2')e^{-\phi} \tag{27}$$

The values of b' and γ' are obtained as follows. Placing equations (14) through (21) into the scale invariant conditions given by equations (22) and (23) yields the following results

$$b' = b + \frac{E\left(bV \frac{\partial\psi}{\partial V} - T \frac{\partial\psi}{\partial T}\right)}{E + V \frac{\partial E}{\partial V} - EV \frac{\partial\psi}{\partial V}} \tag{28}$$

$$\gamma' = \gamma + \frac{P\left(V \frac{\partial\phi}{\partial V} - \gamma T \frac{\partial\phi}{\partial T}\right)}{P - T \frac{\partial P}{\partial T} + PT \frac{\partial\phi}{\partial T}} \tag{29}$$

Now since b' and γ' are associated E' and P' respectively, while b and γ are associated with E and P respectively, one can write

$$b' = b + \frac{db}{dE}(E' - E) + \cdots \tag{30}$$

$$\gamma' = \gamma + \frac{d\gamma}{dP}(P' - P) + \cdots \tag{31}$$

while from $E' = Ee^{-\psi}$ and $P' = Pe^{-\phi}$ it follows that

$$E' - E = E(e^{-\psi} - 1) \tag{32}$$

$$P' - P = P(e^{-\phi} - 1) \tag{33}$$

so that combining equations (28) through (33) yields the following differential equations for ψ and ϕ,

$$\frac{db}{dE} = \frac{e^{\psi}}{e^{\psi} - 1}\left(\frac{T \frac{\partial\psi}{\partial T} - bV \frac{\partial\psi}{\partial V}}{E + V \frac{\partial E}{\partial V} - EV \frac{\partial\psi}{\partial V}}\right) \tag{34}$$

$$\frac{d\gamma}{dP} = \frac{e^{\phi}}{e^{\phi} - 1}\left(\frac{\gamma T \frac{\partial \phi}{\partial T} - V \frac{\partial \phi}{\partial V}}{P - T \frac{\partial P}{\partial T} + PT \frac{\partial \phi}{\partial T}}\right) \tag{35}$$

Therefore it is always possible to find a ψ and ϕ such that the trace equation (13) is locally scale invariant.

Consider now the infinitesimal local scale transformation where ψ and ϕ are small quantities, then equations (32) and (33) become

$$E' - E = -\psi E \tag{36}$$

$$P' - P = -\phi P \tag{37}$$

and equations (34) and (35) become

$$\frac{db}{dE} = \frac{\frac{T}{\psi}\frac{\partial \psi}{\partial T} - b\frac{V}{\psi}\frac{\partial \psi}{\partial V}}{E + V\frac{\partial E}{\partial V} - EV\frac{\partial \psi}{\partial V}} \tag{38}$$

$$\frac{d\gamma}{dP} = \frac{\gamma\frac{T}{\phi}\frac{\partial \phi}{\partial T} - \frac{V}{\phi}\frac{\partial \phi}{\partial V}}{P - T\frac{\partial P}{\partial T} + PT\frac{\partial \phi}{\partial T}} \tag{39}$$

Therefore it is always possible to find a ψ and ϕ such that the trace equation (13) is invariant under an infinitesimal local scale transformation. Note that for the case when b and γ are slowly varying functions, equations (38) and (39) yield

$$b = \frac{T\frac{\partial \psi}{\partial T}}{V\frac{\partial \psi}{\partial V}} \tag{40}$$

$$\gamma = \frac{V\frac{\partial \phi}{\partial V}}{T\frac{\partial \phi}{\partial T}} \tag{41}$$

It turns out that the solutions to equations (40) and (41) are very simple. Choose ψ and ϕ as follows

$$\psi = \frac{PV}{P_i V_i} \tag{42}$$

$$\phi = \frac{\theta_D}{T} \tag{43}$$

where $P_i V_i$ = an initial value of PV. Then from equations (40) and (41) it follows that

$$b = \frac{T \frac{\partial P}{\partial T}}{(P - K_T)} \tag{44}$$

$$\gamma = \frac{-\frac{V}{\theta_D}\frac{\partial \theta_D}{\partial V}}{1 - \frac{T}{\theta_D}\frac{\partial \theta_D}{\partial T}} \tag{45}$$

Equation (44) is just the basic definition of the quantity b given in equation (10), while equation (45) is just the standard result for the Grüneisen parameter given in equation (6). Therefore the values of ψ and ϕ given by equations (42) and (43) are the proper potential functions for the infinitesimal local scale transformation of relativistic thermodynamics for the case of slowly varying values of b and γ. The explicit determination of ψ and ϕ for the general case of local scale invariance given by equations (34) and (35) is much more complicated, and has not been accomplished.

The scale invariance of the trace equations (1) or (9) requires the presence of the parameters b and γ, although as mentioned earlier, b and γ are not independent. This means that for the zero-temperature case, the zero-temperature values of the pressure and the Grüneisen parameter must be determined simultaneously as shown in equations (2) and (3). The establishment of local scale invariance of the relativistic trace equation gives one confidence to use equations (2) and (3) to determine the equations governing the propagation of small amplitude waves in a thermodynamic medium. But first the ground state of the thermodynamic system must be described.

3. GROUND STATE. The nonrelativistic state equation of the ground state of a thermal system is assumed to have the following form[5]

$$E^a = E^a_o + E^a_j T^j + \cdots \tag{46}$$

$$P^a = P^a_o + P^a_j T^j + \cdots \tag{47}$$

where $\mathcal{E}^a$ and P^a = nonrelativistic energy density and pressure respectively, $\mathcal{E}_o^a$ and P_o^a = nonrelativistic zero-temperature values of the energy density and pressure respectively, $\mathcal{E}_j^a$ and P_j^a = nonrelativistic thermal coefficients for the energy density and pressure respectively, T = absolute temperature of the system (°K), and j = numerical index having values characteristic of the type of physical system. Typical examples of systems that are described by equations (46) and (47) are[5]

j = 1	high temperature solid
j = 2	low temperature Fermi gas
j = 5/2	low temperature molecular Bose gas
j = 4	low temperature solid

A commonly used descriptor of the thermal state equations given by equations (46) and (47) is the nonrelativistic zero-temperature value of the Grüneisen parameter that is defined by[5]

$$\gamma_o^a = \frac{P_j^a}{\mathcal{E}_j^a} = \frac{1}{(j-1)} \frac{1}{\mathcal{E}_j^a} \frac{d}{dV} (V\mathcal{E}_j^a) \tag{48}$$

except for j = 1 . Here γ_o^a = nonrelativistic zero-temperature value of the Grüneisen parameter, and V = volume of the material system. When j = 1 , $\gamma_o^a = 2/3$.

The corresponding relativistic state equations will be written as

$$\mathcal{E} = \mathcal{E}_o + \mathcal{E}_j T^j + \cdots \tag{49}$$

$$P = P_o + P_j T^j + \cdots \tag{50}$$

$$\gamma_o = \frac{P_j}{\mathcal{E}_j} = \frac{1}{(j-1)} \frac{1}{\mathcal{E}_j} \frac{d}{dV} (V\mathcal{E}_j) \tag{51}$$

except for j = 1 , when $\mathcal{E}_1 = \mathcal{E}_1^a$, and where $\mathcal{E}_o$ and P_o = relativistic zero-temperature energy density and pressure respectively, $\mathcal{E}_j$ and P_j = relativistic thermal coefficients for the energy density and pressure respectively, and γ_o = relativistic zero-temperature Grüneisen parameter.

The relativistic values of the zero-temperature energy density and Grüneisen parameter for the ground state thermodynamic system are given by the solution of the following two coupled equations[3]

$$\mathcal{E}_o - 3[(1 + \gamma_o)P_o - K_o] = \mathcal{E}_o^a \tag{52}$$

$$\mathcal{E}_j \left(1 + j + \frac{j\gamma_o P_o}{P_o - K_o} + 3n \frac{d\gamma_o}{dn}\right) = \mathcal{E}^a_j \left(1 + j + \frac{j\gamma^a_o P^a_o}{P^a_o - K^a_o}\right) \tag{53}$$

where $n = 1/V$ = reciprocal volume

$K_o = n \frac{dP_o}{dn}$ = relativistic zero-temperature bulk modulus

$K^a_o = n \frac{dP^a_o}{dn}$ = nonrelativistic zero-temperature bulk modulus

From equations (48) through (51) it follows that[3]

$$\mathcal{E}_j = nC_j \exp \left[- (j-1) \int^n \gamma_o \frac{dn}{n} \right] \tag{54}$$

$$\mathcal{E}^a_j = nC^a_j \exp \left[- (j-1) \int^n \gamma^a_o \frac{dn}{n} \right] \tag{55}$$

so that

$$\frac{\mathcal{E}_j}{\mathcal{E}^a_j} = \frac{C_j}{C^a_j} \exp \left[- (j-1) \int^n (\gamma_o - \gamma^a_o) \frac{dn}{n} \right] \tag{56}$$

When $\gamma_o = \gamma^a_o$, it must follow that $\mathcal{E}_j = \mathcal{E}^a_j$ so that quite generally $C_j = C^a_j$. Combining equation (56) with equation (53) shows that equations (52) and (53) are two nonlinear coupled equations for determining $\mathcal{E}_o$ and γ_o in terms of the known values of $\mathcal{E}^a_o$ and γ^a_o. These equations give the recipe for calculating the relativistic ground state of a thermal system in terms of the corresponding nonrelativistic description of the ground state. The calculation of the relativistic excited states will now be given.

4. EXCITATIONS. The excitations in thermal media that are considered in this paper are mechanical radiation and electromagnetic waves. Only waves of small amplitude are treated. The thermal state equations of the radiation are assumed to have a form similar to the ground state equations (46), (47), (49) and (50) and are written as

$$\mathcal{E}^a_r = \mathcal{E}^a_{or} + \mathcal{E}^a_{jr} T^j + \cdots \tag{57}$$

$$P^a_r = P^a_{or} + P^a_{jr} T^j + \cdots \tag{58}$$

and

$$E_r = E_{or} + E_{jr}T^j + \cdots \tag{59}$$

$$P_r = P_{or} + P_{jr}T^j + \cdots \tag{60}$$

where

E^a_{or} and P^a_{or} = nonrelativistic zero-temperature radiation energy density and pressure respectively

E^a_{jr} and P^a_{jr} = nonrelativistic thermal coefficients for the radiation energy density and pressure respectively

E_{or} and P_{or} = relativistic zero-temperature radiation energy density and pressure respectively

E_{jr} and P_{jr} = relativistic thermal coefficients for the radiation energy density and pressure respectively

It will be shown in this paper that the relativistic state equations for radiation given in equations (59) and (60) must have nonzero thermal components even when the corresponding nonrelativistic thermal components that appear in equations (57) and (58) are taken to be zero. Finally, the radiation terms are assumed to be much smaller than the ground state terms, i.e., $E_r \ll E$ and $P_r \ll P$.

When radiation is present in a system whose ground state is described by P_o , K_o , and γ_o , these three parameters become $P_o + P_{or}$, $K_o + K_{or}$, and $\gamma_o + \delta_{or}$, where K_{or} = relativistic zero-temperature incompressibility associated with the radiation, and δ_{or} = incremental change in the relativistic Grüneisen parameter of the system due to the presence of radiation. The bulk modulus (incompressibility) associated with the radiation is given by

$$K_{or} = n \frac{dP_{or}}{dn} \tag{61}$$

The increment in the zero-temperature Grüneisen parameter of the system due to the presence of small amplitude radiation is obtained from the defining equation (51) by noting that when radiation is present this equation becomes

$$\gamma_o + \delta_{or} = \frac{P_j + P_{jr}}{E_j + E_{jr}} = \frac{P_j + P_{jr}}{E_j\left(1 + \frac{E_{jr}}{E_j}\right)} \tag{62}$$

Expanding the denominator in equation (62), keeping only first order terms, and finally subtracting equation (51) gives

$$\delta_{or} = \frac{E_{jr}}{E_j}\left(\frac{P_{jr}}{E_{jr}} - \frac{P_j}{E_j}\right) \tag{63}$$

$$= \frac{E_{jr}}{E_j}(\gamma_{or} - \gamma_o)$$

where γ_{or} = zero-temperature Grüneisen parameter for the radiation field itself, and is defined by

$$\gamma_{or} = \frac{P_{jr}}{E_{jr}} \tag{64}$$

The corresponding expressions for the nonrelativistic bulk modulus and Grüneisen parameters associated with the radiation field in matter are

$$K^a_{or} = n\frac{dP^a_{or}}{dn} \tag{65}$$

$$\delta^a_{or} = \frac{E^a_{jr}}{E^a_j}\left(\gamma^a_{or} - \gamma^a_o\right) \tag{66}$$

$$\gamma^a_{or} = \frac{P^a_{jr}}{E^a_{jr}} \tag{67}$$

From equations (63) and (66) it follows that if E_{jr} and E^a_{jr} are small quantities then so also are δ_{or} and δ^a_{or}. However, the radiation Gruneisen parameters themselves, γ_{or} and γ^a_{or}, are generally not small quantities being the ratio of two small quantities, and at low pressures have the value 1/3 for isotropic radiation.[5] The quantities E_{or}, P_{or}, K_{or}, E_{jr}, P_{jr}, and δ_{or}, and their nonrelativistic analogs, are taken to be small quantities. Finally, from equation (64) and the law of energy conservation (see Appendix A), it follows that the ratio of thermal terms that occurs in equation (63) can be written as

$$\frac{\mathcal{E}_{jr}}{\mathcal{E}_j} = \frac{D_{jr}}{C_j} \exp\left[- (j-1) \int^{n} (\gamma_{or} - \gamma_o) \frac{dn}{n} \right] \tag{68}$$

where D_{jr} = constant associated with radiation field.

5. DERIVATION OF WAVE EQUATIONS. When radiation is present in the thermodynamic systems being considered, equation (52) becomes

$$\mathcal{E}_o + \mathcal{E}_{or} - 3[(1 + \gamma_o + \delta_{or})(P_o + P_{or}) - (K_o + K_{or})] = \mathcal{E}^a_o + \mathcal{E}^a_{or} \tag{69}$$

Similarly, when radiation is present, equation (53) becomes

$$(\mathcal{E}_j + \mathcal{E}_{jr})\left[1 + j + \frac{j(\gamma_o + \delta_{or})(P_o + P_{or})}{P_o + P_{or} - K_o - K_{or}} + 3n \frac{d}{dn}(\gamma_o + \delta_{or})\right] \tag{70}$$

$$= (\mathcal{E}^a_j + \mathcal{E}^a_{jr})\left[1 + j + \frac{j(\gamma^a_o + \delta^a_{or})(P^a_o + P^a_{or})}{P^a_o + P^a_{or} - K^a_o - K^a_{or}}\right]$$

Equations (69) and (70) are the relativistic equations for waves in matter, however they are too complex for simple solutions to be obtained. Simplification can be achieved by assuming the radiation components to be small quantities.

Considering only first order terms in equation (69), and subtracting equation (52), gives

$$\mathcal{E}_{or} - 3[(1 + \gamma_o)P_{or} - K_{or}] - 3P_o\delta_{or} = \mathcal{E}^a_{or} \tag{71}$$

In a similar fashion, by expanding the denominators in equation (70), using $1/(1 + x) \sim 1 - x$, and keeping only first order terms, and finally subtracting equation (53) yields the following result

$$jE_j(\alpha K_{or} - \beta P_{or}) + 3E_j n \frac{d\delta_{or}}{dn} + \frac{jE_j P_o \delta_{or}}{P_o - K_o} \tag{72}$$

$$+ E_{jr}\left(1 + j + \frac{j\gamma_o P_o}{P_o - K_o} + 3n \frac{d\gamma_o}{dn}\right)$$

$$= jE_j^a\left(\alpha^a K^a_{or} - \beta^a P^a_{or}\right) + \frac{jE^a_j P^a_o \delta^a_{or}}{P^a_o - K^a_o} + E^a_{jr}\left(1 + j + \frac{j\gamma^a_o P^a_o}{P^a_o - K^a_o}\right)$$

where

$$\alpha = \frac{\gamma_o P_o}{(P_o - K_o)^2} \qquad \alpha^a = \frac{\gamma^a_o P^a_o}{(P^a_o - K^a_o)^2} \tag{73}$$

$$\beta = \frac{\gamma_o K_o}{(P_o - K_o)^2} \qquad \beta^a = \frac{\gamma^a_o K^a_o}{(P^a_o - K^a_o)^2} \tag{74}$$

The radiation equations (71) and (72) will now be written in a much simpler form.

Equation (63) can be used for δ_{or} that occurs in equations (71) and (72), while the first derivative of δ_{or} that occurs in equation (72) is evaluated by using equations (63) and (68) and can be written as (see Appendix B)

$$n \frac{d\delta_{or}}{dn} = \frac{E_{jr}}{E_j}\left[n \frac{d\gamma_{or}}{dn} - n \frac{d\gamma_o}{dn} - (j-1)(\gamma_{or} - \gamma_o)^2\right] \tag{75}$$

Substituting the results of equations (63) and (75) into the two basic equations (71) and (72) yields

$$E_{or} - 3[(1 + \gamma_o)P_{or} - K_{or}] - 3 \frac{E_{jr}}{E_j} P_o(\gamma_{or} - \gamma_o) = E^a_{or} \tag{76}$$

and

$$jE_j(\alpha K_{or} - \beta P_{or}) + E_{jr}\left[1 + j + \frac{jP_o\gamma_{or}}{P_o - K_o} + 3n\,\frac{d\gamma_{or}}{dn} - 3(j-1)(\gamma_{or} - \gamma_o)^2\right] \quad (77)$$

$$= jE^a_j\left(\alpha^a K^a_{or} - \beta^a P^a_{or}\right) + E^a_{jr}\left(1 + j + \frac{jP^a_o\gamma^a_{or}}{P^a_o - K^a_o}\right) .$$

These two radiation equations can be further simplified by using some basic mathematical properties of the radiation field.

For radiation, the zero-temperature pressure is related to the zero-temperature energy density through the radiation Grüneisen parameter as follows (see Appendix A)

$$P_{or} = \gamma_{or}E_{or} \quad (78)$$

This proportionality of the pressure and energy density is characteristic of radiation fields. The bulk modulus of the radiation field can be written using equation (78) as

$$K_{or} = n\,\frac{dP_{or}}{dn} = \gamma_{or}\,n\,\frac{dE_{or}}{dn} + E_{or}\,n\,\frac{d\gamma_{or}}{dn} \quad (79)$$

Equations analogous to (78) and (79) hold for the nonrelativistic quantities P^a_{or} and K^a_{or} respectively. Placing equations (78) and (79) into equation (76) and (77) and then dividing equation (77) by E_j yields the two fundamental first order coupled nonlinear differential equations for E_{or} and γ_{or} ,

$$In\,\frac{dE_{or}}{dn} + JE_{or} + g = E^a_{or} \quad (80)$$

$$Gn\,\frac{dE_{or}}{dn} + RE_{or} + f = \psi^a_{or} \quad (81)$$

where $I = 3\gamma_{or}$ (82)

$$J = 3n \frac{d\gamma_{or}}{dn} - 3(1 + \gamma_o)\gamma_{or} + 1 \quad (83)$$

$$g = 3 \frac{E_{jr}}{E_j} P_o(\gamma_o - \gamma_{or}) \quad (84)$$

$$G = j\alpha\gamma_{or} \quad (85)$$

$$R = j\left(\alpha n \frac{d\gamma_{or}}{dn} - \beta\gamma_{or}\right) \quad (86)$$

$$f = \frac{E_{jr}}{E_j}\left[1 + j + \frac{jP_o\gamma_{or}}{P_o - K_o} + n\frac{d\gamma_{or}}{dn} - 3(j-1)(\gamma_{or} - \gamma_o)^2\right] \quad (87)$$

$$\psi^a_{or} = \frac{E^a_{jr}}{E_j}\left[1 + j + \frac{jP^a_o\gamma^a_{or}}{P^a_o - K^a_o}\right] + G^a n \frac{dE^a_{or}}{dn} + R^a E^a_{or} \quad (88)$$

$$G^a = j \frac{E^a_j}{E_j} \alpha^a \gamma^a_{or} \quad (89)$$

$$R^a = j \frac{E^a_j}{E_j}\left(\alpha^a n \frac{d\gamma^a_{or}}{dn} - \beta^a \gamma^a_{or}\right) \quad (90)$$

The ratio E_{jr}/E_j that occurs in equation (84) and (87) is a function of γ_{or} as shown in equation (68), while the ratio E^a_j/E_j that appears in equations (89) and (90) is given by equation (56). Equations (80) and (81) are a set of coupled relativistic wave equations that determine the relativistic energy density E_{or} and Grüneisen parameter γ_{or} for radiation in terms of the corresponding nonrelativistic radiation parameters E^a_{or} and γ^a_{or}, and in terms of the ground state material parameters P_o, K_o, γ_o, P^a_o, K^a_o, and γ^a_o.

6. MECHANICAL WAVES IN A SOLID. The temperature independent part of the nonrelativistic energy density for mechanical waves in a solid is given by[10]

$$E^a_{or} = \frac{1}{4} K^a_o k_a^2 A_a^2 \tag{91}$$

where

k_a = nonrelativistic wave number

A_a = nonrelativistic wave amplitude

The nonrelativistic Grüneisen parameter (diffuse radiation factor) for the radiation itself is given by[10]

$$\gamma^a_{or} = \frac{1}{3} + \frac{n}{W_a}\frac{dW_a}{dn} = \frac{1}{3} - \frac{n}{k_a}\frac{dk_a}{dn} \tag{92}$$

where $W_a = \omega/k_a$ = nonrealtivistic phase velocity of the waves, and $\omega = 2\pi f$ where f = frequency of the waves. The corresponding relativistic expressions are written

$$E_{or} = \frac{1}{4} K_o k^2 A^2 \tag{93}$$

$$\gamma_{or} = \frac{1}{3} + \frac{n}{W}\frac{dW}{dn} = \frac{1}{3} - \frac{n}{k}\frac{dk}{dn} \tag{94}$$

where

k = relativistic wave number

A = relativistic wave amplitude

$W = \omega/k$ = relativistic phase velocity

As before the relativistic bulk modulus K_o is obtained from a solution of the ground state material equations (52) and (53).

The solution of the coupled equations (80) and (81) for the case of waves in a solid, described by equations (91) through (94), determines the relativistic radiation parameters k and A in terms of the nonrelativistic radiation parameters k_a and A_a and in terms of the material parameters in the following general forms

$$k = k(k_a, A_a, P_o, K_o, \gamma_o, P^a_o, K^a_o, \gamma^a_o, \omega) \tag{95}$$

$$A = A(k_a, A_a, P_o, K_o, \gamma_o, P^a_o, K^a_o, \gamma^a_o, \omega) \tag{96}$$

The relativistic phase velocity is $W = \omega/k$, so that it also is given by a function of the form

$$W = W(k_a, A_a, P_o, K_o, \gamma_o, P_o^a, K_o^a, \gamma_o^a, \omega) \tag{97}$$

The coupled relativistic wave equations (80) and (81) are complicated differential equations that are difficult to solve analytically, even for the simple case of a solid. However, some qualitative results can be obtained for the case of wave propagation in a solid by combining equations (80) and (93) and assuming that k and A (and hence W) are not density dependent. Actually, there is a density dependence of the phase velocity.[9] But within the crude approximation that k and A are not density dependent, equations (80) and (93) yield

$$\frac{1}{4} k^2 A^2 \left(\ln \frac{dK_o}{dn} + JK_o \right) + g = \frac{1}{4} K_o^a k_a^2 A_a^2 \tag{98}$$

Then assuming $\gamma_{or} = 1/3$ and $\gamma_o = 1$ in equation (82) and (83) gives $I = 1$ and $J = -1$; and taking $K_o \sim n^\sigma$, where σ = adiabatic index, and finally $g \sim 2E_{or}$, gives the following approximate result for equation (98)

$$k^2 A^2 \sim \frac{K_o^a k_a^2 A_a^2}{(\sigma + 1) K_o} \tag{99}$$

But the solution of the ground state equations (52) and (53) is known to give $K_o \sim K_o^a$ at low pressures and $K_o << K_o^a$ at high pressures.[3] Therefore within the limits of the approximations made to obtain equation (99), it follows that $kA \sim k_a A_a / \sqrt{\sigma + 1}$ for very low pressures, $kA \sim k_a A_a$ for moderate pressures, and $kA >> k_a A_a$ for high pressures. The values of σ can range from zero to about two depending on the density of the system.

Finally, an explicit expression is given for the relativistic phase velocity of mechanical waves in a thermodynamic solid. The source terms for equations (80) and (81) are E_{or}^a and γ_{or}^a , where E_{or}^a is given by equation (91) while γ_{or}^a is obtained from equation (92) with[3]

$$\left(\frac{W_a}{c} \right)^2 = \frac{K_o^a}{E_o^a + P_o^a + K_o^a} \tag{100}$$

where c = light speed. The simultaneous solution of (80) and (81) gives $\mathcal{E}_{or}$ and γ_{or}. Then equation (94) is integrated to obtain the relativistic sound speed as follows

$$\frac{W}{c} = \exp\left[-\int_n^{\infty}(\gamma_{or} - \frac{1}{3})\frac{dn}{n}\right] \tag{101}$$

Equation (94) shows that if W is an increasing function of density it follows that $\gamma_{or} > 1/3$ for mechanical waves, and therefore $W \leq c$ which is required by special relativity.

7. ELECTROMAGNETIC WAVES IN MATTER. An analogous calculation can be done for the case of the propagation of electromagnetic waves in matter. For this case the radiation energy density and pressure can also be written in the form of equations (57) through (60) except that now

$$\mathcal{E}_{or} = \frac{1}{2}\left(\varepsilon E^2 + \mu H^2\right) \tag{102}$$

$$\mathcal{E}^a_{or} = \frac{1}{2}\left(\varepsilon_a E_a^2 + \mu_a H_a^2\right) \tag{103}$$

where E and H = relativistic electric and magnetic radiation fields respectively; E_a and H_a = nonrelativistic electric and magnetic radiation fields respectively; ε, μ and ε_a, μ_a = relativistic and nonrelativistic permittivities and permeabilities respectively. Therefore the energy density is a function of many variables. The electroweak case includes even more parameters. Therefore it is not possible to obtain relativistic values of each parameter by solving the two coupled relativistic wave equations (80) and (81) unless the assumption is made that the electric and magnetic fields are unaffected by equations (80) and (81), and only the permittivity and permeability need to be considered. The solution of these two equations gives only the relativistic energy density and Grüneisen parameter for electromagnetic radiation. However an explicit solution for the relativistic phase velocity can be obtained.

The fundamental waves equations (80) and (81) determine $\mathcal{E}_{or}$ and γ_{or} in terms of $\mathcal{E}^a_{or}$ and γ^a_{or}, where $\mathcal{E}^a_{or}$ is given by equation (103) and γ^a_{or} is given by equation (92) with[11]

$$W_a^2 = \frac{1}{\varepsilon_a \mu_a} \tag{104}$$

Equation (94) is then integrated to obtain the relativistic phase velocity of electromagnetic waves in a thermodynamic medium as follows

$$\frac{W}{c} = \exp\left[-\int_{o}^{n}\left(\frac{1}{3} - \gamma_{or}\right)\frac{dn}{n}\right] \tag{105}$$

Equation (94) shows that if W is a decreasing function of density (as is the case for electromagnetic waves in a solid or low temperature quantum system) it follows that $\gamma_{or} < 1/3$ and $W \leq c$ for electromagnetic waves in matter. Note that the relativistic phase velocity depends on the material parameters P_o , K_o , γ_o , P_o^a , K_o^a and γ_o^a .

8. CONCLUSION. Scale invariance has been demonstrated for a theory of relativistic thermodynamics that is based on a trace equation. The Grüneisen parameter must be introduced in addition to the pressure to insure local scale invariance. For solids and low temperature quantum systems, this means that the zero-temperature values of the pressure and Grüneisen parameter must be determined simultaneously through the solution of two coupled second order differential equations. The equations governing small amplitude waves in thermodynamic media can be obtained by a perturbation calculation on these ground state equations. Two coupled first order differential equations have been derived that describe relativistic wave propagation in solids and low temperature quantum systems. The solution of these wave equations determines the relativistic energy density and Grüneisen parameter for radiation in terms of the corresponding nonrelativistic radiation energy density and Grüneisen parameter and in terms of material parameters of the ground state. For mechanical waves the solution of the wave equations determines the relativistic amplitude and phase velocity.

Possible observable effects may occur in a number of high density systems. For instance, unusual dispersion effects and anomalously large wave amplitudes may be observed in the high pressure states of solids and liquids that occur in the interior of planets, stars, and stellar compact objects.[12,13] Measurable effects may also occur in the vibrations of atomic nuclei that are associated with the giant nuclear resonances.[14] Practical effects may also be noticed in the effects of nuclear explosions and in the interaction of high energy laser beams with solids.[15] Finally, it should be pointed out that because the phase velocity and the wave amplitude must be calculated simultaneously from a pair of coupled nonlinear equations, it follows that the phase velocity depends on the wave amplitude, and this characteristic of nonlinear wave propagation may lead to the formation of shocks and solitons under certain conditions.[16]

APPENDIX A - RADIATION CONDITIONS. For radiation, the pressure is linearly related to the energy density as follows

$$P_r = \Gamma_r E_r \tag{A1}$$

where Γ_r = diffuse radiation factor given by[10]

$$\Gamma_r = \frac{1}{3} + \frac{n}{W}\frac{dW}{dn} \tag{A2}$$

where the phase velocity W is generally density dependent. Combining equation (A1) with equations (59) and (60) gives

$$P_{or} = \Gamma_r E_{or} \tag{A3}$$

$$P_{jr} = \Gamma_r E_{jr} \tag{A4}$$

But from the definition of the radiation Grüneisen parameter given by equation (64), it follows from equation (A4) that

$$\Gamma_r = \gamma_{or} \tag{A5}$$

which means that the radiation Grüneisen parameter is equal to the diffuse radiation factor.

If the pressure and the internal energy of the radiation are written as

$$P_r = P_{or} + P_{jr}T^j \tag{A6}$$

$$U_r = VE_r = U_{or} + U_{jr}T^j \tag{A7}$$

then the application of the Gibbs-Helmholtz equation of thermodynamics

$$\left(\frac{\partial U_r}{\partial V}\right)_T = T\left(\frac{\partial P_r}{\partial T}\right)_V - P_r \tag{A8}$$

yields the following completely general equations

$$P_{or} = - \frac{dU_{or}}{dV} = - \frac{d}{dV}(VE_{or}) \quad \text{(A9)}$$

$$(j-1)P_{jr} = \frac{dU_{jr}}{dV} = \frac{d}{dV}(VE_{jr}) \quad \text{(A10)}$$

Combining equations (A3) and (A4) with equations (A9) and (A10) respectively gives

$$\frac{d}{dV}(VE_{or}) + \gamma_{or}E_{or} = 0 \quad \text{(A11)}$$

$$\frac{d}{dV}(VE_{jr}) - (j-1)\gamma_{or}E_{jr} = 0 \quad \text{(A12)}$$

or equivalently

$$\frac{V}{U_{or}} \frac{dU_{or}}{dV} = - \gamma_{or} \quad \text{(A13)}$$

$$\frac{V}{U_{jr}} \frac{dU_{jr}}{dV} = (j-1)\gamma_{or} \quad \text{(A14)}$$

The radiation equations (A13) and (A14) can be immediately integrated to give

$$U_{or} = D_{or} \exp\left(\int^{n} \gamma_{or} \frac{dn}{n} \right) \quad \text{(A15)}$$

$$U_{jr} = D_{jr} \exp\left[- (j-1) \int^{n} \gamma_{or} \frac{dn}{n} \right] \quad \text{(A16)}$$

and finally,

$$E_{or} = nD_{or} \exp\left(\int^{n} \gamma_{or} \frac{dn}{n} \right) \quad \text{(A17)}$$

$$E_{jr} = nD_{jr} \exp\left[- (j-1) \int^{n} \gamma_{or} \frac{dn}{n} \right] \tag{A18}$$

Then combining equation (A18) with equation (54) gives

$$\frac{E_{jr}}{E_j} = \frac{D_{jr}}{C_j} \exp\left[- (j-1) \int^{n} (\gamma_{or} - \gamma_o) \frac{dn}{n} \right] \tag{A19}$$

which is the desired result.

APPENDIX B - EVALUATION OF $d\delta_{or}/dn$. The derivative that occurs in equation (72) can be evaluated from equation (63) as follows,

$$n \frac{d\delta_{or}}{dn} = (\gamma_{or} - \gamma_o)\, n \frac{d}{dn}\left(\frac{E_{jr}}{E_j}\right) + \frac{E_{jr}}{E_j}\left(n \frac{d\gamma_{or}}{dn} - n \frac{d\gamma_o}{dn}\right) \tag{B1}$$

In order to evaluate the derivative of the energy ratio term in equation (B1) one uses equation (A19) of Appendix A to get the following result

$$n \frac{d}{dn}\left(\frac{E_{jr}}{E_j}\right) = - (j-1) \frac{E_{jr}}{E_j} (\gamma_{or} - \gamma_o) \tag{B2}$$

Combining equations (B1) and (B2) gives the result

$$n \frac{d\delta_{or}}{dn} = \frac{E_{jr}}{E_j}\left[n \frac{d\gamma_{or}}{dn} - n \frac{d\gamma_o}{dn} - (j-1)(\gamma_{or} - \gamma_o)^2 \right] \tag{B3}$$

which is the desired result.

REFERENCES

1. Aitchinson, I. and Hey, A., Gauge Theories in Particle Physics, Adam Hilger Ltd., Bristol, United Kingdom, 1982.

2. Leader, E. and Predazzi, E., Gauge Theories and the 'New Physics', Cambridge University Press, London, p. 29, 1982.

3. Weiss, R. A., Relativistic Thermodynamics, Vols. 1 and 2, Exposition Press, New York, 1976.

4. Aitchinson, I., An Informal Introduction to Gauge Field Theories, Cambridge University Press, p. 155, 1982.

5. Zharkov, V. N. and Kalinin, V. A., Equations of State for Solids at High Pressures and Temperatures, Consultants Bureau, New York, 1971.

6. Varley, J., "The Thermal Expansion of Pure Metals and the Possibility of Negative Coefficients of Volume Expansion," Proc. R. Soc. A237, 413, 1956.

7. Barron, T., Collins, J. and White, G., "Thermal Expansion of Solids at Low Temperatures," Adv. in Physics, 29, 609, 1980.

8. Grot, R. A. and Eringen, A. C., "Relativistic Continuum Mechanics, Part I - Mechanics and Thermodynamics; and Part II - Electromagnetic Interactions with Matter," Int. J. Engng. Sci. Vol 4, pp. 611-670, Pergamon Press, New York, 1966.

9. Tolstoy, I., "On Elastic Waves in Prestressed Solids," Journal of Geophysical Research, Vol 87, No. B8, pp. 6823-6827, Aug 10 1982.

10. Brillouin, L., Tensors in Mechanics and Elasticity, Academic Press, New York, 1964.

11. Born, M. and Wolf, E., Principles of Optics, Pergamon Press, New York, 1959.

12. Zeldovich, Ya. B. and Novikov, I. D., Relativistic Astrophysics, Vol. I Stars and Relativity, University of Chicago Press, 1978.

13. Misner, C. W., Thorne, K. S. and Wheeler, J. A., Gravitation, W. H. Freeman, San Francisco, 1973.

14. Vazquez, A., "Giant Resonances as Oscillations of Two Elastically Coupled Fluids," Phys. Rev. Lett., Vol 50, No. 22, pp. 1756-1758, 30 May 1983.

15. Rodean, H. C., Nuclear-Explosion Seismology, U. S. Atomic Energy Commission, AEC Critical Review Series, 1971.

16. Dodd, R. K., Eilbeck, J. C., Gibbon, J. D. and Morris, H. C., Solitons and Nonlinear Wave Equations, Academic Press, New York, 1982.

B

SCALE INVARIANT EQUATIONS FOR RELATIVISTIC WAVES

Richard A. Weiss
Environmental Laboratory
U. S. Army Engineer Waterways Experiment Station
Vicksburg, Mississippi 39180

ABSTRACT. The basic trace equation of relativistic thermodynamics is decoupled into two Callan-Symanzik type renormalization group equations that connect the matter fields with the thermodynamic gauge parameters. These equations determine two characteristic curves along which the solution to the trace equation assumes a simple form. The differential equation describing the variation of the Grüneisen parameter with pressure is derived. A perturbation procedure is applied to the potential form of the renormalization group equations in order to develop the corresponding potential form of the renormalization equations for waves in a relativistic medium. A method for calculating the Debye temperature for the excited states of solids and quantum liquids is developed. The amplitude and spectrum of waves in thermodynamic media are calculated. A simple equation is derived that scales the wave amplitudes for different material densities (pressures). The results of this paper will have applications to nuclear blast loadings, the interaction of directed energy beams with matter, and to various high density geophysical and astrophysical phenomena.

1. INTRODUCTION. The renormalization group was originally developed for problems in quantum field theory.[1,2] But over the years it has become an important technique in many areas of physics including, phase transitions, critical phenomena, hydrodynamics, and statistical mechanics.[3-5] The renormalization group consists of a set of continuous transformations that establish a correspondence between sets of parameters that define physically different states. In particular, the renormalization group gives a correspondence between systems having different correlation lengths. The correlation length of a physical system is the distance over which local particle densities are correlated. Ordinarily the correlation length is approximately equal to the range of interaction between two component particles, however, near the critical point of a fluid the correlation length is much greater than the range of pair interactions.[6] The renormalization group is commonly described by a set of differential equations for the physical state parameters.[3]

A set of renormalization group equations in potential form has been developed for the ground state parameters of a relativistic thermodynamic system.[7] In this case the correspondence between sets of parameters refers to a change of the local scale (gauge), and this change of scale is equivalent to a change in the correlation length. The potential form of the renormalization group equations consists of a set of differential equations for the two gauge parameters of relativistic thermodynamics.[7] These equations are obtained by requiring the basic trace equation of relativistic thermodynamics to be invariant under a local scale transformation that corresponds to a change in the correlation length of the system.

The trace equation of relativistic thermodynamics is written as[8]

$$U + T\left(\frac{dU}{dT}\right)_{PV} - 3V\frac{d}{dV}(PV)_U = U^a + T\left(\frac{dU^a}{dT}\right)_{P^aV} \tag{1}$$

where U = relativistic internal energy, P = relativistic pressure, T = absolute temperature, V = volume of substance, and U^a and P^a = corresponding nonrelativistic internal energy and pressure. Throughout this paper the index "a" will refer to nonrelativistic calculations. The trace equation (1) can be rewritten as[7]

$$\left(1 - b + T\frac{\partial}{\partial T} - bV\frac{\partial}{\partial V}\right)\mathcal{E} - 3\left(1 + \gamma + V\frac{\partial}{\partial V} - \gamma T\frac{\partial}{\partial T}\right)P \tag{2}$$

$$= \left(1 - b^a + T\frac{\partial}{\partial T} - b^aV\frac{\partial}{\partial V}\right)\mathcal{E}^a$$

where $\mathcal{E}$ = relativistic energy density = U/V, $\mathcal{E}^a$ = nonrelativistic energy density, and where[7]

$$\gamma = \frac{V}{C_V}\left(\frac{\partial P}{\partial T}\right)_V \tag{3}$$

$$b = \frac{T(\partial P/\partial T)_V}{(P - K_T)} \tag{4}$$

$$b^a = \frac{T(\partial P^a/\partial T)_V}{(P^a - K_T^a)} \tag{5}$$

where γ = Grüneisen parameter, C_V = heat capacity at constant volume, and

$$K_T = -V\left(\frac{\partial P}{\partial V}\right)_T \tag{6}$$

$$K_T^a = -V\left(\frac{\partial P^a}{\partial V}\right)_T \tag{7}$$

are the relativistic and nonrelativistic values of the bulk modulus respectively. The parameters b and γ are the gauge parameters of relativistic thermodynamics.

For a solid or low temperature quantum system the nonrelativistic state equation of the ground state is assumed to have the following form[8,9]

$$E^a = E^a_o + E^a_j T^j + \cdots \tag{8}$$

$$P^a = P^a_o + P^a_j T^j + \cdots \tag{9}$$

where E^a and P^a = nonrelativistic energy density and pressure respectively, E^a_o and P^a_o = nonrelativistic zero-temperature values of the energy density and pressure respectively, E^a_j and P^a_j = nonrelativistic thermal coefficients for the energy density and pressure respectively, T = absolute temperature of the system (°K), and j = numerical index having values characteristic of the type of physical system. Typical examples of systems that are described by equations (8) and (9) are[8]

j = 1	high temperature solid
j = 2	low temperature Fermi gas
j = 5/2	low temperature molecular Bose gas
j = 4	low temperature solid

A commonly used descriptor of the thermal state equations given by equations (8) and (9) is the nonrelativistic zero-temperature value of the Grüneisen parameter that is defined by[8,9]

$$\gamma^a_o = \frac{P^a_j}{E^a_j} = \frac{1}{(j-1)} \; \frac{1}{E^a_j} \frac{d}{dV}(VE^a_j) \tag{10}$$

except for j = 1. Here γ^a_o = nonrelativistic zero-temperature value of the Grüneisen parameter, and V = volume of the material system. When j = 1 , $\gamma^a_o = 2/3$. The zero temperature value of the nonrelativistic bulk modulus is given by $K^a_o = n dP^a_o/dn$, where n = N/V = number of moles per unit volume, and N = number of moles of a substance.

The corresponding relativistic state equations will be written as[8,9]

$$E = E_o + E_j T^j + \cdots \tag{11}$$

$$P = P_o + P_j T^j + \cdots \tag{12}$$

$$\gamma_o = \frac{P_j}{E_j} = \frac{1}{(j-1)} \frac{1}{E_j} \frac{d}{dV}(VE_j) \tag{13}$$

except for j = 1 , when $E_1 = E_1^a$, and where E_o and P_o = relativistic zero-temperature energy density and pressure respectively, E_j and P_j = relativistic thermal coefficients for the energy density and pressure respectively, and γ_o = relativistic zero-temperature Grünesien parameter. The relativistic value of the zero temperature bulk modulus is given by $K_o = n dP_o/dn$. Combining equation (2) with the state equations (8) through (13) yields the following ground state equations[8]

$$E_o - 3[(1 + \gamma_o)P_o - K_o] = E_o^a \tag{14}$$

$$E_j \left(1 + j + \frac{j\gamma_o P_o}{P_o - K_o} + 3n \frac{d\gamma_o}{dn}\right) = E_j^a \left(1 + j + \frac{j\gamma_o^a P_o^a}{P_o^a - K_o^a}\right) \tag{15}$$

The potential forms of the ground state renormalization group equations for the gauge parameters b and γ are determined from the requirement of local scale invariance of equation (2).[7] A form of the renormalization group equations that is commonly used in quantum field theory and the theory of critical phenomena are the Callan-Symanzik equations.[1-3] In this paper the Callan-Symanzik form of the renormalization group equations for the relativistic ground state of thermal media will be obtained directly from equation (2).

Two forms of the renormalization group equations for radiation in matter are obtained in this paper. The potential form of these equations is obtained by a perturbation procedure that is applied to the potential form of the renormalization group equations for the ground state. The Callan-Symanzik form of the renormalization group equations for radiation will be obtained by a perturbation procedure applied directly to equation (2).

An important task of modern physics is the determination of the effects of gauge (scale) invariance on the ground state and excitations of matter. These effects have been treated for the ground state of a thermodynamic system.[8] An approximate treatment has been developed for waves in solids and quantum liquids by assuming that the diffuse radiation factor and the radiation Grüneisen parameter are equal.[7] This paper presents a completely general procedure for calculating the relativistic amplitude and spectrum of elastic waves in solids and quantum liquids. A set of coupled second order radiation equations is developed that determines the relativistic energy density and Grüneisen parameter for radiation in solids and Fermi and Bose liquids.

2. RENORMALIZATION GROUP EQUATIONS FOR THE GROUND STATE. The ground state of a relativistic thermodynamic medium is described by equation (2) where γ and b are gauge parameters.[7] Equation (2) can be decoupled into two independent equations by noting that $\mathcal{E}$ and P are related by the Gibbs-Helmholtz relation as follows

$$\left(\frac{\partial U}{\partial V}\right)_T = \mathcal{E} + V\left(\frac{\partial \mathcal{E}}{\partial V}\right)_T = T\left(\frac{\partial P}{\partial T}\right)_V - P \tag{16}$$

With the introduction of a Lagrange undetermined multiplier η, equation (16) can be rewritten as

$$\eta\left(1 + V\frac{\partial}{\partial V}\right)\mathcal{E} + \eta\left(1 - T\frac{\partial}{\partial T}\right)P = 0 \tag{17}$$

Combining equations (2) and (17) yields the following decoupled equations

$$[T\frac{\partial}{\partial T} + (\eta - b)V\frac{\partial}{\partial V} + \eta + 1 - b]\mathcal{E} = (T\frac{\partial}{\partial T} - b^a V\frac{\partial}{\partial V} + 1 - b^a)\mathcal{E}^a \tag{18}$$

$$[V\frac{\partial}{\partial V} - (\gamma - \frac{\eta}{3})T\frac{\partial}{\partial T} - \frac{\eta}{3} + \gamma + 1]P = 0 \tag{19}$$

The undetermined multiplier η is in general a function of V and T. From equation (19) it follows that

$$\frac{\eta}{3} = \frac{V\frac{\partial P}{\partial V} - \gamma T\frac{\partial P}{\partial T} + (\gamma + 1)P}{P - T\frac{\partial P}{\partial T}} \tag{20}$$

except when the denominator is zero (as in the case of an ideal gas for which $P = nRT$). For $T = 0$, equations (18) through (20) become

$$(\eta_o V\frac{\partial}{\partial V} + \eta_o + 1)\mathcal{E}_o = \mathcal{E}_o^a \tag{21}$$

$$(V\frac{\partial}{\partial V} - \frac{\eta_o}{3} + \gamma_o + 1)P_o = 0 \tag{22}$$

$$\frac{\eta_o}{3} = \frac{V}{P_o}\frac{dP_o}{dV} + \gamma_o + 1 \tag{23}$$

where η_o = the Lagrange undetermined multiplier for T = 0 . It is easy to show that combining equation (21) through (23) by eliminating the Lagrange multiplier and using the T = 0 form of equation (16) which is

$$P_o = -V \frac{dE_o}{dV} - E_o \tag{24}$$

gives the T = 0 ground state equation (14).

Equations (18) and (19) can be rewritten as

$$(T \frac{\partial}{\partial T} + f \frac{\partial}{\partial v} + M)E = \psi^a \tag{25}$$

$$(T \frac{\partial}{\partial T} + h \frac{\partial}{\partial v} + N)P = 0 \tag{26}$$

where

$$V = e^v \tag{27}$$

$$f = \eta - b \tag{28}$$

$$h = \frac{1}{\eta/3 - \gamma} \tag{29}$$

$$M = f + 1 \tag{30}$$

$$N = h - 1 \tag{31}$$

$$\psi^a = (T \frac{\partial}{\partial T} - b^a \frac{\partial}{\partial v} + 1 - b^a)E^a \tag{32}$$

Equations (25) and (26) are immediately recognized to be similar in form to the Callan-Symanzik equations that describe the renormalization group.[1,3] The physical meaning of equations (25) and (26) is that T and V can be considered to be arbitrary parameters and that the trace equation of relativistic thermodynamics is form invariant for any choices of values for T and V, i.e., arbitrary values of temperature and volume are acceptable in equation (2). Equations (25) and (26) connect the matter fields E and P to the gauge fields γ and b. Using equations (25) and (26) allows f and h to be written as

$$f = \frac{T \frac{\partial E}{\partial T} + E - \psi^a}{P - T \frac{\partial P}{\partial T}} \tag{33}$$

$$h = \frac{P - T \frac{\partial P}{\partial T}}{P - K_T} \tag{34}$$

which for the case T = 0 become

$$f_o = \frac{E_o - E_o^a}{P_o} \tag{35}$$

$$h_o = \frac{P_o}{P_o - K_o} \tag{36}$$

The functions f and h are the thermodynamic analogs of the Gell-Mann-Low functions.[1]

In analogy to equation (27) the introduction of

$$T = e^t \tag{37}$$

allows equations (25) and (26) to be rewritten in simpler form by evaluating the derivatives along two characteristic curves as follows

$$\frac{dE}{dt} + ME = \psi^a \tag{38}$$

$$\frac{dP}{dt} + NP = 0 \tag{39}$$

The two characteristic curves are defined by

$$f(V,T) = \frac{dv}{dt} = \frac{T}{V}\frac{dV}{dT} \tag{40}$$

$$h(V,T) = \frac{dv}{dt} = \frac{T}{V}\frac{dV}{dT} \tag{41}$$

The characteristic equations (38) and (39) can be solved formally as

$$E = Ce^{-\int Mdt} + e^{-\int Mdt}\int \psi_a e^{\int Mdt}\, dt \tag{42}$$

$$P = De^{-\int Ndt}$$

where C and D are constants. Therefore, along the characteristic curves the solutions to equations (25) and (26) assume a simple form.

The requirement of local scale (gauge) invariance demands that equation (2) be invariant under transformations of the form $P \rightarrow P' = Pe^{-\phi}$ and $E \rightarrow E' = Ee^{-\psi}$ where ϕ and ψ are functions of V and T.[7] These transformations describe a correspondence between physical states having different correlation lengths. The scale invariance condition is taken to be analogous to the condition of local gauge invariance, and yields the following differential equations for the gauge parameters using the $e^{-\phi}$ and $e^{-\psi}$ transformations[7]

$$\left(\frac{d\gamma}{dP}\right)^{-} = \frac{\gamma \dfrac{T}{\phi}\dfrac{\partial\phi}{\partial T} - \dfrac{V}{\phi}\dfrac{\partial\phi}{\partial V}}{P - T\dfrac{\partial P}{\partial T} + PT\dfrac{\partial\phi}{\partial T}} \tag{44}$$

$$\left(\frac{db}{dE}\right)^{-} = \frac{\dfrac{T}{\psi}\dfrac{\partial\psi}{\partial T} - b\dfrac{V}{\psi}\dfrac{\partial\psi}{\partial V}}{E + V\dfrac{\partial E}{\partial V} - EV\dfrac{\partial\psi}{\partial V}} \tag{45}$$

These are the potential forms of the renormalization group equations for the relativistic ground state of a thermodynamic system; they correspond to scale transformations with negative signs in front of the potential functions ϕ and ψ.

From equations (44) and (45) it is clear that the denominators of these equations are not symmetrical under $\phi \rightarrow -\phi$ and $\psi \rightarrow -\psi$. In fact, the requirement of scale invariance for equation (2) under the transformations $P \rightarrow P' = Pe^{+\phi}$ and $E \rightarrow E' = Ee^{+\psi}$ yields the following result

$$\left(\frac{d\gamma}{dP}\right)^{+} = \frac{\gamma \dfrac{T}{\phi}\dfrac{\partial\phi}{\partial T} - \dfrac{V}{\phi}\dfrac{\partial\phi}{\partial V}}{P - T\dfrac{\partial P}{\partial T} - PT\dfrac{\partial\phi}{\partial T}} \tag{46}$$

$$\left(\frac{db}{dE}\right)^{+} = \frac{\frac{T}{\psi}\frac{\partial\psi}{\partial T} - b\,\frac{V}{\psi}\frac{\partial\psi}{\partial V}}{E + V\frac{\partial E}{\partial V} + EV\frac{\partial\psi}{\partial V}} \tag{47}$$

However, the physical derivatives of the gauge parameters must be independent of the signs of the potential functions that appear in the scale transformations, and the simplest way to accomplish this is to assume that the physical derivatives are given by the following symmetric forms

$$\frac{d\gamma}{dP} = \frac{1}{2}\left[\left(\frac{d\gamma}{dP}\right)^{-} + \left(\frac{d\gamma}{dP}\right)^{+}\right] \tag{48}$$

$$\frac{db}{dE} = \frac{1}{2}\left[\left(\frac{db}{dE}\right)^{-} + \left(\frac{db}{dE}\right)^{+}\right] \tag{49}$$

Equations (48) and (49) can be rewritten as

$$\frac{d\gamma}{dP} = \frac{\left(P - T\frac{\partial P}{\partial T}\right)\left(\gamma\,\frac{T}{\phi}\frac{\partial\phi}{\partial T} - \frac{V}{\phi}\frac{\partial\phi}{\partial V}\right)}{\left(P - T\frac{\partial P}{\partial T} + PT\frac{\partial\phi}{\partial T}\right)\left(P - T\frac{\partial P}{\partial T} - PT\frac{\partial\phi}{\partial T}\right)} \tag{50}$$

$$\frac{db}{dE} = \frac{\left(E + V\frac{\partial E}{\partial V}\right)\left(\frac{T}{\psi}\frac{\partial\psi}{\partial T} - b\,\frac{V}{\psi}\frac{\partial\psi}{\partial V}\right)}{\left(E + V\frac{\partial E}{\partial V} - EV\frac{\partial\psi}{\partial V}\right)\left(E + V\frac{\partial E}{\partial V} + EV\frac{\partial\psi}{\partial V}\right)} \tag{51}$$

Equations (50) and (51) can be rewritten as

$$\frac{d\gamma}{dP} = \frac{\left(P - T\frac{\partial P}{\partial T}\right)\left(\gamma\,\frac{T}{\phi}\frac{\partial\phi}{\partial T} - \frac{V}{\phi}\frac{\partial\phi}{\partial V}\right)}{\left(P - T\frac{\partial P}{\partial T}\right)^{2} - P^{2}\left(T\frac{\partial\phi}{\partial T}\right)^{2}} \tag{50a}$$

$$\frac{db}{dE} = \frac{\left(E + V\frac{\partial E}{\partial V}\right)\left(\frac{T}{\psi}\frac{\partial\psi}{\partial T} - b\,\frac{V}{\psi}\frac{\partial\psi}{\partial V}\right)}{\left(E + V\frac{\partial E}{\partial V}\right)^{2} - E^{2}\left(V\frac{\partial\psi}{\partial V}\right)^{2}} \tag{51a}$$

Equation (50) has the obvious property that for an ideal gas $d\gamma/dP = 0$ because in this case $P = nRT$, and this agrees with the fact that the Grüneisen parameter for an ideal gas is given by $\gamma = 2/3$. Equations (50) and (51) are the potential forms of the renormalization group equations for the relativistic thermodynamic ground state.

The potential function ϕ is related to the Debye temperature θ_D for high temperatures ($T > \theta_D$) by $\phi = \theta_D/T$, and for low temperatures ($T < \theta_D$) by $\phi = T/\theta_D$. For high temperatures, the use of $\phi = \theta_D/T$ in equation (50) gives

$$\frac{d\gamma}{dP} = \left(T \frac{\partial P}{\partial T} - P\right)\left[\gamma\left(1 - \frac{T}{\theta_D}\frac{\partial\theta_D}{\partial T}\right) + \frac{V}{\theta_D}\frac{\partial\theta_D}{\partial V}\right]/(A_H B_H) \tag{52}$$

where

$$A_H = T\frac{\partial P}{\partial T} - P + P\frac{\theta_D}{T}\left(1 - \frac{T}{\theta_D}\frac{\partial\theta_D}{\partial T}\right) \tag{53}$$

$$B_H = T\frac{\partial P}{\partial T} - P - P\frac{\theta_D}{T}\left(1 - \frac{T}{\theta_D}\frac{\partial\theta_D}{\partial T}\right) \tag{54}$$

Equation (52) can be rewritten as

$$\frac{d\gamma}{dP} = \frac{\left(T\frac{\partial P}{\partial T} - P\right)\left[\gamma\left(1 - \frac{T}{\theta_D}\frac{\partial\theta_D}{\partial T}\right) + \frac{V}{\theta_D}\frac{\partial\theta_D}{\partial V}\right]}{\left(T\frac{\partial P}{\partial T} - P\right)^2 - P^2\left(\frac{\theta_D}{T}\right)^2\left(1 - \frac{T}{\theta_D}\frac{\partial\theta_D}{\partial T}\right)^2} \tag{52a}$$

Since in general the following conditions hold for ordinary materials at high temperature[11]

$$T\frac{\partial P}{\partial T} > P \tag{55}$$

$$\left|\frac{V}{\theta_D}\frac{\partial\theta_D}{\partial V}\right| \gtrsim \gamma\left(1 - \frac{T}{\theta_D}\frac{\partial\theta_D}{\partial T}\right) \tag{56}$$

it follows from equation (52) that in general $d\gamma/dP < 0$. The slow varying Grüneisen parameter condition ($d\gamma/dP \sim 0$) in equation (52) yields[7]

$$\gamma = \frac{-\frac{V}{\theta_D}\left(\frac{\partial\theta_D}{\partial V}\right)_T}{\left[1 - \frac{T}{\theta_D}\left(\frac{\partial\theta_D}{\partial T}\right)_V\right]} \tag{57}$$

which is a standard equation of high pressure physics.[7] If θ_D does not depend appreciably on temperature it follows from equation (57) that

$$\gamma \approx - \frac{V}{\theta_D} \frac{\partial \theta_D}{\partial V} \tag{58}$$

For low temperatures the use of $\phi = T/\theta_D$ in equation (50) yields

$$\frac{d\gamma}{dP} = \left(P - T \frac{\partial P}{\partial T}\right)\left[\gamma\left(1 - \frac{T}{\theta_D}\frac{\partial \theta_D}{\partial T}\right) + \frac{V}{\theta_D}\frac{\partial \theta_D}{\partial V}\right] / (A_L B_L) \tag{59}$$

where

$$A_L = P - T\frac{\partial P}{\partial T} + P\frac{T}{\theta_D}\left(1 - \frac{T}{\theta_D}\frac{\partial \theta_D}{\partial T}\right) \tag{60}$$

$$B_L = P - T\frac{\partial P}{\partial T} - P\frac{T}{\theta_D}\left(1 - \frac{T}{\theta_D}\frac{\partial \theta_D}{\partial T}\right) \tag{61}$$

Equation (59) can be rewritten as

$$\frac{d\gamma}{dP} = \frac{\left(P - T\frac{\partial P}{\partial T}\right)\left[\gamma\left(1 - \frac{T}{\theta_D}\frac{\partial \theta_D}{\partial T}\right) + \frac{V}{\theta_D}\frac{\partial \theta_D}{\partial V}\right]}{\left(P - T\frac{\partial P}{\partial T}\right)^2 - P^2\left(\frac{T}{\theta_D}\right)^2\left(1 - \frac{T}{\theta_D}\frac{\partial \theta_D}{\partial T}\right)^2} \tag{59a}$$

The slow varying Grüneisen parameter condition ($d\gamma/dP \sim 0$) for equation (59) also yields the result in equation (57). For the case $T = 0$, equation (59) becomes

$$\frac{d\gamma_o}{dP_o} = \left(\gamma_o + \frac{V}{\theta_D^o}\frac{\partial \theta_D^o}{\partial V}\right) / P_o \tag{62}$$

where θ_D^o = the $T = 0$ value of the Debye temperature. The slow varying Grüneisen parameter approximation applied to equation (62) yields

$$\gamma_o \approx - \frac{V}{\theta_D^o}\frac{\partial \theta_D^o}{\partial V} \tag{63}$$

3. LOCAL GAUGE INVARIANCE FOR PHASE ROTATIONS. The sign of the zero temperature derivative $d\gamma_o/dP_o$ that appears in equation (62) is not obtained from the previous analysis. An indication of what this sign is can be obtained by introducing complex local gauge transformations for the pressure and energy density. This corresponds to phase rotations of the pressure and energy dendity relative to the gauge parameters as follows, $\bar{P}' = \bar{P}e^{\pm i\phi}$ and $\bar{E}' = \bar{E}e^{\pm i\psi}$, where $\bar{P}$, $\bar{E}$, $\bar{\gamma}$ and $\bar{b}$ now represent complex numbers: $\bar{P} = Pe^{i\theta_P}$, $\bar{E} = Ee^{i\theta_E}$, $\bar{\gamma} = \gamma e^{i\theta_\gamma}$ and $\bar{b} = be^{i\theta_b}$ whose magnitudes are P, E, γ and b. Thus whereas the real exponentials $e^{\pm\phi}$ and $e^{\pm\psi}$ correspond to changes in the pressure and energy density, the complex exponentials correspond to phase rotations where the magnitudes P, E, γ and b are held fixed.

The phase angles θ_γ and θ_b are determined from the condition of gauge invariance (local phase invariance) of the fundamental trace equation (2) under internal phase rotations of $\bar{P}$ and $\bar{E}$. In a manner analogous to that used to determine equations (50a) and (51a) for scale invariance, the local phase invariance condition gives

$$\frac{d\bar{\gamma}}{d\bar{P}} = \frac{\left(\bar{P} - T\frac{\partial\bar{P}}{\partial T}\right)\left(\bar{\gamma}\frac{T}{\phi}\frac{\partial\phi}{\partial T} - \frac{V}{\phi}\frac{\partial\phi}{\partial V}\right)}{\left(\bar{P} - T\frac{\partial\bar{P}}{\partial T}\right)^2 + \bar{P}^2\left(T\frac{\partial\phi}{\partial T}\right)^2} \tag{63a}$$

$$\frac{d\bar{b}}{d\bar{E}} = \frac{\left(\bar{E} + V\frac{\partial\bar{E}}{\partial V}\right)\left(\frac{T}{\psi}\frac{\partial\psi}{\partial T} - \bar{b}\frac{V}{\psi}\frac{\partial\psi}{\partial V}\right)}{\left(\bar{E} + V\frac{\partial\bar{E}}{\partial V}\right)^2 + \bar{E}^2\left(V\frac{\partial\psi}{\partial V}\right)^2} \tag{63b}$$

Note the positive sign in the denominators. When γ, P, b and E are held fixed and neglecting the phase angles of $\bar{\gamma}$, $\bar{P}$, $\bar{b}$ and $\bar{E}$ it follows that

$$\frac{d\bar{\gamma}}{d\bar{P}} \sim \frac{\gamma}{P}\frac{d\theta_\gamma}{d\theta_P} \qquad \frac{d\bar{b}}{d\bar{E}} \sim \frac{b}{E}\frac{d\theta_b}{d\theta_E} \tag{63c}$$

Then it follows that the local gauge invariance conditions and the symmetrization equations (48) and (49) yield the following renormalization group equations for phase rotations in analogy to equations (50a) and (51a) and after neglecting complex number phase factors

$$\frac{\gamma}{P}\frac{d\theta_\gamma}{d\theta_P} \sim \frac{\left(P - T\frac{\partial P}{\partial T}\right)\left(\gamma\frac{T}{\phi}\frac{\partial\phi}{\partial T} - \frac{V}{\phi}\frac{\partial\phi}{\partial V}\right)}{\left(P - T\frac{\partial P}{\partial T}\right)^2 + P^2\left(T\frac{\partial\phi}{\partial T}\right)^2} \tag{63d}$$

$$\frac{b}{E}\frac{d\theta_b}{d\theta_E} \sim \frac{\left(E + v\frac{\partial E}{\partial V}\right)\left(\frac{T}{\psi}\frac{\partial\psi}{\partial T} - b\frac{V}{\psi}\frac{\partial\psi}{\partial V}\right)}{\left(E + v\frac{\partial E}{\partial V}\right)^2 + E^2\left(v\frac{\partial\psi}{\partial V}\right)^2} \tag{63e}$$

Equations (63d) and (63e) are the approximate scalar equations corresponding to the exact complex number equations (63a) and (63b).

For high temperatures $\phi = \theta_D/T$ and for this case equation (63d) becomes

$$\frac{\gamma}{P}\frac{d\theta_\gamma}{d\theta_P} \sim \frac{\left(T\frac{\partial P}{\partial T} - P\right)\left[\gamma\left(1 - \frac{T}{\theta_D}\frac{\partial\theta_D}{\partial T}\right) + \frac{V}{\theta_D}\frac{\partial\theta_D}{\partial V}\right]}{\left(T\frac{\partial P}{\partial T} - P\right)^2 + P^2\left(\frac{\theta_D}{T}\right)^2\left(1 - \frac{T}{\theta_D}\frac{\partial\theta_D}{\partial T}\right)^2} \tag{63f}$$

where

$$\frac{d\theta_\gamma}{d\theta_P} = \frac{\frac{\partial\theta_\gamma}{\partial V}dV + \frac{\partial\theta_\gamma}{\partial T}dT}{\frac{\partial\theta_P}{\partial V}dV + \frac{\partial\theta_P}{\partial T}dT} \tag{63g}$$

An examination of equations (52a) and (63f) shows that for high temperatures $d\gamma/dP < 0$ and $d\theta_\gamma/d\theta_P < 0$.

For low temperatures $\phi = T/\theta_D$ and equation (63d) gives

$$\frac{\gamma}{P}\frac{d\theta_\gamma}{d\theta_P} \sim \frac{\left(P - T\frac{\partial P}{\partial T}\right)\left[\gamma\left(1 - \frac{T}{\theta_D}\frac{\partial\theta_D}{\partial T}\right) + \frac{V}{\theta_D}\frac{\partial\theta_D}{\partial V}\right]}{\left(P - T\frac{\partial P}{\partial T}\right)^2 + P^2\left(\frac{T}{\theta_D}\right)^2\left(1 - \frac{T}{\theta_D}\frac{\partial\theta_D}{\partial T}\right)^2} \tag{63h}$$

For T = 0 equation (63h) becomes

$$\frac{\gamma_o}{P_o}\frac{d\theta^o_\gamma}{d\theta^o_P} \sim \left(\gamma_o + \frac{V}{\theta^o_D}\frac{\partial\theta^o_D}{\partial V}\right)/P_o \tag{63i}$$

Note that $P_o < 0$ for bound systems such as solids at standard pressure. A comparison of equations (62) and (63i) shows that

$$\frac{d\gamma_o}{dP_o} = \frac{\gamma_o}{P_o}\frac{d\theta^o_\gamma}{d\theta^o_P} \qquad (63j)$$

and therefore it follows that if $d\theta^o_\gamma/d\theta^o_P > 0$ then $d\gamma_o/dP_o > 0$, and from equation (62) one concludes that

$$\gamma_o > \left|\frac{V}{\theta^o_D}\frac{\partial\theta^o_D}{\partial V}\right| \qquad (63k)$$

Since in general $d\gamma/dP < 0$ at high temperatures, while equation (63j) shows that $d\gamma/dP > 0$ at $T = 0$, it follows that as the temperature of a solid or quantum liquid is lowered a value of the temperature is finally reached at which point the sign of $d\gamma/dP$ changes from a negative to a positive value.

It should be pointed out that a similar analysis is not possible for the angles ψ and θ_b (associated with $\mathcal{E}$ and b respectively) that appear in equation (63e) because the gauge parameter $b \to 0$ when $T \to 0$. This condition combined with equation (63e) suggests that ψ has the following low temperature form

$$\psi(V,T) = A(V)\exp\left(\int G(V,T)dT\right) \qquad (63\ell)$$

where $G(V,T)$ = some polynomial function of T .

4. RENORMALIZATION GROUP EQUATIONS FOR EXCITATIONS. When electromagnetic or mechanical waves of small amplitude are present in a thermodynamic medium, the renormalization group equations for the excitations can be obtained either directly as a perturbation on the ground state equation (2) for the energy density, or as a perturbation on the potential forms of the ground state renormalization group equations given by equations (44) through (49). When excitations are present the pressure, energy density, bulk modulus, and heat capacity become $P + P_r$, $\mathcal{E} + \mathcal{E}_r$, $K_T + K_{Tr}$, and $C_V + C_{Vr}$ respectively, where P_r = radiation pressure, $\mathcal{E}_r$ = radiation energy density, and where

$$K_{Tr} = n\left(\frac{\partial P_r}{\partial n}\right)_T = \text{radiation bulk modulus}$$

$$C_{Vr} = V\left(\frac{\partial \mathcal{E}_r}{\partial T}\right)_V = \text{radiation heat capacity}$$

The Grüneisen parameter for a thermodynamic medium with excitations is obtained from equation (3) as follows

$$\gamma + \delta_r = \frac{V}{C_V + C_{Vr}} \frac{\partial}{\partial T}(P + P_r) \tag{64}$$

where δ_r = change in the system Grüneisen parameter due to the presence of radiation. Expanding equation (64), subtracting equation (3), and keeping only first order terms gives

$$\delta_r = \frac{V}{C_V} \frac{\partial E_r}{\partial T} (\gamma_r - \gamma) = \frac{C_{Vr}}{C_V} (\gamma_r - \gamma) \tag{65}$$

where γ_r = Grüneisen parameter for the radiation field itself and is given by

$$\gamma_r = \frac{\partial P_r}{\partial T} \Big/ \frac{\partial E_r}{\partial T} \tag{66}$$

The gauge parameter b for an excited thermodynamic medium is obtained from equation (4) to be

$$b + \beta_r = \frac{T\left(\frac{\partial P}{\partial T} + \frac{\partial P_r}{\partial T}\right)}{P - K_T + P_r - K_{Tr}} \tag{67}$$

where β_r = change in b parameter due to the presence of radiation in the system. Expanding equation (67), keeping only first order terms, and subtracting equation (4) gives

$$\beta_r = \frac{T \frac{\partial P_r}{\partial T}}{P - K_T} - \frac{T \frac{\partial P}{\partial T} (P_r - K_{Tr})}{(P - K_T)^2} \tag{68}$$

$$= b_r \frac{P_r - K_{Tr}}{P - K_T} - \frac{T \frac{\partial P}{\partial T} (P_r - K_{Tr})}{(P - K_T)^2}$$

where b_r = radiation gauge parameter given by

$$b_r = \frac{T \frac{\partial P_r}{\partial T}}{P_r - K_{Tr}} \tag{68a}$$

The parameters γ_r and b_r are the two radiation gauge parameters of the thermal medium.

The renormalization group equations for radiation can be put into a Callan-Symanzik form by first combining equation (2) with equations (64) and (67) as follows

$$\left[1 - (b + \beta_r) + T\frac{\partial}{\partial T} - (b + \beta_r)V\frac{\partial}{\partial V}\right](E + E_r) \tag{69}$$

$$- 3\left[1 + \gamma + \delta_r + V\frac{\partial}{\partial V} - (\gamma + \delta_r)T\frac{\partial}{\partial T}\right](P + P_r)$$

$$= \left[1 - (b^a + \beta_r^a) + T\frac{\partial}{\partial T} - (b^a + \beta_r^a)V\frac{\partial}{\partial V}\right](E^a + E_r^a)$$

where β_r^a is given by the nonrelativistic analog of equation (68). Subtracting equation (2) from equation (69), and keeping only first order radiation terms yields the following radiation equation

$$\left(1 - b + T\frac{\partial}{\partial T} - bV\frac{\partial}{\partial V}\right)E_r - \beta_r\left(T\frac{\partial P}{\partial T} - P\right) \tag{70}$$

$$- 3\left[\left(1 + \gamma + V\frac{\partial}{\partial V} - \gamma T\frac{\partial}{\partial T}\right)P_r - \delta_r\left(T\frac{\partial P}{\partial T} - P\right)\right]$$

$$= \left(1 - b^a + T\frac{\partial}{\partial T} - b^a V\frac{\partial}{\partial V}\right)E_r^a - \beta_r^a\left(T\frac{\partial P^a}{\partial T} - P^a\right)$$

where the following standard thermodynamic relation was used

$$\frac{\partial U}{\partial V} = E + V\frac{\partial E}{\partial V} = T\frac{\partial P}{\partial T} - P \tag{71}$$

Equation (70) is a first order radiation equation that can be applied to any thermodynamic system such as gases, solids, and quantum liquids.

Equation (70) can be separated into two radiation equations each of which is similar in form to the Callan-Symanzik equation. This is done by using the Gibbs-Helmholtz equation (71), which for radiation becomes

$$\frac{\partial U_r}{\partial V} = E_r + V\frac{\partial E_r}{\partial V} = T\frac{\partial P_r}{\partial T} - P_r \tag{72}$$

Introducing a radiation Lagrange multiplier η_r as follows

$$\eta_r\left[E_r + V\frac{\partial E_r}{\partial V} + P_r - T\frac{\partial P_r}{\partial T}\right] = 0 \tag{73}$$

allows equation (70) to be separated as follows

$$\left(T\frac{\partial}{\partial T} + f_r\frac{\partial}{\partial v} + M_r\right)E_r - \beta_r\left(T\frac{\partial P}{\partial T} - P\right) = \psi_r^a \tag{74}$$

$$\left(T\frac{\partial}{\partial T} + h_r\frac{\partial}{\partial v} + N_r\right)P_r - h_r\delta_r\left(T\frac{\partial P}{\partial T} - P\right) = 0 \tag{75}$$

where v is defined in equation (27), and where

$$f_r = \eta_r - b \tag{76}$$

$$h_r = (\eta_r/3 - \gamma)^{-1} \tag{77}$$

$$M_r = f_r + 1 \tag{78}$$

$$N_r = h_r - 1 \tag{79}$$

$$\psi_r^a = \left(T\frac{\partial}{\partial T} - b^a V\frac{\partial}{\partial V} + 1 - b^a\right)E_r^a - \beta_r^a\left(T\frac{\partial P^a}{\partial T} - P^a\right) \tag{80}$$

Equations (72), (74), and (75) are coupled radiation equations that give E_r, P_r, and η_r. These equations simplify somewhat for solids and quantum liquids.[7]

The potential form of the renormalization group equations for radiation will now be obtained by a perturbation procedure that is applied to equations (44) through (49). When excitations are present the renormalization group equations (44) through (47) become

$$\frac{\left(\frac{d\gamma}{dP}\right)^- + \left(\frac{d\delta_r}{dP}\right)^-}{\left(1 + \frac{dP_r}{dP}\right)} = \frac{(\gamma + \delta_r)\frac{T}{(\phi + \phi_r)}\left(\frac{\partial\phi}{\partial T} + \frac{\partial\phi_r}{\partial T}\right) - \frac{V}{(\phi + \phi_r)}\left(\frac{\partial\phi}{\partial V} + \frac{\partial\phi_r}{\partial V}\right)}{P + P_r - T\left(\frac{\partial P}{\partial T} + \frac{\partial P_r}{\partial T}\right) + (P + P_r)T\left(\frac{\partial\phi}{\partial T} + \frac{\partial\phi_r}{\partial T}\right)} \tag{81}$$

$$\frac{\left(\frac{d\gamma}{dP}\right)^+ + \left(\frac{d\delta_r}{dP}\right)^+}{\left(1 + \frac{dP_r}{dP}\right)} = \frac{(\gamma + \delta_r)\frac{T}{(\phi + \phi_r)}\left(\frac{\partial\phi}{\partial T} + \frac{\partial\phi_r}{\partial T}\right) - \frac{V}{(\phi + \phi_r)}\left(\frac{\partial\phi}{\partial V} + \frac{\partial\phi_r}{\partial V}\right)}{P + P_r - T\left(\frac{\partial P}{\partial T} + \frac{\partial P_r}{\partial T}\right) - (P + P_r)T\left(\frac{\partial\phi}{\partial T} + \frac{\partial\phi_r}{\partial T}\right)} \tag{82}$$

$$\frac{\left(\frac{db}{dE}\right)^- + \left(\frac{d\beta_r}{dE}\right)^-}{\left(1 + \frac{dE_r}{dE}\right)} = \frac{\frac{T}{(\psi + \psi_r)}\left(\frac{\partial\psi}{\partial T} + \frac{\partial\psi_r}{\partial T}\right) - (b + \beta_r)\frac{V}{(\psi + \psi_r)}\left(\frac{\partial\psi}{\partial V} + \frac{\partial\psi_r}{\partial V}\right)}{E + E_r + V\left(\frac{\partial E}{\partial V} + \frac{\partial E_r}{\partial V}\right) - (E + E_r)V\left(\frac{\partial\psi}{\partial V} + \frac{\partial\psi_r}{\partial V}\right)} \tag{83}$$

$$\frac{\left(\frac{db}{dE}\right)^+ + \left(\frac{d\beta_r}{dE}\right)^+}{\left(1 + \frac{dE_r}{dE}\right)} = \frac{\frac{T}{(\psi + \psi_r)}\left(\frac{\partial\psi}{\partial T} + \frac{\partial\psi_r}{\partial T}\right) - (b + \beta_r)\frac{V}{(\psi + \psi_r)}\left(\frac{\partial\psi}{\partial V} + \frac{\partial\psi_r}{\partial V}\right)}{E + E_r + V\left(\frac{\partial E}{\partial V} + \frac{\partial E_r}{\partial V}\right) + (E + E_r)V\left(\frac{\partial\psi}{\partial V} + \frac{\partial\psi_r}{\partial V}\right)} \tag{84}$$

Expanding equations (81) through (84), keeping only first order terms, and subtracting equations (44) through (47) gives

$$\left(\frac{d\delta_r}{dP}\right)^- = \left(\frac{d\gamma}{dP}\right)^- \frac{dP_r}{dP} + \frac{A_r - \left(\frac{d\gamma}{dP}\right)^- B_r^-}{P - T\frac{\partial P}{\partial T} + PT\frac{\partial\phi}{\partial T}} \tag{85}$$

$$\left(\frac{d\delta_r}{dP}\right)^+ = \left(\frac{d\gamma}{dP}\right)^+ \frac{dP_r}{dP} + \frac{A_r - \left(\frac{d\gamma}{dP}\right)^+ B_r^+}{P - T\frac{\partial P}{\partial T} - PT\frac{\partial\phi}{\partial T}} \tag{86}$$

$$\frac{dP_r}{dP} = \frac{K_{Tr} + \frac{\partial P_r}{\partial T} n \frac{dT}{dn}}{K_T + \frac{\partial P}{\partial T} n \frac{dT}{dn}} \tag{86a}$$

$$A_r = \frac{T}{\phi}\left(\delta_r \frac{\partial\phi}{\partial T} + \gamma \frac{\partial\phi_r}{\partial T}\right) + \frac{\phi_r}{\phi}\left(\frac{V}{\phi}\frac{\partial\phi}{\partial V} - \gamma \frac{T}{\phi}\frac{\partial\phi}{\partial T}\right) - \frac{V}{\phi}\frac{\partial\Phi_r}{\partial V} \tag{87}$$

$$B_r^- = P_r - T\frac{\partial P_r}{\partial T} + P_r T \frac{\partial\phi}{\partial T} + PT\frac{\partial\phi_r}{\partial T} \tag{88}$$

$$B_r^+ = P_r - T\frac{\partial P_r}{\partial T} - P_r T \frac{\partial\Phi}{\partial T} - PT\frac{\partial\phi_r}{\partial T} \tag{89}$$

and

$$\left(\frac{d\beta_r}{dE}\right)^- = \left(\frac{db}{dE}\right)^- \frac{dE_r}{dE} + \frac{C_r - \left(\frac{db}{dE}\right)^- D_r^-}{E + V\frac{\partial E}{\partial V} - EV\frac{\partial\psi}{\partial V}} \tag{90}$$

$$\left(\frac{d\beta_r}{dE}\right)^+ = \left(\frac{db}{dE}\right)^+ \frac{dE_r}{dE} + \frac{C_r - \left(\frac{db}{dE}\right)^+ D_r^+}{E + V\frac{\partial E}{\partial V} + EV\frac{\partial\psi}{\partial V}} \tag{91}$$

$$\frac{dE_r}{dE} = \frac{E_r + P_r - T\frac{\partial P_r}{\partial T} + \frac{\partial E_r}{\partial T}\, n\, \frac{dT}{dn}}{E + P - T\frac{\partial P}{\partial T} + \frac{\partial E}{\partial T}\, n\, \frac{dT}{dn}} \tag{91a}$$

$$C_r = \frac{T}{\psi}\frac{\partial\psi_r}{\partial T} + \frac{\psi_r}{\psi}\left(b\frac{V}{\psi}\frac{\partial\psi}{\partial V} - \frac{T}{\psi}\frac{\partial\psi}{\partial T}\right) - \frac{V}{\psi}\left(b\frac{\partial\psi_r}{\partial V} + \beta_r\frac{\partial\psi}{\partial V}\right) \tag{92}$$

$$D_r^- = E_r + V\frac{\partial E_r}{\partial V} - E_r V\frac{\partial\psi}{\partial V} - EV\frac{\partial\psi_r}{\partial V} \tag{93}$$

$$D_r^+ = E_r + V\frac{\partial E_r}{\partial V} + E_r V\frac{\partial\psi}{\partial V} + EV\frac{\partial\psi_r}{\partial V} \tag{94}$$

where $(d\gamma/dP)^-$ and $(d\gamma/dP)^+$ are given by equations (44) and (46), while $(db/dE)^-$ and $(db/dE)^+$ are given by equations (45) and (47).

The symmetric equations that describe the variation of the radiation gauge parameters are then given by

$$\frac{d\delta_r}{dP} = \frac{1}{2}\left[\left(\frac{d\delta_r}{dP}\right)^- + \left(\frac{d\delta_r}{dP}\right)^+\right] \tag{95}$$

$$\frac{d\beta_r}{dE} = \frac{1}{2}\left[\left(\frac{d\beta_r}{dE}\right)^- + \left(\frac{d\beta_r}{dE}\right)^+\right] \tag{96}$$

Equations (95) and (96) are the potential forms of the renormalization group equations for a thermodynamic system that contains radiation. These equations determine the radiation potentials ϕ_r and ψ_r . The potential function ϕ_r is related to the radiative change in the Debye temperature θ_{Dr} by $\phi_r = \theta_{Dr}/T$ for high temperatures and by $\phi_r = T/\theta_{Dr}$ for low temperatures. The derivatives on the left side of equations (95) and (96) can be evaluated using equations (65) and (68) and the following simple relationships

$$\frac{d\delta_r}{dP} = \frac{n\dfrac{\partial\delta_r}{\partial n} + \dfrac{\partial\delta_r}{\partial T}\, n\, \dfrac{dT}{dn}}{K_T + \dfrac{\partial P}{\partial T}\, n\, \dfrac{dT}{dn}} \tag{97}$$

$$\frac{d\beta_r}{dE} = \frac{n\dfrac{\partial\beta_r}{\partial n} + \dfrac{\partial\beta_r}{\partial T}\, n\, \dfrac{dT}{dn}}{E + P - T\dfrac{\partial P}{\partial T} + \dfrac{\partial E}{\partial T}\, n\, \dfrac{dT}{dn}} \tag{98}$$

5. WAVES IN SOLIDS AND QUANTUM LIQUIDS. Excitations in relativistic solids and quantum liquids have already been considered using some simplifying assumptions.[7] A general procedure for calculating the amplitude and spectrum of relativistic waves in solids and quantum liquids will be outlined here. The energy density and pressure for radiation in these systems is written as

$$E^a_r = E^a_{or} + E^a_{jr}T^j + \cdots \tag{99}$$

$$P^a_r = P^a_{or} + P^a_{jr}T^j + \cdots \tag{100}$$

and

$$E_r = E_{or} + E_{jr}T^j + \cdots \tag{101}$$

$$P_r = P_{or} + P_{jr}T^j + \cdots \tag{102}$$

where

$\mathcal{E}^a_{or}$ and P^a_{or} = nonrelativistic zero-temperature radiation energy density and pressure respectively

$\mathcal{E}^a_{jr}$ and P^a_{jr} = nonrelativistic thermal coefficients for the radiation energy density and pressure respectively

$\mathcal{E}_{or}$ and P_{or} = relativistic zero-temperature radiation energy density and pressure respectively

$\mathcal{E}_{jr}$ and P_{jr} = relativistic thermal coefficients for the radiation energy density and pressure respectively

The zero temperature value of the radiation Grüneisen parameter is obtained from equations (66) and (99) through (102) to be

$$\gamma^a_{or} = \frac{P^a_{jr}}{\mathcal{E}^a_{jr}} \qquad \gamma_{or} = \frac{P_{jr}}{\mathcal{E}_{jr}} \tag{103}$$

The zero temperature values of the nonrelativistic and relativistic radiation bulk modulus is written as $K^a_{or} = ndP^a_{or}/dn$ and $K_{or} = ndP_{or}/dn$ respectively.

The basic relativistic equations describing excitations in solids and quantum liquids are written as (equations 76 and 77 of Reference 7)

$$\mathcal{E}_{or} - 3[(1 + \gamma_o)P_{or} - K_{or}] - 3\,\frac{\mathcal{E}_{jr}}{\mathcal{E}_j}\,P_o(\gamma_{or} - \gamma_o) = \mathcal{E}^a_{or} \tag{104}$$

$$j\mathcal{E}_j(\alpha K_{or} - \beta P_{or}) + \mathcal{E}_{jr}S_{jr} = j\mathcal{E}^a_j\left(\alpha^a K^a_{or} - \beta^a P^a_{or}\right) + \mathcal{E}^a_{jr}T^a_{jr} \tag{105}$$

where

$$S_{jr} = 1 + j + \frac{jP_o\gamma_{or}}{P_o - K_o} + 3n\,\frac{d\gamma_{or}}{dn} - 3(j-1)(\gamma_{or} - \gamma_o)^2 \tag{106}$$

$$T^a_{jr} = 1 + j + \frac{jP^a_o\gamma^a_{or}}{P^a_o - K^a_o} \tag{107}$$

and where

$$\alpha = \frac{\gamma_o P_o}{(P_o - K_o)^2} \qquad \alpha^a = \frac{\gamma_o^a P_o^a}{(P_o^a - K_o^a)^2} \tag{108}$$

$$\beta = \frac{\gamma_o K_o}{(P_o - K_o)^2} \qquad \beta^a = \frac{\gamma_o^a K_o^a}{(P_o^a - K_o^a)^2} \tag{109}$$

Equations (104) and (105) can be deduced directly from equation (70) by using equations (11) and (101). For example, the expression for δ_r that appears in equation (65) can be evaluated for the zero temperature case of solids and quantum liquids to be

$$\delta_{or} = \frac{E_{jr}}{E_j}(\gamma_{or} - \gamma_o) \tag{110}$$

Using the following basic relationships

$$P_{or} = n\frac{dE_{or}}{dn} - E_{or} \tag{111}$$

$$K_{or} = n\frac{dP_{or}}{dn} = n^2\frac{d^2E_{or}}{dn^2} \tag{112}$$

allows equations (104) and (105) to be written as

$$3n^2\frac{d^2E_{or}}{dn^2} - 3(1 + \gamma_o)n\frac{dE_{or}}{dn} + (3\gamma_o + 4)E_{or} + 3\frac{E_{jr}}{E_j}P_o(\gamma_o - \gamma_{or}) = E^a_{or} \tag{113}$$

$$\alpha n^2\frac{d^2E_{or}}{dn^2} - \beta n\frac{dE_{or}}{dn} + \beta E_{or} + \frac{E_{jr}S_{jr}}{jE_j} \tag{114}$$

$$= \frac{E^a_j}{E_j}\left[\alpha^a n^2\frac{d^2E^a_{or}}{dn^2} - \beta^a\left(n\frac{dE^a_{or}}{dn} - E^a_{or}\right)\right] + \frac{E^a_{jr}T^a_{jr}}{jE_j}$$

The quantities S_{jr} and T^a_{jr} are functions of γ_{or} and γ^a_{or}, while the ratios E_{jr}/E_j, E^a_j/E_j, and E^a_{jr}/E_j are functions of γ_{or}, γ^a_{or}, γ_o and γ^a_o.[7] Equations (113) and (114) are simultaneous second order differential equations that determine E_{or} and γ_{or} in terms of E^a_{or}, γ^a_{or}, and in terms of the parameters of the ground state.

Equations (113) and (114) must be solved simultaneously to obtain the relativistic wave amplitude and the relativistic wave number in terms of the corresponding nonrelativistic values. However, in order to obtain an approximate value for the relativistic wave amplitude and wave number, only equation (113) will be used. The zero temperature values of the nonrelativistic and relativistic radiation energy densities are respectively written as[10]

$$E^a_{or} = \frac{1}{4} K^a_o k^2_a A^2_a \qquad (115)$$

$$E_{or} = \frac{1}{4} K_o k^2 A^2 \qquad (116)$$

where k_a and A_a = nonrelativistic wave number and wave amplitude respectively, and k and A = relativistic wave number and wave amplitude respectively. Placing equations (115) and (116) into equation (113) gives

$$\frac{3}{4} K_o n^2 \frac{d^2}{dn^2}(k^2A^2) + \frac{3}{4}\left[2n \frac{dK_o}{dn} - (1 + \gamma_o)K_o\right] n \frac{d}{dn}(k^2A^2) \qquad (117)$$

$$+ \frac{1}{4} k^2A^2\left[3n^2 \frac{d^2K_o}{dn^2} - 3(1 + \gamma_o)n \frac{dK_o}{dn} + (3\gamma_o + 4)K_o\right] + g = \frac{1}{4} k^2_a A^2_a K^a_o$$

where

$$g = 3 \frac{E_{jr}}{E_j} P_o(\gamma_o - \gamma_{or}) \qquad (118)$$

As a crude approximation take $\gamma_{or} = 1/3$, $E_{jr}/E_j = E_{or}/E_o$, and $P_o \sim n^{\sigma_o}$ where σ_o = adiabatic index, and assume kA is not explicitly density dependent, and get

$$g \sim 3E_{or}(\sigma_o - 1)(\gamma_o - \tfrac{1}{3}) \qquad (119)$$

and

$$k^2A^2\left[3n^2 \frac{d^2K_o}{dn^2} - 3(1 + \gamma_o)n \frac{dK_o}{dn} + (3\sigma_o\gamma_o - \sigma_o + 5)K_o\right] = k^2_a A^2_a K^a_o \qquad (120)$$

Using $K_o \sim n^{\sigma_o}$ in equation (120) gives

$$k^2A^2 = \frac{k_a^2A_a^2K_o^a/K_o}{3\sigma_o^2 - 7\sigma_o + 5} \tag{121}$$

$$\approx k_a^2A_a^2\left(\frac{3\sigma_o^2 - 3\sigma_o(2 + \gamma_o) + 3\gamma_o + 4}{3\sigma_o^2 - 7\sigma_o + 5}\right)$$

The relative values of kA and k_aA_a at low and high densities depend on the values of σ_o and γ_o at these densities. For a low density Fermi gas where $\sigma_o = 5/3$ and $\gamma_o = 2/3$ one has $kA = 0.77\ k_aA_a$, for a solid where $\sigma_o \sim 8$ and $\gamma_o \sim 3.83$ the result is $kA = 0.69\ k_aA_a$. The high density limit of equation (121) is somewhat more delicate. If the high density limit is associated with asymptotic freedom, then $\sigma_o = 4/3$ and $\gamma_o = 1/3$ and $kA = k_aA_a$. On the other hand, if at high densities the interactions increase without limit and $\sigma_o \to \infty$, but with γ_o = constant, then $kA = k_aA_a$. However, γ_o is probably a function of σ_o and may be written as $\gamma_o = \sigma_o - 7/3$.[8] In this case equation (121) goes as $\sigma_o/(3\sigma_o^2)$ as $\sigma_o \to \infty$ so that $kA/(k_aA_a) \to 0$. This behaviour contrasts with the results of the first order radiation differential equation approximation of Reference 7, where $kA/(k_aA_a) \to 3\sigma_o^2/\sigma_o$ and is large for $\sigma_o \to \infty$ with γ_o = constant while $kA/(k_aA_a) \to \sigma_o/\sigma_o = 1$ for $\sigma_o \to \infty$ with $\gamma_o = \sigma_o - 7/3$. Finally for the case of asymptotic freedom with $\sigma_o = 4/3$ and $\gamma_o = 1/3$ the first order differential equation approximation gives $kA = 0.65\ k_aA_a$. Thus the effect of relativistic thermodynamics on wave amplitudes is system and model dependent.

6. <u>RELATIVISTIC PHASE VELOCITY.</u> A general procedure is given for determining the relativistic phase velocity for waves in solids and quantum liquids. The procedure will be first to determine E_{or}^a and γ_{or}^a from the values of the nonrelativistic sound speed, then to solve for the relativistic quantities E_{or} and γ_{or} using quations (113) and (114), and then finally working backward to determine the relativistic sound speed from the relativistic energy density and Grüneisen parameter for the radiation.

The nonrelativistic expression for the phase velocity of mechanical waves is given by[8]

$$\left(\frac{W^a}{c}\right)^2 = \frac{K_T^a + \gamma^a T\,\frac{\partial P^a}{\partial T}}{\Sigma^a + \gamma^a\Theta^a} \tag{122}$$

where

$$\Sigma^a = E^a + P^a + K_T^a - T\,\frac{\partial P^a}{\partial T} \tag{123}$$

$$\Theta^a = T\frac{\partial E^a}{\partial T} + T\frac{\partial P^a}{\partial T} \tag{124}$$

where W^a = nonrelativistic sound speed, and c = light speed. The zero temperature limit of equation (122) is[7,8]

$$\left(\frac{W_o^a}{c}\right)^2 = \frac{K_o^a}{E_o^a + P_o^a + K_o^a} \tag{125}$$

where W_o^a = sound speed in a T = 0 solid or quantum liquid.

The nonrelativistic diffuse radiation factor is defined to be

$$\Gamma_r^a = \frac{P_r^a}{E_r^a} \tag{126}$$

For isotropic radiation the diffuse radiation factor can be expressed as follows[10]

$$\Gamma_r^a = \frac{1}{3} + \frac{n}{W^a}\frac{dW^a}{dn} \tag{127}$$

The phase velocity that appears in equation (122) through (124) can be expanded in powers of the temperature, so that the diffuse radiation factor can be written in the following general form

$$\Gamma_r^a = \Gamma_{or}^a + \Gamma_{jr}^a T^j + \cdots \tag{128}$$

Therefore the coefficients of the diffuse radiation are expressed, through the phase velocity W^a , in terms of the ground state functions E_o^a , P_o^a , E_j^a , and P_j^a .

Using equations (99) and (100) to represent the radiation pressure and energy density that appear in the defining equation (126) for the diffuse radiation factor, expanding the denominator, and keeping only first order terms yields the following results

$$\Gamma^a_{or} = \frac{P^a_{or}}{\mathcal{E}^a_{or}} \qquad (129)$$

$$\Gamma^a_{jr} = \frac{\mathcal{E}^a_{jr}}{\mathcal{E}^a_{or}} \left(\gamma^a_{or} - \Gamma^a_{or} \right) \qquad (130)$$

where the left hand side of these equations are obtained from the sound speed.

Equations (129) and (130) can be used to determine $\mathcal{E}^a_{or}$, $\mathcal{E}^a_{jr}$, and γ^a_{or} . For instance, placing equation (111) into equation (129) gives

$$\Gamma^a_{or} = \frac{n}{\mathcal{E}^a_{or}} \frac{d\mathcal{E}^a_{or}}{dn} - 1 \qquad (131)$$

which is a differential equation that can be solved for $\mathcal{E}^a_{or}$, since Γ^a_{or} is known from the ground state parameters through equations (122) and (127). In fact, the solution of equation (131) is

$$\mathcal{E}^a_{or} = nD^a_{or} \exp \left(\int^n \Gamma^a_{or} \frac{dn}{n} \right) \qquad (132)$$

where D^a_{or} = constant. The determination of $\mathcal{E}^a_{jr}$ and γ^a_{or} from equation (130) goes as follows. It is easily shown that[7]

$$\mathcal{E}^a_{jr} = nD^a_{jr} \exp \left[- (j-1) \int^n \gamma^a_{or} \frac{dn}{n} \right] \qquad (133)$$

where D^a_{jr} = constant. Placing equation (133) into equation (130) gives an integral equation which can be solved for γ^a_{or} . In this way the nonrelativistic radiation energy density and Grüneisen parameter, $\mathcal{E}^a_{or}$ and γ^a_{or} respectively, can be determined from the phase velocity.

The corresponding relativistic values of the radiation energy density $\mathcal{E}_{or}$ and Grüneisen parameter γ_{or} are obtained by the solution of the simultaneous equations (113) and (114). The relativistic thermal energy density coefficient is then determined by[7]

$$E_{jr} = nD_{jr} \exp\left[- (j-1) \int^{n} \gamma_{or} \frac{dn}{n} \right] \tag{134}$$

where D_{jr} = constant. The relativistic diffuse radiation factor coefficients are then calculated by

$$\Gamma_{or} = \frac{P_{or}}{E_{or}} = \frac{n}{E_{or}} \frac{dE_{or}}{dn} - 1 \tag{135}$$

$$\Gamma_{jr} = \frac{E_{jr}}{E_{or}} (\gamma_{or} - \Gamma_{or}) \tag{136}$$

The relativistic diffuse radiation factor is then written as

$$\Gamma_r(n,T) = \Gamma_{or} + \Gamma_{jr} T^j + \cdots \tag{137}$$

$$= \frac{P_r}{E_r}$$

Finally the relativistic phase velocity is obtained as a solution to the following equation

$$\Gamma_r = \frac{1}{3} + \frac{n}{W} \frac{dW}{dn} \tag{138}$$

which can be written as

$$\frac{W}{c} = \exp\left[- \int_{n}^{\infty} (\Gamma_r - \frac{1}{3}) \frac{dn}{n} \right] \tag{139}$$

If it is assumed that the diffuse radiation factor is independent of temperature it follows from equations (136) and (137) that

$$\Gamma_r = \Gamma_{or} = \gamma_{or} \tag{140}$$

In this case it follows that $P_{or} = \gamma_{or}\mathcal{E}_{or}$, and the wave equations (104) and (105) reduce to a set of coupled first order differential equations instead of the second order differential equations that appear in equations (113) and (114).[7]

7. SCALING THE WAVE AMPLITUDE. This section obtains a simple expression for the wave amplitude in a T = 0 system. Combining equation (135) with the T = 0 form of equation (138) yields

$$\frac{n}{\mathcal{E}_{or}}\frac{d\mathcal{E}_{or}}{dn} = \frac{4}{3} + \frac{n}{W_o}\frac{dW_o}{dn} \tag{141}$$

where W_o = zero temperature value of the relativistic phase velocity. An equation analogous to (141) holds for the corresponding nonrelativistic quantities. The solution of equation (141) is easily obtained to be

$$\mathcal{E}_{or} = G_r W_o n^{4/3} \tag{142}$$

where G_r = constant independent of n. The relationship between wave number and phase velocity is $W_o = \omega/k$, where ω = angular frequency. Using this in equation (116) gives the following expression for the radiation energy density

$$\mathcal{E}_{or} = \frac{1}{4}\frac{\omega^2 A^2 K_o}{W_o^2} \tag{143}$$

Combining equation (142) and (143) gives the wave amplitude as

$$A^2 = \frac{4G_r W_o^3 n^{4/3}}{\omega^2 K_o} \tag{144}$$

Let n and n_1 be two particle number densities, then it follows from equation (144) that

$$\left[\frac{A(n)}{A(n_1)}\right]^2 = \left[\frac{W_o(n)}{W_o(n_1)}\right]^3 \left(\frac{n}{n_1}\right)^{4/3} \frac{K_o(n_1)}{K_o(n)} \tag{145}$$

which is the scaling equation for wave amplitudes under a change in density in a T = 0 system. Equation (145) is expected to be a good approximation for finite temperature systems if the corresponding finite temperature parameters are used.

8. ELECTROMAGNETIC WAVES IN MATTER. The relativistic calculation of the energy density and phase velocity of electromagnetic waves in matter proceeds in a manner analogous to the case of mechanical waves. The relativistic and nonrelativistic electromagnetic energy densities at zero temperature are given by

$$\mathcal{E}_{or} = \frac{1}{2}(\varepsilon_o E^2 + \mu_o H^2) \tag{146}$$

$$\mathcal{E}^a_{or} = \frac{1}{2}(\varepsilon^a_o E^2_a + \mu^a_o H^2_a) \tag{147}$$

where E and H = relativistic electric and magnetic radiation fields respectively; E_a and H_a = nonrelativistic electric and magnetic radiation fields respectively; ε_o , μ_o and ε^a_o , μ^a_o = zero temperature values of the relativistic and nonrelativistic permittivities and permeabilities respectively. The thermal part of the radiation energy density and pressure is written in the form of equations (101) through (103), and therefore the determination of $\mathcal{E}_{or}$ and γ_{or} is necessary for a relativistic description of electromagnetic waves in matter. The crude approximation $\gamma_{or} = 1/3$ is made in this section, and the problem is to determine ε_o , μ_o , E, and H .

It will be assumed that the nonrelativistic values of the zero temperature values of the permittivity and permeability can be represented by some theoretical expressions in terms of the density, pressure and Grüneisen parameters as follows

$$\varepsilon^a_o = X[n, P^a_o(n), \gamma^a_o(n)] \tag{148}$$

$$\mu^a_o = Y[n, P^a_o(n), \gamma^a_o(n)] \tag{149}$$

Then the relativistic values ε_o and μ_o are determined using the same functional relationships but now evaluated for the relativistic values of the pressure and Grüneisen parameter as follows

$$\varepsilon_o = X[n, P_o(n), \gamma_o(n)] \tag{150}$$

$$\mu_o = Y[n, P_o(n), \gamma_o(n)] \tag{151}$$

The relativistic values of P_o and γ_o are obtained from the solution of the simultaneous equations (14) and (15). Thus the relativistic values of ε_o and μ_o are obtained indirectly from the ground state solution of the relativistic trace equation (1).

A complete relativistic thermodynamic description of electromagnetic waves in matter requires the determination of $\mathcal{E}_{or}$ and γ_{or} by the simultaneous solution of equations (113) and (114). But for an approximate solution only equation (113) can be used with $\gamma_{or} = 1/3$. Placing equation (146) and (147) into equation (113) gives the following differential equations for E and H

$$\frac{3}{2}\,\varepsilon_o n^2 \frac{d^2}{dn^2}(E^2) + \frac{3}{2}\left[2n \frac{d\varepsilon_o}{dn} - (1 + \gamma_o)\varepsilon_o\right]n \frac{d}{dn}(E^2) \tag{152}$$

$$+ \frac{1}{2}\,E^2\left[3n^2 \frac{d^2\varepsilon_o}{dn^2} - 3(1 + \gamma_o)n \frac{d\varepsilon_o}{dn} + (3\gamma_o + 4)\varepsilon_o\right] + g_E = \frac{1}{2}\,\varepsilon_o^a E_a^2$$

$$\frac{3}{2}\,\mu_o n^2 \frac{d^2}{dn^2}(H^2) + \frac{3}{2}\left[2n \frac{d\mu_o}{dn} - (1 + \gamma_o)\mu_o\right]n \frac{d}{dn}(H^2) \tag{153}$$

$$+ \frac{1}{2}\,H^2\left[3n^2 \frac{d^2\mu_o}{dn^2} - 3(1 + \gamma_o)n \frac{d\mu_o}{dn} + (3\gamma_o + 4)\mu_o\right] + g_H = \frac{1}{2}\,\mu_o^a H_a^2$$

where g_E and g_H are obtained from equations (118) and (119) to be given approximately as

$$g_E \sim \frac{3}{2}\,\varepsilon_o E^2(\sigma_o - 1)(\gamma_o - \frac{1}{3}) \tag{154}$$

$$g_H \sim \frac{3}{2}\,\mu_o H^2(\sigma_o - 1)(\gamma_o - \frac{1}{3}) \tag{155}$$

Combining equations (152) through (155) yields the following equations for the relativistic values of the electric and magnetic fields in matter assuming that E and H are not explicitly density dependent

$$E^2\left[3n^2 \frac{d^2\varepsilon_o}{dn^2} - 3(1 + \gamma_o)n \frac{d\varepsilon_o}{dn} + (3\sigma_o\gamma_o - \sigma_o + 5)\varepsilon_o\right] = \varepsilon_o^a E_a^2 \tag{156}$$

$$H^2\left[3n^2 \frac{d^2\mu_o}{dn^2} - 3(1 + \gamma_o)n \frac{d\mu_o}{dn} + (3\sigma_o\gamma_o - \sigma_o + 5)\mu_o\right] = \mu_o^a H_a^2 \tag{157}$$

Assuming $\varepsilon_o \sim n^{\rho_o}$, $\mu_o \sim n^{\nu_o}$, and $P_o \sim n^{\sigma_o}$ in equations (156) and (157) gives the following approximate equations

$$E^2 = \frac{E_a^2 \varepsilon_o^a / \varepsilon_o}{3\rho_o^2 - 6\rho_o - \sigma_o + 3(\sigma_o - \rho_o)\gamma_o + 5} \tag{158}$$

$$H^2 = \frac{H_a^2 \mu_o^a / \mu_o}{3\nu_o^2 - 6\nu_o - \sigma_o + 3(\sigma_o - \nu_o)\gamma_o + 5} \tag{159}$$

where

$$\sigma_o = \frac{n}{P_o}\frac{dP_o}{dn} = \frac{K_o}{P_o} \tag{160}$$

$$\rho_o = \frac{n}{\varepsilon_o}\frac{d\varepsilon_o}{dn} \tag{161}$$

$$\nu_o = \frac{n}{\mu_o}\frac{d\mu_o}{dn} \tag{162}$$

The determination of the relativistic phase velocity for electromagnetic waves in matter proceeds in a manner similar to that for the case of mechanical waves that was treated in equations (122) through (139) except that the temperature dependent nonrelativistic phase velocity is given by

$$\left(\frac{W^a}{c}\right)^2 = (\varepsilon^a \mu^a)^{-1} \tag{163}$$

where ε^a and μ^a = nonrelativistic permittivity and permeability respectively. The zero temperature limit of equation (163) is written as

$$\left(\frac{W_o^a}{c}\right)^2 = (\varepsilon_o^a \mu_o^a)^{-1} \tag{164}$$

From the phase velocities given in equations (163) and (164) one can calculate the nonrelativistic radiation energy density and Grüneisen parameters, E^a_{or} and γ^a_{or} respectively, by the procedure outlined in equations (126) through (133). Then the solution of equations (113) and (114) yields the corresponding relativistic radiation energy density and Grüneisen parameter, E_{or} and γ_{or} respectively. From E_{or} and γ_{or} one obtains an estimate of the relativistic diffuse radiation factor Γ_r by the procedure outlined in equations (134) through (138). Finally the relativistic phase velocity for electromagnetic waves in matter is given by

$$\frac{W}{c} = \exp\left[- \int_o^n \left(\frac{1}{3} - \Gamma_r\right) \frac{dn}{n} \right] \tag{165}$$

If it is assumed that the diffuse radiation factor is independent of temperature the substitution $\Gamma_r = \gamma_{or}$ can be made in equation (165).

9. CONCLUSION. The trace equation of the relativistic thermodynamic ground state is reduced to two Callan-Symanzik type renormalization group equations that connect the matter fields E and P with the thermodynamic gauge fields γ and b . The gauge parameters are necessary to insure that the trace equation is invariant under a local scale transformation. The assumption of local scale invariance under changes of the correlation length of the system leads in a natural way to a set of differential equations for the gauge parameters. These are the potential forms of the renormalization group equations for the ground state. The renormalization group equations for radiation in matter can be written in terms of radiation potentials or in the form of radiative Callan-Symanzik equations. The radiation equations for a general thermodynamic system are applied to waves in solids and quantum liquids, and a set of coupled second order differential equations are developed that determine the relativistic radiation energy density and Grüneisen parameter. Finally, a simple scaling relation is developed for the amplitude of waves propagating through materials of different density.

No mass or energy scale occurs in the equations of relativistic thermodynamics, but the temperature and volume scales that appear in these equations is similar to the mass cutoff parameter that appears in the Callan-Symanzik equations of quantum field theory.[1] Therefore in analogy to the dimensional transmutation of Coleman and Weinberg there may appear a mass associated with the gauge bosons that correspond with the gauge parameters γ_r and b_r .[12] On the other hand, the ground state of a relativistic thermodynamic system may exhibit a broken symmetry in which case the gauge bosons can become massive by the Higgs mechanism.[1] In either case massive thermal gauge bosons should exist that are associated with the thermodynamic gauge parameters γ_r and b_r . The gauge boson associated with the Grüneisen parameter should exist even for T = 0 solids or quantum liquids. Therefore new physical phenomena are expected to occur in bulk matter that is subjected to high pressures. In addition, the results of this paper should have engineering and geophysics applications.

REFERENCES

1. Huang, K., Quarks Leptons and Gauge Fields, World Scientific, Singapore, 1982.

2. Moriyasu, K., An Elementary Primer for Gauge Theory, Heyden & Son, Philadelphia, 1983.

3. Pfeuty, P. and Toulouse, G., Introduction to the Renormalization Group and to Critical Phenomena, John Wiley, New York, 1977.

4. Wilson, K. G. and Kogut, J., "The Renormalization Group and the ε Expansion", Phys. Rep. 12C, p. 75, 1974.

5. Wilson, K. G., "The Renormalization Group and Critical Phenomena", Rev. Mod. Phys. Vol. 55, No. 3, July 1983.

6. Sengers, A. L., Hocken, R., and Sengers, J. V., "Critical Point Universality and Fluids", Physics Today, pp. 42, Dec. 1977.

7. Weiss, R. A., "Relativistic Wave Equations for Solids and Low Temperature Quantum Systems", Third Army Conference on Applied Mathematics and Computing, Georgia Institute of Technology, ARO 86-1, May 13-16 1985, p. 717.

8. Weiss, R. A., Relativistic Thermodynamics, Vols 1 and 2, Exposition Press, New York, 1976.

9. Zharkov, V. N. and Kalinin, V. A., Equations of State for Solids at High Pressures and Temperatures, Consultants Bureau, New York, 1971.

10. Brillouin, L., Tensors in Mechanics and Elasticity, Academic Press, New York, 1964.

11. Boehler, R., "Adiabats of Quartz, Coesite, Olivine, and Magnesium Oxide to 50 KBAR and 1000 K, and the Adiabatic Gradient in the Earth's Mantle", Journal of Geophysical Research, Vol 87, No. B7, 5501-5506, July 10, 1982.

12. Coleman, S. and Weinberg, E., "Radiative Corrections as the Origin of Spontaneous Symmetry Breaking", Phys. Rev., D7, p. 1888, 1974.

C

RELATIVISTIC WAVE EQUATIONS FOR REAL GASES

Richard A. Weiss
Environmental Laboratory
U. S. Army Engineer Waterways Experiment Station
Vicksburg, Mississippi 39180

ABSTRACT. The relativistic wave equation for a generalized thermodynamic system is developed. The solution of this equation is obtained for the real gases whose pressure is described by a virial expansion. A procedure is given for calculating the relativistic amplitude and phase velocity for mechanical waves propagating in real gases. The relativistic wave amplitude is calculated by a virial expansion whose coefficients are determined from the wave equation. The relativistic effects on wave propagation in gases manifest themselves only through the third virial coefficient, and therefore these effects are expected to be observed only at high pressures such as found in atmospheric nuclear explosions, the interaction of directed energy beams with the atmosphere, stellar atmospheres, and in high-pressure-physics laboratory experiments. The effects of curvature waves in spacetime on the pressure of real gases are also considered, and applications to the detection of gravitational radiation are suggested.

1. INTRODUCTION. Local gauge (scale) invariance plays a fundamental role in the description of diverse physical phenomena.[1-3] The requirement of local scale invariance suggests that relativistic thermodynamics can be formulated on the basis of a relativistic trace equation that relates the pressure and internal energy fields to a set of gauge parameters.[4,5] The trace equation for a thermodynamic system can be written as a partial differential equation involving the energy density, pressure, and two gauge parameters.[5] The scale transformations refer to changes in the correlation length of the system, and the scale invariance establishes a correspondence between different physical states of a relativistic thermodynamic system. This correspondence is encompassed by the renormalization group differential equations that describe the variation of the gauge parameters with the magnitude of ambient matter fields such as pressure and energy density.

For the case where the thermodynamic system has a well defined zero temperature state, such as is the case for solids and quantum liquids, the trace equation leads to a set of coupled second order differential equations for the simultaneous determination of the zero temperature values of the pressure and Grüneisen parameter.[4] For real gases whose pressure is described by a virial expansion, the trace equation yields a relativistic expression for the third virial coefficient.[4] This paper derives the relativistic equation for radiation in a generalized thermodynamic system, and then derives the equations that are necessary to calculate the wave amplitudes and phase velocity for waves in real gases. This is done by a perturbation procedure that is applied to the basic trace equation that describes the ground state of a relativistic thermodynamic system.

The trace equation of relativistic thermodynamics is written as[4]

$$U + T\left(\frac{dU}{dT}\right)_{PV} - 3V\,\frac{d}{dV}\,(PV)_U = U^a + T\left(\frac{dU^a}{dT}\right)_{P^aV} \tag{1}$$

where U = relativistic internal energy, P = relativistic pressure, T = absolute temperature, V = volume of substance, and U^a and P^a = corresponding nonrelativistic internal energy and pressure. Throughout this paper the index "a" will refer to nonrelativistic calculations. It is easy to show that equation (1) can be written as follows[4]

$$\frac{\partial}{\partial T}(TU) - bV\,\frac{\partial U}{\partial V} - 3V\left[\frac{\partial}{\partial V}(PV) - \gamma\,\frac{\partial U}{\partial V}\right] \tag{2}$$

$$= \frac{\partial}{\partial T}(TU^a) - b^aV\,\frac{\partial U^a}{\partial V}$$

where

$$\gamma = \frac{V}{C_V}\left(\frac{\partial P}{\partial T}\right)_V \tag{3}$$

$$b = \frac{T(\partial P/\partial T)_V}{(P - K_T)} \tag{4}$$

$$b^a = \frac{T(\partial P^a/\partial T)_V}{(P^a - K_T^a)} \tag{5}$$

and where γ = relativistic Grüneisen parameter, C_V = relativistic heat capacity at constant volume, and where

$$K_T = -\,V\left(\frac{\partial P}{\partial V}\right)_T \tag{6}$$

$$K_T^a = -V\left(\frac{\partial P^a}{\partial V}\right)_T \tag{7}$$

are the relativistic and nonrelativistic values of the bulk modulus respectively. The nonrelativistic Grüneisen parameter is defined as follows

$$\gamma^a = \frac{V}{C_V^a}\left(\frac{\partial P^a}{\partial T}\right)_V \tag{8}$$

where C_V^a = nonrelativistic heat capacity and constant volume. Equation (2) can be rewritten in terms of the energy density as follows[5]

$$\left(1 - b + T\frac{\partial}{\partial T} - bV\frac{\partial}{\partial V}\right)E - 3\left(1 + \gamma + V\frac{\partial}{\partial V} - \gamma T\frac{\partial}{\partial T}\right)P \tag{9}$$

$$= \left(1 - b^a + T\frac{\partial}{\partial T} - b^a V\frac{\partial}{\partial V}\right)E^a$$

where $E = U/V$ = relativistic energy density, and $E^a = U^a/V$ = nonrelativistic energy density. The parameters γ and b are the two gauge parameters of relativistic thermodynamics.[5]

Wave motion in relativistic gases can be of two types. The first corresponds to mechanical vibrations of the gas which results in pressure changes in time and space. This type of wave motion is described by a relativistic wave equation for real gases. Such an equation can be developed by first considering relativistic waves in a completely general thermodynamic medium and then specializing to the case of real gases. It is required to find both the relativistic amplitude and sound speed for waves in real gases. In order to do this the nonrelativistic wave amplitude and phase velocity must first be determined. The relativistic effects appear only in the third and higher virial coefficients of the real gas state equation.[4] Therefore it is necessary to solve the relativistic radiation equation for real gases to determine the relativistic value of the third virial coefficient for radiation in the real gas system. The relativistic diffuse radiation factor and the relativistic sound speed in real gases are then determined from the values of the relativistic third virial coefficient for radiation. These effects should be important only for real gases at high pressures where the third virial coefficient contributes significantly to the equation of state.

A second kind of wave motion in gases can be induced by the coupling of the wave motions in spacetime with some characteristic parameter of real gases. The wave motions in spacetime are gravitational waves. The attempts at detecting gravitational radiation over the past twenty years by various methods including the use of solid body resonance detectors have not been successful.[6–8] This may be due to the lack of adequate sensitivity of present day detectors because the cosmic sources of gravity waves are thought to be very weak.[9–11] On the other hand the lack of positive experimental results using solid body detectors may be due to a basic insensitivity of this type of detector, and it has been suggested that real gases and liquids may be better suited for a detector material because the third virial coefficient is expected to be sensitive to changes in the metric of spacetime.[12] This paper calculates the adiabatic changes in gas pressure that are expected to occur in a detector that is subjected to the tidal effects of gravity waves.

The procedure followed in this paper is to: a) review the theory of the relativistic ground state of real gases, b) determine the equations that describe relativistic waves in a generalized thermodynamic medium, c) develop a simple nonrelativistic calculation of the amplitude of waves in real gases, d) determine the solution of the wave equation for real gases by performing a perturbation calculation on the relativistic ground state equation for real gases, e) determine the relativistic values of the wave amplitude and phase velocity, f) determine the adiabatic changes of pressure, volume, and temperature for a real gas that is interacting with gravitational radiation.

2. RELATIVISTIC GROUND STATE OF REAL GASES. The form of the solution of the trace equation (1) depends on the type of physical system being considered. For real gases the nonrelativistic and relativistic pressure, energy density, bulk modulus, and molar heat capacity (specific heat) are written in virial form respectively as[13,14]

$$P^a = nR^aT[1 + nB^a(T) + n^2C^a(T) + \cdots] \tag{10}$$

$$\mathcal{E}^a = nR^aT\left[\frac{3}{2} - nT\,\frac{\partial B^a}{\partial T} - \frac{1}{2}\,n^2T\,\frac{\partial C^a}{\partial T} - \cdots\right] \tag{11}$$

$$K_T^a = nR^aT[1 + 2nB^a(T) + 3n^2C^a(T) + \cdots] \tag{12}$$

$$\tilde{C}_V^a = R^a\left[\frac{3}{2} - n\left(T^2\,\frac{\partial^2B^a}{\partial T^2} + 2T\,\frac{\partial B^a}{\partial T}\right) - \frac{1}{2}\,n^2\left(T^2\,\frac{\partial^2C^a}{\partial T^2} + 2T\,\frac{\partial C^a}{\partial T}\right) - \cdots\right] \tag{13}$$

and

$$P = nRT[1 + nB(T) + n^2C(T) + \cdots] \tag{14}$$

$$E = nRT\left[\frac{3}{2} - nT\frac{\partial B}{\partial T} - \frac{1}{2}n^2T\frac{\partial C}{\partial T} - \cdots\right] \qquad (15)$$

$$K_T = nRT[1 + 2nB(T) + 3n^2C(T) + \cdots] \qquad (16)$$

$$\tilde{C}_V = R\left[\frac{3}{2} - n\left(T^2\frac{\partial^2 B}{\partial T^2} + 2T\frac{\partial B}{\partial T}\right) - \frac{1}{2}n^2\left(T^2\frac{\partial^2 C}{\partial T^2} + 2T\frac{\partial C}{\partial T}\right) - \cdots\right] \qquad (17)$$

where

$$n = N/V = 1/\tilde{V} \qquad (18)$$

where N = number of moles, $\tilde{V}$ = molar volume; R^a, $B^a(T)$ and $C^a(T)$ = nonrelativistic values of the gas constant, second virial coefficient, and third virial coefficient respectively; R, B(T), and C(T) = relativistic values of the gas constant, second virial coefficient, and third virial coefficient respectively; and $\tilde{C}_V^a$ and $\tilde{C}_V$ = nonrelativistic and relativistic values of the molar heat capacity (specific heat) respectively. The relationship between the extensive, intensive, and molar quantities that are used in this paper is as follows

$$C_V^a = N\tilde{C}_V^a = \left(\frac{\partial U^a}{\partial T}\right)_V \qquad (19)$$

$$C_V = N\tilde{C}_V = \left(\frac{\partial U}{\partial T}\right)_V \qquad (20)$$

$$E^a = n\tilde{U}^a = U^a/V \qquad (21)$$

$$E = n\tilde{U} = U/V \qquad (22)$$

where $\tilde{U}^a$ and $\tilde{U}$ = nonrelativistic and relativistic internal energy per mole.

The relationship between the relativistic and the nonrelativistic functions that appear in equations (10) through (17) are given by[4]

$$R = R^a \tag{23}$$

$$B(T) = B^a(T) \tag{24}$$

$$C(T) = C^a(T) - 3[B^a(T)]^2 \ln \psi^a \tag{25}$$

where

$$\psi^a = \frac{T}{T_R} \left| \frac{B^a(T)}{B^a(T_R)} \right|^{2/3} = \frac{T}{T_{CR}} \left| \frac{B^a(T)}{B^a(T_{CR})} \right|^{2/3} \tag{26}$$

and where T_R = relativity temperature constant, and T_{CR} = conjugate relativity temperature constant. The relationship between T_R and T_{CR} is shown in Figure 1. Thus the relativistic effects enter the real gas state equation only through the third and higher virial coefficients; the ideal gas term and the second virial coefficient are unaffected.

The relativity temperature T_R and the conjugate relativity temperature T_{CR} are related to the critical temperature of a real gas. The conditions for the critical point can be expressed in terms of the second and third virial coefficients as follows[15]

$$B(T_{crit}) = -\tilde{V}(T_{crit}) \tag{27}$$

$$3C(T_{crit}) = \tilde{V}^2(T_{crit}) \tag{28}$$

or equivalently

$$3C(T_{crit}) = B^2(T_{crit}) \tag{29}$$

where T_{crit} = critical temperature. Equations (24), (25), and (29) give the critical point condition as[16]

$$C^a = \frac{1}{3}[B^a]^2(1 + 9 \ln \psi^a) \tag{30}$$

and gives the relationship between T_{crit} and T_R (or T_{CR}) that is shown in Figure 2. Figure 3 gives the dependence of the critical molar volume on the relativity temperature.

The Grüneisen function can be evaluated for real gases using equation (14) which gives

$$\left(\frac{\partial P}{\partial T}\right)_n = nR[1 + nf_1(T) + n^2 f_2(T) + \cdots] \tag{31}$$

where

$$f_1(T) = T\frac{\partial B}{\partial T} + B \tag{32}$$

$$f_2(T) = T\frac{\partial C}{\partial T} + C \tag{33}$$

and equation (17) which gives

$$\frac{1}{\tilde{C}_V} = \frac{2}{3R}[1 + ng_1(T) + n^2 g_2(T) + \cdots] \tag{34}$$

where

$$g_1(T) = \frac{2}{3}\left(T^2\frac{\partial^2 B}{\partial T^2} + 2T\frac{\partial B}{\partial T}\right) \tag{35}$$

$$g_2(T) = \frac{1}{3}\left(T^2\frac{\partial^2 C}{\partial T^2} + 2T\frac{\partial C}{\partial T}\right) + \frac{4}{9}\left(T^2\frac{\partial^2 B}{\partial T^2} + 2T\frac{\partial B}{\partial T}\right)^2 \tag{36}$$

Then equations (3), (31), and (34) give the relativistic Grüneisen parameter as

$$\gamma = \frac{2}{3} [1 + n\gamma_1(T) + n^2\gamma_2(T) + \cdots] \tag{37}$$

where

$$\gamma_1(T) = f_1(T) + g_1(T) \tag{38}$$

$$\gamma_2(T) = f_2(T) + g_2(T) + f_1(T)g_1(T) \tag{39}$$

Expressions analogous to equations (31) through (39) hold for the nonrelativistic Grüneisen parameter.

3. EXCITATIONS IN THERMODYNAMIC SYSTEMS. This section considers mechanical radiation in thermal media. Only small amplitude vibrations are considered. When radiation is present in a thermal system the relativistic energy density, pressure, bulk modulus, and heat capacity are written as, $E + E_r$, $P + P_r$, $K_T + K_{Tr}$, and $C_V + C_{Vr}$ respectively, while the corresponding nonrelativistic quantities become $E^a + E^a_r$, $P^a + P^a_r$, $K^a_T + K^a_{Tr}$, and $C^a_V + C^a_{Vr}$ respectively, where

E^a_r and E_r = nonrelativistic and relativistic radiation energy density respectively

P^a_r and P_r = nonrelativistic and relativistic radiation pressure respectively

$K_{Tr} = n\left(\frac{\partial P_r}{\partial n}\right)_T$ = relativistic bulk modulus of the radiation

$K^a_{Tr} = n\left(\frac{\partial P^a_r}{\partial n}\right)_T$ = nonrelativistic bulk modulus of the radiation

$C_{Vr} = V\left(\frac{\partial E_r}{\partial T}\right)_V$ = relativistic heat capacity of radiation

$C^a_{Vr} = V\left(\frac{\partial E^a_r}{\partial T}\right)_V$ = nonrelativistic heat capacity of radiation

The radiation terms are assumed to be much smaller than the ground state terms, i.e., $E_r \ll E$ and $P_r \ll P$.

The Grüneisen parameter γ and the gauge parameter b become $\gamma + \delta_r$ and $b + \beta_r$, where δ_r and β_r = incremental changes in the parameters γ and b when radiation is present in the system. The increment in the Grüneisen parameter of the system due to the presence of small amplitude radiation is obtained from the defining equation (3) by noting that

$$\gamma + \delta_r = \frac{V}{C_V + C_{Vr}} \frac{\partial}{\partial T} (P + P_r) = \frac{V}{C_V\left(1 + \frac{C_{Vr}}{C_V}\right)} \frac{\partial}{\partial T} (P + P_r) \tag{40}$$

Expanding the denominator in equation (40), keeping only first order terms, and finally substracting equation (3) gives

$$\delta_r = \frac{V}{C_V} \frac{\partial E_r}{\partial T} (\gamma_r - \gamma) \tag{41}$$

$$= \frac{V}{C_V} \left[E_r \frac{\partial \Gamma_r}{\partial T} + (\Gamma_r - \gamma) \frac{\partial E_r}{\partial T} \right] \tag{42}$$

where γ_r = relativistic Grüneisen parameter of the radiation itself, and is defined as

$$\gamma_r = \frac{V}{C_{Vr}} \frac{\partial P_r}{\partial T} = \frac{\partial P_r}{\partial T} \Big/ \frac{\partial E_r}{\partial T} \tag{43}$$

and where Γ_r = relativistic diffuse radiation factor which is defined by

$$\Gamma_r = \frac{P_r}{E_r} \tag{44}$$

Note that a comparison of equations (41) and (42) shows that if Γ_r is independent of temperature, then $\Gamma_r = \gamma_r$.

Similarly, the increment in the gauge parameter b due to the presence of radiation in the medium is obtained from equation (4) by observing that

$$b + \beta_r = \frac{T\left(\frac{\partial P}{\partial T} + \frac{\partial P_r}{\partial T}\right)}{P - K_T + P_r - K_{Tr}} = \frac{T\left(\frac{\partial P}{\partial T} + \frac{\partial P_r}{\partial T}\right)}{\left(P - K_T\right)\left(1 + \frac{P_r - K_{Tr}}{P - K_T}\right)} \tag{45}$$

Expanding the denominator in equation (45), keeping only first order terms, and subtracting equation (4) gives

$$\beta_r = \frac{T \frac{\partial P_r}{\partial T}}{P - K_T} - \frac{T \frac{\partial P}{\partial T}\left(P_r - K_{Tr}\right)}{\left(P - K_T\right)^2} \tag{46}$$

$$= \frac{\gamma_r T \frac{\partial E_r}{\partial T}}{P - K_T} - \frac{\gamma T \frac{\partial E}{\partial T}\left(P_r - K_{Tr}\right)}{\left(P - K_T\right)^2}$$

$$= \frac{T}{P - K_T} \frac{\partial}{\partial T}\left(\Gamma_r E_r\right) - \frac{T \frac{\partial P}{\partial T}}{\left(P - K_T\right)^2}\left[\Gamma_r E_r + V \frac{\partial}{\partial V}\left(\Gamma_r E_r\right)\right]$$

Note that equation (46) can be rewritten as

$$\beta_r = b_r \frac{P_r - K_{Tr}}{P - K_T} - \frac{T \frac{\partial P}{\partial T}\left(P_r - K_{Tr}\right)}{\left(P - K_T\right)^2} \tag{47}$$

where b_r = radiation gauge parameter given by

$$b_r = \frac{T \frac{\partial P_r}{\partial T}}{P_r - K_{Tr}} \tag{48}$$

The parameters γ_r and b_r are the two gauge parameters for radiation in a thermal medium.

The corresponding nonrelativistic values of the δ_r and β_r are given by

$$\delta_r^a = \frac{V}{C_V^a} \frac{\partial E_r^a}{\partial T} (\gamma_r^a - \gamma^a) \tag{49}$$

$$\beta_r^a = \frac{T}{P^a - K_T^a} \frac{\partial P_r^a}{\partial T} - \frac{T \frac{\partial P^a}{\partial T}}{\left(P^a - K_T^a\right)^2} \left[\Gamma_r^a E_r^a + V \frac{\partial}{\partial V} \left(\Gamma_r^a E_r^a\right)\right] \tag{50}$$

where γ_r^a is given by the nonrelativistic analog of equation (43), and Γ_r^a is given by the nonrelativistic analog of equation (44). Note also that δ_r and β_r are small quantities because E_r is assumed to be small compared to E. But the radiation gauge parameters γ_r and b_r, and the diffuse radiation factor Γ_r, are not small quantities since they are defined as the ratio of two small numbers.

When radiation is present in a general thermodynamic system, equation (9) can be written as

$$\begin{aligned} &\left[1 - \left(b + \beta_r\right) + T \frac{\partial}{\partial T} - \left(b + \beta_r\right) V \frac{\partial}{\partial V}\right] \left(E + E_r\right) \\ &\quad - 3\left[1 + \gamma + \delta_r + V \frac{\partial}{\partial V} - \left(\gamma + \delta_r\right) T \frac{\partial}{\partial T}\right]\left(P + P_r\right) \\ &\quad = \left[1 - \left(b^a + \beta_r^a\right) + T \frac{\partial}{\partial T} - \left(b^a + \beta_r^a\right) V \frac{\partial}{\partial V}\right]\left(E^a + E_r^a\right) \end{aligned} \tag{51}$$

Subtracting equation (9) from equation (51) and keeping only first order terms yields the following first order radiation equation

$$\begin{aligned} &\left(1 - b + T \frac{\partial}{\partial T} - bV \frac{\partial}{\partial V}\right) E_r - \beta_r\left(T \frac{\partial P}{\partial T} - P\right) \\ &\quad - 3\left[\left(1 + \gamma + V \frac{\partial}{\partial V} - \gamma T \frac{\partial}{\partial T}\right) P_r - \delta_r\left(T \frac{\partial P}{\partial T} - P\right)\right] \\ &\quad = \left(1 - b^a + T \frac{\partial}{\partial T} - b^a V \frac{\partial}{\partial V}\right) E_r^a - \beta_r^a\left(T \frac{\partial P^a}{\partial T} - P^a\right) \end{aligned} \tag{52}$$

where the following standard thermodynamic relationship was used

$$\frac{\partial U}{\partial V} = E + V\frac{\partial E}{\partial V} = T\frac{\partial P}{\partial T} - P \tag{53}$$

Equation (52) can also be written as

$$\frac{\partial}{\partial T}\left(TU_r\right) - bV\frac{\partial U_r}{\partial V} - \beta_r V\frac{\partial U}{\partial V} \tag{54}$$

$$- 3V\left[\frac{\partial}{\partial V}\left(VP_r\right) - \gamma\frac{\partial U_r}{\partial V} - \delta_r\frac{\partial U}{\partial V}\right]$$

$$= \frac{\partial}{\partial T}\left(TU_r^a\right) - b^a V\frac{\partial U_r^a}{\partial V} - \beta_r^a V\frac{\partial U^a}{\partial V}$$

where $U_r = VE_r$ = relativistic radiation internal energy, and $U_r^a = VE_r^a$ = nonrelativistic radiation internal energy. Equation (52) or equation (54) can serve as the basic first order relativistic thermodynamic equation governing radiation in a thermal medium. Equations (52) and (54) are completely general and can be used to derive the radiation equations for real gases. To do this it is first necessary to develop a nonrelativistic theory of mechanical radiation in real gases so that the terms on the right hand side of equations (52) or (54) can be evaluated.

4. NONRELATIVISTIC THEORY OF WAVES IN REAL GASES. A simple nonlinear nonrelativistic calculation of the radiation energy density and pressure for waves in real gases is presented that will allow the calculation of the nonrelativistic amplitude of the waves. The nonrelativistic radiation pressure and energy density are written in a virial form analogous to the ground state equations (10) and (11) as follows

$$P_r^a = \Gamma_{ro}^a nR_r^a T + n^2 R^a TB_r^a(T) + n^3 R^a TC_r^a(T) + \cdots \tag{55}$$

$$E_r^a = nR_r^a T - n^2 R^a T^2\frac{\partial B_r^a}{\partial T} - \frac{1}{2}n^3 R^a T^2\frac{\partial C_r^a}{\partial T} - \cdots \tag{56}$$

where $\Gamma_{ro}^a = 1/3$ = diffuse radiation factor for an ideal gas, and where the nonrelativistic radiation coefficients R_r^a , $B_r^a(T)$, and $C_r^a(T)$ are to be determined from a simple model that describes the vibrations in a real gas.

The form of the energy density in equation (56) follows from equation (55) by the requirement that

$$\frac{\partial U_r^a}{\partial V} = T\frac{\partial P_r^a}{\partial T} - P_r^a = \mathcal{E}_r^a - n\frac{\partial \mathcal{E}_r^a}{\partial n} \qquad (57)$$

The functions δ_r^a, β_r^a, γ_r^a, and Γ_r^a that appear in the right hand side of the wave equation (52) can be calculated in terms of the nonrelativistic radiation virial coefficients from equations (55) and (56) by using equations (49) and (50) and the nonrelativistic analogs of equations (43) and (44).

The nonrelativistic vibrations in a mechanical medium have an energy density given by[17]

$$\mathcal{E}_r^a = \frac{1}{4}k_a^2A_a^2K_T^a \qquad (58)$$

where k_a = nonrelativistic wave number, and A_a = nonrelativistic wave amplitude. The wave number and wave amplitude that appear in equation (58) are also expected to have a virial expansion of the form

$$k_a^2A_a^2 = k_o^2A_o^2\left[1 + n\alpha_1^a(T) + n^2\alpha_2^a(T) + \cdots\right] \qquad (59)$$

where α_1^a and α_2^a are unknown functions of temperature that are to be determined, and where k_o and A_o = known wave number and wave amplitude respectively associated with waves in an ideal gas. Combining equations (12), (58) and (59) gives

$$\mathcal{E}_r^a = \frac{1}{4}k_o^2A_o^2nR^aT\left[1 + n\left(2B^a + \alpha_1^a\right) + n^2\left(3C^a + 2\alpha_1^aB^a + \alpha_2^a\right) + \cdots\right] \qquad (60)$$

Comparing equations (56) and (60) gives

$$R_r^a = \frac{1}{4}k_o^2A_o^2R^a \qquad (61)$$

$$-T\frac{\partial B_r^a}{\partial T} = \frac{1}{4}k_o^2A_o^2\left(2B^a + \alpha_1^a\right) \qquad (62)$$

$$-T \frac{\partial C_r^a}{\partial T} = \frac{1}{2} k_o^2 A_o^2 \left(3C^a + 2\alpha_1^a B^a + \alpha_2^a\right) \tag{63}$$

Equation (61) immediately determines the value of the radiation coefficient R_r^a, but further equations in addition to equations (62) and (63) are needed to determine the radiation virial coefficients B_r^a and C_r^a. This is so because the functions α_1^a and α_2^a are also unknown and need to be determined.

The additional equations needed in conjunction with equations (62) and (63) are those involving the nonrelativistic diffuse radiation factor Γ_r^a defined by

$$P_r^a = \Gamma_r^a E_r^a \tag{64}$$

Combining equations (55), (56) and (64) gives the following expression for Γ_r^a

$$\Gamma_r^a = \Gamma_{ro}^a + n\Gamma_{r1}^a + n^2\Gamma_{r2}^a + \cdots \tag{65}$$

where as before $\Gamma_{ro}^a = 1/3$ and

$$\Gamma_{r1}^a = \frac{R^a}{R_r^a}\left[B_r^a + \Gamma_{ro}^a T \frac{\partial B_r^a}{\partial T}\right] \tag{66}$$

$$\Gamma_{r2}^a = \frac{R^a}{R_r^a}\left[C_r^a + \frac{1}{2}\Gamma_{ro}^a T \frac{\partial C_r^a}{\partial T} + \Gamma_{r1}^a T \frac{\partial B_r^a}{\partial T}\right] \tag{67}$$

where R^a/R_r^a is given by equation (61). But it is well known that the general expression for the diffuse radiation factor is[17]

$$\Gamma_r^a = \frac{1}{3} + \frac{n}{W^a}\frac{dW^a}{dn} \tag{68}$$

where W^a = phase velocity of mechanical waves in a thermodynamic medium. The phase velocity of waves in a general thermodynamic medium is given by[4]

$$\left(\frac{W^a}{c}\right)^2 = \frac{K_T^a + \gamma^a T \frac{\partial P^a}{\partial T}}{\Sigma^a + \gamma^a \Theta^a} \tag{69}$$

where

$$\Sigma^a = E^a + P^a + K_T^a - T\frac{\partial P^a}{\partial T} \tag{70}$$

$$\Theta^a = T\frac{\partial E^a}{\partial T} + T\frac{\partial P^a}{\partial T} \tag{71}$$

Thus in general $W^a = W^a(n,T)$ and substitution of equation (69) into equation (68) and expanding in terms of the ground state virial coefficients automatically determines the expansion coefficients for the diffuse radiation factor given in equation (65). Therefore it will be assumed that Γ_{ro}^a , Γ_{r1}^a , Γ_{r2}^a , and so on, can be obtained from the sound speed and are known functions of temperature through the ground state virial expansion coefficients. Then equations (66) and (67) can be integrated to obtain the nonrelativistic radiation virial coefficients $B_r^a(T)$ and $C_r^a(T)$. Finally equations (62) and (63) can be used to calculate α_1^a and α_2^a as follows

$$\alpha_1^a(T) = -\frac{4}{k_o^2 A_o^2} T \frac{\partial B_r^a}{\partial T} - 2B^a \tag{72}$$

$$\alpha_2^a(T) = -\frac{2}{k_o^a A_o^2} T \frac{\partial C_r^a}{\partial T} - 3C^a - 2\alpha_1^a B^a \tag{73}$$

Then $k_a A_a$ can be determined from equation (59). It will be assumed that $k_a = k_o$ and therefore equation (59) gives the nonrelativistic wave amplitude.

5. SOLUTION OF THE WAVE EQUATION FOR REAL GASES. The solution of the radiation equation (52) for the real gases can be most easily obtained from equations (14), (15) and (23) through (26) that describe the relativistic ground state solution of equation (1) for the real gases. The relativistic expressions for the radiation pressure and energy density are written in a form analogous to equations (55) and (56) as follows

$$P_r = \Gamma_{ro} n R_r T + n^2 RTB_r(T) + n^3 RTC_r(T) + \cdots \tag{74}$$

$$E_r = nR_r T - n^2 RT^2 \frac{\partial B_r}{\partial T} - \frac{1}{2} n^3 RT^2 \frac{\partial C_r}{\partial T} - \cdots \tag{75}$$

where the relativistic radiation parameters Γ_{ro} and R_r , and the relativistic radiation virial coefficients B_r and C_r, are to be determined from the solution of the radiation equation (52). The functions δ_r , β_r , γ_r , and Γ_r that appear in equation (52) can be calculated from equations (74) and (75) by using equations (42), (43), (44), and (46).

The solution of the radiation equation (52) for the real gases can be immediately obtained from the ground state solution of equation (1), as given by equations (23) through (26) for the real gases, by a simple perturbation method applied to this solution. Thus when mechanical radiation is present in a real gas, equations (23) through (26) become

$$R + R_r = R^a + R_r^a \tag{76}$$

$$B(T) + B_r(T) = B^a(T) + B_r^a(T) \tag{77}$$

$$C(T) + C_r(T) = C^a(T) + C_r^a(T) - 3\left[B^a(T) + B_r^a(T)\right]^2 \ell n \left(\psi^a + \psi_r^a\right) \tag{78}$$

Subtracting equations (23) through (25) from equations (76) through (78) respectively and keeping only first order terms yields

$$R_r = R_r^a \tag{79}$$

$$B_r = B_r^a \tag{80}$$

$$C_r = C_r^a - 3[2B^a B_r^a + (B_r^a)^2] \, \ell n \, \psi^a - 3(B^a + B_r^a)^2 \, \ell n \left(1 + \frac{\psi_r^a}{\psi^a}\right) \tag{81}$$

$$C_r \approx C_r^a - 6B^a B_r^a \, \ell n \, \psi^a - 3(B^a)^2 \psi_r^a / \psi^a \tag{81a}$$

where the following first order approximation has been used

$$\ell n \left(\psi^a + \psi_r^a \right) = \ell n \ \psi^a + \ell n \left(1 + \frac{\psi_r^a}{\psi^a} \right) \approx \ell n \ \psi^a + \frac{\psi_r^a}{\psi^a} \tag{82}$$

to obtain the result in equation (81a). The small radiation term ψ_r^a that occurs in equation (81a) is obtained from the defining equation (26) as follows

$$\psi^a + \psi_r^a = \frac{T}{T_R} \left| \frac{B^a(T) + B_r^a(T)}{B^a(T_R) + B_r^a(T_R)} \right|^{2/3} \tag{83}$$

Expanding the denominator in equation (83), and subtracting equation (26), and finally dividing by ψ^a yields the following first order approximation

$$\frac{\psi_r^a}{\psi^a} \approx \frac{2}{3} \left[\frac{B_r^a(T)}{B^a(T)} - \frac{B_r^a(T_R)}{B^a(T_R)} \right] \tag{84}$$

$$\approx \frac{2}{3} \left[\frac{B_r^a(T)}{B^a(T)} - \frac{B_r^a(T_{CR})}{B^a(T_{CR})} \right]$$

Equations (79) through (81) give the relativistic radiation virial coefficients in terms of the corresponding nonrelativistic radiation virial coefficients and in terms of the second order ground state virial coefficient $B^a(T)$. Note that at the Boyle temperature T_B , at which $B^a(T_B) = 0$, it follows from equation (81a) that $C_r(T_B) = C_r^a(T_B)$. Also note that at the relativity temperature T_R (or at the conjugate relativity temperature T_{CR}) it follows from equation (26) that $\psi^a = 1$, and from equation (84) that $\psi_r^a = 0$, so that $C_r(T_R) = C_r^a(T_R)$ and $C_r(T_{CR}) = C_r^a(T_{CR})$. Therefore any experimental test that is conducted to determine the difference between $C_r(T)$ and $C_r^a(T)$ should exclude the temperature regions around T_B , T_R and T_{CR} . A similar result is already known for the ground state third virial coefficient.[4]

6. RELATIVISTIC WAVE AMPLITUDE AND PHASE VELOCITY.

Relativistic effects on waves in real gases will manifest themselves in the amplitude and dispersive properties of the waves. Therefore it is important to be able to calculate the relativistic amplitude and phase velocity of waves in real gases and to compare them with their corresponding nonrelativistic values. The relativistic energy density for mechanical waves in a real gas is written in analogy to equation (58) as[17]

$$E_r = \frac{1}{4} k^2 A^2 K_T \tag{85}$$

where k = relativistic wave number, A = relativistic amplitude, and where K_T is given by equation (16). In a form similar to that of equation (59), the product k^2A^2 is written as

$$k^2 A^2 = k_o^2 A_o^2 \left[1 + n\alpha_1(T) + n^2\alpha_2(T) + \cdots\right] \tag{86}$$

where the relativistic functions $\alpha_1(T)$ and $\alpha_2(T)$ need to be determined. Combining equations (16), (75), (85), and (86) gives

$$R_r = \frac{1}{4} k_o^2 A_o^2 R = R_r^a \tag{87}$$

$$-T \frac{\partial B_r}{\partial T} = \frac{1}{4} k_o^2 A_o^2 (2B + \alpha_1) \tag{88}$$

$$-T \frac{\partial C_r}{\partial T} = \frac{1}{2} k_o^2 A_o^2 (3C + 2\alpha_1 B + \alpha_2) \tag{89}$$

Because $B(T) = B^a(T)$ and $B_r(T) = B_r^a(T)$ it follows from equations (62) and (88) that $\alpha_1 = \alpha_1^a$. The value of α_2 is obtained from equation (89) to be

$$\alpha_2 = -\frac{2}{k_o^2 A_o^2} T \frac{\partial C_r}{\partial T} - 3C - 2\alpha_1^a B^a \tag{90}$$

where C_r is given by equation (81) and C is given by equation (25). In this way the relativistic expression for k^2A^2 given by equation (86) can be calculated in terms of nonrelativistic quantities. Since $\alpha_1 = \alpha_1^a$, it is clear that relativistic effects affect only the second order and higher terms in equation (86). Essentially this is due to the fact that only the third and higher virial coefficients of the ground state are affected by relativity as shown in equations (24) and (25).

The relativistic phase velocity can be obtained by first noting that the relativistic diffuse radiation factor is obtained from equations (44), (74), and (75) to be

$$\Gamma_r = \Gamma_{ro} + n\Gamma_{r1} + n^2\Gamma_{r2} + \cdots \tag{91}$$

where

$$\Gamma_{ro} = \Gamma^a_{ro} = 1/3 \tag{92}$$

$$\Gamma_{r1} = \Gamma^a_{r1} \tag{93}$$

$$\Gamma_{r2} = \frac{R}{R_r}\left[C_r + \frac{1}{2}\Gamma_{ro}T\frac{\partial C_r}{\partial T} + \Gamma_{r1}T\frac{\partial B_r}{\partial T}\right] \tag{94}$$

where C_r is given by equation (81). Therefore $\Gamma_r(n,T)$ can be evaluated in terms of nonrelativistic quantities. The relativistic sound speed can then be calculated by solving the following equation

$$\Gamma_r(n,T) = \frac{1}{3} + \frac{n}{W}\frac{dW}{dn} \tag{95}$$

or

$$\frac{W}{c} = \exp\left[-\int_n^\infty \left(\Gamma_r - \frac{1}{3}\right)\frac{dn}{n}\right] \tag{96}$$

where c = light speed.

7. GRAVITATIONAL WAVES IN REAL GASES. It has been suggested that real gases can possibly be used in a gravity wave detector.[12] This is possible because the relativity temperature parameter T_R that occurs in the state equation of relativistic real gases is a measure of the interaction of the real gases with the vacuum state, and gravity waves are oscillations of the vacuum, i.e., waves of curvature in spacetime. Gravity waves are shear-like in nature and are not expected to directly change the volume, pressure, or temperature of a gas, liquid or solid. Thus in the case of the Weber bar design for a gravity wave detector, only the surface shear strain is attempted to be measured, but no success has been reported.[6-11]

The interactions of real gases are of dipole-dipole, dipole-quadrupole, and quadrupole-quadrupole types.[13] These interactions depend on the separation and shape of the molecules through their dipole, quadrupole, and higher moments.[13] The values of T_R and T_{crit} depend on these multipole moments as both temperatures are species dependent.[4] The tidal nature of gravity waves will alter the multipole moments across a volume of gas, and will produce a gradient of T_R across the volume of gas in a detector. Gravity wave detector calculations must be done in conjunction with the relativistic state equations of the materials used in a detector. Real gases and liquids exhibit a critical point, and the critical temperature is related to the relativity temperature by equation (30). Solids, on the other hand, do not have a parameter akin to T_R in their relativistic state equations.[12] Gases and liquids are expected to be sensitive to gravity waves while solids are not expected to show any response.

The values of the relativity temperature T_R and the critical temperature T_{crit} are expected to vary across the volume of a gaseous gravitational wave detector due to the tidal effects of gravity waves. Heat exchange in the detector gas will tend to produce a uniform change in temperature. The tidal effects of gravitation can be described by the difference between the metric $g_{\mu\nu}$ for gravitational waves and the Minkowski metric $g^o_{\mu\nu} = (1, 1, 1, -1)$ which is written as[18]

$$h_{\mu\nu} = g_{\mu\nu} - g^o_{\mu\nu} \equiv h \tag{97}$$

where the values of the small dimensionless number h give a measure of the strength of gravitational radiation at the detector.

The gravitational potential that is associated with this weak gravitational field is $\chi = hc^2$. In the presence of a gravitational field the energy of a body is altered by the following quasi-static factor[19]

$$\left(1 + \frac{2\chi}{c^2}\right)^{1/2} \tag{98}$$

so that the effects of a gravitational wave on the relativity temperature is to give it the value

$$\begin{aligned} T_{RG} &= \left(1 + \frac{2\chi}{c^2}\right)^{1/2} T_R \\ &= (1 + 2h)^{1/2} T_R \\ &\approx (1 + h) T_R \end{aligned} \tag{99}$$

where T_{RG} = value of relativity temperature in the presence of gravity waves. The change in the value of T_R due to ambient gravity waves is therefore[12]

$$\delta T_R = hT_R \tag{100}$$

A similar analysis holds for the conjugate relativity temperature T_{CR}. The order of magnitude change in the value of the relativity temperature depends on the value of h at the detector.

Many studies have been done on the relative strengths of possible astronomical sources of gravity waves.[9-11] These sources include pulsars $10^{-27} < h < 10^{-24}$, supernovae $10^{-22} < h < 10^{-19}$, and binary stars $h < 10^{-21}$. It is possible that the galactic center radiates gravity waves with $h < 10^{-16}$. Taking $T_R \sim 100°K$ gives $10^{-25} < \delta T_R < 10^{-14}$ as a likely range for the change in the relativity temperature of a gas due to astronomical sources of gravity waves. The corresponding changes in pressure, temperature, and volume in a gaseous gravitational wave detector will now be calculated.

8. GENERALIZED FORCE ASSOCIATED WITH RELATIVITY TEMPERATURE. The generalized work done during a change of volume and a change of the relativity temperature of the system is given by

$$dW = PdV + S_R dT_R \tag{101}$$

$$= -\frac{PN}{n^2}\,dn + S_R dT_R$$

where S_R = generalized force associated with T_R. Clearly S_R has the dimensions of an entropy. The generalized force associated with dV is clearly the system pressure P. The generalized forces can be calculated using the Gibbs-Helmholtz equation which states that if a generalized work is written as $E\,dq$, where E = generalized force associated with a physical variable q, then[20]

$$\left(\frac{\partial U}{\partial q}\right)_{T,V} = T\left(\frac{\partial E}{\partial T}\right)_{q,V} - E \tag{102}$$

For instance E might be an electric field and q an electric charge. The Gibbs-Helmholtz equations associated with the situation in equation (101) are

$$\left(\frac{\partial U}{\partial V}\right)_{T,T_R} = T\left(\frac{\partial P}{\partial T}\right)_{V,T_R} - P \tag{103}$$

$$\left(\frac{\partial U}{\partial T_R}\right)_{T,V} = T\left(\frac{\partial S_R}{\partial T}\right)_{T_R,V} - S_R \tag{104}$$

Equation (104) can be used to determine the function S_R .

An expression for S_R can easily be obtained from equation (104) by making the substitution $S_R = Ts_R$ because then equation (104) becomes

$$\left(\frac{\partial U}{\partial T_R}\right)_{T,V} = T^2\left(\frac{\partial s_R}{\partial T}\right)_{T_R,V} \tag{105}$$

Combining equation (15) and (18) with equation (105) gives

$$-\frac{1}{2}\, NRT^2 n^2 \,\frac{\partial^2 C}{\partial T \partial T_R} = T^2\left(\frac{\partial s_R}{\partial T}\right)_{T_R,V} \tag{106}$$

which reduces immediately to

$$s_R = -\frac{1}{2}\, NRn^2 \left(\frac{\partial C}{\partial T_R}\right)_T \tag{107}$$

Finally $S_R = Ts_R$ gives

$$S_R = -\frac{1}{2}\, NRTn^2\left(\frac{\partial C}{\partial T_R}\right)_T \tag{108}$$

which can be written per unit volume as

$$S_R/V = -\frac{1}{2}\, RTn^3\left(\frac{\partial C}{\partial T_R}\right)_T \tag{108A}$$

or in molar quantities as

$$\tilde{S}_R = -\frac{1}{2} RTn^2\left(\frac{\partial C}{\partial T_R}\right)_T \tag{108B}$$

where C = relativistic third virial coefficient. This generalized force (entropy) will be used subsequently to calculate the changes in volume, temperature, and pressure in a gas due to gravity waves.

The derivative in equation (108) can be evaluated by using equations (25) and (26) which give

$$T_R \frac{\partial C}{\partial T_R} = -3[B^a(T)]^2 \frac{T_R}{\psi^a}\frac{\partial \psi^a}{\partial T_R} \tag{109}$$

and

$$\frac{T_R}{\psi^a}\frac{\partial \psi^a}{\partial T_R} = -\left[1 + \frac{2}{3}\frac{T_R}{B^a(T_R)}\frac{\partial B^a(T_R)}{\partial T_R}\right] \tag{110}$$

For reference it is noted also that

$$\frac{T}{\psi^a}\frac{\partial \psi^a}{\partial T} = 1 + \frac{2}{3}\frac{T}{B^a(T)}\frac{\partial B^a(T)}{\partial T} \tag{111}$$

Combining equations (108B), (109), and (110) gives the final result as

$$\tilde{S}_R(n,T,T_R) = -\frac{3}{2} R \frac{T}{T_R} n^2[B^a(T)]^2\left[1 + \frac{2}{3}\frac{T_R}{B^a(T_R)}\frac{\partial B^a(T_R)}{\partial T_R}\right] \tag{112}$$

The entropy $\tilde{S}_R$ is a purely relativistic quantity that is associated with the variation of T_R and is related to the interaction of the vacuum state with the molecules of a real gas. Equations (23) through (26) and equation (112) represent a relativistic thermodynamic analog of the Casimir effect of quantum electrodynamics.[21]

9. ADIABATIC CHANGES OF TEMPERATURE, VOLUME, AND PRESSURE. The first law of thermodynamics for the relativistic real gas can be written as follows

$$dU = dQ - PdV - S_R dT_R \tag{113}$$

$$= dQ + \frac{PN}{n^2} dn - S_R dT_R$$

where dQ = TdS = increment of heat associated with the absorption of gravity waves by a real gas, and dS = corresponding increase in entropy. Because the internal energy is a state function it has a perfect differential which can be written as

$$dU = \left(\frac{\partial U}{\partial T}\right)_{V,T_R} dT + \left(\frac{\partial U}{\partial V}\right)_{T,T_R} dV + \left(\frac{\partial U}{\partial T_R}\right)_{V,T} dT_R \tag{114}$$

Using the Gibbs-Helmholtz equations (103) and (104) brings equation (114) into the following form

$$dU = \left(\frac{\partial U}{\partial T}\right)_{V,T_R} dT + \left[T\left(\frac{\partial P}{\partial T}\right)_{V,T_R} - P\right] dV + \left[T\left(\frac{\partial S_R}{\partial T}\right)_{V,T_R} - S_R\right] dT_R \tag{115}$$

Placing equation (115) into equation (113) gives the following expression for the heat increment

$$dQ = \left(\frac{\partial U}{\partial T}\right)_{V,T_R} dT + T\left(\frac{\partial P}{\partial T}\right)_{V,T_R} dV + T\left(\frac{\partial S_R}{\partial T}\right)_{V,T_R} dT_R \tag{116}$$

The condition for adiabatic processes is given by dQ = 0 or

$$C_V dT + T\left(\frac{\partial P}{\partial T}\right)_{V,T_R} dV + T\left(\frac{\partial S_R}{\partial T}\right)_{V,T_R} dT_R = 0 \tag{117}$$

where C_V is given by equations (17) and (20). It will be assumed that gravity wave interactions with the real gases are sufficiently rapid that they can be described as adiabatic processes. The general expression for the change of pressure in a gas due to the passage of a gravity wave will be written as

$$dP = \left(\frac{\partial P}{\partial T}\right)_{V,T_R} dT + \left(\frac{\partial P}{\partial V}\right)_{T,T_R} dV + \left(\frac{\partial P}{\partial T_R}\right)_{T,V} dT_R \tag{118}$$

Combining equations (14), (16), and (118) gives

$$dP = nR[1 + nf_1(T) + n^2 f_2(T) + \cdots]\ dT \tag{119}$$

$$+ K_T \frac{dn}{n} + RTn^3 \frac{\partial C}{\partial T_R}\ dT_R$$

where f_1 and f_2 are given by equations (32) and (33) respectively. Several special cases will now be examined.

Using equation (117) allows several interesting adiabatic situations to be considered.

<u>Case a.</u> Adiabatic Change of Temperature at Constant Volume.

For this case equation (117) gives

$$dT\big|_{S,V} = -\frac{T}{\tilde{C}_V}\left(\frac{\partial \tilde{S}_R}{\partial T}\right)_{V,T_R} dT_R \tag{120}$$

Combining equations (108) and (120) gives

$$dT\big|_{S,V} = \frac{Rn^2 CJT}{2\tilde{C}_V T_R}\ dT_R \tag{121}$$

where the dimensionless quantity J is given by

$$J = \frac{T_R}{C}\frac{\partial C}{\partial T_R} + \frac{TT_R}{C}\frac{\partial^2 C}{\partial T \partial T_R} \tag{122}$$

The second derivative that occurs in equation (122) is calculated using equation (109) as follows

$$TT_R \frac{\partial^2 C}{\partial T \partial T_R} = - 6B^a(T)T \frac{\partial B^a}{\partial T} \frac{T_R}{\psi^a} \frac{\partial \psi^a}{\partial T_R} \tag{123}$$

The result in equation (121) can be rewritten using equation (34) as follows

$$dT|_{S,V} = \frac{CJn^2 T}{3T_R} \left(1 + g_1 n + g_2 n^2 + \cdots \right) dT_R \tag{124}$$

where g_1 and g_2 are given by equations (35) and (36) respectively. The sign of the temperature change given by equation (124) depends on the sign of the product CJ which is temperature dependent and can be positive or negative according to the value of temperature being considered.

Case b. Adiabatic Change of Volume at Constant Temperature.

By using the definition of the Grüneisen function given in equation (3) it follows from equations (117) and (120) that

$$d\tilde{V}|_{S,T} = - \frac{\tilde{V}}{\gamma \tilde{C}_V} \left(\frac{\partial \tilde{S}_R}{\partial T} \right)_{V,T_R} dT_R \tag{125}$$

$$= \frac{\tilde{V}}{\gamma T} dT|_{S,V}$$

$$= - \left. \frac{dn}{n^2} \right|_{S,T}$$

Combining equations (34), (37), (121), and (125) gives

$$d\tilde{V}|_{S,T} = - \left. \frac{dn}{n^2} \right|_{S,T} = \frac{RnCJ}{2\gamma \tilde{C}_V T_R} dT_R \tag{126}$$

$$= \frac{nCJ}{2T_R} [1 - f_1 n + (f_1^2 - f_2)n^2 - \cdots] dT_R \tag{127}$$

where f_1 and f_2 are given by equations (32) and (33) respectively.

Case c. Adiabatic Change in Pressure at Constant Volume.

Placing equation (124) into equation (119) with dn = 0 gives

$$dP|_{S,V} = \frac{RTCn^3}{3T_R}(F_o + F_1 n + F_2 n^2 + \cdots)\, dT_R \tag{128}$$

where

$$F_o = J + \frac{3T_R}{C}\frac{\partial C}{\partial T_R} \tag{129}$$

$$F_1 = \gamma_1 J \tag{130}$$

$$F_2 = \gamma_2 J \tag{131}$$

and where γ_1 and γ_2 are defined in equation (38) and (39) respectively. An equivalent expression for dP can also be written in terms of the Grüneisen parameter as follows

$$dP|_{S,V} = \frac{RTCn^3}{T_R}\left[\frac{\gamma J}{2} + \frac{T_R}{C}\frac{\partial C}{\partial T_R}\right] dT_R \tag{132}$$

Substituting the power series expansion for γ given by equation (37) into equation (132) yields the result given in equation (128). Thus $dP \sim n^3$ for low densities.

Case d. Adiabatic Change in Pressure at Constant Temperature.

Combining equation (126) with equation (119) for dT = 0 yields

$$dP|_{S,T} = \frac{RTCn^3}{T_R}\left[\frac{T_R}{C}\frac{\partial C}{\partial T_R} - \frac{JK_T}{2n\gamma\tilde{C}_V T}\right] dT_R \tag{133}$$

Using equations (16), (34), and (35) allows equation (133) to be rewritten as

$$dP|_{S,T} = \frac{RTCn^3}{2T_R} (G_o + G_1 n + G_2 n^2 + \cdots)\, dT_R \tag{134}$$

where

$$G_o = 2 \frac{T_R}{C} \frac{\partial C}{\partial T_R} - J \tag{135}$$

$$G_1 = J(f_1 - 2B) \tag{136}$$

$$G_2 = J(f_2 - f_1^2 + 2f_1 B - 3C) \tag{137}$$

where J is given by equation (122) and f_1 and f_2 are given by equations (32) and (33) respectively. Therefore at low densities $dP \sim n^3$. Because in general $dP/P \sim n^2$ for low densities, the efficiency of a gaseous gravitational wave detector can be improved by increasing the density of the gas in the detector.

Consider now the case of a constant pressure system. From equation (118) it follows that the constant pressure condition is written as

$$\left(\frac{\partial P}{\partial V}\right)_{T,T_R} dV + \left(\frac{\partial P}{\partial T}\right)_{V,T_R} dT + \left(\frac{\partial P}{\partial T_R}\right)_{V,T} dT_R = 0 \tag{138}$$

Two cases of the constant pressure system are of interest.

<u>Case e.</u> Change in volume at Constant Pressure and Temperature.

From equation (138) and equations (6), (14), and (23) through (26) it follows that

$$d\tilde{V}|_{P,T} = \frac{\tilde{V}}{K_T} \left(\frac{\partial P}{\partial T_R}\right)_{T,V} dT_R \tag{139}$$

$$= \frac{RTn^2}{K_T} \frac{\partial C}{\partial T_R} dT_R$$

$$= -\frac{2\tilde{S}_R}{K_T} dT_R$$

where $\tilde{S}_R$ is given by equation (108B). Equation (139) can be rewritten using equation (16) as follows

$$d\tilde{V}\Big|_{P,T} = n[1 - 2nB + n^2(4B^2 - 3C) - \cdots] \frac{\partial C}{\partial T_R} dT_R \tag{140}$$

Case f. Change in Temperature at Constant Pressure and Volume.

From equation (138) and equations (3), (14), and (23) through (26) it follows that

$$dT\Big|_{P,V} = -\frac{\tilde{V}}{\gamma \tilde{C}_V} \left(\frac{\partial P}{\partial T_R}\right)_{T,V} dT_R \tag{141}$$

$$= -\frac{RTn^2}{\gamma \tilde{C}_V} \frac{\partial C}{\partial T_R} dT_R$$

Using equations (34) and (37) allows equation (141) to be rewritten as

$$dT\Big|_{P,V} = -Tn^2[1 - f_1 n + (f_1^2 - f_2)n^2 - \cdots] \frac{\partial C}{\partial T_R} dT_R \tag{142}$$

where f_1 and f_2 are given by equations (32) and (33) respectively.

10. CONCLUSION. The description of relativistic wave motion in real gases must include the coupling of the matter and radiation fields with the thermodynamic gauge parameters for matter and radiation. This means that the quantities P, γ, b and P_r, γ_r, b_r are coupled as shown in equation (52). This is true for wave motion in any relativistic physical system. The form of the relativistic third virial coefficient of the ground state of a real gas is affected by the ground state gauge parameters. When mechanical radiation is present in real gases, the third virial coefficient of the radiation itself is correspondingly affected by both the ground state and radiation gauge parameters. Because only the third and higher virial coefficients are affected by the gauge parameters, measurable relativistic effects should be observed only at high pressures such as can occur in nuclear explosions in the atmosphere, during the interaction of directed energy beams with the atmosphere, or in high pressure laboratory experiments. The tidal effects of gravitational radiation are expected to appear in the third and higher virial coefficients of the real gases, and therefore these gases under high pressure can serve as suitable materials for a gravitational wave detector.

REFERENCES

1. Quigg, C., Gauge Theories of the Strong, Weak, and Electromagnetic Interactions, Addison-Wesley, New York, 1983.

2. Moriyasu, K., An Elementary Primer for Gauge Theory, World Scientific, Singapore, 1983.

3. Cheng, R. P. and Li, L. F., Gauge Theory of Elementary Particle Physics, Oxford, New York, 1984.

4. Weiss, R. A., Relativistic Thermodynamics, Vols. 1 and 2, Exposition Press, New York, 1976.

5. Weiss, R. A., "Relativistic Wave Equations for Solids and Low Temperature Quantum Systems", Third Army Conference on Applied Mathematics and Computing, Georgia Institute of Technology, ARO Report 86-1, May 13-16 1985, p. 717.

6. Weber, J., "Gravitational Radiation", Phys. Rev. Lett., 18, 498, 1967.

7. Papini, G., "Gravitational Radiation and its Detection", Can. J. Phys., 52, 880, 1973.

8. Braginsky, V. B. and Manukin, A. B., Measurement of Weak Forces in Physics Experiments, Chicago Univ. Press, p. 98, 1977.

9. Press, W. H. and Thorne, K. S., Annual Reviews of Astronomy and Astrophysics, 10, 335 (1972).

10. Ostriker, J. P., article in Sources of Gravitational Radiation, edited by L. Smarr (Cambridge Univ. Press, 1979), p. 461.

11. Thorne, K. S., "Gravitational Wave Research: Current Status and Future Prospects", Revs. Mod. Phys., Vol 52, No. 2, Part 1, April 1980.

12. Weiss, R. A., "A Gaseous Gravitational Wave Detector", article in After Einstein, edited by P. Barker and C. G. Shugart, Memphis State University Press, 1981, pp. 103.

13. Hirschfelder, J. O., Curtiss, C. F. and Bird, R. B., Molecular Theory of Gases and Liquids, John Wiley, New York, 1954.

14. Beattie, J. A., "Thermodynamic Properties of Real Gases and Mixtures of Real Gases", article in Thermodynamics and Physics of Matter, edited by F. D. Rossini, Princeton University Press, 1955, pp. 240.

15. Rice, O. K., "Critical Phenomena", article in Thermodynamics and Physics of Matter, edited by F. D. Rossini, Princeton University Press, 1955, pp. 438.

16. Weiss, R. A., "Relativistic Effects on the Critical Point of Gases and Liquids", Bulletin Amer, Phys. Soc., NF 17, 1980.

17. Brillouin, L., Tensors in Mechanics and Elasticity, Academic Press, New York, 1964.

18. Weinberg, S., Gravitation and Cosmology: Principles and Applications of the General Theory of Relativity, John Wiley, New York, 1972.

19. Møller, C., The Theory of Relativity, Oxford, London, 1955.

20. Epstein, P., Textbook of Thermodynamics, John Wiley, New York, 1937.

21. Brevik, I. and Kolbenstvedt, H., "Attractive Casimir Stress on a Thin Spherical Shell", Can. J. Phys., Vol. 63, 1985.

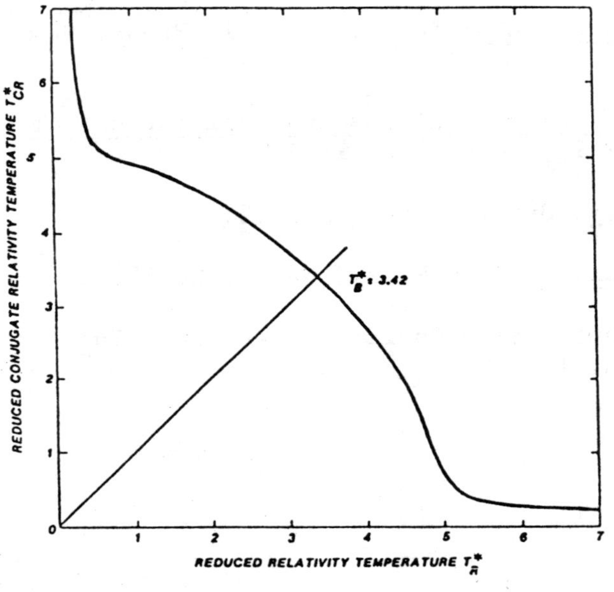

(1)

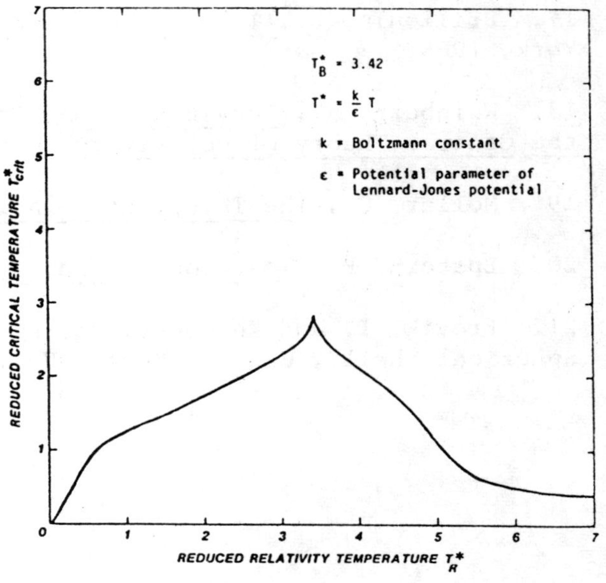

(2)

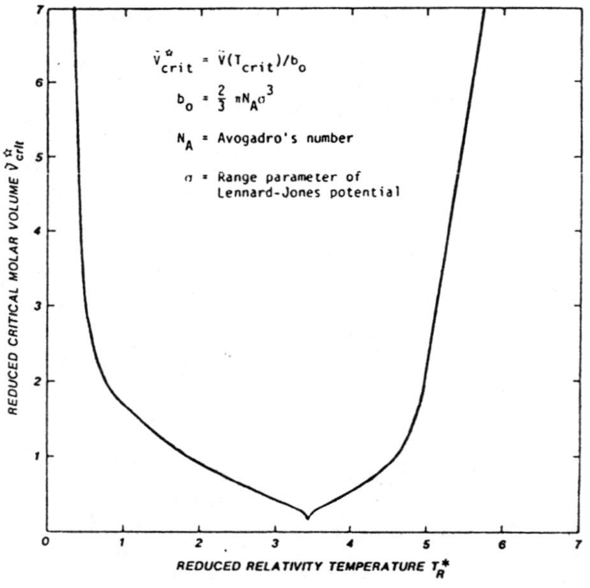

(3)

Figure 1. Relationship between the relativity temperature and the conjugate relativity temperature.

Figure 2. Dependence of reduced critical temperature on the reduced relativity temperature. Note that $T^*_{crit} < T^*_B$ and $B^a(T_{crit}) < 0$.

Figure 3. Dependence of the reduced critical molar volume on the reduced relativity temperature. Note that $\tilde{V}_{crit} = -B(T_{crit})$.

D

LAGRANGIAN FORMULATION OF RELATIVISTIC THERMODYNAMICS

Richard A. Weiss
U. S. Army Engineer Waterways Experiment Station
Vicksburg, Mississippi 39180

ABSTRACT. Matter and radiation Lagrangians are developed from which the renormalization group equations of locally gauge invariant relativistic thermodynamics can be obtained by the Euler-Lagrange equations. The Noether current tensor and conservation equations of relativistic thermodynamics can be derived from these Lagrangians. These Lagrangians exhibit a minimum value when expressed in terms of the fractal dimension of a physical system, and are locally symmetric about this minimum value. This suggests that all matter and radiation in matter is fractal in nature. Gases, liquids, solids, quantum liquids, and the mechanical waves that propagate in these systems, have fractal properties. Equations for calculating the fractal dimensions of matter and radiation in matter are presented, and general expressions for the void ratio of gases, condensed matter, and radiation are derived. These results will have applications to matter and radiation at high densities such as occur in neutron stars, nuclear explosions, and the interaction of directed energy beams with matter.

1. INTRODUCTION. Lagrangian formulations of the theory of continuous systems are common in the classical mechanics of particles and fields.[1,2] But it is in the quantum theory of fields that the Lagrangian formalisms have exhibited their unique power to describe new physical effects in addition to yielding the dynamical equations of motion.[3-5] For instance, the properties of a chiral Lagrangian yield the left-right asymmetries of the electroweak force.[6,7] The spontaneously broken symmetry of a Lagrangian gives rise to such diverse phenomena as mass generation of gauge bosons, the existance of Goldstone bosons, the ferromagnetic ground state, the Meissner effect for superconductors, and many other subtle effects.[6,7] These results suggest that other locally gauge invariant systems, such as relativistic thermodynamics, may have a simple Lagrangian description.

A set of relativistic thermodynamic renormalization group equations for the ground and excited states of matter and radiation has been derived using the local scale (gauge) invariance of relativistic thermodynamics.[8,9] These renormalization group equations are partial differential equations for the energy and gauge parameters, and are similar in form to the Callan-Symanzik equations of relativistic quantum field theory.[7] These equations are derived from a relativistic trace equation that accounts for the vacuum interactions of matter and radiation in a four-dimensional formalism.[10] The trace equation is locally gauge invariant under the U(1) group in the sense that the values of the gauge transformation functions depend on the local density and temperature of a system.[9]

This paper develops a Lagrangian formulation of relativistic thermodynamics that is shown to be equivalent to the renormalization group equations. The Lagrangian density can be used to determine the fractal dimension (Hausdorf number) of bulk matter and radiation in bulk matter. Fractal matter systems are discussed extensively in the literature.[11-17] For relativistic thermodynamics, the fractal dimension is related to the state equation of a system and its deviation from the homogeneous case is due to the vacuum interactions of matter and radiation. In this paper the Gibbs-Helmholtz equation is used to estimate the void ratios of fractal matter and radiation.

2. RENORMALIZATION GROUP EQUATIONS FOR A FRACTAL GROUND STATE. The locally gauge invariant interaction of the vacuum state with uniform bulk matter and radiation is described by a relativistic trace equation.[10] The question arises, however, whether the vacuum interaction will produce a uniform system of matter and radiation or whether it will result in a fractal state. This question can be answered by developing the renormalization group equations for the fractal states of matter and radiation. The fractal analog of the relativistic trace equation is written as[10]

$$U + T\left(\frac{dU}{dT}\right)_{PV} - DV\frac{d}{dV}(PV)_U = U^a + T\left(\frac{dU^a}{dT}\right)_{P^aV} \tag{1}$$

where D = fractal dimension = Hausdorf number.[11-17] The vacuum state (space-time) has D = 3 to a very high degree of accuracy.[18] On the other hand, matter and radiation in matter need not have D = 3 , and in general the fractal dimension of a system will depend on volume and temperature, D = D(V,T). In equation (1), U = relativistic internal energy, P = relativistic pressure, T = absolute temperature, V = volume of substance, and U^a and P^a = corresponding nonrelativistic internal energy and pressure. Throughout this paper the index "a" will refer to nonrelativistic calculations.

For a fractal system with Hausdorf number D, equation (1) becomes[9]

$$\left(1 - b + T\frac{\partial}{\partial T} - bV\frac{\partial}{\partial V}\right)E - D\left(1 + \gamma + V\frac{\partial}{\partial V} - \gamma T\frac{\partial}{\partial T}\right)P \tag{2}$$

$$= \left(1 - b^a + T\frac{\partial}{\partial T} - b^aV\frac{\partial}{\partial V}\right)E^a$$

where E = relativistic energy density = U/V , E^a = nonrelativistic energy density, and where[9]

$$\gamma = \frac{V}{C_V}\left(\frac{\partial P}{\partial T}\right)_V \tag{3}$$

$$b = \frac{T(\partial P/\partial T)_V}{(P - K_T)} \tag{4}$$

$$b^a = \frac{T(\partial P^a/\partial T)_V}{(P^a - K_T^a)} \qquad (5)$$

where γ = relativistic Grüneisen parameter; C_V = relativistic heat capacity at constant volume, and C_V^a = nonrelativistic heat capacity at constant volume, given respectively by

$$C_V = \left(\frac{\partial U}{\partial T}\right)_V \qquad (6)$$

$$C_V^a = \left(\frac{\partial U^a}{\partial T}\right)_V \qquad (7)$$

and where

$$K_T = -V\left(\frac{\partial P}{\partial V}\right)_T \qquad (8)$$

$$K_T^a = -V\left(\frac{\partial P^a}{\partial V}\right)_T \qquad (9)$$

are the relativistic and nonrelativistic values of the bulk modulus respectively. The parameters b and γ are the gauge parameters of relativistic thermodynamics.

Equation (2) can be rewritten as the following two renormalization group equations[9]

$$(T \frac{\partial}{\partial T} + fV \frac{\partial}{\partial V} + M)E = \psi^a \qquad (10)$$

$$(T \frac{\partial}{\partial T} + hV \frac{\partial}{\partial V} + N)P = 0 \qquad (11)$$

where

$$f = \eta - b \qquad (12)$$

$$h = \frac{1}{\eta/D - \gamma} \qquad (13)$$

$$M = f + 1 \qquad (14)$$

$$N = h - 1 \tag{15}$$

$$\psi^a = (T \frac{\partial}{\partial T} - b^a V \frac{\partial}{\partial V} + 1 - b^a) E^a \tag{16}$$

and where η = Lagrange multiplier given by[9]

$$\frac{\eta}{D} = \frac{V \frac{\partial P}{\partial V} - \gamma T \frac{\partial P}{\partial T} + (\gamma + 1)P}{P - T \frac{\partial P}{\partial T}} \tag{17}$$

Equations (10) through (17) reduce to equation (25) through (32) of Reference 9 for the case D = 3 . Thus f , h , M , and N for the fractal ground state are now explicit functions of the fractal dimension D, and therefore E , P , and γ are also explicit functions of D. Finally, it will be assumed that $f \neq 0$ and $h \neq 0$ so that equations (10) and (11) can be rewritten as

$$V \frac{\partial E}{\partial V} + \frac{1}{f} T \frac{\partial E}{\partial T} + \frac{M}{f} E = \psi^a / f \tag{18}$$

$$V \frac{\partial P}{\partial V} + \frac{1}{h} T \frac{\partial P}{\partial T} + \frac{N}{h} P = 0 \tag{19}$$

For a solid or low temperature quantum system the nonrelativistic state equation of the ground state is assumed to have the following form[10]

$$E^a = E^a_o + E^a_j T^j + \cdots \tag{20}$$

$$P^a = P^a_o + P^a_j T^j + \cdots \tag{21}$$

where E^a and P^a = nonrelativistic energy density and pressure respectively, E^a_o and P^a_o = nonrelativistic zero-temperature values of the energy density and pressure respectively, E^a_j and P^a_j = nonrelativistic thermal coefficients for the energy density and pressure respectively, T = absolute temperature of the system (°K), and j = numerical index having values characteristic of the type of physical system. A commonly used descriptor of the thermal state equations given by equations (20) and (21) is the nonrelativistic zero-temperature value of the Grüneisen parameter that is defined by

$$\gamma^a_o = \frac{P^a_j}{E^a_j} = \frac{1}{(j-1)} \frac{1}{E^a_j} \frac{d}{dV} (V E^a_j) \tag{22}$$

except for j = 1. When j = 1 , $\gamma_o^a = 2/3$. The zero temperature value of the nonrelativistic bulk modulus is given by $K_o^a = ndP_o^a/dn$, where n = N/V = number of moles per unit volume, and N = number of moles of a substance.

The corresponding relativistic state equations will be written as[10]

$$E = E_o + E_j T^j + \cdots \tag{23}$$

$$P = P_o + P_j T^j + \cdots \tag{24}$$

$$\gamma_o = \frac{P_j}{E_j} = \frac{1}{(j-1)} \frac{1}{E_j} \frac{d}{dV}(VE_j) \tag{25}$$

except for j = 1 , when $E_1 = E_1^a$, and where E_o and P_o = relativistic zero-temperature energy density and pressure respectively, E_j and P_j = relativistic thermal coefficients for the energy density and pressure respectively, and γ_o = relativistic zero-temperature Grüneisen parameter. The relativistic value of the zero temperature bulk modulus is given by $K_o = ndP_o/dn$.

Combining equation (2) with the state equations (20) through (25) yields the following ground state equations for fractal solids and low temperature quantum systems[10]

$$E_o - D_o[(1 + \gamma_o)P_o - K_o] = E_o^a \tag{26}$$

$$E_j\left(1 + j + \frac{j\gamma_o P_o}{P_o - K_o} + D_o n \frac{d\gamma_o}{dn}\right) = E_j^a\left(1 + j + \frac{j\gamma_o^a P_o^a}{P_o^a - K_o^a}\right) \tag{27}$$

where $D_o = D_o(n)$ is the T = 0 value of the fractal dimension, and where[10]

$$\frac{E_j}{E_j^a} = \exp[-(j - 1)\int^n (\gamma_o - \gamma_o^a) \frac{dn}{n}] \tag{28}$$

Note that in the derivation of equation (27) it is assumed that any explicit temperature dependence of D in the for $D = D_o + D_j T^j + \cdots$ can be neglected and that essentially $D_j = 0$. If this is not assumed, an additional term $-D_j[(1 + \gamma_o)P_o - K_o]$ has to be inserted into the left hand side of equation (27).

Equation (26) can be rewritten as follows

$$D_o n^2 \frac{d^2E_o}{dn^2} - D_o(1 + \gamma_o)n \frac{dE_o}{dn} + [D_o(1 + \gamma_o) + 1]E_o = E_o^a \tag{29}$$

Equivalent forms of equation (29) are

$$D_o n^2 \frac{d^2P_o}{dn^2} - D_o(1 + \gamma_o)n \frac{dP_o}{dn} + \left[D_o\left(\gamma_o - n \frac{d\gamma_o}{dn}\right) + D_o + 1\right]P_o = P_o^a \tag{30}$$

$$D_o V^2 \frac{d^2P_o}{dV^2} + D_o(3 + \gamma_o)V \frac{dP_o}{dV} + \left[D_o\left(\gamma_o + V \frac{d\gamma_o}{dV}\right) + D_o + 1\right]P_o = P_o^a \tag{31}$$

Thus in general $E_o = E_o(E_o^a , \gamma_o^a , D_o , V)$ and $\gamma_o = \gamma_o(E_o^a , \gamma_o^a , D_o , V)$. It is possible that originally the unrenormalized state is fractal in nature so that $E_o^a = E_o^a(D_o^a , V)$ and $\gamma_o^a = \gamma_o^a(D_o^a , V)$. If in equation (29) one takes $E_o \sim n^{\sigma_o}$ and $E_o^a \sim n^{\sigma_o}$, where σ_o = adiabatic index, and γ_o = constant, one gets

$$E_o = \frac{E_o^a}{G(D_o)} \tag{32}$$

where

$$G(D_o) = D_o\sigma_o^2 - D_o\sigma_o(2 + \gamma_o) + D_o(1 + \gamma_o) + 1 \tag{33}$$

It then follows that

$$P_o = \frac{P_o^a}{G(D_o)} \qquad K_o = \frac{K_o^a}{G(D_o)} \tag{34}$$

A general expression for the void ratio in a fractal ground state can be obtained from the Gibbs-Helmholtz equation[10]

$$E + V \frac{\partial E}{\partial V} = T \frac{\partial P}{\partial T} - P \tag{35}$$

which can be rewritten as

$$\frac{dV}{V} = \frac{dE}{T \frac{\partial P}{\partial T} - P - E} \tag{36}$$

Equation (36) will be written in finite difference form corresponding to a change of fractal dimension from the uniform D = 3 case to the general case of arbitrary fractal dimension as follows

$$\frac{\Delta V}{V} \sim \frac{E(D) - E(3)}{P(3) + E(3) - T\,\dfrac{\partial P(3)}{\partial T}} \tag{37}$$

where the notation, $E(D)$ = energy density associated with a fractal dimension D, and $E(3)$ and $P(3)$ = energy density and pressure respectively for the homogeneous case of D = 3 , is introduced for calculating void ratios. The T = 0 limit of equation (37) is given by

$$\left(\frac{\Delta V}{V}\right)_o = \frac{E_o(D_o) - E_o(3)}{P_o(3) + E_o(3)} = \frac{E_o(D_o) - E_o(3)}{n\,\dfrac{dE_o(3)}{dn}} \tag{38}$$

Note that the energy density for a fractal ground state is greater than that of the homogeneous state.[19]

3. RENORMALIZATION GROUP EQUATIONS FOR FRACTAL RADIATION IN FRACTAL MATTER. The renormalization group equation for fractal radiation in fractal matter can be written as a simple extension of the corresponding equation for homogeneous matter, using the same notation as in equation (70) of Reference 9, as follows

$$\begin{aligned}&\left(1 - b + T\frac{\partial}{\partial T} - bV\frac{\partial}{\partial V}\right)E_r - \beta_r\left(T\frac{\partial P}{\partial T} - P\right)\\
&\quad - D\left[\left(1 + \gamma + V\frac{\partial}{\partial V} - \gamma T\frac{\partial}{\partial T}\right)P_r - \delta_r\left(T\frac{\partial P}{\partial T} - P\right)\right]\\
&\quad + d_r\left(1 + \gamma + V\frac{\partial}{\partial V} - \gamma T\frac{\partial}{\partial T}\right)P\\
&\quad = \left(1 - b^a + T\frac{\partial}{\partial T} - b^aV\frac{\partial}{\partial V}\right)E_r^a - \beta_r^a\left(T\frac{\partial P^a}{\partial T} - P^a\right)\end{aligned} \tag{39}$$

where the fractal dimension of the radiation in matter is $D_r = D - d_r$, and where $d_r > 0$ is the incremental change in the fractal dimension due to the presence of radiation in the system, and where[9]

$$\beta_r = b_r\frac{P_r - K_{Tr}}{P - K_T} - \frac{T\dfrac{\partial P}{\partial T}(P_r - K_{Tr})}{(P - K_T)^2} \tag{40}$$

$$b_r = \frac{T\partial P_r/\partial T}{P_r - K_{Tr}} \quad (41)$$

$$K_{Tr} = -V \frac{\partial P_r}{\partial V} \quad (42)$$

$$\delta_r = \frac{\partial E_r/\partial T}{\partial E/\partial T} (\gamma_r - \gamma) \quad (43$$

$$\gamma_r = \frac{\partial P_r/\partial T}{\partial E_r/\partial T} \quad (44)$$

The functions γ_r and b_r are the radiation gauge parameters, and K_{Tr} = radiation bulk modulus. Note that β_r , δ_r , and d_r are generally small quantities, while γ_r , b_r , and D_r refer to the radiation itself and are not small quantities. For D = 3 and $d_r = 0$, equation (39) reduces to equation (70) of Reference 9.

Equation (39) can be decoupled into two independent radiation renormalization group equations as follows

$$(T \frac{\partial}{\partial T} + f_r V \frac{\partial}{\partial V} + M_r)E_r - \beta_r(T \frac{\partial P}{\partial T} - P) = \psi_r^a \quad (45)$$

$$(T \frac{\partial}{\partial T} + h_r V \frac{\partial}{\partial V} + N_r)P_r - h_r\delta_r(T \frac{\partial P}{\partial T} - P) \quad (46)$$

$$- \frac{h_r d_r}{D}\left(1 + \gamma + V \frac{\partial}{\partial V} - \gamma T \frac{\partial}{\partial T}\right) P = 0$$

$$f_r = \eta_r - b \quad (47)$$

$$h_r = (\eta_r/D - \gamma)^{-1} \quad (48)$$

$$M_r = f_r + 1 \quad (49)$$

$$N_r = h_r - 1 \quad (50)$$

$$\psi_r^a = (T \frac{\partial}{\partial T} - b^a V \frac{\partial}{\partial V} + 1 - b^a)E_r^a - \beta_r^a(T \frac{\partial P^a}{\partial T} - P^a) \quad (51)$$

where η_r = Lagrange multiplier. For the case D = 3 and $d_r = 0$, equations (45) through (51) reduce to equations (74) through (80) of Reference 9. It will be assumed that $f_r \neq 0$ and $h_r \neq 0$ so that equations (45) and (46) can be written as

$$V \frac{\partial E_r}{\partial V} + \frac{T}{f_r} \frac{\partial E_r}{\partial T} + \frac{M_r}{f_r} E_r - \frac{\beta_r}{f_r} \left(T \frac{\partial P}{\partial T} - P\right) = \psi^a_r / f_r \tag{52}$$

$$V \frac{\partial P_r}{\partial V} + \frac{T}{h_r} \frac{\partial P_r}{\partial T} + \frac{N_r}{h_r} P_r - \delta_r \left(T \frac{\partial P}{\partial T} - P\right) \tag{53}$$

$$- \frac{d_r}{D} \left(1 + \gamma + V \frac{\partial}{\partial V} - \gamma T \frac{\partial}{\partial T}\right) P = 0$$

The energy density and pressure for radiation in solids and quantum liquids is written as[9]

$$E^a_r = E^a_{or} + E^a_{jr} T^j + \cdots \tag{54}$$

$$P^a_r = P^a_{or} + P^a_{jr} T^j + \cdots \tag{55}$$

and

$$E_r = E_{or} + E_{jr} T^j + \cdots \tag{56}$$

$$P_r = P_{or} + P_{jr} T^j + \cdots \tag{57}$$

where

E^a_{or} and P^a_{or} = nonrelativistic zero-temperature radiation energy density and pressure respectively

E^a_{jr} and P^a_{jr} = nonrelativistic thermal coefficients for the radiation energy density and pressure respectively

E_{or} and P_{or} = relativistic zero-temperature radiation energy density and pressure respectively

E_{jr} and P_{jr} = relativistic thermal coefficients for the radiation energy density and pressure respectively

The zero temperature value of the radiation Grüneisen parameter is obtained from equations (44) and (54) through (57) to be

$$\gamma^a_{or} = \frac{P^a_{jr}}{E^a_{jr}} \qquad \gamma_{or} = \frac{P_{jr}}{E_{jr}} \tag{58}$$

The zero temperature values of the nonrelativistic and relativistic radiation bulk modulus is written as $K^a_{or} = ndP^a_{or}/dn$ and $K_{or} = ndP_{or}/dn$ respectively.

When radiation is present in a fractal solid or low temperature quantum liquid, the T = 0 fractal dimension of the radiation will be written as $D_{or} = D_o - d_{or}$ where $d_{or} > 0$ is the small change in fractal dimension associated with the addition of radiation to the material system. The excitation equations for such a system are obtained from equations (45) and (46) and are an extension of equations (104) and (105) of Reference 9. They are written as

$$E_{or} - D_o[(1 + \gamma_o)P_{or} - K_{or}] - D_o \frac{E_{jr}}{E_j} P_o(\gamma_{or} - \gamma_o) + d_{or}[(1 + \gamma_o)P_o - K_o] = E^a_{or} \tag{59}$$

$$jE_j(\alpha K_{or} - \beta P_{or}) + E_{jr}S_{jr} - d_{or}E_j n \frac{d\gamma_o}{dn} = jE^a_j(\alpha^a K^a_{or} - \beta^a P^a_{or}) + E^a_{jr}T^a_{jr} \tag{60}$$

where

$$S_{jr} = 1 + j + \frac{jP_o\gamma_{or}}{P_o - K_o} + D_o n \frac{d\gamma_{or}}{dn} - D_o(j - 1)(\gamma_{or} - \gamma_o)^2 \tag{61}$$

$$T^a_{jr} = 1 + j + jP^a_o\gamma^a_{or}/(P^a_o - K^a_o) \tag{62}$$

$$\alpha = \gamma_o P_o/(P_o - K_o)^2 \qquad \alpha^a = \gamma^a_o P^a_o/(P^a_o - K^a_o)^2 \tag{63}$$

$$\beta = \gamma_o K_o/(P_o - K_o)^2 \qquad \beta^a = \gamma^a_o K^a_o/(P^a_o - K^a_o)^2 \tag{64}$$

In the derivation of equation (60) it is assumed that one can neglect a temperature dependence of D_r of the form $D_r = D_{or} + D_{jr}T^j + \cdots$ (or equivalently, $d_r = D - D_r = D_o - D_{or} + (D_j - D_{jr})T^j + \cdots = d_{or} + d_{jr}T^j + \cdots$) and that $D_j = 0$, $D_{jr} = 0$, and $d_{jr} = 0$. If this is not the case and $D_j \neq 0$, then an additional term $+d_{jr}[(1 + \gamma_o)P_o - K_o]$ has to be inserted into the left hand side of equation (60). For this case both d_{or} and d_{jr} must be determined. Equation (59) can be rewritten as

$$D_o n^2 \frac{d^2 E_{or}}{dn^2} - D_o(1 + \gamma_o)n \frac{dE_{or}}{dn} + [D_o(1 + \gamma_o) + 1]E_{or} \tag{65}$$

$$+ D_o \frac{E_{jr}}{E_j} P_o(\gamma_o - \gamma_{or}) + d_{or}[(1 + \gamma_o)P_o - K_o] = E^a_{or}$$

Equations (59) through (65) reduce to equations (104) through (113) of Reference 9 for the case $D_o = 3$ and $d_{or} = 0$. The values of d_{or} and d_{jr} can be obtained from an appropriate Lagrangian formalism.

From the Gibbs-Helmholtz equation for a radiation system

$$E_r + V \frac{\partial E_r}{\partial V} = T \frac{\partial P_r}{\partial T} - P_r \tag{66}$$

one obtains the following estimate for the void ratio of a fractal mechanical radiation system in matter

$$\left(\frac{\Delta V}{V}\right)_r = \frac{E_r(D,d_r) - E_r(3,0)}{P_r(3,0) + E_r(3,0) - T \dfrac{\partial P_r(3,0)}{\partial T}} \tag{67}$$

where the notation $E_r(D,d_r)$ = fractal radiation energy density, and $E_r(3,0)$ and $P_r(3,0)$ = homogeneous radiation energy density and pressure respectively, will be introduced for calculating the void ratios of the fractal radiation field. The T = 0 limit of equation (67) is given by

$$\left(\frac{\Delta V}{V}\right)_{or} = \frac{E_{or}(D_o,d_{or}) - E_{or}(3,0)}{P_{or}(3,0) + E_{or}(3,0)} \tag{68}$$

Specific examples of the use of these equations for radiation in gases and condensed matter are given in Sections 7 and 8.

4. <u>GROUND STATE LAGRANGIAN.</u> Lagrangian formulations of nonlocal gauge field theories have been used to describe the four basic interactions that occur in nature.[6,7] One is tempted to write a similar Lagrangian formulation of the effects of the vacuum state on bulk matter. This section develops a Lagrangian description of a nonlocal gauge theory of relativistic thermodynamics. Let the Lagrangian function of a relativistic thermodynamic system be written as

$$L = \int \mathcal{L}\left(\frac{\partial \phi}{\partial v}, \frac{\partial \phi}{\partial t}, \phi, v, t\right)dv \tag{69}$$

and the thermodynamic action I be written as

$$I = \int \mathcal{L}\left(\frac{\partial \phi}{\partial v}, \frac{\partial \phi}{\partial t}, \phi, v, t\right)dvdt = \int L dt \tag{69A}$$

where $\mathcal{L}$ = Lagrangian density, $\phi = \phi(v,t)$ is an appropriately selected field, and where

$$v = \ell n\ V \tag{70}$$

$$t = \ell n\ T \tag{71}$$

The introduction of the variables in equations (70) and (71) is made because it simplifies the ground state renormalization group equations (18) and (19) which now become

$$\frac{\partial E}{\partial v} + \frac{1}{f}\frac{\partial E}{\partial t} + \frac{M}{f} E = \psi^a/f \tag{72}$$

$$\frac{\partial P}{\partial v} + \frac{1}{h}\frac{\partial P}{\partial t} + \frac{N}{h} P = 0 \tag{73}$$

The Euler-Lagrange field equations derived from $\delta I = 0$ are[1,3]

$$\frac{d}{dt}\left(\frac{\partial \mathcal{L}}{\partial \phi_{,t}}\right) + \frac{d}{dv}\left(\frac{\partial \mathcal{L}}{\partial \phi_{,v}}\right) = \frac{\partial \mathcal{L}}{\partial \phi} \tag{74}$$

where the following notation was introduced

$$\phi_{,t} = \frac{\partial \phi}{\partial t} \qquad \phi_{,v} = \frac{\partial \phi}{\partial v} \tag{75}$$

The Lagrangian density $\mathcal{L}(\phi_{,t}, \phi_{,v}, \phi, v, t)$ and the field $\phi(v,t)$ are selected in an appropriate way for relativistic thermodynamics so that the Euler-Lagrange equations (74) will reproduce the ground state renormalization group equations (72) and (73). In order to reproduce equation (72) one takes $\phi = \xi$ where

$$\xi = \int E dv = \int E \frac{dV}{V} = \xi(V,T,D) \tag{76}$$

The corresponding Lagrangian density is

$$\mathcal{L}_1 = \frac{1}{2} A\xi_{,v}^2 + I\xi_{,t} + B\xi + \frac{1}{2} C\xi^2 \tag{77}$$

where

$$A = 1 + Z \tag{78}$$

$$Z = \int \frac{M}{f}\, dv \tag{79}$$

$$I = \int \frac{1}{f} \frac{\partial E}{\partial t} \, dt \tag{80}$$

$$B = Z(\xi_{,v} + \xi_{,vv}) + \frac{\psi^a}{f} = Z(E + E_{,v}) + \frac{\psi^a}{f} \tag{81}$$

$$C = \frac{M}{f} \tag{82}$$

where $\xi_{,vv} = E_{,v}$ is treated as a parameter dependent of v and t but independent of ξ, $\xi_{,v}$ or $\xi_{,t}$. In order to see that equation (72) can be derived from $\mathcal{L}_1$ and equation (74) it is noted that

$$\frac{\partial \mathcal{L}_1}{\partial \xi_{,t}} = I = \int \frac{1}{f} \frac{\partial E}{\partial t} \, dt \tag{83}$$

$$\frac{d}{dt}\left(\frac{\partial \mathcal{L}_1}{\partial \xi_{,t}}\right) = \frac{1}{f} \frac{\partial E}{\partial t} \tag{84}$$

$$\frac{\partial \mathcal{L}_1}{\partial \xi_{,v}} = A\xi_{,v} + \xi Z = AE + \xi Z \tag{85}$$

$$\frac{d}{dv}\left(\frac{\partial \mathcal{L}_1}{\partial \xi_{,v}}\right) = AE_{,v} + C(\xi + E) + ZE \tag{86}$$

$$\frac{\partial \mathcal{L}_1}{\partial \xi} = B + C\xi \tag{87}$$

Placing these quantities in equation (74) yields equation (72).

In a similar fashion, in order to reproduce equation (73) one takes $\phi = \zeta$

$$\zeta = \int P dv = \int P \frac{dV}{V} = \zeta(V,T,D) \tag{88}$$

The corresponding Lagrangian density is

$$\mathcal{L}_2 = \frac{1}{2} F\zeta_{,v}^2 + J\zeta_{,t} + G\zeta + \frac{1}{2} H\zeta^2 \tag{89}$$

where

$$F = 1 + X \tag{90}$$

$$X = \int \frac{N}{h} \, dv \tag{91}$$

$$J = \int \frac{1}{h} \frac{\partial P}{\partial t} dt \tag{92}$$

$$G = X(\zeta_{,v} + \zeta_{,vv}) = X(P + P_{,v}) \tag{93}$$

$$H = \frac{N}{h} \tag{94}$$

where $\zeta_{,vv} = P_{,v}$ is treated as a parameter dependent on v and t but independent of ζ , $\zeta_{,v}$ or $\zeta_{,t}$. The following relationships hold

$$\frac{\partial \mathcal{L}_2}{\partial \zeta_{,t}} = J = \int \frac{1}{h} \frac{\partial P}{\partial t} dt \tag{95}$$

$$\frac{d}{dt}\left(\frac{\partial \mathcal{L}_2}{\partial \zeta_{,t}}\right) = \frac{1}{h} \frac{\partial P}{\partial t} \tag{96}$$

$$\frac{\partial \mathcal{L}_2}{\partial \zeta_{,v}} = F\zeta_{,v} + \zeta X = FP + \zeta X \tag{97}$$

$$\frac{d}{dv}\left(\frac{\partial \mathcal{L}_2}{\partial \zeta_{,v}}\right) = FP_{,v} + H(\zeta + P) + XP \tag{98}$$

$$\frac{\partial \mathcal{L}_2}{\partial \zeta} = G + H\zeta \tag{99}$$

Placing equations (95) through (99) into equation (74) shows that $\mathcal{L}_2$ is a proper Lagrangian density for the pressure renormalization group equation (73).

It should be pointed out that the Lagrangian densities $\mathcal{L}_1$ and $\mathcal{L}_2$ have a proper T = 0 limit and are given by

$$\mathcal{L}_1^o = \frac{1}{2} A_o\xi_{o,v}^2 + B_o\xi_o + \frac{1}{2} C_o\xi_o^2 \tag{100}$$

$$\mathcal{L}_2^o = \frac{1}{2} F_o\zeta_{o,v}^2 + G_o\zeta_o + \frac{1}{2} H_o\zeta_o^2 \tag{101}$$

where

$$\xi_o = \int E_o dv \qquad \zeta_o = \int P_o dv \tag{102}$$

and where

$$A_o = 1 + Z_o \tag{103}$$

$$Z_o = \int \frac{M_o}{f_o}\, dv = \int C_o dv \tag{104}$$

$$B_o = A_o(\mathcal{E}_o + \mathcal{E}_{o,v}) + \frac{\mathcal{E}^a_o}{f_o} \tag{105}$$

$$C_o = \frac{M_o}{f_o} \tag{106}$$

$$F_o = 1 + X_o \tag{107}$$

$$X_o = \int \frac{N_o}{h_o}\, dv = \int H_o dv \tag{108}$$

$$G_o = X_o(P_o + P_{o,v}) \tag{109}$$

$$H_o = \frac{N_o}{h_o} \tag{110}$$

where f_o , h_o , M_o and N_o are defined in Reference 9. Equations (100) and (101) yield the following T = 0 ground state equations

$$\frac{d\mathcal{E}_o}{dv} + \frac{M_o}{f_o}\mathcal{E}_o = \frac{\mathcal{E}^a_o}{f_o} \tag{111}$$

$$\frac{dP_o}{dv} + \frac{N_o}{h_o}P_o = 0 \tag{112}$$

The two potential functions associated with $\mathcal{L}_1$ and $\mathcal{L}_2$ are obtained from equations (77) and (89) to be

$$V_1 = B\xi + \frac{1}{2}C\xi^2 \tag{113}$$

$$V_2 = G\zeta + \frac{1}{2} H\zeta^2 \tag{114}$$

If $B < 0$, $G < 0$ and $C > 0$, $H > 0$ the potentials have a minimum at certain values of ξ and ζ which are determined from

$$\frac{\partial V_1}{\partial \xi} = 0 \qquad \frac{\partial V_2}{\partial \zeta} = 0 \tag{115}$$

where ξ and ζ are varied by changing the fractal dimension D for fixed V and T. The conditions in equation (115) are equivalent to

$$B + C\xi_M = 0 \tag{116}$$

$$G + H\zeta_M = 0 \tag{117}$$

or

$$\xi_M = - \frac{B}{C} = - \frac{1}{C} [Z(E + E_{,v}) + \psi^a/f] \tag{118}$$

$$\zeta_M = - \frac{G}{H} = - \frac{X}{H} (P + P_{,v}) \tag{119}$$

where $\xi_M(V,T)$ and $\zeta_M(V,T)$ = values of ξ and ζ for which V_1 and V_2 are respectively minimum. Taking the derivatives of equations (118) and (119) with respect to v yields respectively

$$ZE_{,vv} + \left(Z + C - Z \frac{C_{,v}}{C}\right)E_{,v} + \left(2C - Z \frac{C_{,v}}{C}\right)E = \frac{C_{,v}}{C} \frac{\psi^a}{f} - \frac{\partial}{\partial v}\left(\frac{\psi^a}{f}\right) \tag{120}$$

$$XP_{,vv} + \left(X + H - X \frac{H_{,v}}{H}\right)P_{,v} + \left(2H - X \frac{H_{,v}}{H}\right)P = 0 \tag{121}$$

Either equation (120) or (121) can be solved for $D = D_M(V,T)$ that makes the potential V_1 and V_2 have minimum values. About the minimum, the potentials have the form

$$V_1(\xi) - V_1(\xi_M) = \frac{1}{2} C(\Delta\xi)^2 \tag{122}$$

$$V_2(\zeta) - V_2(\zeta_M) = \frac{1}{2} H(\Delta\zeta)^2 \tag{123}$$

where $\xi = \xi_M + \Delta\xi$ and $\zeta = \zeta_M + \Delta\zeta$. Thus the potentials are locally symmetric about the minimum points. The value of D at the minimum points will be designated as $D_M(V,T)$, and in general $D_M < 3$ so that matter will have voids and is fractal in nature for specified values of V and T. The fractal ground state of matter occurs only for limited regions of V and T corresponding to the conditions $B < 0$, $G < 0$ and $C > 0$, $H > 0$.

It should be pointed out that the Lagrangians $\mathcal{L}_1$ and $\mathcal{L}_2$ are not unique in the sense that the following two Lagrangians also yield the desired renormalization group equations

$$\mathcal{L}_1' = \frac{1}{2} A\xi_{,v}^2 + I\xi_{,t} + \xi(\xi_{,v} + Z\xi_{,vv} + \psi^a/f) \tag{124}$$

$$\mathcal{L}_2' = \frac{1}{2} F\zeta_{,v}^2 + J\zeta_{,t} + \zeta(\zeta_{,v} + X\zeta_{,vv}) \tag{125}$$

These Lagrangians are linear in ξ and ζ respectively and do not contain quadratis potential terms as in the cases of equations (77) and (89). The quadratic Lagrangians are chosen instead of the linear potential Lagrangians because they are symmetrical about their minimum values.

5. <u>EXCITED STATE LAGRANGIAN.</u> The radiation renormalization group equations (45) and (46) can also be obtained from a Lagrangian formulation. The Lagrangian density that gives equation (45) is

$$\mathcal{L}_{1r} = \frac{1}{2} A_r\xi_{r,v}^2 + I_r\xi_{r,t} + B_r\xi_r + \frac{1}{2} C_r\xi_r^2 \tag{126}$$

where

$$\xi_r = \int E_r dv = \int E_r \frac{dV}{V} = \xi_r(V,T,D,D_r) \tag{127}$$

and

$$A_r = 1 + Z_r \tag{128}$$

$$Z_r = \int \frac{M_r}{f_r} dv \tag{129}$$

$$I_r = \int \frac{1}{f_r} \frac{\partial E_r}{\partial t} dt \tag{130}$$

$$B_r = Z_r(\xi_{r,v} + \xi_{r,vv}) + \frac{\beta_r}{f_r}\left(T \frac{\partial P}{\partial T} - P\right) + \frac{\psi_r^a}{f_r} \tag{131}$$

$$C_r = \frac{M_r}{f_r} \tag{132}$$

where $\xi_{r,vv} = E_{r,v}$ is treated as parameter which is dependent on v and t but is independent of ξ_r, $\xi_{r,v}$ or $\xi_{r,t}$. To show that the Euler-Lagrange equation (74) yields equation (45) when $\phi = \xi_r$ it is noted that

$$\frac{d}{dt}\left(\frac{\partial \mathcal{L}_{1r}}{\partial \xi_{r,t}}\right) = \frac{1}{f_r}\frac{\partial E_r}{\partial t} \tag{133}$$

$$\frac{\partial \mathcal{L}_{1r}}{\partial \xi_{r,v}} = A_r \xi_{r,v} + \xi_r Z_r = A_r E_r + \xi_r Z_r \tag{134}$$

$$\frac{d}{dv}\left(\frac{\partial \mathcal{L}_{1r}}{\partial \xi_{r,v}}\right) = A_r E_{r,v} + C_r(\xi_r + E_r) + Z_r E_r \tag{135}$$

$$\frac{\partial \mathcal{L}_{1r}}{\partial \xi_r} = B_r + C_r \xi_r \tag{136}$$

Combining equations (133) through (136) with equation (74) yields equation (45).

The Lagrangian density that yields equation (46) is found by choosing $\phi = \zeta_r$ in equation (74) with $\mathcal{L} = \mathcal{L}_{2r}$ where

$$\mathcal{L}_{2r} = \frac{1}{2} F_r \zeta_{r,v}^2 + J_r \zeta_{r,t} + G_r \zeta_r + \frac{1}{2} H_r \zeta_r^2 \tag{137}$$

where

$$\zeta_r = \int P_r dv = \zeta_r(V,T,D,d_r) \tag{138}$$

and

$$F_r = 1 + X_r \tag{139}$$

$$X_r = \int \frac{N_r}{h_r}\, dv = \int H_r dv \tag{140}$$

$$J_r = \int \frac{1}{h_r} \frac{\partial P_r}{\partial t}\, dt \tag{141}$$

$$G_r = X_r(\zeta_{r,v} + \zeta_{r,vv}) + \delta_r(T \frac{\partial P}{\partial T} - P) + \frac{d_r}{D}\left(1 + \gamma + V \frac{\partial}{\partial V} - \gamma T \frac{\partial}{\partial T}\right)P \tag{142}$$

$$H_r = \frac{N_r}{h_r} \tag{143}$$

If $\zeta_{r,vv}$ is taken to be a parameter only dependent on v and t, one has

$$\frac{d}{dt}\left(\frac{\partial \mathcal{L}_{2r}}{\partial \zeta_{r,t}}\right) = \frac{1}{h_r} \frac{\partial P_r}{\partial t} \tag{144}$$

$$\frac{\partial \mathcal{L}_{2r}}{\partial \zeta_{r,v}} = F_r P_r + \zeta_r X_r \tag{145}$$

$$\frac{d}{dv}\left(\frac{\partial \mathcal{L}_{2r}}{\partial \zeta_{r,v}}\right) = F_r P_{r,v} + H_r(\zeta_r + P_r) + X_r P_r \tag{146}$$

$$\frac{\partial \mathcal{L}_{2r}}{\partial \zeta_r} = G_r + H_r \zeta_r \tag{147}$$

which combine with equation (74) to give equation (46). It should be noted that the Lagrangians $\mathcal{L}_{1r}$ and $\mathcal{L}_{2r}$ have natural extensions to the case T = 0 .

In a form similar to that in equations (113) and (114), the potentials associated with radiation in matter are given by

$$V_{1r} = B_r \xi_r + \frac{1}{2} C_r \xi_r^2 \tag{148}$$

$$V_{2r} = G_r \zeta_r + \frac{1}{2} H_r \zeta_r^2 \tag{149}$$

If $B_r < 0$, $G_r < 0$, and $C_r > 0$, $H_r > 0$ these potentials will have minimum values at specific values of ξ_r and ζ_r given by

$$\frac{\partial V_{1r}}{\partial \xi_r} = B_r + C_r \xi_{rM} = 0 \tag{150}$$

$$\frac{\partial V_{2r}}{\partial \zeta_r} = G_r + H_r \zeta_{rM} = 0 \tag{151}$$

where $\xi_{rM}(V,T,D)$ and $\zeta_{rM}(V,T,D)$ = values of ξ_r and ζ_r for which V_{1r} and V_{2r} are respectively minimum, and where the variation of ξ_r and ζ_r in equations (150) and (151) corresponds to a change in d_r for fixed values of V, T, and D. The value of the fractal dimension of radiation in fractal matter is $D_r = D - d_r$ generally, where $d_r > 0$ so that in general there will be voids in a mechanical radiation field. If the value $d_r = d_{rM}(V,T,D)$ minimizes the potentials V_{1r} and V_{2r} , then the fractal dimension of mechanical radiation in fractal matter is $D_{rM} = D_M - d_{rM}$, and the fractal dimension of radiation in homogeneous matter is $D_{rM} = 3 - d_{rM}$. Mechanical radiation in matter is fractal in nature, and this includes waves in gases, liquids, and solids for limited regions of temperature and density where $B_r < 0$, $G_r < 0$ and $C_r > 0$, $H_r > 0$. Note that in general $D_M < 3$ so that $D_{rM} < 3$. Finally, when V_{1r} and V_{2r} have minimum values they can be expanded about these minimum values by writing $\xi_r = \xi_{rM} + \Delta\xi_r$ and $\zeta_r = \zeta_{rM} + \Delta\zeta_r$ as follows

$$V_{1r}(\xi_r) - V_{1r}(\xi_{rM}) = \frac{1}{2} C_r (\Delta\xi_r)^2 \tag{152}$$

$$V_{2r}(\zeta_r) - V_{2r}(\zeta_{rM}) = \frac{1}{2} H_r (\Delta\zeta_r)^2 \tag{153}$$

Electromagnetic radiation in matter will be treated in another paper.

6. THERMODYNAMIC NOETHER CURRENT TENSOR. Because the renormalization group equations can be derived from a variational principle, there exists a formal procedure for determining the conservation laws as a result of the form invariance of the Lagrangian density under continuous transformations. This procedure is given by Noether's theorem.[20] If the coordinates of a field $\phi(x_\mu)$ undergo a continuous translation of the form

$$x_\mu \rightarrow x'_\mu = x_\mu + \Delta x_\mu \tag{154}$$

then the Noether tensor

$$\Phi_{\mu\nu} = \frac{\partial \mathcal{L}}{\partial \phi_{,\nu}} \phi_{,\mu} - g_{\mu\nu}\mathcal{L} \tag{155}$$

satisfies the conservation law

$$\frac{\partial \Phi_{\mu\nu}}{\partial x_{\mu}} = 0 \tag{156}$$

In this paper it will be assumed that the thermodynamic Lagrangian densities are form invariant under continuous changes of volume and temperature of the form

$$t \rightarrow t' = t + \Delta t \tag{157}$$

$$v \rightarrow v' = v + \Delta v \tag{158}$$

This is true for the Lagrangian densities of relativistic thermodynamics because they are ultimately expressed in terms of the energy density and pressure, and these quantities are form invariant under continuous changes of volume and temperature.

The Noether current tensor for the energy density of the ground state of a thermodynamic system is given by

$$\Phi^{E}_{vv} = \frac{\partial \mathcal{L}_1}{\partial \xi_{,v}} \xi_{,v} - \mathcal{L}_1 \tag{159}$$

$$\Phi^{E}_{tt} = \frac{\partial \mathcal{L}_1}{\partial \xi_{,t}} \xi_{,t} - \mathcal{L}_1 \tag{160}$$

$$\Phi^{E}_{vt} = \frac{\partial \mathcal{L}_1}{\partial \xi_{,t}} \xi_{,v} \tag{161}$$

$$\Phi^{E}_{tv} = \frac{\partial \mathcal{L}}{\partial \xi_{,v}} \xi_{,t} \tag{162}$$

which satisfy the following conservation equations

$$\frac{\partial \Phi^{E}_{tv}}{\partial t} + \frac{\partial \Phi^{E}_{vv}}{\partial v} = 0 \tag{163}$$

$$\frac{\partial \Phi^{E}_{vt}}{\partial v} + \frac{\partial \Phi^{E}_{tt}}{\partial t} = 0 \tag{164}$$

The components of the Noether tensor for the ground state energy density are obtained from equations (159) through (162) by using the expression for $\mathcal{L}_1$ given in equation (77) as follows

$$\Phi^E_{vv} = \frac{1}{2} A\xi^2_{,v} - I\xi_{,t} - \xi(B - Z\xi_{,v}) - \frac{1}{2} C\xi^2 \tag{165}$$

$$\Phi^E_{tt} = -\frac{1}{2} A\xi^2_{,v} - B\xi - \frac{1}{2} C\xi^2 \tag{166}$$

$$\Phi^E_{vt} = I\xi_{,v} \tag{167}$$

$$\Phi^E_{tv} = \xi_{,t}(A\xi_{,v} + \xi Z) \tag{168}$$

The Noether tensor for the ground state pressure of a thermodynamic system is written as

$$\Phi^P_{vv} = \frac{\partial \mathcal{L}_2}{\partial \zeta_{,v}} \zeta_{,v} - \mathcal{L}_2 \tag{169}$$

$$\Phi^P_{tt} = \frac{\partial \mathcal{L}_2}{\partial \zeta_{,t}} \zeta_{,t} - \mathcal{L}_2 \tag{170}$$

$$\Phi^P_{vt} = \frac{\partial \mathcal{L}_2}{\partial \zeta_{,t}} \zeta_{,v} \tag{171}$$

$$\Phi^P_{tv} = \frac{\partial \mathcal{L}_2}{\partial \zeta_{,v}} \zeta_{,t} \tag{172}$$

which satisfy the following conservation equations

$$\frac{\partial \Phi^P_{tv}}{\partial t} + \frac{\partial \Phi^P_{vv}}{\partial v} = 0 \tag{173}$$

$$\frac{\partial \Phi^P_{vt}}{\partial v} + \frac{\partial \Phi^P_{tt}}{\partial t} = 0 \tag{174}$$

where

$$\Phi^{P}_{vv} = \frac{1}{2} F\zeta^{2}_{,v} - J\zeta_{,t} - \zeta(G - X\zeta_{,v}) - \frac{1}{2} H\zeta^{2} \tag{175}$$

$$\Phi^{P}_{tt} = -\frac{1}{2} F\zeta^{2}_{,v} - G\zeta - \frac{1}{2} H\zeta^{2} \tag{176}$$

$$\Phi^{P}_{vt} = J\zeta_{,v} \tag{177}$$

$$\Phi^{P}_{tv} = \zeta_{,t}(F\zeta_{,v} + X\zeta) \tag{178}$$

The solution of the conservation equations (163), (164), (173), and (174) yields the conserved quantities of the ground state of renormalized relativistic thermodynamics.

The Noether tensor for the radiation energy density in a thermodynamic system is given by

$$\Phi^{Er}_{vv} = \frac{\partial \mathcal{L}_{lr}}{\partial \xi_{r,v}} \xi_{r,v} - \mathcal{L}_{lr} \tag{179}$$

$$\Phi^{Er}_{tt} = \frac{\partial \mathcal{L}_{lr}}{\partial \xi_{r,t}} \xi_{r,t} - \mathcal{L}_{lr} \tag{180}$$

$$\Phi^{Er}_{vt} = \frac{\partial \mathcal{L}_{lr}}{\partial \xi_{r,t}} \xi_{r,v} \tag{181}$$

$$\Phi^{Er}_{tv} = \frac{\partial \mathcal{L}_{lr}}{\partial \xi_{r,v}} \xi_{r,t} \tag{182}$$

which satisfy

$$\frac{\partial \Phi^{Er}_{tv}}{\partial t} + \frac{\partial \Phi^{Er}_{vv}}{\partial v} = 0 \tag{183}$$

$$\frac{\partial \Phi^{Er}_{vt}}{\partial v} + \frac{\partial \Phi^{Er}_{tt}}{\partial t} = 0 \tag{184}$$

where

$$\phi_{vv}^{Er} = \frac{1}{2} A_r \xi_{r,v}^2 - I_r \xi_{r,t} - \xi_r (B_r - Z_r \xi_{r,v}) - \frac{1}{2} C_r \xi_r^2 \quad (185)$$

$$\phi_{tt}^{Er} = -\frac{1}{2} A_r \xi_{r,v}^2 - B_r \xi_r - \frac{1}{2} C_r \xi_r^2 \quad (186)$$

$$\phi_{vt}^{Er} = I_r \xi_{r,v} \quad (187)$$

$$\phi_{tv}^{Er} = \xi_{r,t} (A_r \xi_{r,v} + Z_r \xi_r) \quad (188)$$

The Noether tensor for the radiation pressure in a thermodynamic system is given by

$$\phi_{vv}^{Pr} = \frac{\partial \mathcal{L}_{2r}}{\partial \zeta_{r,v}} \zeta_{r,v} - \mathcal{L}_{2r} \quad (189)$$

$$\phi_{tt}^{Pr} = \frac{\partial \mathcal{L}_{2r}}{\partial \zeta_{r,t}} \zeta_{r,t} - \mathcal{L}_{2r} \quad (190)$$

$$\phi_{vt}^{Pr} = \frac{\partial \mathcal{L}_{2r}}{\partial \zeta_{r,t}} \zeta_{r,v} \quad (191)$$

$$\phi_{tv}^{Pr} = \frac{\partial \mathcal{L}_{2r}}{\partial \zeta_{r,v}} \zeta_{r,t} \quad (192)$$

while the conservation equations are

$$\frac{\partial \phi_{tv}^{Pr}}{\partial t} + \frac{\partial \phi_{vv}^{Pr}}{\partial v} = 0 \quad (193)$$

$$\frac{\partial \phi_{vt}^{Pr}}{\partial v} + \frac{\partial \phi_{tt}^{Pr}}{\partial t} = 0 \quad (194)$$

where

$$\Phi^{Pr}_{vv} = \frac{1}{2} F_r \zeta^2_{r,v} - J_r \zeta_{r,t} - \zeta_r (G_r - X_r \zeta_{r,v}) - \frac{1}{2} H_r \zeta^2_r \tag{195}$$

$$\Phi^{Pr}_{tt} = - \frac{1}{2} F_r \zeta^2_{r,v} - G_r \zeta_r - \frac{1}{2} H_r \zeta^2_r \tag{196}$$

$$\Phi^{Pr}_{vt} = J_r \zeta_{r,v} \tag{197}$$

$$\Phi^{Pr}_{tv} = \zeta_{r,t} (F_r \zeta_{r,v} + X_r \zeta_r) \tag{198}$$

The simultaneous solution of equations (183), (184), (193), and (194) yield the conserved quantities for radiation in matter.

7. <u>VOID RATIOS FOR THE REAL GASES.</u> In order to use the expression for the ground state void ratio given in equation (37) it is necessary to calculate the energy density and pressure of real gases for both the fractal and homogeneous cases. For the homogeneous case with D = 3, the renormalized pressure and energy density are given by [10,21]

$$P(3) = nRT[1 + nB^a + n^2 C(3) + \cdots] \tag{199}$$

$$E(3) = nRT[\frac{3}{2} - nT \frac{\partial B^a}{\partial T} - \frac{1}{2} n^2 T \frac{\partial C(3)}{\partial T} - \cdots] \tag{200}$$

where[10]

$$C(3) = C^a - 3(B^a)^2 \ln \psi^a \tag{201}$$

where B^a and C^a = unrenormalized second and third virial coefficients respectively, and C(3) = renormalized third virial coefficient for the uniform D = 3 real gas, and where ψ^a [not to be confused with equation (16)] is a function of the second virial coefficient given by[10,21]

$$\psi^a = \frac{T}{T_R} \left| \frac{B^a(T)}{B^a(T_R)} \right|^{2/3} \tag{202}$$

where T_R = species dependent relativity temperature of real gases.[10] The corresponding state equations for a fractal real gas are written as

$$P(D) = nRT[1 + nB^a + n^2 C(D) + \cdots] \tag{203}$$

$$E(D) = nRT\left[\frac{3}{2} - nT\frac{\partial B^{a}}{\partial T} - \frac{1}{2}n^{2}T\frac{\partial C(D)}{\partial T} - \cdots\right] \tag{204}$$

where

$$C(D) = C^{a} - D(B^{a})^{2}\, \ell n\, \psi^{a} \tag{205}$$

In order to obtain equation (205) it is necessary to assume that D is independent of T and n.

Combining equations (199), (200), (203), and (204) with equation (37) gives

$$\frac{\Delta V}{V} = \frac{1}{3}n^{2}T\frac{\partial}{\partial T}\left[C(3) - C(D)\right] \tag{206}$$

where the following approximation was used in equation (37)

$$E(3) + P(3) - T\frac{\partial P(3)}{\partial T} \sim \frac{3}{2}nRT \tag{207}$$

Combining equations (201), (205), and (206) gives

$$\frac{\Delta V}{V} = -\frac{n^{2}}{3}(3 - D)T\frac{\partial}{\partial T}\left[(B^{a})^{2}\,\ell n\,\psi^{a}\right] + \cdots \tag{208}$$

$$= \frac{n^{2}}{9}(3 - D)T\frac{\partial}{\partial T}\left[C(3) - C^{a}\right] + \cdots$$

Note that in general

$$C(3) - C(D) = \frac{1}{3}(3 - D)\left[C(3) - C^{a}\right] \tag{209}$$

Thus voids will exist in the ground state of real gases only in the temperature intervals for which $\Delta V/V > 0$ in equation (208). This condition gives

$$\frac{\partial}{\partial T}\left[(B^{a})^{2}\,\ell n\,\psi^{a}\right] < 0 \tag{210}$$

or equivalently

$$\frac{\partial}{\partial T}\left[C(3) - C^{a}\right] > 0 \tag{211}$$

as shown in Figure 1. From Figure 1 it is clear that the largest size voids will occur at low temperatures. There is also a narrow fractal region just above the Boyle temperature, and a broad fractal region at high temperatures.

Ultimately, the voids that occur in real gases are due to the interaction of these gases with the vacuum state, which manifests itself in the third and higher virial coefficients. The ideal gas is not fractal.

Consider now mechanical radiation in a real gas. For the homogeneous case with D = 3 and $d_r = 0$ the renormalized radiation pressure and energy density are[21]

$$P_r(3,0) = nRT[\frac{1}{12}k_o^2A_o^2 + nB_r^a + n^2C_r(3,0) + \cdots] \tag{212}$$

$$E_r(3,0) = nRT[\frac{1}{4}k_o^2A_o^2 - nT\frac{\partial B_r^a}{\partial T} - \frac{1}{2}n^2T\frac{\partial C_r(3,0)}{\partial T} - \cdots] \tag{213}$$

where k_o and A_o = wave number and amplitude of waves in an ideal gas, and where B_r^a = nonrelativistic (unrenormalized) second radiation virial coefficient, and $C_r(3,0)$ = relativistic third virial coefficient for a homogeneous (D = 3 and d_r = 0) mechanical radiation field given by[21]

$$C_r(3,0) = C_r^a - 3[2B^aB_r^a + (B_r^a)^2]\,\ln\psi^a - 3(B^a + B_r^a)^2\,\ln\left(1 + \frac{\psi_r^a}{\psi^a}\right) \tag{214}$$

where ψ^a is given by equation (202) and ψ_r^a [not to be confused with equation (51)] is given by[21]

$$\psi^a + \psi_r^a = \frac{T}{T_R}\left|\frac{B^a(T) + B_r^a(T)}{B^a(T_R) + B_r^a(T_R)}\right|^{2/3} \tag{215}$$

The procedure for calculating B_r^a and C_r^a is given in Reference 21. The corresponding state equations for fractal radiation in a fractal real gas are given by

$$P_r(D,d_r) = nRT[\frac{1}{12}k_o^2A_o^2 + nB_r^a + n^2C_r(D,d_r) + \cdots] \tag{216}$$

$$E_r(D,d_r) = nRT[\frac{1}{4}k_o^2A_o^2 - nT\frac{\partial B_r^a}{\partial T} - \frac{1}{2}n^2T\frac{\partial C_r(D,d_r)}{\partial T} - \cdots] \tag{217}$$

where $C_r(D,d_r)$ = relativistic third virial coefficient for a fractal mechanical radiation field in a fractal real gas and is given by

$$C_r(D,d_r) = C_r^a - D[2B^aB_r^a + (B_r^a)^2]\,\ell n\,\psi^a - D(B^a + B_r^a)^2\,\ell n\left(1 + \frac{\psi_r^a}{\psi^a}\right) \quad (218)$$

$$+ d_r(B^a)^2\,\ell n\,\psi^a$$

where the fractal dimension of the mechanical radiation in a real gas is now $D_r = D - d_r$ which is lower than the fractal dimension D of the ground state of a real gas. Note that in order to obtain equation (218) it is assumed that D and d_r are independent of T and V.

The calculation of the void ratio for radiation in a real gas then proceeds from equation (67). Note first the following approximation

$$E_r(3,0) + P_r(3,0) - T\frac{\partial P_r(3,0)}{\partial T} \sim \frac{1}{4}\,nRTk_o^2A_o^2 \quad (219)$$

Combining equations (213) and (217) gives

$$E_r(D,d_r) - E_r(3,0) = \frac{1}{2}\,n^3RT^2\,\frac{\partial}{\partial T}\,[C_r(3,0) - C_r(D,d_r)] \quad (220)$$

Placing equation (219) and (220) into equation (67) yields

$$\left(\frac{\Delta V}{V}\right)_r = \frac{2n^2}{k_o^2A_o^2}\,T\,\frac{\partial}{\partial T}\,[C_r(3,0) - C_r(D,d_r)] + \cdots \quad (221)$$

where

$$C_r(3,0) - C_r(D,d_r) = -(3 - D)\left\{[2B^aB_r^a + (B_r^a)^2]\,\ell n\,\psi^a + (B^a + B_r^a)^2\,\ell n\left(1 + \frac{\psi_r^a}{\psi^a}\right)\right\}$$

$$- d_r(B^a)^2\,\ell n\,\psi^a \quad (222)$$

Placing equations (201) and (214) into equation (222) yields

$$C_r(3,0) - C_r(D,d_r) = \frac{1}{3}\,(3 - D)[C_r(3,0) - C_r^a] + \frac{d_r}{3}[C(3) - C^a] \quad (223)$$

Combining equations (221) and (223) gives the following expression for void ratio of mechanical radiation

$$\left(\frac{\Delta V}{V}\right)_r = \frac{2n^2}{3k_o^2A_o^2}\left\{(3 - D)T\frac{\partial}{\partial T}[C_r(3,0) - C_r^a] + d_rT\frac{\partial}{\partial T}[C(3) - C^a]\right\} + \cdots \qquad (224)$$

It has been shown that the coefficients B_r^a , B_r , C_r^a , and $C_r(3,0)$ are all proportional to $k_o^2A_o^2$.[21] In addition, the radiation fractal decrement d_r is proportional to the radiation energy density so that

$$d_r = \tau_r E_r(D,d_r) \qquad (225)$$

$$= \frac{1}{4}\tau_r nRTk_o^2A_o^2 + \cdots$$

where τ_r is independent of k_o and A_o . It then follows that

$$C_r(3,0) = k_o^2A_o^2C_r'(3,0) \qquad (226)$$

$$C_r(D,d_r) = k_o^2A_o^2C_r'(D,d_r) \qquad (227)$$

$$C_r^a = k_o^2A_o^2C_r^{a'} \qquad (228)$$

where $C_r^{a'}$, $C_r'(3,0)$, and $C_r'(D,d_r)$ are independent of k_o and A_o . Therefore equation (224) can be written as

$$\left(\frac{\Delta V}{V}\right)_r = \frac{2}{3}n^2\left\{(3 - D)T\frac{\partial}{\partial T}[C_r'(3,0) - C_r^{a'}] + \frac{1}{4}\tau_r nRT^2\frac{\partial}{\partial T}[C(3) - C^a]\right\} + \cdots \quad (229)$$

The void ratio for mechanical radiation in real gases is independent of wave amplitude and frequency, and depends only on temperature and density. For the case of fractal mechanical radiation in a homogeneous (D = 3) ground state, equation (229) becomes

$$\left(\frac{\Delta V}{V}\right)_r = \frac{1}{6}\tau_r n^3RT^2\frac{\partial}{\partial T}[C(3) - C^a] + \cdots \qquad (230)$$

Equation (230) is somewhat similar to the result for the ground state in equation (208).

Even if the ground state is homogeneous (D = 3) the mechanical radiation state can be fractal with fractal dimension $D_r = 3 - d_r$. In general, however, the ground state may be fractal, and the fractal dimension of the radiation field will be written as $D_r = D - d_r$. Only in limited temperature regions will mechanical radiation in real gases have fractal properties, i.e., those regions for which $(\Delta V/V)_r > 0$ in equation (229). The fractal dimension D_r of mechanical radiation can be calculated from the general radiation Lagrangian formalism given in Section 5. The voids in a mechanical radiation field are due to the interaction of the excitations of a real gas with the vacuum state. Mechanical radiation in an ideal gas is not fractal.

8. VOID RATIOS FOR FRACTAL SOLIDS AND QUANTUM LIQUIDS. In this section the void ratios for the fractal ground state and excited states of solids and quantum liquids are calculated. For simplicity only the T = 0 state will be considered. In order to use equation (38) for the calculation of the ground state void ratio, the T = 0 energy density and pressure must be calculated for both the fractal and homogeneous states. It will be assumed that

$$E_o^a = An^{\sigma_o} \tag{231}$$

$$P_o^a = (\sigma_o - 1)E_o^a \tag{232}$$

where A and σ_o = constants independent of density. Using equation (29) with γ_o and D_o taken to be constants independent of density gives the renormalized energy densities of the fractal and homogeneous systems respectively as

$$E_o(D_o) = \frac{E_o^a}{G(D_o)} \tag{233}$$

$$E_o(3) = \frac{E_o^a}{G(3)} \tag{234}$$

where

$$G(D_o) = D_o\sigma_o^2 - D_o\sigma_o(2 + \gamma_o) + D_o(1 + \gamma_o) + 1 \tag{235}$$

$$G(3) = 3\sigma_o^2 - 3(2 + \gamma_o)\sigma_o + 3\gamma_o + 4 \tag{236}$$

Then the difference in energy densities for the fractal and homogeneous states is given by

$$E_o(D_o) - E_o(3) = \frac{(3 - D_o)F_oE_o^a}{G(3)G(D_o)} \tag{237}$$

where

$$F_o = \sigma_o^2 - (2 + \gamma_o)\sigma_o + \gamma_o + 1 \tag{238}$$

which satisfies

$$G(D_o) = D_o F_o + 1 \tag{239}$$

$$G(3) = 3F_o + 1 \tag{240}$$

Combining equations (231) and (234) gives

$$n \frac{dE_o(3)}{dn} = \frac{\sigma_o E_o^a}{G(3)} \tag{241}$$

and therefore equation (38) gives the following expression for the T = 0 ground state void ratio for solids and quantum liquids

$$\left(\frac{\Delta V}{V}\right)_o = \frac{[E_o(D_o) - E_o(3)]G(3)}{\sigma_o E_o^a} = \frac{(3 - D_o)F_o}{\sigma_o G(D_o)} \tag{242}$$

In addition to the obvious dependence of D_o , the void ratio depends on the constants σ_o and γ_o .

Often in the literature the average energy per particle is introduced as follows[10]

$$\varepsilon_o^a = E_o^a/n = An^{\sigma_o - 1} = Ax^{\kappa} \tag{243}$$

where x is defined by $n = x^3$. This gives

$$\sigma_o = \kappa/3 + 1 \tag{244}$$

Using the parameter κ gives[10]

$$G(3) = \frac{\kappa^2}{3} - \kappa\gamma_o + 1 \tag{245}$$

$$G(D_o) = \frac{D_o\kappa^2}{9} - \frac{\kappa D_o \gamma_o}{3} + 1 \tag{246}$$

$$F_o = \frac{\kappa}{3}\left(\frac{\kappa}{3} - \gamma_o\right) \qquad (247)$$

These functions are expressed in terms of both κ and γ_o .

Of particular interest are the cases of the non-relativistic and ultra-relativistic non-interacting degenerate T = 0 Fermi gases. The non-relativistic Fermi gas has the following properties

$$\gamma_o = 2/3 \qquad \kappa = 2 \qquad \sigma_o = 5/3 \qquad (248)$$

$$G(3) = 1 \qquad G(D_o) = 1 \qquad F_o = 0 \qquad (249)$$

For the ultra-relativistic case these quantities are

$$\gamma_o = 1/3 \qquad \kappa = 1 \qquad \sigma_o = 4/3 \qquad (250)$$

$$G(3) = 1 \qquad G(D_o) = 1 \qquad F_o = 0$$

Because $F_o = 0$ for the ideal non-relativistic and ultra-relativistic Fermi gases, it is clear from equation (242) that no voids exist for these cases. Ideal non-relativistic and ultra-relativistic Fermi gases are homogeneous with $D_o = 3$.

In general the Grüneisen parameter γ_o is a function of the index κ . For instance, the effective mass approximation for an interacting system gives the following expression for the zero-temperature Grüneisen parameter[10]

$$\gamma_o \sim \frac{\kappa - 4}{3} = \sigma_o - \frac{7}{3} \qquad (252)$$

Using this relationship gives[10]

$$G(3) \sim 1 + \frac{4\kappa}{3} = 4\sigma_o - 3 > 0 \qquad (253)$$

$$G(D_o) \sim 1 + \frac{4\kappa D_o}{9} = \frac{4}{3} D_o\sigma_o - \left(\frac{4}{3}D_o - 1\right) > 0 \qquad (254)$$

$$F_o \sim \frac{4\kappa}{9} = \frac{4}{3}(\sigma_o - 1) > 0 \qquad (255)$$

Equation (252) is valid only for $\kappa \geqslant 5$ or $\sigma_o \geqslant 8/3$. In general the value of D_o for solids or quantum liquids may be determined experimentally or possibly theoretically from a T = 0 Lagrangian formulation outlined in Section 4. In general T = 0 solids and quantum liquids are fractal for all physical densities because equation (242) shows that $(\Delta V/V)_o > 0$.

Consider now the void ratio of mechanical waves in a solid or quantum liquid. For simplicity only waves in a T = 0 system will be considered. The wave equation (59) can be simplified by using equations (231) through (234) and the approximation $E_{jr}/E_j = E_{or}/E_o$ with the result that

$$D_o \frac{E_{jr}}{E_j} P_o(\gamma_o - \gamma_{or}) = D_o E_{or}(\sigma_o - 1)(\gamma_o - \gamma_{or}) \tag{256}$$

and therefore equation (65) can be rewritten as

$$D_o n^2 \frac{d^2 E_{or}}{dn^2} - D_o(1 + \gamma_o) n \frac{dE_{or}}{dn} + C_o E_{or} = E^a_{or} \tag{257}$$

where

$$C_o = D_o \sigma_o(\gamma_o - \gamma_{or}) + D_o(1 + \gamma_{or}) + 1 + \tau_{or}[(1 + \gamma_o)P_o - K_o] \tag{258}$$

and where the relation $d_{or} = \tau_{or} E_{or}$ is used which is similar to equation (225) that was used for the real gases. Assume now that[9]

$$E_{or} = \frac{1}{4} K_o k^2 A^2 \qquad E^a_{or} = \frac{1}{4} K^a_o k^2_a A^2_a \tag{259}$$

where k_a and A_a = nonrelativistic wave number and wave amplitude respectively, and k and A = relativistic wave number and wave amplitude respectively. Placing equation (259) into equation (257) gives

$$\frac{D_o}{4} K_o n^2 \frac{d^2}{dn^2}(k^2 A^2) + \frac{D_o}{4}\left[2n \frac{dK_o}{dn} - (1 + \gamma_o)K_o\right] n \frac{d}{dn}(k^2 A^2)$$
$$+ \frac{1}{4} k^2 A^2 K_o G_r(D_o, \tau_{or}) = \frac{1}{4} k^2_a A^2_a K^a_o \tag{260}$$

where

$$G_r(D_o, \tau_{or}) = D_o \frac{n^2}{K_o} \frac{d^2 K_o}{dn^2} - D_o(1 + \gamma_o) \frac{n}{K_o} \frac{dK_o}{dn} + C_o \tag{261}$$

Using $K_o \sim n^{\sigma_o}$ in equation (261) gives

$$G_r(D_o,\tau_{or}) = D_o\sigma_o^2 - D_o\sigma_o(2 + \gamma_o) + C_o \tag{262}$$

$$= D_o\sigma_o^2 - D_o\sigma_o(2 + \gamma_{or}) + D_o(1 + \gamma_{or}) + 1 - \tau_{or}[K_o - (1 + \gamma_o)P_o]$$

If kA is taken to be independent of density it follows from equation (260) that

$$k^2A^2 = \frac{k_a^2A_a^2K_o^a/K_o}{G_r(D_o,\tau_{or})} \tag{263}$$

Combining equations (34) and (263) gives

$$k^2A^2 = k_a^2A_a^2G(D_o)/G_r(D_o,\tau_{or}) \tag{264}$$

Similarly, from equation (260) it follows that

$$E_{or}(D_o,\tau_{or}) = E_{or}^a/G_r(D_o,\tau_{or}) \tag{265}$$

$$E_{or}(3,0) = E_{or}^a/G_r(3,0) \tag{266}$$

where

$$G_r(3,0) = 3\sigma_o^2 - 3\sigma_o(2 + \gamma_{or}) + 3\gamma_{or} + 4 \tag{267}$$

Taking $D_o = 3$, $\tau_{or} = 0$, and $\gamma_{or} = 1/3$ in equations (235), (262), and (264) yields equation (121) of Reference 9.

The calculation of the void ratio for mechanical radiation in a fractal solid or quantum liquid follows from equation (68). First determine the enhanced energy of the fractal radiation state from equations (265) and (266) as follows

$$E_{or}(D_o,\tau_{or}) - E_{or}(3,0) = \frac{E_{or}^a[G_r(3,0) - G_r(D_o,\tau_{or})]}{G_r(3,0)G_r(D_o,\tau_{or})} \tag{268}$$

where

$$G_r(3,0) - G_r(D_o,\tau_{or}) = (3 - D_o)F_{or} - \tau_{or}[(1 + \gamma_o)P_o - K_o] \tag{269}$$

$$F_{or} = \sigma_o^2 - \sigma_o(2 + \gamma_{or}) + \gamma_{or} + 1 \tag{270}$$

Calculate also the following expression

$$E_{or}(3,0) + P_{or}(3,0) = n\,\frac{dE_{or}(3,0)}{dn} = \frac{\sigma_o E^a_{or}}{G_r(3,0)} \tag{271}$$

Then equation (68) gives the void ratio for mechanical waves as follows

$$\left(\frac{\Delta V}{V}\right)_{or} = \frac{(3 - D_o)F_{or} + \tau_{or}[K_o - (1 + \gamma_o)P_o]}{\sigma_o G_r(D_o,\tau_{or})} \tag{272}$$

Note that

$$G_r(3,0) = 3F_{or} + 1 \tag{273}$$

$$G_r(D_o,\tau_{or}) = D_o F_{or} + 1 + \tau_{or}[(1 + \gamma_o)P_o - K_o] \tag{274}$$

A homogeneous ground state ($D_o = 3$) can have a fractal radiation excited state described by

$$\left(\frac{\Delta V}{V}\right)_{or} = \frac{\tau_{or}[K_o - (1 + \gamma_o)P_o]}{\sigma_o G_r(3,\tau_{or})} \tag{275}$$

where

$$G_r(3,\tau_{or}) = 3\sigma_o^2 - 3\sigma_o(2 + \gamma_{or}) + 3\gamma_{or} + 4 - \tau_{or}[K_o - (1 + \gamma_o)P_o] \tag{276}$$

Both equations (272) and (275) are valid only for $(\Delta V/V)_{or} > 0$ and this restricts the density regions in which radiation voids are possible. Only in the unphysical regions below equilibrium density are equations (272) or (275) negative. In fact at the equilibrium density of a T = 0 solid or quantum liquid one has $P_o = 0$ so that $(\Delta V/V)_{or} > 0$ at this point. Radiation voids are also possible in the high density regions beyond the equilibrium density of an interacting system. Note that for ideal non-relativistic and ultra-relativistic T = 0 Fermi gases $(1 + \gamma_o)P_o - K_o = 0$, and $D_o = 3$, and from equation (272) the radiation state is homogeneous.

9. CONCLUSION. The renormalization group equations for fractal matter and fractal mechanical radiation are presented and a Lagrangian formulation of these equations is developed. For limited temperature and density regions the equilibrium fractal dimensions of the ground and excited states of real gases, solids, and quantum liquids may possibly be obtained by minimizing the Lagrangian density with respect to the fractal dimension of the system. Ideal non-relativistic and ultra-relativistic quantum thermodynamic systems are not fractal. For interacting thermodynamic systems the ground and excited states are fractal in nature. The ground and excited states of real gases are fractal only in limited temperature regions. Although in general $D \neq 3$ it has not been shown that the voids in matter and mechanical radiation fields are self similar, which is a basic characteristic of fractal systems.

ACKNOWLEDGEMENT

The author wishes to thank Elizabeth K. Klein for typing this paper.

REFERENCES

1. Goldstein, H., Classical Mechanics, Addison-Wesley, New York, 1980.

2. Lanczos, C., The Variational Principles of Mechanics, University of Toronto Press, Toronto, 1960.

3. Yourgrau, W. and Mandelstam, S., Variational Principles in Dynamics and Quantum Theory, Pitman, London, 1955.

4. Mandl, F. and Shaw, G., Quantum Field Theory, John Wiley, New York, 1984.

5. Bogoliubov, N. and Shirkov, D., Introduction to the Theory of Quantized Fields, Interscience, New York, 1959.

6. Lee, T., Particle Physics and Introduction to Field Theory, Harwood, New York, 1981.

7. Aitchison, I. and Hey, A., Gauge Theories in Particle Physics, Adam Hilger, Bristol, 1982.

8. Weiss, R., "Relativistic Wave Equations for Solids and Low Temperature Quantum Systems," Third Army Conference on Applied Mathematics and Computing, Georgia Institute of Technology, ARO 86-1, May 13-16, 1985, p. 717.

9. Weiss, R., "Scale Invariant Equations for Relativistic Waves," Fourth Army Conference on Applied Mathematics and Computing, Cornell University, ARO 87-1, May 27-30, 1986, p. 307.

10. Weiss, R., Relativistic Thermodynamics, Vols. 1 and 2, Exposition Press, New York, 1976.

11. Pietronero, L. and Tosatti, E., editors, Fractals in Physics, North-Holland-Elsevier, New York, 1986.

12. Stanley, H. and Ostrowsky, N., editors, On Growth and Form, Martinus Nijhoff, Boston, 1986.

13. Family, F. and Landau, D., editors, Kinetics of Aggregation and Gelation, North-Holland, Amsterdam, 1984.

14. Pynn, R. and Skjeltorp, A., editors, Scaling Phenomena in Disordered Systems, Plenum, New York, 1985.

15. Boccara, N. and Daoud, M., editors, Physics of Finely Divided Matter, Springer, Berlin, 1985.

16. de Gennes, P., Scaling Concepts in Polymer Physics, Cornell Univ. Press, Ithaca, 1979.

17. Mandelbrot, B., The Fractal Geometry of Nature, Freeman, San Francisco, 1983.

18. Müller, B. and Schäfer, A., "Improved Bounds on the Dimension of Space-Time, Phys. Rev. Lett. 56, 1215, 1986.

19. Hirth, J. and Lothe, J., Theory of Dislocations, McGraw-Hill, New York, 1968.

20. Quigg, C., Gauge Theories of the Strong, Weak, and Electromagnetic Interactions, Addison-Wesley, New York, 1983.

21. Weiss, R., "Relativistic Wave Equations for Real Gases," Fourth Army Conference on Applied Mathematics and Computing, Cornell University, ARO 87-1, May 27-30, 1986, p. 341.

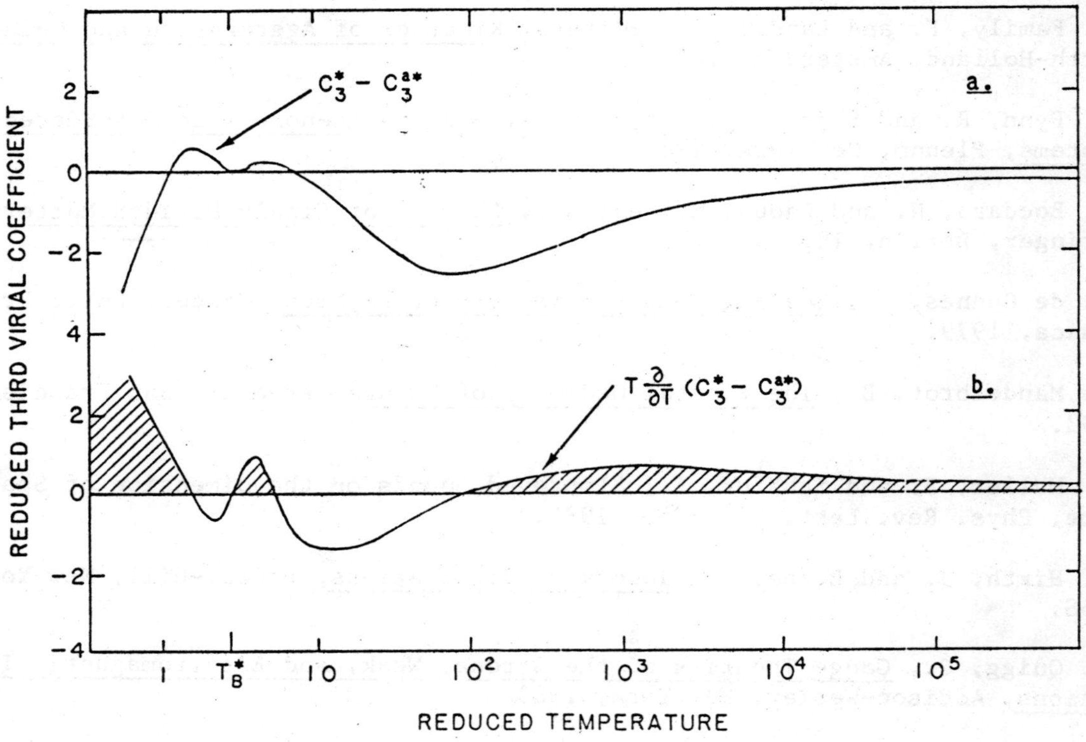

Figure 1. a) $C_3^* - C_3^{a*}$ as a function of reduced temperature; b) $T\partial/\partial T(C_3^* - C_3^{a*})$ versus reduced temperature. When positive, this indicates the regions (shaded areas) where voids are possible for real gases.

E

THERMODYNAMIC GAUGE THEORY OF SOLIDS AND QUANTUM LIQUIDS WITH INTERNAL PHASE

Richard A. Weiss
U. S. Army Engineer Waterways Experiment Station
Vicksburg, Mississippi 39180

ABSTRACT. The local gauge invariance of relativistic thermodynamics under phase rotations suggests that bulk matter systems have density and temperature dependent internal phase angles associated with the state functions. A procedure for determining the internal phase angles associated with energy, pressure, entropy, thermodynamic potentials, and gauge parameters of solids and quantum liquids is presented in terms of the renormalization group equations of relativistic thermodynamics. The calculated magnitudes of the thermodynamic state functions depend on the values of the internal phase angles. It is suggested that the external angular momentum of systems may be coupled to the angular momenta associated with the internal space of thermodynamic phase angles. Applications to mechanical waves in matter with internal phase are considered. These effects are expected to be found in high density and pressure systems such as atomic nuclei, neutron stars, nuclear explosions, and the interaction of directed energy beams with matter.

1. INTRODUCTION. The complete understanding of matter and radiation at high densities requires a locally scale and gauge invariant theory of the forces and fields that determine the properties of a physical system.[1,2] The basic forces in a physical system are associated with a local gauge group, as for example the gauge group of the standard model of the strong and electroweak interactions is $SU(3)_c \times SU(2)_L \times U(1)_Y$. For simple systems, such as electromagnetism, the gauge group is U(1) the group of phase rotations.[3] Local gauge symmetry has unified the interactions of nature, and it is only natural because of such success to attempt a similar synthesis in other areas of physics such as thermodynamics and mechanics.

The vacuum state plays an important role in the development of local gauge theories of the four fundamental interactions. It produces observable effects in quantum electrodynamic calculations of the fermion self-energy, vertex modification, and vacuum polarization as manifested in the Lamb shift.[4,5] In quantum flavordynamics the nonzero expectation values of the vacuum Higgs field produces the spontaneous symmetry breaking that gives rise to the massive intermediate vector bosons that mediate the weak interactions.[1] In quantum chromodynamics the vacuum polarization leads to the concepts of a running coupling constant and asymptotic freedom for the non-Abelian gauge theories.[1-3] The question then arises as to whether vacuum effects appear in systems at the macroscopic level, and whether a synthesis of thermodynamics and continuum mechanics can be based on a locally gauge invariant theory that includes the effects of the vacuum state.

As part of a general program to determine the state equation of systems at high densities, a local gauge theory of matter and radiation has been developed that is based on a gauge and scale invariant relativistic trace equation written to include vacuum effects as follows[6]

$$U + T\left(\frac{dU}{dT}\right)_{PV} - 3V\,\frac{d}{dV}(PV)_U = U^a + T\left(\frac{dU^a}{dT}\right)_{P^aV} \tag{1}$$

where U = relativistic (renormalized) internal energy, P = relativistic pressure, T = absolute temperature, V = volume of substance, and U^a and P^a = corresponding nonrelativistic internal energy and pressure. Throughout this paper the index "a" will refer to nonrelativistic (unrenormalized) calculations. The trace equation (1) can be rewritten as[7,8]

$$\left(1 - b + T\frac{\partial}{\partial T} - bV\frac{\partial}{\partial V}\right)E - 3\left(1 + \gamma + V\frac{\partial}{\partial V} - \gamma T\frac{\partial}{\partial T}\right)P \tag{2}$$

$$= \left(1 - b^a + T\frac{\partial}{\partial T} - b^aV\frac{\partial}{\partial V}\right)E^a$$

Equation (1) can also be written as

$$E + \frac{T}{V}C_V - 3(P - K_T) + (T\frac{\partial P}{\partial T} - P)(3\gamma - b) = E^a + \frac{T}{V}C_V^a - b^a(T\frac{\partial P^a}{\partial T} - P^a) \tag{3}$$

where E = relativistic energy density = U/V , E^a = nonrelativistic energy density, and where[7,8]

$$\gamma = \frac{V}{C_V}\left(\frac{\partial P}{\partial T}\right)_V \tag{4}$$

$$b = \frac{T(\partial P/\partial T)_V}{(P - K_T)} \tag{5}$$

$$b^a = \frac{T(\partial P^a/\partial T)_V}{(P^a - K_T^a)} \tag{6}$$

where γ = Grüneisen parameter, C_V = relativistic heat capacity at constant volume, and C_V^a = nonrelativistic heat capacity at constant volume, given respectively by

$$C_V = \left(\frac{\partial U}{\partial T}\right)_V \tag{7}$$

$$C_V^a = \left(\frac{\partial U^a}{\partial T}\right)_V \tag{8}$$

and where

$$K_T = -V\left(\frac{\partial P}{\partial V}\right)_T \tag{9}$$

$$K_T^a = -V\left(\frac{\partial P^a}{\partial V}\right)_T \tag{10}$$

are the relativistic and nonrelativistic values of the bulk modulus respectively. The parameters b and γ are the gauge parameters of relativistic thermodynamics. Equation (2) can be decoupled into two independent equations by noting that E and P are related by the Gibbs-Helmholtz relation as follows[9]

$$\left(\frac{\partial U}{\partial V}\right)_T = E + V\left(\frac{\partial E}{\partial V}\right)_T = T\left(\frac{\partial P}{\partial T}\right)_V - P \tag{11}$$

With the introduction of a Lagrange undetermined multiplier η , equation (11) can be rewritten as

$$\eta\left(1 + V\frac{\partial}{\partial V}\right)E + \eta\left(1 - T\frac{\partial}{\partial T}\right)P = 0 \tag{12}$$

Using equation (12) allows equation (2) to be decoupled as follows[8]

$$\left(T\frac{\partial}{\partial T} + fV\frac{\partial}{\partial V} + M\right)E = \psi^a \tag{13}$$

$$\left(T\frac{\partial}{\partial T} + hV\frac{\partial}{\partial V} + N\right)P = 0 \tag{14}$$

where

$$f = \eta - b \tag{15}$$

$$h = \frac{1}{\eta/3 - \gamma} \tag{16}$$

$$M = f + 1 \tag{17}$$

$$N = h - 1 \tag{18}$$

$$\psi^a = (T \frac{\partial}{\partial T} - b^a V \frac{\partial}{\partial V} + 1 - b^a) E^a \tag{19}$$

Equation (13) and (14) are the ground state renormalization group equations of relativistic thermodynamics.

For a solid or low temperature quantum system the nonrelativistic, scalar state equation of the ground state is assumed to have the following form[6-8]

$$E^a = E^a_o + E^a_j T^j + \cdots \tag{20}$$

$$P^a = P^a_o + P^a_j T^j + \cdots \tag{21}$$

where E^a and P^a = nonrelativistic energy density and pressure respectively, E^a_o and P^a_o = nonrelativistic zero-temperature values of the energy density and pressure respectively, E^a_j and P^a_j = nonrelativistic thermal coefficients for the energy density and pressure respectively, T = absolute temperature of the system (°K), and j = numerical index having values characteristic of the type of physical system. Note that $U^a = VE^a$ and $U^a_o = VE^a_o$ where U^a_o = zero-temperature value of the unrenormalized internal energy.

A commonly used descriptor of the thermal state equations given by equations (20) and (21) is the nonrelativistic zero-temperature value of the Grüneisen parameter that is defined by[6-8]

$$\gamma^a_o = \frac{P^a_j}{E^a_j} = \frac{1}{(j-1)} \frac{1}{E^a_j} \frac{d}{dV}(VE^a_j) \tag{22}$$

except for j = 1. Here γ^a_o = nonrelativistic zero-temperature value of the Grüneisen parameter, and V = volume of the material system. When j = 1 , $\gamma^a_o = 2/3$. The zero temperature value of the nonrelativistic bulk modulus is given by $K^a_o = n dP^a_o/dn$, where n = N/V = number of moles per unit volume, and N = number of moles of a substance.

The corresponding relativistic scalar state equation will be written as[6-8]

$$E = E_o + E_j T^j + \cdots \tag{23}$$

$$P = P_o + P_j T^j + \cdots \tag{24}$$

$$\gamma_o = \frac{P_j}{E_j} = \frac{1}{(j-1)} \frac{1}{E_j} \frac{d}{dV}(VE_j) \tag{25}$$

except for $j = 1$, when $E_1 = E_1^a$, where E_o and P_o = relativistic zero-temperature energy density and pressure respectively, E_j and P_j = relativistic thermal coefficients for the energy density and pressure respectively, and γ_o = relativistic zero-temperature Grüneisen parameter. The relativistic value of the zero temperature bulk modulus is given by $K_o = ndP_o/dn$. Note that $U_o = VE_o$, where U_o = zero temperature value of the renormalized internal energy. Combining equation (2) with the state equations (20) through (25) yields the following ground state equations[6]

$$E_o - 3[(1 + \gamma_o)P_o - K_o] = E_o^a \tag{26}$$

$$E_j\left(1 + j + \frac{j\gamma_o P_o}{P_o - K_o} - 3V\frac{d\gamma_o}{dV}\right) = E_j^a\left(1 + j + \frac{j\gamma_o^a P_o^a}{P_o^a - K_o^a}\right) \tag{27}$$

where the internal energy coefficients are given by

$$\frac{E_j^a}{E_j} = \exp\left[(j - 1)\int^V (\gamma_o^a - \gamma_o)\frac{dV}{V}\right] \tag{28}$$

and where P_o, K_o and γ_o = zero temperature values of the relativistic pressure, incompressibility ($-VdP_o/dV$) and Gruneisen parameter respectively, and P_o^a, K_o^a and γ_o^a are the corresponding nonrelativistic values of these quantities. Eqs. (26) and (27) are a set of coupled nonlinear differential equations for P_o and γ_o . Equation (26) is equivalent to[6]

$$3n^2\frac{d^2E_o}{dn^2} - 3(1 + \gamma_o)n\frac{dE_o}{dn} + (3\gamma_o + 4)E_o = E_o^a \tag{29}$$

The trace equation for radiation in matter can be derived by a perturbation technique applied to equation (2) with the result[7,8]

$$\left(1 - b + T\frac{\partial}{\partial T} - bV\frac{\partial}{\partial V}\right)E_r - \beta_r\left(T\frac{\partial P}{\partial T} - P\right) \tag{30}$$

$$- 3\left[\left(1 + \gamma + V\frac{\partial}{\partial V} - \gamma T\frac{\partial}{\partial T}\right)P_r - \delta_r\left(T\frac{\partial P}{\partial T} - P\right)\right] = \psi_r^a$$

where $\mathcal{E}_r$ and P_r = radiation energy density and pressure respectively, and where[7,8]

$$\beta_r = b_r \frac{P_r - K_{Tr}}{P - K_T} - \frac{T \frac{\partial P}{\partial T}(P_r - K_{Tr})}{(P - K_T)^2} \tag{31}$$

$$b_r = \frac{T \frac{\partial P_r}{\partial T}}{P_r - K_{Tr}} \tag{32}$$

$$\delta_r = \frac{\partial \mathcal{E}_r/\partial T}{\partial \mathcal{E}/\partial T} (\gamma_r - \gamma) \tag{33}$$

$$\gamma_r = \frac{\partial P_r/\partial T}{\partial \mathcal{E}_r/\partial T}$$

$$K_{Tr} = - V \frac{\partial P_r}{\partial V} \tag{35}$$

$$\psi_r^a = \left(T \frac{\partial}{\partial T} - b^a V \frac{\partial}{\partial V} + 1 - b^a\right)\mathcal{E}_r^a - \beta_r^a\left(T \frac{\partial P^a}{\partial T} - P^a\right) \tag{36}$$

The parameters γ_r and b_r are the two radiation gauge parameters of the thermal medium.

Equation (30) can be separated into two radiation equations each of which is similar in form to the Callan-Symanzik equation. This is done by using the Gibbs-Helmholtz equation which for radiation becomes[9]

$$\frac{\partial U_r}{\partial V} = \mathcal{E}_r + V \frac{\partial \mathcal{E}_r}{\partial V} = T \frac{\partial P_r}{\partial T} - P_r \tag{37}$$

Introducing a radiation Lagrange multiplier η_r as follows

$$\eta_r\left[\mathcal{E}_r + V \frac{\partial \mathcal{E}_r}{\partial V} + P_r - T \frac{\partial P_r}{\partial T}\right] = 0 \tag{38}$$

allows equation (30) to be separated as follows[8]

$$\left(T\frac{\partial}{\partial T} + f_r V\frac{\partial}{\partial V} + M_r\right)E_r - \beta_r\left(T\frac{\partial P}{\partial T} - P\right) = \psi_r^a \tag{39}$$

$$\left(T\frac{\partial}{\partial T} + h_r V\frac{\partial}{\partial V} + N_r\right)P_r - h_r\delta_r\left(T\frac{\partial P}{\partial T} - P\right) = 0 \tag{40}$$

where

$$f_r = \eta_r - b \tag{41}$$

$$h_r = (\eta_r/3 - \gamma)^{-1} \tag{42}$$

$$M_r = f_r + 1 \tag{43}$$

$$N_r = h_r - 1 \tag{44}$$

Local gauge and scale invariance has unified continuum mechanics and thermodynamics.[6-8] In particular, it has been shown that local scale invariance for thermodynamics requires the introduction of two gauge parameters b and γ which must be determined simultaneously with the energy density. It has been shown that the Lie group $e^{\pm\phi}$ is the scale invariance group of relativistic thermodynamics that is based on a trace equation.[7] The group U(1) of phase rotations $e^{\pm i\phi}$ is the gauge invariance group of this theory.[8]

The invariance of the trace equation under scale transformations of the form $P \to P' = Pe^{\pm\phi}$ and $E \to E' = Ee^{\pm\psi}$, and under phase rotations of the form $\overline{P} \to \overline{P}' = \overline{P}e^{\pm i\phi}$, $\overline{E} \to \overline{E}' = \overline{E}e^{\pm i\psi}$, with $\overline{\gamma} = \gamma e^{i\theta_\gamma}$ and $\overline{b} = be^{i\theta_b}$, leads to the renormalization group equations of relativistic thermodynamics.[7,8] For instance, gauge invariance for phase rotations yields the following renormalization group equations for the complex gauge parameters[8]

$$\left(\frac{d\overline{\gamma}}{d\overline{P}}\right)^{\pm} = \frac{\overline{\gamma}\,\frac{T}{\phi}\frac{\partial\phi}{\partial T} - \frac{V}{\phi}\frac{\partial\phi}{\partial V}}{\overline{P} - T\frac{\partial\overline{P}}{\partial T} \mp i\overline{P}T\frac{\partial\phi}{\partial T}} \tag{45}$$

$$\left(\frac{d\overline{b}}{d\overline{E}}\right)^{\pm} = \frac{\frac{T}{\psi}\frac{\partial\psi}{\partial T} - \overline{b}\,\frac{V}{\psi}\frac{\partial\psi}{\partial V}}{\overline{E} + V\frac{\partial\overline{E}}{\partial V} \pm i\overline{E}V\frac{\partial\psi}{\partial V}} \tag{46}$$

where P and E are the magnitudes of the pressure and energy density respectively. Symmetrization gives the following results for gauge invariance under phase rotations[8]

$$\frac{d\overline{\gamma}}{d\overline{\overline{P}}} = \frac{1}{2}\left[\left(\frac{d\overline{\gamma}}{d\overline{\overline{P}}}\right)^{-} + \left(\frac{d\overline{\gamma}}{d\overline{\overline{P}}}\right)^{+}\right] = \frac{\left(\overline{P} - T\frac{\partial\overline{P}}{\partial T}\right)\left(\overline{\gamma}\frac{T}{\phi}\frac{\partial\phi}{\partial T} - \frac{V}{\phi}\frac{\partial\phi}{\partial V}\right)}{\left(\overline{P} - T\frac{\partial\overline{P}}{\partial T}\right)^2 + \overline{P}^2\left(T\frac{\partial\phi}{\partial T}\right)^2} \tag{47}$$

$$\frac{d\overline{b}}{d\overline{\overline{E}}} = \frac{1}{2}\left[\left(\frac{d\overline{b}}{d\overline{\overline{E}}}\right)^{-} + \left(\frac{d\overline{b}}{d\overline{\overline{E}}}\right)^{+}\right] = \frac{\left(\overline{E} + V\frac{\partial\overline{E}}{\partial V}\right)\left(\frac{T}{\psi}\frac{\partial\psi}{\partial T} - \overline{b}\frac{V}{\psi}\frac{\partial\psi}{\partial V}\right)}{\left(\overline{E} + V\frac{\partial\overline{E}}{\partial V}\right)^2 + \overline{E}^2\left(V\frac{\partial\psi}{\partial V}\right)^2} \tag{48}$$

Similar equations result from the scale invariance condition which gives $d\gamma/dP$ and db/dE for changes in the magnitudes of the pressure and energy density.[7,8]

These results suggest that the pressure, energy density and the gauge paramaters may themselves be intrinsically complex numbers that are associated with internal phase angles. Accordingly the relativistic trace equation (1) will be written as

$$\overline{U} + T\left(\frac{d\overline{U}}{dT}\right)_{\overline{P}V} - 3V\frac{d}{dV}(\overline{P}V)_{\overline{U}} = U^a + T\left(\frac{dU^a}{dT}\right)_{P^aV} \tag{49}$$

or equivalently as

$$\left(1 - \overline{b} + T\frac{\partial}{\partial T} - \overline{b}V\frac{\partial}{\partial V}\right)\overline{E} - 3\left(1 + \overline{\gamma} + V\frac{\partial}{\partial V} - \overline{\gamma}T\frac{\partial}{\partial T}\right)\overline{P} = \psi^a \tag{50}$$

where $\overline{U}$, $\overline{E}$, $\overline{P}$, $\overline{\gamma}$, and $\overline{b}$ are complex number representations of the internal energy, energy density, pressure, and the gauge parameters. The corresponding equation for radiation in matter with internal phases is derived from equation (50) to be

$$\left(1 - \overline{b} + T\frac{\partial}{\partial T} - \overline{b}V\frac{\partial}{\partial V}\right)\overline{E}_r - \overline{\beta}_r\left(T\frac{\partial\overline{P}}{\partial T} - \overline{P}\right)$$
$$- 3\left[\left(1 + \overline{\gamma} + V\frac{\partial}{\partial V} - \overline{\gamma}T\frac{\partial}{\partial T}\right)\overline{P}_r - \overline{\delta}_r\left(T\frac{\partial\overline{P}}{\partial T} - \overline{P}\right)\right] = \psi_r^a \tag{51}$$

where $\overline{E}_r$, $\overline{P}_r$, $\overline{\beta}_r$, and $\overline{\delta}_r$ are the complex number generalizations of the functions that appear in equations (30) through (36).

This paper presents a theory of the relativistic thermodynamics of solids and quantum liquids with internal phase. The renormalization group equations for systems with internal phase are derived, and a procedure for solving the complex number relativistic trace equation is presented that allows the determination of the internal phase angles associated with the pressure, energy density, and gauge parameters. The non-zero values of the phase angles represent a spontaneously broken symmetry.

2. THERMODYNAMIC STATE FUNCTIONS FOR SYSTEMS WITH INTERNAL PHASE. In order to solve the complex number trace equation (50) it is first necessary to determine the relations between the complex thermodynamic state functions and to determine their connection to the internal phase angles. This will be done using the first and second laws of thermodynamics. The complex number thermodynamic state functions that appear in equations (49) and (50) will be written in terms of their internal phase angles as follows

$$\overline{U} = Ue^{i\theta_u} \tag{52}$$

$$\overline{E} = \overline{U}/V = Ee^{i\theta_u} \tag{53}$$

$$\overline{P} = Pe^{i\theta_p} \tag{54}$$

$$\overline{\gamma} = \gamma e^{i\theta_\gamma} \tag{55}$$

$$\overline{b} = be^{i\theta_b} \tag{56}$$

where θ_u, θ_p, θ_γ, and θ_b = internal phase angles of the internal energy, pressure, Grüneisen parameter, and b gauge parameter respectively. In addition the complex number entropy will be written as

$$\overline{S} = Se^{i\theta_s} \tag{57}$$

where θ_s = internal phase angle of the entropy. In general all of the phase angles are functions of V and T. The quantities U, E, P, γ, b, and S are the magnitudes of the complex thermodynamic state functions, and are also functions of V and T.

The complex number bulk modulus is obtained from equation (54) as follows

$$\overline{K}_T = -V\left(\frac{\partial \overline{P}}{\partial V}\right)_T = K_T e^{i\theta_K} = -e^{i\theta_p}\left(V\frac{\partial P}{\partial V} + iPV\frac{\partial \theta_p}{\partial V}\right) \tag{58}$$

$$= e^{i\theta_p}\left(n\frac{\partial P}{\partial n} + iPn\frac{\partial \theta_p}{\partial n}\right)$$

where

$$K_T = \sqrt{\left(V\frac{\partial P}{\partial V}\right)^2 + P^2\left(V\frac{\partial \theta_p}{\partial V}\right)^2} \tag{59}$$

$$K_T \cos\omega = n\,\partial P/\partial n \tag{59A}$$

$$K_T \sin\omega = Pn\,\partial\theta_p/\partial n \tag{59B}$$

$$\theta_K = \theta_p + \omega \tag{60}$$

$$\tan \omega = \frac{Pn \dfrac{\partial \theta_P}{\partial n}}{n \dfrac{\partial P}{\partial n}} \tag{61}$$

Equation (52) immediately gives the complex number heat capacity as

$$\overline{C}_V = \left(\frac{\partial \overline{U}}{\partial T}\right)_V = C_V e^{i\theta_{C_V}} \tag{62}$$

where

$$C_V = \sqrt{\left(\frac{\partial U}{\partial T}\right)^2 + U^2\left(\frac{\partial \theta_u}{\partial T}\right)^2} \tag{63}$$

$$C_V \cos \rho = \partial U/\partial T \tag{63A}$$

$$C_V \sin \rho = U\, \partial \theta_u/\partial T \tag{63B}$$

$$\theta_{C_V} = \theta_u + \rho \tag{64}$$

$$\tan \rho = \frac{U \dfrac{\partial \theta_u}{\partial T}}{\dfrac{\partial U}{\partial T}} = \frac{T \dfrac{\partial \theta_u}{\partial T}}{\dfrac{T}{U} \dfrac{\partial U}{\partial T}} \tag{65}$$

Thus the renormalized values of C_V and K_T include the effects of the internal phase angles θ_u and θ_p respectively.

The relationships between the various state functions and their internal phase angles are determined from the First and Second laws of thermodynamics which can be written for matter and radiation with internal phase angles as follows

$$Td\overline{S} = Te^{i\theta_s}(dS + iSd\theta_s) = d\overline{U} + \overline{P}dV \tag{66}$$

or equivalently as

$$T \frac{\partial \overline{S}}{\partial V} = \frac{\partial \overline{U}}{\partial V} + \overline{P} \tag{67}$$

$$T \frac{\partial \overline{S}}{\partial T} = \frac{\partial \overline{U}}{\partial T} \tag{68}$$

Combining equations (52), (53) and (57) with equations (67) and (68), and separating into real and imaginary parts, yields the following equations

$$T\left(\cos\theta_s \frac{\partial S}{\partial V} - S\sin\theta_s \frac{\partial\theta_s}{\partial V}\right) = \cos\theta_u \frac{\partial U}{\partial V} + P\cos\theta_p - U\sin\theta_u \frac{\partial\theta_u}{\partial V} \quad (69)$$

$$T\left(\sin\theta_s \frac{\partial S}{\partial V} + S\cos\theta_s \frac{\partial\theta_s}{\partial V}\right) = \sin\theta_u \frac{\partial U}{\partial V} + P\sin\theta_p + U\cos\theta_u \frac{\partial\theta_u}{\partial V} \quad (70)$$

$$T\left(\cos\theta_s \frac{\partial S}{\partial T} - S\sin\theta_s \frac{\partial\theta_s}{\partial T}\right) = \cos\theta_u \frac{\partial U}{\partial T} - U\sin\theta_u \frac{\partial\theta_u}{\partial T} \quad (71)$$

$$T\left(\sin\theta_s \frac{\partial S}{\partial T} + S\cos\theta_s \frac{\partial\theta_s}{\partial T}\right) = \sin\theta_u \frac{\partial U}{\partial T} + U\cos\theta_u \frac{\partial\theta_u}{\partial T} \quad (72)$$

Squaring and adding equations (69) and (70) gives

$$T^2\left[\left(\frac{\partial S}{\partial V}\right)^2 + S^2\left(\frac{\partial\theta_s}{\partial V}\right)^2\right] = \left(\frac{\partial U}{\partial V}\right)^2 + P^2 + U^2\left(\frac{\partial\theta_u}{\partial V}\right)^2 + 2P\frac{\partial U}{\partial V}\cos(\theta_u - \theta_p) \quad (73)$$

$$+ 2PU\frac{\partial\theta_u}{\partial V}\sin(\theta_p - \theta_u)$$

Squaring and adding equations (71) and (72) gives

$$T^2\left[\left(\frac{\partial S}{\partial T}\right)^2 + S^2\left(\frac{\partial\theta_s}{\partial T}\right)^2\right] = \left(\frac{\partial U}{\partial T}\right)^2 + U^2\left(\frac{\partial\theta_u}{\partial T}\right)^2 \quad (74)$$

The Gibbs-Helmholtz equation for matter with internal phase is written as

$$\frac{\partial\overline{U}}{\partial V} = T\frac{\partial\overline{P}}{\partial T} - \overline{P} \quad (75)$$

Using the Gibbs-Helmholtz equation allows Equation (67) to be rewritten as[9]

$$\frac{\partial\overline{S}}{\partial V} = \frac{\partial\overline{P}}{\partial T} \quad (76)$$

which allows equations (69) and (70) to be rewritten as

$$\cos\theta_s \frac{\partial S}{\partial V} - S\sin\theta_s \frac{\partial\theta_s}{\partial V} = \cos\theta_p \frac{\partial P}{\partial T} - P\sin\theta_p \frac{\partial\theta_p}{\partial T} \quad (77)$$

$$\sin\theta_s \frac{\partial S}{\partial V} + S\cos\theta_s \frac{\partial\theta_s}{\partial V} = \sin\theta_p \frac{\partial P}{\partial T} + P\cos\theta_p \frac{\partial\theta_p}{\partial T} \quad (78)$$

Squaring and adding equations (77) and (78) gives

$$\left(\frac{\partial S}{\partial V}\right)^2 + S^2\left(\frac{\partial \theta_s}{\partial V}\right)^2 = \left(\frac{\partial P}{\partial T}\right)^2 + P^2\left(\frac{\partial \theta_p}{\partial T}\right)^2 \tag{79}$$

Also, from Maxwell's relationship it follows that[9]

$$T = \left(\frac{\partial \overline{U}}{\partial \overline{S}}\right)_V \tag{80}$$

From which it follows that

$$T = \frac{\partial U}{\partial S} \cos\left(\theta_u - \theta_s\right) - U \frac{\partial \theta_u}{\partial S} \sin\left(\theta_u - \theta_s\right) \tag{81}$$

$$TS \frac{\partial \theta_s}{\partial S} = \frac{\partial U}{\partial S} \sin\left(\theta_u - \theta_s\right) + U \frac{\partial \theta_u}{\partial S} \cos\left(\theta_u - \theta_s\right) \tag{82}$$

and

$$T^2 + T^2S^2\left(\frac{\partial \theta_s}{\partial S}\right)^2 = \left(\frac{\partial U}{\partial S}\right)^2 + U^2\left(\frac{\partial \theta_u}{\partial S}\right)^2 \tag{83}$$

The Gibbs-Helmholtz equation (75) can be separated into real and imaginary components as follows

$$\cos\theta_u \frac{\partial U}{\partial V} - U \sin\theta_u \frac{\partial \theta_u}{\partial V} = \cos\theta_p\left(T\frac{\partial P}{\partial T} - P\right) - TP \sin\theta_p \frac{\partial \theta_p}{\partial T} \tag{84}$$

$$\sin\theta_u \frac{\partial U}{\partial V} + U \cos\theta_u \frac{\partial \theta_u}{\partial V} = \sin\theta_p\left(T\frac{\partial P}{\partial T} - P\right) + TP \cos\theta_p \frac{\partial \theta_p}{\partial T} \tag{85}$$

Equations (84) and (85) can be rewritten in terms of the energy density as follows

$$\left(\cos\theta_u - \sin\theta_u V \frac{\partial \theta_u}{\partial V} + \cos\theta_u V \frac{\partial}{\partial V}\right)E + \left(\cos\theta_p + \sin\theta_p T \frac{\partial \theta_p}{\partial T} - \cos\theta_p T \frac{\partial}{\partial T}\right)P = 0 \tag{86}$$

$$\left(\sin\theta_u + \cos\theta_u V \frac{\partial \theta_u}{\partial V} + \sin\theta_u V \frac{\partial}{\partial V}\right)E + \left(\sin\theta_p - \cos\theta_p T \frac{\partial \theta_p}{\partial T} - \sin\theta_p T \frac{\partial}{\partial T}\right)P = 0 \tag{87}$$

Squaring and adding equations (84) and (85) gives

$$\left(\frac{\partial U}{\partial V}\right)^2 + U^2\left(\frac{\partial \theta_u}{\partial V}\right)^2 = (T\frac{\partial P}{\partial T} - P)^2 + P^2\left(T\frac{\partial \theta_p}{\partial T}\right)^2 \quad (88)$$

Combining equations (73) and (79) gives

$$T^2\left[\left(\frac{\partial P}{\partial T}\right)^2 + P^2\left(\frac{\partial \theta_p}{\partial T}\right)^2\right] = P^2 + \left(\frac{\partial U}{\partial V}\right)^2 + U^2\left(\frac{\partial \theta_u}{\partial V}\right)^2 + 2P\frac{\partial U}{\partial V}\cos(\theta_u - \theta_p) \quad (89)$$

$$+ 2PU\frac{\partial \theta_u}{\partial V}\sin(\theta_p - \theta_u)$$

Expanding the right hand side of equation (88) and using equation (89) gives the following simple result

$$T\frac{\partial P}{\partial T} - P = \frac{\partial U}{\partial V}\cos(\theta_p - \theta_u) + U\frac{\partial \theta_u}{\partial V}\sin(\theta_p - \theta_u) \quad (90)$$

Similarly

$$P^2\left(T\frac{\partial \theta_p}{\partial T}\right)^2 = \left(\frac{\partial U}{\partial V}\right)^2 \sin^2(\theta_p - \theta_u) + U^2\left(\frac{\partial \theta_u}{\partial V}\right)^2 \cos^2(\theta_p - \theta_u) \quad (91)$$

$$- 2U\frac{\partial U}{\partial V}\frac{\partial \theta_u}{\partial V}\cos(\theta_p - \theta_u)\sin(\theta_p - \theta_u)$$

The three basic thermodynamic potentials will now be considered.

The enthalpy of a substance with internal phase is written as

$$\overline{H} = He^{i\theta_H} = \overline{U} + \overline{P}V \quad (92)$$

where $\overline{H}$ = complex number enthalpy, H = enthalpy magnitude, and θ_H = internal phase angle of the enthalpy. Combining equation (92) with equations (52) and (53) gives

$$H^2 = (U\cos\theta_u + PV\cos\theta_p)^2 + (U\sin\theta_u + PV\sin\theta_p)^2 \quad (93)$$

$$= U^2 + P^2V^2 + 2UPV\cos(\theta_u - \theta_p)$$

$$\tan \theta_H = \frac{U \sin \theta_u + PV \sin \theta_p}{U \cos \theta_u + PV \cos \theta_p} \tag{94}$$

The differential of the vector enthalpy is

$$d\overline{H} = e^{i\theta_H}(dH + iHd\theta_H) = Td\overline{S} + Vd\overline{P} \tag{95}$$

which yields

$$\cos \theta_H \frac{\partial H}{\partial S} - H \sin \theta_H \frac{\partial \theta_H}{\partial S} = T \cos \theta_s - TS \sin \theta_s \frac{\partial \theta_s}{\partial S} - PV \sin \theta_p \frac{\partial \theta_p}{\partial S} \tag{96}$$

$$\sin \theta_H \frac{\partial H}{\partial S} + H \cos \theta_H \frac{\partial \theta_H}{\partial S} = T \sin \theta_s + TS \cos \theta_s \frac{\partial \theta_s}{\partial S} + PV \cos \theta_p \frac{\partial \theta_p}{\partial S} \tag{97}$$

$$\cos \theta_H \frac{\partial H}{\partial P} - H \sin \theta_H \frac{\partial \theta_H}{\partial P} = V \cos \theta_p - TS \sin \theta_s \frac{\partial \theta_s}{\partial P} - PV \sin \theta_p \frac{\partial \theta_p}{\partial P} \tag{98}$$

$$\sin \theta_H \frac{\partial H}{\partial P} + H \cos \theta_H \frac{\partial \theta_H}{\partial P} = V \sin \theta_p + TS \cos \theta_s \frac{\partial \theta_s}{\partial P} + PV \cos \theta_p \frac{\partial \theta_p}{\partial P} \tag{99}$$

where S and P are taken to be the two independent variables. Combining equations (96) and (97) gives

$$\left(\frac{\partial H}{\partial S}\right)^2 + H^2\left(\frac{\partial \theta_H}{\partial S}\right)^2 = T^2 + T^2S^2\left(\frac{\partial \theta_s}{\partial S}\right)^2 + P^2V^2\left(\frac{\partial \theta_p}{\partial S}\right)^2 \tag{100}$$

$$+ 2PVT \frac{\partial \theta_p}{\partial S} \sin (\theta_s - \theta_p) + 2TSPV \frac{\partial \theta_s}{\partial S} \frac{\partial \theta_p}{\partial S} \cos (\theta_s - \theta_p)$$

while combining equations (98) and (99) gives

$$\left(\frac{\partial H}{\partial P}\right)^2 + H^2\left(\frac{\partial \theta_H}{\partial P}\right)^2 = V^2 + T^2S^2\left(\frac{\partial \theta_s}{\partial P}\right)^2 + P^2V^2\left(\frac{\partial \theta_p}{\partial P}\right)^2 \tag{101}$$

$$+ 2TSV \frac{\partial \theta_s}{\partial P} \sin (\theta_p - \theta_s) + 2TSPV \frac{\partial \theta_s}{\partial P} \frac{\partial \theta_p}{\partial P} \cos (\theta_s - \theta_p)$$

The complex free energy is written as

$$\bar{A} = Ae^{i\theta_A} = \bar{U} - T\bar{S} \tag{102}$$

where $\bar{A}$ = complex number free energy, A = magnitude of the free energy, and θ_A = internal phase angle of the free energy. Combining equations (52) and (57) with equation (102) yields

$$A^2 = (U \cos \theta_u - TS \cos \theta_s)^2 + (U \sin \theta_u - TS \sin \theta_s)^2 \tag{103}$$

$$= U^2 + T^2S^2 - 2UTS \cos (\theta_u - \theta_s)$$

$$\tan \theta_A = \frac{U \sin \theta_u - TS \sin \theta_s}{U \cos \theta_u - TS \cos \theta_s} \tag{104}$$

The differential of the vector free energy in equation (102) is

$$d\bar{A} = e^{i\theta_A}(dA + iAd\theta_A) = -\bar{P}dV - \bar{S}dT \tag{105}$$

from which it follows that

$$\cos \theta_A \frac{\partial A}{\partial V} - A \sin \theta_A \frac{\partial \theta_A}{\partial V} = - P \cos \theta_p \tag{106}$$

$$\sin \theta_A \frac{\partial A}{\partial V} + A \cos \theta_A \frac{\partial \theta_A}{\partial V} = - P \sin \theta_p \tag{107}$$

$$\cos \theta_A \frac{\partial A}{\partial T} - A \sin \theta_A \frac{\partial \theta_A}{\partial T} = - S \cos \theta_s \tag{108}$$

$$\sin \theta_A \frac{\partial A}{\partial T} + A \cos \theta_A \frac{\partial \theta_A}{\partial T} = - S \sin \theta_s \tag{109}$$

and

$$\left(\frac{\partial A}{\partial V}\right)_T^2 + A^2\left(\frac{\partial \theta_A}{\partial V}\right)_T^2 = P^2 \tag{110}$$

$$\left(\frac{\partial A}{\partial T}\right)_V^2 + A^2\left(\frac{\partial \theta_A}{\partial T}\right)_V^2 = S^2 \tag{111}$$

The complex number form of the Gibbs-Helmholtz equation for the free energy is written as[9]

$$\overline{A} - T\left(\frac{\partial \overline{A}}{\partial T}\right)_V = \overline{U} \tag{112}$$

which gives immediately

$$\cos\theta_A\left(A - T\frac{\partial A}{\partial T}\right) + A\sin\theta_A\, T\frac{\partial\theta_A}{\partial T} = U\cos\theta_u \tag{113}$$

$$\sin\theta_A\left(A - T\frac{\partial A}{\partial T}\right) - A\cos\theta_A\, T\frac{\partial\theta_A}{\partial T} = U\sin\theta_u \tag{114}$$

and

$$\left(A - T\frac{\partial A}{\partial T}\right)^2 + A^2\left(T\frac{\partial\theta_A}{\partial T}\right)^2 = U^2 \tag{115}$$

Combining equations (103), (111) and (115) gives

$$A\frac{\partial A}{\partial T} = S[TS - U\cos(\theta_u - \theta_s)] \tag{116}$$

The complex number form of the Gibbs free energy is given by

$$\overline{G} = Ge^{i\theta_G} = \overline{U} + \overline{P}V - T\overline{S} = \overline{A} + \overline{P}V \tag{117}$$

where $\overline{G}$ = complex number Gibbs free energy, G = magnitude of the Gibbs free energy, and θ_G = internal phase angle of the Gibbs free energy. It follows immediately from equation (117) that

$$G\cos\theta_G = U\cos\theta_u + PV\cos\theta_p - TS\cos\theta_s \tag{118}$$

$$G\sin\theta_G = U\sin\theta_u + PV\sin\theta_p - TS\sin\theta_s \tag{119}$$

which gives

$$G^2 = U^2 + P^2V^2 + T^2S^2 + 2UPV\cos(\theta_p - \theta_u) - 2UTS\cos(\theta_s - \theta_u) - 2PVTS\cos(\theta_s - \theta_p) \tag{120}$$

and

$$\tan \theta_G = \frac{U \sin \theta_u + PV \sin \theta_p - TS \sin \theta_s}{U \cos \theta_u + PV \cos \theta_p - TS \cos \theta_s} \tag{121}$$

The differential of the vector Gibbs function in equation (117) is

$$d\overline{G} = e^{i\theta_G}(dG + iGd\theta_G) = -\overline{S}dT + Vd\overline{P} \tag{122}$$

from which it follows that

$$\cos \theta_G \frac{\partial G}{\partial T} - G \sin \theta_G \frac{\partial \theta_G}{\partial T} = -S \cos \theta_s - PV \sin \theta_p \frac{\partial \theta_p}{\partial T} \tag{123}$$

$$\sin \theta_G \frac{\partial G}{\partial T} + G \cos \theta_G \frac{\partial \theta_G}{\partial T} = -S \sin \theta_s + PV \cos \theta_p \frac{\partial \theta_p}{\partial T} \tag{124}$$

$$\cos \theta_G \frac{\partial G}{\partial P} - G \sin \theta_G \frac{\partial \theta_G}{\partial P} = V \cos \theta_p - PV \sin \theta_p \frac{\partial \theta_p}{\partial P} \tag{125}$$

$$\sin \theta_G \frac{\partial G}{\partial P} + G \cos \theta_G \frac{\partial \theta_G}{\partial P} = V \sin \theta_p + PV \cos \theta_p \frac{\partial \theta_p}{\partial P} \tag{126}$$

Combining equation (123) and (124) gives

$$\left(\frac{\partial G}{\partial T}\right)^2 + G^2\left(\frac{\partial \theta_G}{\partial T}\right)^2 = S^2 + 2PVS \sin (\theta_p - \theta_s) \frac{\partial \theta_p}{\partial T} + P^2V^2\left(\frac{\partial \theta_p}{\partial T}\right)^2 \tag{127}$$

while equations (125) and (126) give

$$\left(\frac{\partial G}{\partial P}\right)^2 + G^2\left(\frac{\partial \theta_G}{\partial P}\right)^2 = V^2 + P^2V^2\left(\frac{\partial \theta_p}{\partial P}\right)^2 \tag{128}$$

Further relationships can be obtained from the vector form of the Gibbs-Helmholtz equation for the Gibbs function which is written as

$$\overline{A} = \overline{G} - \overline{P}V = \overline{G} - \overline{P}\left(\frac{\partial \overline{G}}{\partial \overline{P}}\right)_T \tag{129}$$

When the T = 0 limit exists (as in the case of solids and quantum liquids) for matter with internal phase, the following equations corresponding to equations (52) through (57) can be written

$$\overline{U}_o = U_o e^{i\theta_u^o} \tag{130}$$

$$\overline{E}_o = \overline{U}_o/V = E_o e^{i\theta_u^o} \tag{131}$$

$$\overline{P}_o = P_o e^{i\theta_p^o} \tag{132}$$

$$\overline{\gamma}_o = \gamma_o e^{i\theta_\gamma^o} \tag{133}$$

$$\overline{b}_o = 0 \tag{134}$$

$$\overline{S}_o = 0 \tag{135}$$

where θ_u^o , θ_p^o , and θ_γ^o are the T = 0 values of θ_u , θ_p , and θ_γ respectively. Note from the definition of b in equation (5) it follows that $b_o = 0$ so that $\overline{b}_o = 0$ also, however it will be shown later that $\theta_b^o \neq 0$. From the Third law of thermodynamics it follows that $S_o = 0$ and $\overline{S}_o = 0$.

For solids and quantum liquids one can take the T = 0 limit of equations (84) and (85) and get

$$\cos\theta_u^o \frac{dU_o}{dV} - U_o \frac{d\theta_u^o}{dV} \sin\theta_u^o = -P_o \cos\theta_p^o \tag{136}$$

$$\sin\theta_u^o \frac{dU_o}{dV} + U_o \frac{d\theta_u^o}{dV} \cos\theta_u^o = -P_o \sin\theta_p^o \tag{137}$$

From equations (136) and (137) one gets immediately

$$\tan\theta_p^o = \frac{\sin\theta_u^o \frac{dU_o}{dV} + U_o \frac{d\theta_u^o}{dV} \cos\theta_u^o}{\cos\theta_u^o \frac{dU_o}{dV} - U_o \frac{d\theta_u^o}{dV} \sin\theta_u^o} \tag{138}$$

$$P_o = \sqrt{\left(\frac{dU_o}{dV}\right)^2 + U_o^2\left(\frac{d\theta_u^o}{dV}\right)^2} \tag{139}$$

$$= \sqrt{\left(n\,\frac{dE_o}{dn} - E_o\right)^2 + E_o^2\left(n\,\frac{d\theta_u^o}{dn}\right)^2}$$

Using the trigonometrical formula for the tangent of the sum of two angles allows equation (138) to be rewritten as

$$\theta_p^o = \theta_u^o + \Phi_o \tag{140}$$

where

$$\tan \Phi^o = \frac{U_o \dfrac{d\theta_u^o}{dV}}{\dfrac{dU_o}{dV}} = \frac{U_o \dfrac{d\theta_u^o}{dn}}{\dfrac{dU_o}{dn}} = \frac{E_o n \dfrac{d\theta_u^o}{dn}}{n \dfrac{dE_o}{dn} - E_o} \tag{141}$$

$$P_o \cos \Phi_o = n\, dE_o/dn - E_o \tag{141A}$$

$$P_o \sin \Phi_o = E_o n\, d\theta_u^o/dn \tag{141B}$$

From equation (141) it follows that

$$\Phi^o = 0 \qquad \text{when } \frac{d\theta_u^o}{dn} = 0 \tag{142A}$$

$$\Phi^o = \pm\frac{\pi}{2} \qquad \text{when } \frac{dU_o}{dn} = 0 \tag{142B}$$

In general for a T = 0 system

$$\theta_p^o = \theta_u^o + \tan^{-1}\left(\frac{E_o n \dfrac{d\theta_u^o}{dn}}{n \dfrac{dE_o}{dn} - E_o}\right) \tag{143}$$

Figure 1 shows the density dependence of θ_p^o and θ_u^o for an unbound interacting system such as a neutron gas. The two possible signs that occur in equation (142B) arise in systems having a saturation density at which $dU_o/dn = 0$, according to the signs of dU_o/dn and $d\theta_u^o/dn$ as $dU_o/dn \to 0$ as seen in Figure 2. Figure 2 shows the density dependence of θ_p^o and θ_u^o for a system such as N = Z infinite nuclear matter which is bound at a saturation density. These figures show the effects of equations (140) through (143). In general $\theta_p^o < \theta_u^o$ in the high density limit of bound or unbound quantum systems. From equations (58) and (59) it follows that

$$\overline{K}_o = n \frac{d\overline{P}_o}{dn} = K_o e^{i(\theta_p^o + \omega_o)} \tag{144}$$

where

$$K_o = \sqrt{\left(n \frac{dP_o}{dn}\right)^2 + P_o^2\left(n \frac{d\theta_p^o}{dn}\right)^2} \tag{145}$$

$$\tan \omega_o = P_o(d\theta_p^o/dn)/(dP_o/dn) \tag{146}$$

$$K_o \cos \omega_o = n\, dP_o/dn \tag{146A}$$

$$K_o \sin \omega_o = P_o n\, d\theta_p^o/dn \tag{146B}$$

The complex number analogs of the scalar thermal state equations given in equations (23) and (24) are

$$\overline{\mathcal{E}} = \overline{\mathcal{E}}_o + \overline{\mathcal{E}}_j T^j = \mathcal{E}_o e^{i\theta_u^o} + \mathcal{E}_j e^{i\theta_u^j} T^j = \mathcal{E} e^{i\theta_u} \tag{147}$$

$$\overline{P} = \overline{P}_o + \overline{P}_j T^j = P_o e^{i\theta_p^o} + P_j e^{i\theta_p^j} T^j = P e^{i\theta_p} \tag{148}$$

where $\mathcal{E}_j$ and P_j = magnitudes of the thermal components of the energy and pressure respectively, and θ_u^j and θ_p^j = phase angles of the thermal components of the energy density and pressure respectively. From equations (147) and (148) it follows immediately that

$$\mathcal{E}^2 = \mathcal{E}_o^2 + 2\mathcal{E}_o\mathcal{E}_j \cos(\theta_u^o - \theta_u^j)T^j + \mathcal{E}_j^2 T^{2j} \tag{149}$$

$$P^2 = P_o^2 + 2P_oP_j \cos(\theta_p^o - \theta_p^j)T^j + P_j^2 T^{2j} \tag{150}$$

$$\tan \theta_u = \frac{\mathcal{E}_o \sin \theta_u^o + \mathcal{E}_j \sin \theta_u^j T^j}{\mathcal{E}_o \cos \theta_u^o + \mathcal{E}_j \cos \theta_u^j T^j} \tag{151}$$

$$\tan \theta_p = \frac{P_o \sin \theta_p^o + P_j \sin \theta_p^j T^j}{P_o \cos \theta_p^o + P_j \cos \theta_p^j T^j} \tag{152}$$

Note that $\overline{U}_o = V\overline{E}_o$ and $\overline{U}_j = V\overline{E}_j$.

3. GAUGE PARAMETERS FOR SYSTEMS WITH INTERNAL PHASE. The two gauge parameters that appear in the basic trace equation (50) are $\overline{\gamma}$ and $\overline{b}$. The complex number Grüneisen parameter is defined as

$$\overline{\gamma} = \frac{V}{\overline{C}_V} \frac{\partial \overline{P}}{\partial T} = \frac{\partial \overline{P}/\partial T}{\partial \overline{E}/\partial T} = \gamma e^{i\theta_\gamma} \tag{153}$$

where γ and θ_γ = magnitude and phase of the Grüneisen parameter respectively. Combining equations (53) and (54) with equation (153) gives

$$\overline{\gamma} = e^{i(\theta_p - \theta_u)} \left(\frac{\dfrac{\partial P}{\partial T} + iP \dfrac{\partial \theta_p}{\partial T}}{\dfrac{\partial E}{\partial T} + iE \dfrac{\partial \theta_u}{\partial T}} \right) \tag{154}$$

$$= \sqrt{\frac{\left(\dfrac{\partial P}{\partial T}\right)^2 + P^2 \left(\dfrac{\partial \theta_p}{\partial T}\right)^2}{\left(\dfrac{\partial E}{\partial T}\right)^2 + E^2 \left(\dfrac{\partial \theta_u}{\partial T}\right)^2}} \; e^{i(\theta_p - \theta_u + \mu - \rho)} \tag{155}$$

where ρ and μ are given by

$$\tan \rho = \frac{E \dfrac{\partial \theta_u}{\partial T}}{\dfrac{\partial E}{\partial T}} \tag{156}$$

$$\tan \mu = \frac{P \dfrac{\partial \theta_p}{\partial T}}{\dfrac{\partial P}{\partial T}} \tag{157}$$

Comparing equations (153) and (155) gives

$$\theta_\gamma = \theta_p - \theta_u + \mu - \rho \tag{158}$$

$$\gamma = \frac{\partial P/\partial T}{\partial E/\partial T}\frac{\cos\rho}{\cos\mu} = \sqrt{\frac{(\partial P/\partial T)^2 + P^2(\partial\theta_p/\partial T)^2}{(\partial E/\partial T)^2 + E^2(\partial\theta_u/\partial T)^2}} \tag{159}$$

The T = 0 limit of equations (156) and (157) can be obtained by noting that from equations (149) and (150) it follows that

$$\left(\frac{\partial E}{\partial T}\right)_{T\to o} = jE_j \cos\,(\theta_u^j - \theta_u^o)\,T^{j-1} \tag{160}$$

$$\left(\frac{\partial P}{\partial T}\right)_{T\to o} = jP_j \cos\,(\theta_p^j - \theta_p^o)\,T^{j-1} \tag{161}$$

and from equation (151) and (152) it follows that

$$\left(E\,\frac{\partial\theta_u}{\partial T}\right)_{T\to o} = jE_j \sin\,(\theta_u^j - \theta_u^o)\,T^{j-1} \tag{162}$$

$$\left(P\,\frac{\partial\theta_p}{\partial T}\right)_{T\to o} = jP_j \sin\,(\theta_p^j - \theta_p^o)\,T^{j-1} \tag{163}$$

It then follows from equations (156) and (157) and (160) through (163) that

$$\rho_o = \theta_u^j - \theta_u^o \tag{164}$$

$$\mu_o = \theta_p^j - \theta_p^o \tag{165}$$

and therefore from equation (158) it follows that

$$\theta_\gamma^o = \theta_p^j - \theta_u^j \tag{166}$$

Finally combining equation (159) with (160) through (163) gives

$$\gamma_o = \frac{P_j}{E_j} \tag{167}$$

which is the same form as in equation (25) for the scalar thermal state equation. The results in equations (166) and (167) can be obtained directly from the T = 0 limit of equation (153) by using the complex number thermal state equation (147) to get

$$\overline{\gamma}_o = \frac{\overline{P}_j}{\overline{E}_j} = \frac{P_j}{E_j} e^{i(\theta_p^j - \theta_u^j)} \tag{168}$$

The gauge parameter $\overline{b}$ is defined as follows

$$\overline{b} = \frac{T \dfrac{\partial \overline{P}}{\partial T}}{\overline{P} - \overline{K}_T} = b e^{i\theta_b} \tag{169}$$

where $\overline{K}_T$ is defined in equation (58). Equation (169) can be rewritten as

$$\overline{b} = \frac{T \dfrac{\partial P}{\partial T} + iPT \dfrac{\partial \theta_p}{\partial T}}{P + V \dfrac{\partial P}{\partial V} + iPV \dfrac{\partial \theta_p}{\partial V}} = \frac{T \dfrac{\partial P}{\partial T} + iPT \dfrac{\partial \theta_p}{\partial T}}{P - n \dfrac{\partial P}{\partial n} - iPn \dfrac{\partial \theta_p}{\partial n}} \tag{170}$$

$$= \sqrt{\frac{\left(T \dfrac{\partial P}{\partial T}\right)^2 + P^2\left(T \dfrac{\partial \theta_p}{\partial T}\right)^2}{\left(P + V \dfrac{\partial P}{\partial V}\right)^2 + P^2\left(V \dfrac{\partial \theta_p}{\partial V}\right)^2}} \; e^{i(\mu+\chi)} \tag{171}$$

where μ is given by equation (157) and χ is given by

$$\tan \chi = \frac{Pn \dfrac{\partial \theta_p}{\partial n}}{P - n \dfrac{\partial P}{\partial n}} \tag{172}$$

Comparing equations (169) and (171) gives

$$\theta_b = \mu + \chi \tag{173}$$

$$b = \sqrt{\frac{(T\partial P/\partial T)^2 + P^2(T\partial\theta_p/\partial T)^2}{(P - n\partial P/\partial n)^2 + P^2(n\partial\theta_p/\partial n)^2}} = \left| \frac{T\partial P/\partial T}{P - n\partial P/\partial n} \right| \frac{\sec \mu}{\sec \chi} \tag{174}$$

The T = 0 limit of equation (173) is

$$\theta_b^o = \mu^o + \chi^o = \theta_p^j - \theta_p^o + \chi^o \tag{175}$$

where

$$\tan \chi^o = \frac{P_o n \dfrac{d\theta_p^o}{dn}}{P_o - n \dfrac{dP_p}{dn}} \tag{176}$$

while the T = 0 limit of b is obtained from equation (174) to be b = 0.

In the past two sections the relationships between the various phase angles and amplitudes of the thermodynamic state functions have been presented. In the next section a method of calculating the phase angles and amplitudes will be presented.

4. RENORMALIZATION GROUP EQUATIONS FOR THE GROUND STATE OF PHASE MATTER. The phase angles and magnitudes of the complex number thermodynamic state functions are calculated from the solution of the vector renormalization group equation (50). Combining equation (50) with equations (52) through (56) gives

$$\left(1 - be^{i\theta_b} + T\frac{\partial}{\partial T} - be^{i\theta_b} V\frac{\partial}{\partial V}\right)Ee^{i\theta_u} - 3\left(1 + \gamma e^{i\theta_\gamma} + V\frac{\partial}{\partial V} - \gamma e^{i\theta_\gamma} T\frac{\partial}{\partial T}\right)Pe^{i\theta_p} = \psi^a \tag{177}$$

where

$$\psi^a = \left(1 - b^a + T\frac{\partial}{\partial T} - b^a V\frac{\partial}{\partial V}\right)E^a \tag{178}$$

Equation (177) can be separated into real and imaginary parts. The real part is given by

$$\cos\theta_u\left(1 - b\cos\theta_b + T\frac{\partial}{\partial T} - b\cos\theta_b V\frac{\partial}{\partial V} + b\sin\theta_b V\frac{\partial\theta_u}{\partial V}\right)E \tag{179}$$

$$- \sin\theta_u\left(- b\sin\theta_b + T\frac{\partial\theta_u}{\partial T} - b\sin\theta_b V\frac{\partial}{\partial V} - b\cos\theta_b V\frac{\partial\theta_u}{\partial V}\right)E$$

$$- 3\cos\theta_p\left(1 + \gamma\cos\theta_\gamma + V\frac{\partial}{\partial V} - \gamma\cos\theta_\gamma T\frac{\partial}{\partial T} + \gamma\sin\theta_\gamma T\frac{\partial\theta_p}{\partial T}\right)P$$

$$+ 3\sin\theta_p\left(\gamma\sin\theta_\gamma + V\frac{\partial\theta_p}{\partial V} - \gamma\sin\theta_\gamma T\frac{\partial}{\partial T} - \gamma\cos\theta_\gamma T\frac{\partial\theta_p}{\partial T}\right)P = \psi^a$$

The imaginary part of equation (177) is written as

$$\sin\theta_u\left(1 - b\cos\theta_b + T\frac{\partial}{\partial T} - b\cos\theta_b V\frac{\partial}{\partial V} + b\sin\theta_b V\frac{\partial\theta_u}{\partial V}\right)E \tag{180}$$

$$+ \cos\theta_u\left(- b\sin\theta_b + T\frac{\partial\theta_u}{\partial T} - b\sin\theta_b V\frac{\partial}{\partial V} - b\cos\theta_b V\frac{\partial\theta_u}{\partial V}\right)E$$

$$- 3\sin\theta_p\left(1 + \gamma\cos\theta_\gamma + V\frac{\partial}{\partial V} - \gamma\cos\theta_\gamma T\frac{\partial}{\partial T} + \gamma\sin\theta_\gamma T\frac{\partial\theta_p}{\partial T}\right)P$$

$$- 3\cos\theta_p\left(\gamma\sin\theta_\gamma + V\frac{\partial\theta_p}{\partial V} - \gamma\sin\theta_\gamma T\frac{\partial}{\partial T} - \gamma\cos\theta_\gamma T\frac{\partial\theta_p}{\partial T}\right)P = 0$$

Equations (179) and (180) can be further simplified by introducing equations (86) and (87) and their two corresponding Lagrange indeterminate multipliers η and τ to get

$$\eta\left(\cos\theta_u - \sin\theta_u V\frac{\partial\theta_u}{\partial V} + \cos\theta_u V\frac{\partial}{\partial V}\right)E \tag{181}$$

$$+ \eta\left(\cos\theta_p + \sin\theta_p T\frac{\partial\theta_p}{\partial T} - \cos\theta_p T\frac{\partial}{\partial T}\right)P = 0$$

$$\tau\left(\sin\theta_u + \cos\theta_u V\frac{\partial\theta_u}{\partial V} + \sin\theta_u V\frac{\partial}{\partial V}\right)E \tag{182}$$

$$+ \tau\left(\sin\theta_p - \cos\theta_p T\frac{\partial\theta_p}{\partial T} - \sin\theta_p T\frac{\partial}{\partial T}\right)P = 0$$

Combining equation (179) and (180) with the constraints in equations (181) and (182) gives the following four independent partial differential equations

$$\left(\ell T\frac{\partial}{\partial T} + fV\frac{\partial}{\partial V} + M\right)E = \psi^a \tag{183}$$

$$\left(mT\frac{\partial}{\partial T} + qV\frac{\partial}{\partial V} + R\right)E = 0 \tag{184}$$

$$\left(wT\frac{\partial}{\partial T} + xV\frac{\partial}{\partial V} + Y\right)P = 0 \tag{185}$$

$$\left(sT\frac{\partial}{\partial T} + zV\frac{\partial}{\partial V} + I\right)P = 0 \tag{186}$$

where

$$\ell = \cos\theta_u \tag{187}$$

$$f = \cos\theta_u(\eta - b\cos\theta_b) + b\sin\theta_u\sin\theta_b \tag{188}$$

$$M = \cos\theta_u\left[1 + \eta - b\cos\theta_b + b\sin\theta_b V\frac{\partial\theta_u}{\partial V}\right] - \sin\theta_u\left[-b\sin\theta_b + T\frac{\partial\theta_u}{\partial T} + (\eta - b\cos\theta_b)V\frac{\partial\theta_u}{\partial V}\right] \tag{189}$$

$$m = \sin\theta_u \tag{190}$$

$$q = \sin\theta_u(\tau - b\cos\theta_b) - b\cos\theta_u\sin\theta_b \tag{191}$$

$$R = \sin\theta_u\left[1 + \tau - b\cos\theta_b + b\sin\theta_b V\frac{\partial\theta_u}{\partial V}\right] + \cos\theta_u\left[-b\sin\theta_b + T\frac{\partial\theta_u}{\partial T} + (\tau - b\cos\theta_b)V\frac{\partial\theta_u}{\partial V}\right] \tag{192}$$

$$w = \cos\theta_p\left(\frac{\eta}{3} - \gamma\cos\theta_\gamma\right) + \gamma\sin\theta_p\sin\theta_\gamma \tag{193}$$

$$x = \cos\theta_p$$

$$Y = \cos\theta_p\left[1 - \frac{\eta}{3} + \gamma\cos\theta_\gamma + \gamma\sin\theta_\gamma T\frac{\partial\theta_p}{\partial T}\right] - \sin\theta_p\left[\gamma\sin\theta_\gamma + V\frac{\partial\theta_p}{\partial V} + \left(\frac{\eta}{3} - \gamma\cos\theta_\gamma\right)T\frac{\partial\theta_p}{\partial T}\right] \tag{195}$$

$$s = \sin\theta_p\left(\frac{\tau}{3} - \gamma\cos\theta_\gamma\right) - \gamma\cos\theta_p\sin\theta_\gamma \tag{196}$$

$$z = \sin\theta_p \tag{197}$$

$$I = \sin\theta_p\left[1 - \frac{\tau}{3} + \gamma\cos\theta_\gamma + \gamma\sin\theta_\gamma T\frac{\partial\theta_p}{\partial T}\right] + \cos\theta_p\left[\gamma\sin\theta_\gamma + V\frac{\partial\theta_p}{\partial V} + \left(\frac{\tau}{3} - \gamma\cos\theta_\gamma\right)T\frac{\partial\theta_p}{\partial T}\right] \tag{198}$$

In the limit of $\theta_i \to 0$ one has

$$\ell = 1 \qquad m = 0 \qquad w = \frac{\eta}{3} - \gamma \qquad s = 0 \tag{199}$$

$$f = \eta - b \qquad q = 0 \qquad x = 1 \qquad z = 0$$

$$M = 1 + \eta - b \qquad R = 0 \qquad Y = 1 - \frac{\eta}{3} + \gamma \qquad I = 0$$

which agree with equations (13) through (18).

Equations (183) through (186) are the renormalization group equations that describe relativistic thermodynamic systems having internal phase angles. There are ten unknown quantities in equations (183) through (186): E , θ_u , P , θ_p , γ , θ_γ , b , θ_b , η , τ . The ten equations required to determine these quantities are: the four renormalization group equations (183) through (186), the two equations (158) and (159) that define $\overline{\gamma}$, the two equations (173) and (174) that define $\overline{b}$, and the two constraint equations (181) and (182). Equations (183) through (186) can be derived from a Lagrangian formalism in a manner similar to that in the accompanying paper.

5. RENORMALIZATION GROUP EQUATIONS FOR RADIATION IN PHASE MATTER. In a manner similar to equations (52) through (56) the state functions and gauge parameters for radiation that appear in equation (51) are written as

$$\overline{U}_r = U_r e^{i\theta_{ur}} \tag{200}$$

$$\overline{E}_r = \overline{U}_r / V = E_r e^{i\theta_{ur}} \tag{201}$$

$$\overline{P}_r = P_r e^{i\theta_{pr}} \tag{202}$$

$$\overline{\gamma}_r = \gamma_r e^{i\theta_{\gamma r}} \tag{203}$$

$$\overline{b}_r = b_r e^{i\theta_{br}} \tag{204}$$

$$\overline{\delta}_r = \delta_r e^{i\theta_{\delta r}} \tag{205}$$

$$\overline{\beta}_r = \beta_r e^{i\theta_{\beta r}} \tag{206}$$

where θ_{ur} = internal phase angle of the radiation internal energy

θ_{pr} = internal phase angle of the radiation pressure

$\theta_{\gamma r}$ = internal phase angle of the radiation Grüneisen gauge parameter

θ_{br} = internal phase angle of the radiation b_r gauge parameter

$\theta_{\delta r}$ = internal phase angle of δ_r gauge function

$\theta_{\beta r}$ = internal phase angle of β_r gauge function

In general all of the phase angles and magnitudes that appear in equations (200) through (206) are functions of V and T. Also all of the equations that appear in Sections 2. and 3. are also valid for radiation and can be carried over into the present calculation by adding a subscript "r".

The complex number form of the functions that appear in equation (51) can be written in a form analogous to that in equations (31) through (35) as follows

$$\overline{\beta}_r = \overline{b}_r \frac{\overline{P}_r - \overline{K}_{Tr}}{\overline{P} - \overline{K}_T} - \frac{T \frac{\partial \overline{P}}{\partial T} (\overline{P}_r - \overline{K}_{Tr})}{(\overline{P} - \overline{K}_T)^2} \tag{207}$$

$$\overline{b}_r = \frac{T \frac{\partial \overline{P}_r}{\partial T}}{\overline{P}_r - \overline{K}_{Tr}} \tag{208}$$

$$\overline{\delta}_r = \frac{\partial \overline{E}_r / \partial T}{\partial \overline{E} / \partial T} (\overline{\gamma}_r - \overline{\gamma}) \tag{209}$$

$$\overline{\gamma}_r = \frac{\partial \overline{P}_r / \partial T}{\partial \overline{E}_r / \partial T} \tag{210}$$

$$\overline{K}_{Tr} = -V \frac{\partial \overline{P}_r}{\partial V} \tag{211}$$

Combining equations (210) and (201) through (203) gives

$$\gamma_r = \sqrt{\frac{\left(\frac{\partial P_r}{\partial T}\right)^2 + P_r^2 \left(\frac{\partial \theta_{pr}}{\partial T}\right)^2}{\left(\frac{\partial E_r}{\partial T}\right)^2 + E_r^2 \left(\frac{\partial \theta_{ur}}{\partial T}\right)^2}} = \frac{\partial P_r / \partial T}{\partial E_r / \partial T} \frac{\sec \mu_r}{\sec \rho_r} \tag{212}$$

$$\theta_{\gamma r} = \theta_{pr} - \theta_{ur} + \mu_r - \rho_r \tag{213}$$

where

$$\tan \mu_r = \frac{P_r \dfrac{\partial \theta_{pr}}{\partial T}}{\dfrac{\partial P_r}{\partial T}} \tag{214}$$

$$\tan \rho_r = \frac{E_r \dfrac{\partial \theta_{ur}}{\partial T}}{\dfrac{\partial E_r}{\partial T}} \tag{215}$$

Combining equations (174), (204), and (208) gives

$$b_r = \sqrt{\frac{\left(T \dfrac{\partial P_r}{\partial T}\right)^2 + P_r^2\left(T \dfrac{\partial \theta_{pr}}{\partial T}\right)^2}{\left(P_r + V \dfrac{\partial P_r}{\partial V}\right)^2 + P_r^2\left(V \dfrac{\partial \theta_{pr}}{\partial V}\right)^2}} = \left| \frac{T \dfrac{\partial P_r}{\partial T}}{P_r + V \dfrac{\partial P_r}{\partial V}} \right| \frac{\sec \mu_r}{\sec \chi_r} \tag{216}$$

$$\theta_{br} = \mu_r + \chi_r \tag{217}$$

where

$$\tan \chi_r = - \frac{P_r V \dfrac{\partial \theta_{pr}}{\partial V}}{P_r + V \dfrac{\partial P_r}{\partial V}} = \frac{P_r n \dfrac{\partial \theta_{pr}}{\partial n}}{P_r - n \dfrac{\partial P_r}{\partial n}} \tag{218}$$

Combining equations (201), (53), (203) with (209) gives

$$\delta_r = \sqrt{\frac{\left(\dfrac{\partial E_r}{\partial T}\right)^2 + E_r^2\left(\dfrac{\partial \theta_{ur}}{\partial T}\right)^2}{\left(\dfrac{\partial E}{\partial T}\right)^2 + E^2\left(\dfrac{\partial \theta_u}{\partial T}\right)^2}} \sqrt{\gamma_r^2 + \gamma^2 - 2\gamma\gamma_r \cos(\theta_{\gamma r} - \theta_\gamma)} \tag{219}$$

$$\theta_{\delta r} = \theta_{ur} - \theta_u + \rho_r - \rho + \pi_r \tag{220}$$

where

$$\tan \pi_r = \frac{\gamma_r \sin \theta_{\gamma r} - \gamma \sin \theta_\gamma}{\gamma_r \cos \theta_{\gamma r} - \gamma \cos \theta_\gamma} \tag{221}$$

and where ρ is defined in equation (156). Combining equation (206) and (207) gives

$$\overline{\beta}_r = A_r e^{i\psi_r} - B_r e^{i\phi_r} \tag{222}$$

where β_r and $\theta_{\beta r}$ of equation (206) are given as

$$\beta_r = \sqrt{A_r^2 + B_r^2 - 2A_rB_r \cos(\psi_r - \phi_r)} \tag{223}$$

$$\tan\theta_{\beta r} = \frac{A_r \sin\psi_r - B_r \sin\phi_r}{A_r \cos\psi_r - B_r \cos\phi_r} \tag{224}$$

where

$$A_r = b_r \sqrt{\frac{\left(P_r + V\dfrac{\partial P_r}{\partial V}\right)^2 + P_r^2\left(V\dfrac{\partial\theta_{pr}}{\partial V}\right)^2}{\left(P + V\dfrac{\partial P}{\partial V}\right)^2 + P^2\left(V\dfrac{\partial\theta_p}{\partial V}\right)^2}} \tag{225}$$

$$B_r = \frac{\sqrt{\left(T\dfrac{\partial P}{\partial T}\right)^2 + P^2\left(T\dfrac{\partial\theta_p}{\partial T}\right)^2}\sqrt{\left(P_r + V\dfrac{\partial P_r}{\partial V}\right)^2 + P_r^2\left(V\dfrac{\partial\theta_{pr}}{\partial V}\right)^2}}{\left(P + V\dfrac{\partial P}{\partial V}\right)^2 + P^2\left(V\dfrac{\partial\theta_p}{\partial V}\right)^2} \tag{226}$$

$$\psi_r = \theta_{br} + \theta_{pr} - \theta_p - \chi_r + \chi = \theta_{pr} - \theta_p + \mu_r + \chi \tag{227}$$

$$\phi_r = \theta_{pr} - \theta_p + \mu - \chi_r + 2\chi \tag{228}$$

Finally the radiation bulk modulus in equation (211) is given by

$$\overline{K}_{Tr} = e^{i\theta_{pr}}\left(n\frac{\partial P_r}{\partial n} + iP_r n\frac{\partial\theta_{pr}}{\partial n}\right) \tag{229}$$

$$= K_{Tr} e^{i(\theta_{pr} + \omega_r)}$$

where

$$\tan \omega_r = \frac{P_r n \dfrac{\partial \theta_{pr}}{\partial n}}{n \dfrac{\partial P_r}{\partial n}} \tag{230}$$

It remains to show how the radiation equation (51) can be decomposed into four radiation equations which combined with the defining relations in equations (200) through (228) can be used to calculate the eight quantities E_r, θ_{ur}, P_r, θ_{pr}, γ_r, $\theta_{\gamma r}$, b_r, and θ_{br}. First note that equation (51) can be rewritten as

$$\left(1 - be^{i\theta_b} + T\frac{\partial}{\partial T} - be^{i\theta_b} V\frac{\partial}{\partial V}\right)E_r e^{i\theta_{ur}} - \beta_r e^{i\theta_{\beta r}}\left[T\frac{\partial}{\partial T}(Pe^{i\theta_p}) - Pe^{i\theta_p}\right] \tag{231}$$

$$- 3\left\{\left(1 + \gamma e^{i\theta_\gamma} + V\frac{\partial}{\partial V} - \gamma e^{i\theta_\gamma} T\frac{\partial}{\partial T}\right)P_r e^{i\theta_{pr}} - \delta_r e^{i\theta_{\delta r}}\left[T\frac{\partial}{\partial T}(Pe^{i\theta_p}) - Pe^{i\theta_p}\right]\right\} = \psi_r^a$$

The simplification of equation (231) can be realized by noting that the complex Gibbs-Helmholtz equation for radiation

$$\frac{\partial \overline{U}_r}{\partial V} = T\frac{\partial \overline{P}_r}{\partial T} - \overline{P}_r \tag{232}$$

yields the following two constraint equations similar to equations (181) and (182)

$$\eta_r\left(\cos\theta_{ur} - \sin\theta_{ur} V\frac{\partial \theta_{ur}}{\partial V} + \cos\theta_{ur} V\frac{\partial}{\partial V}\right)E_r \tag{233}$$

$$+ \eta_r\left(\cos\theta_{pr} + \sin\theta_{pr} T\frac{\partial \theta_{pr}}{\partial T} - \cos\theta_{pr} T\frac{\partial}{\partial T}\right)P_r = 0$$

$$\tau_r\left(\sin\theta_{ur} + \cos\theta_{ur} V\frac{\partial \theta_{ur}}{\partial V} + \sin\theta_{ur} V\frac{\partial}{\partial V}\right)E_r \tag{234}$$

$$+ \tau_r\left(\sin\theta_{pr} - \cos\theta_{pr} T\frac{\partial \theta_{pr}}{\partial T} - \sin\theta_{pr} T\frac{\partial}{\partial T}\right)P_r = 0$$

where two radiation Lagrange indeterminate multipliers, η_r and τ_r, are introduced. Separating equation (231) into real and imaginary parts and using the constraint equations (233) and (234) yields the following four independent partial differential equations

$$(\ell_r T \frac{\partial}{\partial T} + f_r V \frac{\partial}{\partial V} + M_r)E_r - \beta_r J \cos(\theta_{\beta r} + \theta_p + \lambda) = \psi_r^a \tag{235}$$

$$(m_r T \frac{\partial}{\partial T} + q_r V \frac{\partial}{\partial V} + R_r)E_r - \beta_r J \sin(\theta_{\beta r} + \theta_p + \lambda) = 0 \tag{236}$$

$$(w_r T \frac{\partial}{\partial T} + x_r V \frac{\partial}{\partial V} + Y_r)P_r - \delta_r J \cos(\theta_{\delta r} + \theta_p + \lambda) = 0 \tag{237}$$

$$(s_r T \frac{\partial}{\partial T} + z_r V \frac{\partial}{\partial V} + I_r)P_r - \delta_r J \sin(\theta_{\delta r} + \theta_p + \lambda) = 0 \tag{238}$$

where β_r , $\theta_{\beta r}$, δ_r , and $\theta_{\delta r}$ are given by equations (223), (224), (219), and (220), and where

$$\tan\lambda = \frac{PT \frac{\partial \theta_p}{\partial T}}{T \frac{\partial P}{\partial T} - P} \tag{239}$$

$$J = \sqrt{\left(T \frac{\partial P}{\partial T} - P\right)^2 + P^2\left(T \frac{\partial \theta_p}{\partial T}\right)^2} \tag{239A}$$

and where

$$\ell_r = \cos\theta_{ur} \tag{240}$$

$$f_r = \cos\theta_{ur}(\eta_r - b\cos\theta_b) + b\sin\theta_{ur}\sin\theta_b \tag{241}$$

$$M_r = \cos\theta_{ur}\left(1 + \eta_r - b\cos\theta_b + b\sin\theta_b V\frac{\partial\theta_{ur}}{\partial V}\right) - \sin\theta_{ur}\left[- b\sin\theta_b + T\frac{\partial\theta_{ur}}{\partial T} + (\eta_r - b\cos\theta_b)V\frac{\partial\theta_{ur}}{\partial V}\right] \tag{242}$$

$$m_r = \sin\theta_{ur} \tag{243}$$

$$q_r = \sin\theta_{ur}(\tau_r - b\cos\theta_b) - b\cos\theta_{ur}\sin\theta_b \tag{244}$$

$$R_r = \sin\theta_{ur}\left(1 + \tau_r - b\cos\theta_b + b\sin\theta_b V\frac{\partial\theta_{ur}}{\partial V}\right) \tag{245}$$

$$+ \cos\theta_{ur}\left[- b\sin\theta_b + T\frac{\partial\theta_{ur}}{\partial T} + (\tau_r - b\cos\theta_b)V\frac{\partial\theta_{ur}}{\partial V}\right]$$

$$w_r = \cos\theta_{pr}\left(\frac{\eta_r}{3} - \gamma\cos\theta_\gamma\right) + \gamma\sin\theta_{pr}\sin\theta_\gamma \tag{246}$$

$$x_r = \cos\theta_{pr} \tag{247}$$

$$Y_r = \cos\theta_{pr}\left(1 - \frac{\eta_r}{3} + \gamma\cos\theta_\gamma + \gamma\sin\theta_\gamma T\frac{\partial\theta_p}{\partial T}\right) \tag{248}$$

$$- \sin\theta_{pr}\left[\gamma\sin\theta_\gamma + V\frac{\partial\theta_{pr}}{\partial V} + \left(\frac{\eta_r}{3} - \gamma\cos\theta_\gamma\right)T\frac{\partial\theta_{pr}}{\partial T}\right]$$

$$s_r = \sin\theta_{pr}\left(\frac{\tau_r}{3} - \gamma\cos\theta_\gamma\right) - \gamma\cos\theta_{pr}\sin\theta_\gamma \tag{249}$$

$$z_r = \sin\theta_{pr} \tag{250}$$

$$I_r = \sin\theta_{pr}\left(1 - \frac{\tau_r}{3} + \gamma\cos\theta_\gamma + \gamma\sin\theta_\gamma T\frac{\partial\theta_{pr}}{\partial T}\right) \tag{251}$$

$$+ \cos\theta_{pr}\left[\gamma\sin\theta_\gamma + V\frac{\partial\theta_{pr}}{\partial V} + \left(\frac{\tau_r}{3} - \gamma\cos\theta_\gamma\right)T\frac{\partial\theta_{pr}}{\partial T}\right]$$

Equations (235) through (238) are the renormalization group equations for radiation with internal phase angles. Setting $\theta_i = 0$ in equations (240) through (251) reduces equations (235) through (251) to equations (39) through (44). The radiation renormalization group equations (235) through (238) can easily be derived from a Lagrangian formalism in a manner similar to that given in the accompanying paper.

6. <u>GROUND STATE OF SOLIDS AND LOW TEMPERATURE QUANTUM LIQUIDS WITH INTERNAL PHASE.</u> This section considers the calculation of the energy density, pressure, and internal phase angles associated with the relativistic state equation of the form given in equations (147) and (148). The complex number analogs of equations (26) through (28) are written as

$$\overline{E}_o - 3[(1 + \overline{\gamma}_o)\overline{P}_o - \overline{K}_o] = E_o^a \tag{252}$$

$$\overline{E}_j\left(1 + j + \frac{j\overline{\gamma}_o\overline{P}_o}{\overline{P}_o - \overline{K}_o} + 3n\,\frac{d\overline{\gamma}_o}{dn}\right) = E^a_j\left(1 + j + \frac{j\gamma^a_o P^a_o}{P^a_o - K^a_o}\right) \tag{253}$$

$$\frac{\overline{E}_j}{E^a_j} = \exp\left[(j-1)\int(\gamma^a_o - \overline{\gamma}_o)\,\frac{dn}{n}\right] = \frac{E_j}{E^a_j}\,e^{i\theta^j_u} \tag{254}$$

where

$$\overline{K}_o = n\,\frac{d\overline{P}_o}{dn} = e^{i\theta^o_p}\left(n\,\frac{dP_o}{dn} + iP_o n\,\frac{d\theta^o_p}{dn}\right) \tag{255}$$

$$= K_o e^{i(\theta^o_p + \omega_o)}$$

where P_o , K_o , and ω_o are given by equations (139), (145), and (146) respectively. Equation (252) can also be written as

$$3n^2\,\frac{d^2\overline{E}_o}{dn^2} - 3(1 + \overline{\gamma}_o)n\,\frac{d\overline{E}_o}{dn} + (3\overline{\gamma}_o + 4)\overline{E}_o = E^a_o \tag{256}$$

Equations (252) and (253) and the constraint equations (136) and (137) must be solved for E_o , θ^o_u , P_o , θ^o_p , γ_o , and θ^o_γ . Combining equation (256) with equations (131) and (132) and taking the real and imaginary parts yields the following two equations

$$F_1 \cos\theta^o_u + 3G_1 \sin\theta^o_u = E^a_o \tag{257}$$

$$F_1 \sin\theta^o_u - 3G_1 \cos\theta^o_u = 0 \tag{258}$$

where

$$F_1 = 3n^2\,\frac{d^2E_o}{dn^2} - 3(1 + \gamma_o\cos\theta^o_\gamma)n\,\frac{dE_o}{dn} \tag{259}$$

$$+ \left[4 + 3\gamma_o\left(\cos\theta^o_\gamma + \sin\theta^o_\gamma\, n\,\frac{d\theta^o_u}{dn}\right) - 3\left(n\,\frac{d\theta^o_u}{dn}\right)^2\right]E_o$$

$$G_1 = \left(\gamma_o \sin \theta_\gamma^o - 2n \frac{d\theta_u^o}{dn}\right) n \frac{dE_o}{dn} \tag{260}$$

$$+ \left[(1 + \gamma_o \cos \theta_\gamma^o) n \frac{d\theta_u^o}{dn} - n^2 \frac{d^2\theta_u^o}{dn^2} - \gamma_o \sin \theta_\gamma^o\right] E_o$$

Equation (253) can be written as the following two equations

$$\frac{U_j}{U_j^a}\left[\cos \theta_u^j (1 + j + A + C) - \sin \theta_u^j (B + D)\right] = 1 + j + \frac{j\gamma_o^a P_o^a}{P_o^a - K_o^a} \tag{261}$$

$$\sin \theta_u^j (1 + j + A + C) + \cos \theta_u^j (B + D) = 0 \tag{262}$$

where

$$A = \frac{j\gamma_o P_o \cos(\theta_\gamma^o + \chi_o)}{\sqrt{\left(P_o - n \frac{dP_o}{dn}\right)^2 + P_o^2\left(n \frac{d\theta_p^o}{dn}\right)^2}} \tag{263}$$

$$B = \frac{j\gamma_o P_o \sin(\theta_\gamma^o + \chi_o)}{\sqrt{\left(P_o - n \frac{dP_o}{dn}\right)^2 + P_o^2\left(n \frac{d\theta_p^o}{dn}\right)^2}} \tag{264}$$

$$C = 3\left(\cos \theta_\gamma^o \, n \frac{d\gamma_o}{dn} - \sin \theta_\gamma^o \, n \frac{d\theta_\gamma^o}{dn}\right) \tag{265}$$

$$D = 3\left(\sin \theta_\gamma^o \, n \frac{d\gamma_o}{dn} + \cos \theta_\gamma^o \, n \frac{d\theta_\gamma^o}{dn}\right) \tag{266}$$

where

$$\tan \chi_o = \frac{P_o n \frac{d\theta_p^o}{dn}}{P_o - n \frac{dP_o}{dn}} \tag{267}$$

$$\theta_u^j = -(j - 1)\int \gamma_o \sin \theta_\gamma^o \frac{dn}{n} \tag{268}$$

$$\frac{U_j}{U_j^a} = \exp\left[(j - 1)\int (\gamma_o^a - \gamma_o \cos \theta_\gamma^o) \frac{dn}{n}\right] \tag{269}$$

The angle θ_u^j enters the calculations through the relation (254)

$$\frac{\overline{U}_j}{U_j^a} = \frac{U_j}{U_j^a} e^{i\theta_u^j} = \exp\left[(j - 1)\int (\gamma_o^a - \gamma_o e^{i\theta_\gamma^o}) \frac{dn}{n}\right] \tag{270}$$

from which equations (268) and (269) follow immediately. Equivalently one can use

$$\overline{\gamma}_o = \gamma_o e^{i\theta_\gamma^o} = \frac{1}{(j-1)} \frac{V}{\overline{U}_j} \frac{d\overline{U}_j}{dV} = \frac{V\overline{P}_j}{\overline{U}_j} \tag{271}$$

$$= \frac{1}{j-1}\left(\frac{V}{U_j} \frac{dU_j}{dV} + iV \frac{d\theta_u^j}{dV}\right)$$

to obtain

$$\gamma_o = \frac{1}{(j-1)} \sqrt{\left(\frac{V}{U_j} \frac{dU_j}{dV}\right)^2 + \left(V \frac{d\theta_u^j}{dV}\right)^2} = \frac{VP_j}{U_j} \tag{272}$$

$$\tan \theta_\gamma^o = \frac{V \frac{d\theta_u^j}{dV}}{\frac{V}{U_j} \frac{dU_j}{dV}} \tag{273}$$

Note that $\overline{U}_1 = U_1^a$ when $j = 1$. From equation (271) it follows that

$$\gamma_o \sin \theta_\gamma^o = \frac{1}{j-1} V \frac{d\theta_u^j}{dV} = -\frac{1}{j-1} n \frac{d\theta_u^j}{dn} \tag{274}$$

$$\gamma_o \cos\theta^o_\gamma = \frac{1}{j-1}\frac{V}{U_j}\frac{dU_j}{dV} = -\frac{1}{j-1}\frac{n}{U_j}\frac{dU_j}{dn} \qquad (275)$$

which immediately give equations (268) and (269) respectively. From equation (271) it also follows that

$$\theta^o_\gamma = \theta^j_p - \theta^j_u \qquad (276)$$

$$P_j = \frac{\gamma_o U_j}{V} = \frac{1}{(j-1)}\sqrt{\left(\frac{dU_j}{dV}\right)^2 + U_j^2\left(\frac{d\theta^j_u}{dV}\right)^2} \qquad (277)$$

Note that when $j = 1$: $U_1 = 3/2\ NR = U_1^a$, and $P_1 = \gamma_o 3/2\ nR$. The simultaneous solution of equations (257), (258), (261), and (262) along with the constraint equations (136) and (137) give E_o , θ^o_u , P_o , θ^o_p , γ_o , and θ^o_γ . Then equations (272) and (273) give U_j and θ^j_u ; P_j is obtained from equation (277), and finally θ^j_p is obtained from equation (276). In this way all of the elements of the renormalized state equations (147) and (148) can be determined.

7. EXCITED STATES OF SOLIDS AND LOW TEMPERATURE QUANTUM LIQUIDS WITH INTERNAL PHASE. The complex number state equations for the excited states of solids and low temperature quantum liquids are written in analogy to the ground state equations (147) and (148) as follows

$$\overline{E}_r = \overline{E}_{or} + \overline{E}_{jr}T^j = E_{or}e^{i\theta^o_{ur}} + E_{jr}e^{i\theta^j_{ur}}T^j = E_r e^{i\theta_{ur}} \qquad (278)$$

$$\overline{P}_r = \overline{P}_{or} + \overline{P}_{jr}T^j = P_{or}e^{i\theta^o_{pr}} + P_{jr}e^{i\theta^j_{pr}}T^j = P_r e^{i\theta_{pr}} \qquad (279)$$

where the $T = 0$ equivalents of the quantities in equations (200) through (206) are

$$\overline{U}_{or} = U_{or}e^{i\theta^o_{ur}} \qquad (280)$$

$$\overline{E}_{or} = \overline{U}_{or}/V = E_{or}e^{i\theta^o_{ur}} \qquad (281)$$

$$\overline{P}_{or} = P_{or}e^{i\theta^o_{pr}} \qquad (282)$$

$$\overline{\gamma}_{or} = \gamma_{or}e^{i\theta^o_{\gamma r}} \qquad (283)$$

$$\overline{b}_{or} = 0 \tag{284}$$

$$\overline{\delta}_{or} = \delta_{or} e^{i\theta^o_{\delta r}} \tag{285}$$

$$\overline{\beta}_{or} = 0 \tag{286}$$

where θ^o_{ur} , θ^o_{pr} , $\theta^o_{\gamma r}$, and $\theta^o_{\delta r}$ are the T = 0 values of θ_{ur} , θ_{pr} , $\theta_{\gamma r}$, and $\theta_{\delta r}$ respectively; and where θ^j_{ur} and θ^j_{pr} are the phase angles associated with the thermal components of the radiation energy and pressure respectively.

The T = 0 and T^j components of the complex number equation (51) are respectively[8]

$$\overline{E}_{or} - 3[(1 + \overline{\gamma}_o)\overline{P}_{or} - \overline{K}_{or}] - 3 \frac{\overline{E}_{jr}}{\overline{E}_j} \overline{P}_o(\overline{\gamma}_{or} - \overline{\gamma}_o) = E^a_{or} \tag{287}$$

$$j\overline{E}_j(\overline{\alpha K}_{or} - \overline{\beta P}_{or}) + \overline{E}_{jr}\overline{S}_{jr} = jE^a_j\left(\alpha^a K^a_{or} - \beta^a P^a_{or}\right) + E^a_{jr}T^a_{jr} \tag{288}$$

where

$$\overline{P}_{or} = n \frac{d\overline{E}_{or}}{dn} - \overline{E}_{or} = e^{i\theta^o_{ur}}\left(n \frac{dE_{or}}{dn} - E_{or} + iE_{or} n \frac{d\theta^o_{ur}}{dn}\right) \tag{289}$$

$$= P_{or} e^{i(\theta^o_{ur} + \phi^o_r)}$$

where

$$P_{or} = \sqrt{\left(\frac{dU_{or}}{dV}\right)^2 + U^2_{or}\left(\frac{d\theta^o_{ur}}{dV}\right)^2} \tag{290}$$

$$= \sqrt{\left(n \frac{dE_{or}}{dn} - E_{or}\right)^2 + E^2_{or}\left(n \frac{d\theta^o_{ur}}{dn}\right)^2}$$

$$\theta^o_{pr} = \theta^o_{ur} + \phi^o_r \tag{291}$$

$$P_{or} \cos \phi^o_r = n\, dE_{or}/dn - E_{or} \tag{291A}$$

$$P_{or} \sin \phi^o_r = E_{or} n\, d\theta^o_{ur}/dn \tag{291B}$$

where

$$\tan \phi^o_r = \frac{U_{or} \dfrac{d\theta^o_{ur}}{dV}}{\dfrac{dU_{or}}{dV}} = \frac{E_{or} n \dfrac{d\theta^o_{ur}}{dn}}{n \dfrac{dE_{or}}{dn} - E_{or}} \tag{292}$$

The T = 0 radiation bulk modulus is given by

$$\overline{K}_{or} = n \frac{d\overline{P}_{or}}{dn} = e^{i\theta^o_{pr}}\left(n \frac{dP_{or}}{dn} + iP_{or} n \frac{d\theta^o_{pr}}{dn}\right) \tag{293}$$

$$= K_{or} e^{i(\theta^o_{pr} + \omega^o_r)}$$

where

$$K_{or} = \sqrt{\left(n \frac{dP_{or}}{dn}\right)^2 + P^2_{or}\left(n \frac{d\theta^o_{pr}}{dn}\right)^2} \tag{294}$$

$$K_{or} \cos \omega^o_r = n\, dP_{or}/dn \tag{294A}$$

$$K_{or} \sin \omega^o_r = P_{or} n\, d\theta^o_{pr}/dn \tag{294B}$$

$$\tan \omega^o_r = P_{or}(d\theta^o_{pr}/dn)/(dP_{or}/dn) \tag{295}$$

The functions $\overline{S}_{jr}$ and T^a_{jr} that appear in equation (288) are given by[8]

$$\overline{S}_{jr} = 1 + j + \frac{j\overline{P}_o\overline{\gamma}_{or}}{\overline{P}_o - \overline{K}_o} + 3n \frac{d\overline{\gamma}_{or}}{dn} - 3(j - 1)(\overline{\gamma}_{or} - \overline{\gamma}_o)^2 \tag{296}$$

$$T^a_{jr} = 1 + j + \frac{jP^a_o\gamma^a_{or}}{P^a_o - K^a_o} \tag{297}$$

and where

$$\overline{\alpha} = \frac{\overline{\gamma}_o\overline{P}_o}{(\overline{P}_o - \overline{K}_o)^2} \qquad \alpha^a = \frac{\gamma^a_o P^a_o}{(P^a_o - K^a_o)^2} \tag{298}$$

$$\bar{\beta} = \frac{\bar{\gamma}_o \bar{K}_o}{(\bar{P}_o - \bar{K}_o)^2} \qquad \beta^a = \frac{\gamma_o^a K_o^a}{(P_o^a - K_o^a)^2} \tag{299}$$

In analogy with equations (271) through (275) one has the following relations

$$\bar{\gamma}_{or} = \frac{\bar{P}_{jr}}{\bar{E}_{jr}} = \frac{1}{(j-1)} \frac{V}{\bar{U}_{jr}} \frac{d\bar{U}_{jr}}{dV} = \gamma_{or} e^{i\theta^o_{\gamma r}} \tag{300}$$

$$= \frac{1}{(j-1)} \left(\frac{V}{U_{jr}} \frac{dU_{jr}}{dV} + iV \frac{d\theta^j_{ur}}{dV} \right)$$

$$\gamma_{or} = \frac{1}{(j-1)} \sqrt{\left(\frac{V}{U_{jr}} \frac{dU_{jr}}{dV} \right)^2 + \left(V \frac{d\theta^j_{ur}}{dV} \right)^2} \tag{301}$$

$$\tan \theta^o_{\gamma r} = \frac{V \dfrac{d\theta^j_{ur}}{dV}}{\dfrac{V}{U_{jr}} \dfrac{dU_{jr}}{dV}} \tag{302}$$

$$\gamma_{or} \sin \theta^o_{\gamma r} = \frac{1}{j-1} V \frac{d\theta^j_{ur}}{dV} \tag{303}$$

$$\gamma_{or} \cos \theta^o_{\gamma r} = \frac{1}{j-1} \frac{V}{U_{jr}} \frac{dU_{jr}}{dV} \tag{304}$$

$$\theta^o_{\gamma r} = \theta^j_{pr} - \theta^j_{ur} \tag{305}$$

$$P_{jr} = \frac{\gamma_{or} U_{jr}}{V} = \frac{1}{j-1} \sqrt{\left(\frac{dU_{jr}}{dV} \right)^2 + U^2_{jr} \left(\frac{d\theta^j_{ur}}{dV} \right)^2} \tag{306}$$

When j = 1: $E_{1r} = E^a_{1r}$ and $P_{1r} = \gamma_{or} E_{1r}$. Integration of (303) and (304) yields

$$\theta^j_{ur} = -(j - 1)\int \gamma_{or} \sin \theta^o_{\gamma r} \frac{dn}{n} \tag{307}$$

$$\frac{E_{jr}}{E^a_{jr}} = \exp\left[(j - 1)\int (\gamma^a_{or} - \gamma_{or} \cos \theta^o_{\gamma r}) \frac{dn}{n}\right] \tag{308}$$

Using equations (130) through (133) and (280) through (283) in equation (287) and separating into real and imaginary parts yields the following two radiation renormalization group equations

$$F_{1r} \cos \theta^o_{ur} + 3G_{1r} \sin \theta^o_{ur} - 3 \frac{E_{jr}}{E_j} P_o(A_{or} \cos \Gamma^j_{or} - B^j_{or} \sin \Gamma^j_{or}) = E^a_{or} \tag{309}$$

$$F_{1r} \sin \theta^o_{ur} - 3G_{1r} \cos \theta^o_{ur} - 3 \frac{E_{jr}}{E_j} P_o(A_{or} \sin \Gamma^j_{or} + B^j_{or} \cos \Gamma^j_{or}) = 0 \tag{310}$$

where

$$F_{1r} = 3n^2 \frac{d^2E_{or}}{dn^2} - 3(1 + \gamma_o \cos \theta^o_\gamma)\, n \frac{dE_{or}}{dn} + \left[4 + 3\gamma_o\left(\cos \theta^o_\gamma + \sin \theta^o_\gamma\, n \frac{d\theta^o_{ur}}{dn}\right) - 3\left(n \frac{d\theta^o_{ur}}{dn}\right)^2\right] E_{or} \tag{311}$$

$$G_{1r} = \left(\gamma_o \sin \theta^o_\gamma - 2n \frac{d\theta^o_{ur}}{dn}\right) n \frac{dE_{or}}{dn} + \left[(1 + \gamma_o \cos \theta^o_\gamma)\, n \frac{d\theta^o_{ur}}{dn} - n^2 \frac{d^2\theta^o_{ur}}{dn^2} - \gamma_o \sin \theta^o_\gamma\right] E_{or} \tag{312}$$

$$A_{or} = \gamma_{or} \cos \theta^o_{\gamma r} - \gamma_o \cos \theta^o_\gamma \tag{313}$$

$$B_{or} = \gamma_{or} \sin \theta^o_{\gamma r} - \gamma_o \sin \theta^o_\gamma \tag{314}$$

$$\Gamma^j_{or} = \theta^j_{ur} + \theta^o_p - \theta^o_u \tag{315}$$

In order to decouple the complex number equation (288) into two real equations one must first rewrite the expressions for $\overline{\alpha}$, $\overline{\beta}$, and $\overline{S}_{jr}$. From equation (298) it follows that

$$\overline{\alpha} = \frac{\gamma_o P_o}{T_o^2} e^{i(2\chi_o + \theta_\gamma^o - \theta_p^o)} \tag{316}$$

where γ_o , P_o , and χ_o are given by equations (272), (139), and (267) respectively, and T_o is given by

$$T_o^2 = \left(P_o - n\frac{dP_o}{dn}\right)^2 + P_o^2\left(n\frac{d\theta_p^o}{dn}\right)^2 \tag{317}$$

From equation (299) it follows that

$$\overline{\beta} = \frac{\gamma_o K_o}{T_o^2} e^{i(2\chi_o + \theta_\gamma^o + \omega_o - \theta_p^o)} \tag{318}$$

where K_o and ω_o are given by equations (145) and (146) respectively. The expression $\overline{S}_{jr}$ is obtained from equations (296), (132), (133), (144), and (283) to be

$$\overline{S}_{jr} = R_{jr} + iT_{jr} \tag{319}$$

where

$$R_{jr} = 1 + j + \frac{jP_o\gamma_{or}}{T_o^2}\left[\left(P_o - n\frac{dP_o}{dn}\right)\cos\theta_{\gamma r}^o - P_o n\frac{d\theta_p^o}{dn}\sin\theta_{\gamma r}^o\right] \tag{320}$$

$$+ 3\left(n\frac{d\gamma_{or}}{dn}\cos\theta_{\gamma r}^o - \gamma_{or} n\frac{d\theta_{\gamma r}^o}{dn}\sin\theta_{\gamma r}^o\right)$$

$$- 3(j-1)[\gamma_{or}^2\cos(2\theta_{\gamma r}^o) - 2\gamma_o\gamma_{or}\cos(\theta_p^o + \theta_{\gamma r}^o) + \gamma_o^2\cos(2\theta_\gamma^o)]$$

$$T_{jr} = \frac{jP_o\gamma_{or}}{T_o^2}\left[\left(P_o - n\frac{dP_o}{dn}\right)\sin\theta^o_{\gamma r} + P_o n\frac{d\theta^o_p}{dn}\cos\theta^o_{\gamma r}\right] \tag{321}$$

$$+ 3\left(n\frac{d\gamma_{or}}{dn}\sin\theta^o_{\gamma r} + \gamma_{or} n\frac{d\theta^o_{\gamma r}}{dn}\cos\theta^o_{\gamma r}\right)$$

$$- 3(j-1)[\gamma^2_{or}\sin(2\theta^o_{\gamma r}) - 2\gamma_o\gamma_{or}\sin(\theta^o_\gamma + \theta^o_{\gamma r}) + \gamma^2_o\sin(2\theta^o_\gamma)]$$

Using equations (289) through (321) allows the second radiation equation (288) to be separated into real and imaginary parts as follows

$$\frac{jE_j\gamma_o}{T_o^2}(L_{or}\cos\Omega^j_{or} - M_{or}\sin\Omega^j_{or}) + E_{jr}(R_{jr}\cos\theta^j_{ur} - T_{jr}\sin\theta^j_{ur}) \tag{322}$$

$$= jE^a_j(\alpha^a K^a_{or} - \beta^a P^a_{or}) + E^a_{jr}T^a_{jr}$$

$$\frac{jE_j\gamma_o}{T_o^2}(L_{or}\sin\Omega^j_{or} + M_{or}\cos\Omega^j_{or}) + E_{jr}(R_{jr}\sin\theta^j_{ur} + T_{jr}\cos\theta^j_{ur}) = 0 \tag{323}$$

where

$$L_{or} = P_oK_{or}\cos\omega^o_r - K_oP_{or}\cos\omega_o \tag{324}$$

$$M_{or} = P_oK_{or}\sin\omega^o_r - K_oP_{or}\sin\omega_o \tag{325}$$

$$\Omega^j_{or} = \theta^o_{pr} + \theta^j_u + 2\chi_o + \theta^o_\gamma - \theta^o_p \tag{326}$$

Equations (309), (310), (322), and (323) are the four renormalization group equations needed to solve for the four unknown radiation functions E_{or}, θ^o_{ur}, γ_{or}, and $\theta^o_{\gamma r}$. The radiation analogs of equations (136) and (137) relate θ^o_{ur}, θ^o_{pr}, E_{or}, and P_{or}. Equations (307) and (308) relate $\overline{E}_{jr}$ to $\overline{\gamma}_{or}$. These eight equations can be used to solve for the eight quantities $\overline{E}_{or}$, $\overline{P}_{or}$, $\overline{E}_{jr}$, and $\overline{P}_{jr}$ (or $\overline{\gamma}_{or}$).

8. PROCESSES IN PHASE MATTER AND RADIATION. The internal phase angles of matter and radiation allow an extended interpretation of the types of processes that can occur in these systems. Consider for example the change in complex entropy given by equation (57) as

$$d\overline{S} = e^{i\theta_s}(dS + iSd\theta_s) \tag{327}$$

From equation (327) it is clear that $d\overline{S} = 0$ cannot represent a physical process, as the real and imaginary parts both set equal to zero would determine two $V = V(T)$ curves, and these conditions would perhaps hold jointly only at an intersection point. In fact two distinct processes can be obtained from equation (327)

$$dS = 0 \qquad \text{adiabatic process} \tag{328}$$

$$d\theta_s = 0 \qquad \text{entropy isophase process} \tag{329}$$

Thus an adiabatic process corresponds to a rotation of the entropy vector $\overline{S}$ in internal phase angle space. Processes may occur in nature such that changes in volume and temperature cause rotation of the entropy and internal energy vectors. For the case of constant entropy magnitude ($dS = 0$) the heat increment is obtained from equation (327) to be

$$d\overline{Q} = iT\overline{S}d\theta_s \tag{330}$$

For this adiabatic process the conservation of energy is written as

$$iT\overline{S}d\theta_s = d\overline{U} + \overline{P}dV \tag{331}$$

This results in equations (69) through (91) with $\partial S/\partial V = 0$ and $\partial S/\partial T = 0$.

For the case of constant magnitude of the internal energy, $dU = 0$ and the rotation of the internal energy vector is given by

$$d\overline{U} = i\overline{U}d\theta_u \tag{332}$$

and equation (66) gives

$$Td\overline{S} = i\overline{U}d\theta_u + \overline{P}dV \tag{333}$$

This results in equations (69) through (91) with $\partial U/\partial V = 0$ and $\partial U/\partial T = 0$. For the case when both $dS = 0$ and $dU = 0$, corresponding to rotations of both the entropy and internal energy vectors, one has

$$iT\overline{S}d\theta_s = i\overline{U}d\theta_u + \overline{P}dV \tag{334}$$

This results in equations (69) through (91) with $\partial U/\partial V = 0$, $\partial U/\partial T = 0$, $\partial S/\partial V = 0$, and $\partial S/\partial T = 0$. Similar results apply for the thermodynamic potentials $\overline{H}$, $\overline{A}$, and $\overline{G}$.

In general a process will result in a combined stretch and rotation of the thermodynamic functions $\overline{U}$, $\overline{P}$, $\overline{S}$, $\overline{H}$, $\overline{A}$, and $\overline{G}$, and the time rate of change of a thermodynamic quantity will include an angular velocity of the internal phase angles. For example, the rate of pressure change is given by

$$\frac{d\overline{P}}{dt} = e^{i\theta}P\left(\frac{dP}{dt} + iP\,\frac{d\theta_p}{dt}\right) \tag{335}$$

Thus, even if the magnitude of the pressure were held fixed the pressure vector can rotate internally.

Another possibility is the transfer of external angular momentum to internal angular momentum and vice versa. Thus external rotation may in some cases be coupled to the rotation of the internal phase angles. In fact, the Lagrangian of a rotating body may be of the form

$$L = \frac{1}{2}\sum_e I_e\omega_e^2 + \frac{1}{2}\sum_i I_i\omega_i^2 + \sum_{ie} a_{ei}\omega_e\omega_i - V(\theta_e,\theta_i) \tag{336}$$

where I_e and ω_e = external moment of inertia and angular velocity respectively, I_i and ω_i = internal moments of inertia and angular velocity respectively, and where θ_e and θ_i = external and internal angles respectively. The θ_i consists of θ_p , θ_u , θ_s , etc, and $\dot{\theta}_i$ includes $\dot{\theta}_p$, $\dot{\theta}_u$, $\dot{\theta}_s$ and so on. The internal moments of inertia I_p , I_u , I_s , etc are associated with the internal angle coordinates of pressure, internal energy, entropy, etc. It is expected that for such a system macroscopic energy transfers would occur between the internal and external dynamical systems. Such transfers may account for the glitches that appear in the spin-down of pulsars. Similar processes have been suggested to occur at the level of fundamental particles.[10-13]

9. CONCLUSION. The local gauge invariance of relativistic thermodynamics suggests the possibility that the thermodynamic state functions can be represented as complex numbers whose imaginary parts are related to phase angles in an internal space associated with all interacting systems of matter and radiation. Due to vacuum interactions, bulk matter solids and quantum liquids are coherent in internal space. The phase angles and magnitudes of the thermodynamic state functions are calculated from a solution of the renormalization group equations which represent the mathematical description of the interaction of matter and radiation in matter with the vacuum state. The internal phase angles are expected to manifest themselves in the state equations of matter and radiation in matter. In some cases a transfer of energy may occur between external rotations and the rotations of the internal phase angles. The internal phase angles are expected to affect the equations of motion of classical and quantum systems, and should affect the equilibrium configurations of atomic nuclei and the stars.

The renormalized ground state of a relativistic thermodynamic solid or quantum liquid is associated with a broken symmetry manifested by the nonzero values of the internal phases θ_p , θ_u , θ_s , etc. that are obtained as solutions to the relativistic trace equation. A symmetrical ground state would have $\theta_p = 0$, $\theta_u = 0$, $\theta_s = 0$, etc. This broken symmetry should be associated with massive gauge bosons that are connected with the excited states of the internal phases of bulk matter, i.e., internal spin waves of the pressure and entropy. Similar broken symmetries are common in atomic and nuclear systems.[14] As a practical application of these ideas one can conceive of a bulk matter vacuum-induced broken symmetry thermodynamic engine. Such an internal phase engine would utilize the broken symmetry nature of the ground state of bulk matter in a manner analogous to the broken symmetry ferromagnetic state of an iron armature in an electric motor.

ACKNOWLEDGEMENT

The author wishes to thank Elizabeth K. Klein for typing this paper.

REFERENCES

1. Frampton, P., Gauge Field Theories, Benjamin-Cummings, Menlo Park, 1987.

2. Close, F., An Introduction to Quarks and Partons, Academic, New York, 1979.

3. Okun, L., Particle Physics, Harwood, New York, 1985.

4. Akhiezer, A. and Berestetskii, V., Quantum Electrodynamics, Interscience, New York, 1965.

5. Schweber, S., An Introduction to Relativistic Quantum Field Theory, Harper & Row, New York, 1962.

6. Weiss, R. A., Relativistic Thermodynamics, Vols. 1 and 2, Exposition Press, New York, 1976.

7. Weiss, R. A., "Relativistic Wave Equations for Solids and Low Temperature Quantum Systems," Third Army Conference on Applied Mathematics and Computing, Georgia Institute of Technology, ARO 86-1, May 13-16, 1985, p. 717.

8. Weiss, R. A., "Scale Invariant Equations for Relativistic Waves," Fourth Army Conference on Applied Mathematics and Computing, Cornell University, ARO 87-1, May 27-30, 1986, p. 307.

9. Rumer, Y. and Ryvkin, M., Thermodynamics, Statistical Physics, and Kinetics, MIR Publishers, Moscow, 1980.

10. Cheng, T. and Li, L., Gauge Theory of Elementary Particle Physics, Clarendon Press, Oxford, 1984.

11. Jackiw, R. and Rebbi, C., "Vacuum Periodicity in a Yang-Mills Quantum Theory," Phys. Rev. Lett., 37, 172, (1976).

12. Hasenfratz, P. and 't Hooft, G., "Fermion-Boson Puzzle in a Gauge Theory," Phys. Rev. Lett., 36, 1119, (1976).

13. Goldhaber, A., "Connection of Spin and Statistics for Charge-Monopole Composites," Phys. Rev. Lett., 36, 1122, (1976).

14. Dasso, C. H. and Vitturi, A., "Reconstructing the Nuclear Profile in Gauge Space," Phys. Rev. Lett., 59, 634, (1987).

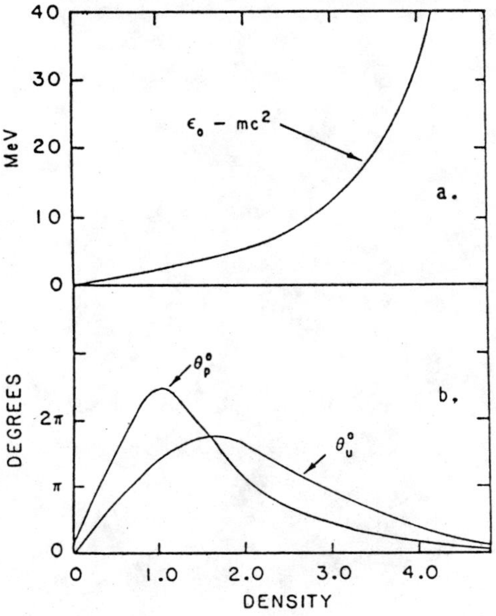

Figure 1. a) Binding energy of T = 0 neutron gas; b) Density dependence of the phase angles for the pressure and internal energy of a T = 0 neutron gas.

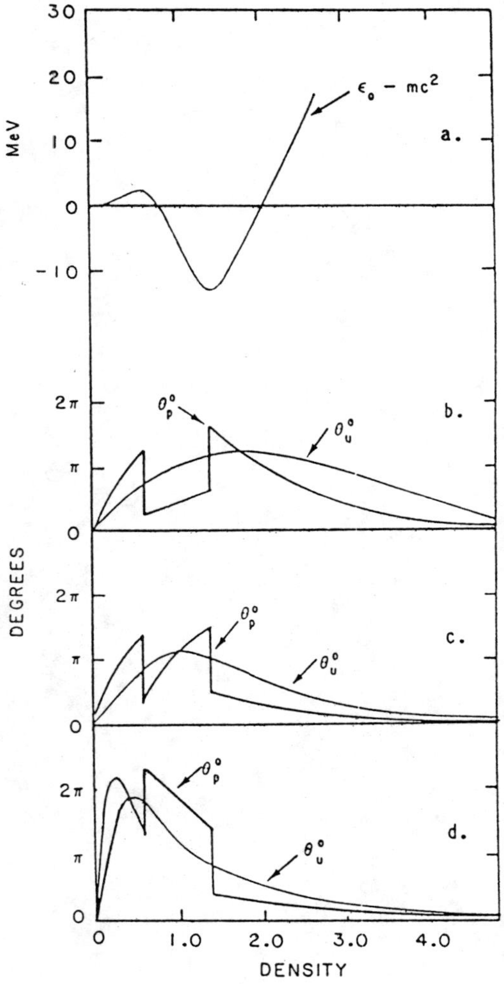

Figure 2. a) Binding energy of T = 0 infinite nuclear matter with N = Z; b), c), d), Three possible cases for the density dependence of the phase angles for the pressure and internal energy of T = 0 infinite nuclear matter with N = Z.

F

RELATIVISTIC THERMODYNAMICS OF REAL GASES WITH BROKEN INTERNAL SYMMETRY

Richard A. Weiss
U. S. Army Engineer Waterways Experiment Station
Vicksburg, Mississippi 39180

ABSTRACT. The relativistic state equation for real gases with broken internal symmetry is developed. This is done by solving the complex form of a relativistic trace equation for the virial state equation of the real gases. The resulting solution affects only the third and higher virial coefficients. The complex relativistic third virial coefficient is given by a solution of two coupled differential equations. An approximate solution is found which is valid in the high temperature region, and an expression for the internal phase angle for the third virial coefficient is obtained. From this it is possible to develop expressions for the internal phase angles of the pressure, internal energy, entropy, enthalpy, and free energy of real gases that exhibit broken internal symmetries. Mixtures of interacting gases with broken internal symmetry are suggested to exhibit an interference phenomenon whereby the total pressure and internal energy will oscillate slightly in magnitude as the density of the system is increased. Accurate high temperature state equations of real gases are important for the description of the equilibrium configurations of gaseous stars, and for the description of nuclear explosions in the atmosphere.

1. INTRODUCTION. Spontaneously broken symmetry is a common phenomenon in physics because it appears in such diverse situations as ferromagnetism, superconductivity, weak interactions, and the vacuum screening currents that produce the asymmetric vacuum.[1-4] It is associated with a phase difference between a free particle in a potential and a particle in a coherent state of particles which forms due to some special system of forces. In superconductivity it is the Cooper pairs of electrons that form a self-coherent system which violates gauge invariance. In a similar fashion the vacuum state is thought to exhibit spontaneous symmetry breaking due to the presence of a Higgs scalar field which has a nonzero value for a minimum potential energy.

In relativistic thermodynamics a similar broken symmetry has been suggested to exist in the renormalized state equations of solids and quantum liquids.[5] This broken symmetry is associated with intrinsic phase angles of the thermodynamic state functions such as internal energy and pressure, and is due to the interaction of spacetime with bulk matter and the vacuum. The internal phase angles of the coherent state must be considered when applying the first and second laws of thermodynamics because the differentials of the state functions, such as the entropy and internal energy, must include a rotation in internal space.[5] This affects the measured state equation of thermodynamic systems such as solids, liquids and gases. This paper considers the vacuum induced broken symmetry of the state functions of the real gases. The broken symmetry appears in the third and higher order virial coefficients of the state equation for real gases.

The effects of the Minkowski metric of spacetime on the equation of state of bulk matter was originally described by the solutions of the scalar trace equation[6]

$$U + T\left(\frac{dU}{dT}\right)_{PV} - 3V\,\frac{d}{dV}(PV)_U = U^a + T\left(\frac{dU^a}{dT}\right)_{P^aV} \quad (1)$$

where U = relativistic (renormalized) internal energy, P = relativistic pressure, T = absolute temperature, V = volume of substance, and U^a and P^a = corresponding nonrelativistic internal energy and pressure. Throughout this paper the index "a" will refer to nonrelativistic (unrenormalized) calculations. It has been suggested that the spacetime induced broken symmetry effects on bulk matter can be represented by the following complex number trace equation[5]

$$\bar{U} + T\left(\frac{d\bar{U}}{dT}\right)_{\bar{P}V} - 3V\,\frac{d}{dV}(\bar{P}V)_{\bar{U}} = U^a + T\left(\frac{dU^a}{dT}\right)_{P^aV} \quad (2)$$

whose solution yields complex numbers, with internal phase angles, for the state functions of relativistic thermodynamics. Complex number solutions arise only in those cases for which there are nonzero gauge terms of the form[5,6]

$$T\left(\frac{d\bar{U}}{dT}\right)_{\bar{P}V} \neq 0 \quad (3)$$

$$\frac{d}{dV}(\bar{P}V)_{\bar{U}} \neq 0 \quad (4)$$

This includes interacting systems and the general case of the noninteracting relativistic Fermi gas.

The unrenormalized pressure and internal energy density for the real gases are given by[7,8]

$$P^a = nR^aT[1 + nB^a(T) + n^2C^a(T) + n^3D^a(T) + \cdots] \quad (5)$$

$$E^a = nR^aT[\frac{3}{2} - nT\,\frac{\partial B^a}{\partial T} - \frac{1}{2}\,n^2T\,\frac{\partial C^a}{\partial T} - \frac{1}{3}\,n^3T\,\frac{\partial D^a}{\partial T} - \cdots] \quad (6)$$

where

$$n = N/V = 1/\tilde{V} \quad (7)$$

where N = number of moles, $\tilde{V}$ = molar volume; R^a, $B^a(T)$, $C^a(T)$ and $D^a(T)$ = nonrelativistic values of the gas constant, second virial coefficient, third virial coefficient, and fourth virial coefficient respectively. The corresponding renormalized pressure and energy density that are obtained from the solutions of the scalar trace equation (1) are written as[6,9]

$$P = nRT[1 + nB(T) + n^2C(T) + n^3D(T) + \cdots] \tag{8}$$

$$\mathcal{E} = nRT[\frac{3}{2} - nT\frac{\partial B}{\partial T} - \frac{1}{2}n^2T\frac{\partial C}{\partial T} - \frac{1}{3}n^3T\frac{\partial D}{\partial T} - \cdots] \tag{9}$$

where

$$R = R^a \tag{9A}$$

$$B(T) = B^a(T) \tag{10}$$

$$C(T) = C^a(T) - 3[B^a(T)]^2 \ln \psi^a \tag{11}$$

$$\psi^a = \frac{T}{T_R}\left|\frac{B^a(T)}{B^a(T_R)}\right|^{2/3} = \frac{T}{T_{CR}}\left|\frac{B^a(T)}{B^a(T_{CR})}\right|^{2/3} \tag{12}$$

and where T_R = relativity temperature constant, and T_{CR} = conjugate relativity temperature constant. An expression for the renormalized fourth virial coefficient D(T) has not been obtained.

The solution to the complex number trace equation (2) for the real gases is written as

$$\bar{P} = Pe^{j\theta_P} = nRT[1 + n\bar{B}(T) + n^2\bar{C}(T) + n^3\bar{D}(T) + \cdots] \tag{13}$$

$$\bar{\mathcal{E}} = \mathcal{E}e^{j\theta_{\mathcal{E}}} = nRT[\frac{3}{2} - nT\frac{\partial\bar{B}}{\partial T} - \frac{1}{2}n^2T\frac{\partial\bar{C}}{\partial T} - \frac{1}{3}n^3T\frac{\partial\bar{D}}{\partial T} - \cdots] \tag{14}$$

where

$$\bar{B} = Be^{j\theta_B} \tag{15}$$

$$\bar{C} = Ce^{j\theta_C} \tag{16}$$

$$\bar{D} = De^{j\theta_D} \tag{16A}$$

are to be determined from a solution of the trace equation (2). Only $\bar{B}$ and $\bar{C}$ are obtained in this paper, and in fact $\bar{C}$ is obtained only through a high temperature approximation. It is the real parts of the complex number virial coefficients, pressure and internal energy that are the measured quantities.

The determination of P and θ_P for real gases with broken internal symmetry is important because the state equation of real gases enters the physical description of such diverse situations as the equilibrium configuration of stars and the latent heat associated with the gas-liquid phase transition. Consider for instance the Clausius-Clapyron equation for a real gas with broken internal symmetry[10-13]

$$\bar{\ell} = T(v_2 - v_1)\frac{d\bar{P}}{dT} \tag{17}$$

where $\bar{\ell}$ = complex number latent heat of vaporization, v_2 = specific volume of vapor, v_1 = specific volume of liquid, and $d\bar{P}/dT$ = slope of the vapor pressure curve. The complex number latent heat of vaporization can be written as

$$\bar{\ell} = \ell e^{j\theta_\ell} \tag{18}$$

where ℓ = magnitude of latent heat, and θ_ℓ = internal phase angle of the latent heat of vaporization. Equation (17) can then be written as the following three equations

$$\ell = T(v_2 - v_1)\sqrt{(\partial P/\partial T)_v^2 + (P\,\partial\theta_P/\partial T)_v^2} \tag{19}$$

$$\theta_\ell = \theta_P + \beta_{P,T} \tag{20}$$

$$\tan\beta_{P,T} = P\,\frac{\partial\theta_P/\partial T}{\partial P/\partial T} \tag{21}$$

Note that if $\theta_P = 0$ the standard form of the Clausius-Clapyron equation is regained. The measured value of the latent heat is $\ell \cos\theta_\ell$.

Accurate state equations for the real gases, including broken symmetry effects, are important for stellar structure calculations because the internal phase angle of the radial coordinate is related to the internal phase angle of the pressure. Therefore the complex values of the third and higher virial coefficients of the real gases will play an important role in stellar equilibrium calculations. In addition, it has been suggested that the third virial coefficient of the real gases can be utilized in the design of a gravitational wave detector.[9,14]

This paper calculates the internal phase angles of the second and third virial coefficients of the real gases. The phase angle of the second virial coefficient is determined to be equal to zero. The evaluation of the internal phase angle of the third virial coefficient is more complicated, and has been determined only in the regions of high temperature. The internal phase angles of the pressure, internal energy, entropy, enthalpy, and free energy are then calculated in terms of the virial coefficients and their internal phases. The heat capacity for a real gas with broken internal symmetry is then evaluated. Finally, a general discussion of the interference effects expected to occur in mixtures of asymmetric gases is given.

2. THIRD VIRIAL COEFFICIENT FOR ASYMMETRIC REAL GASES. This section uses the complex number trace equation (2) to solve for the renormalized values of the second and third virial coefficients of the real gases with broken internal symmetry. It has been shown that the real number trace equation (1) does not change the value of the second virial coefficient as shown in equation (10).[6] In a similar fashion it is easy to show by substituting equations (13) and (14) into the complex number trace equation (2) that the second virial coefficient satisfies the following relation[6]

$$\frac{1}{\bar{\beta}}\frac{d\bar{\beta}}{dT} = \frac{1}{\beta^a}\frac{d\beta^a}{dT} \qquad (22)$$

where

$$\bar{\beta} = RT\bar{B} \qquad (23)$$

$$\beta^a = RTB^a \qquad (24)$$

Equations (22) through (24) imply that

$$\bar{B} = Be^{j\theta_B} = B_R + jB_I = B^a = \text{real number} \qquad (25)$$

which gives

$$B_I = 0 \qquad (26)$$

$$\theta_B = 0 \qquad (27)$$

$$B_R = B^a \qquad (28)$$

Therefore the relativsitic value of the second virial coefficient, as determined from a solution of equation (2), is a real number which is equal to the unrenormalized value of the second virial coefficient.

The calculation of the complex number values of the third virial coefficients follows in a more complicated fashion from a solution of equation (2) for the real gases. An expedient way of doing the caluclation is to make use of the results for real gases that have been obtained for the relativistic form of the third virial coefficient from a solution of the scalar relativistic trace equation (1).[6] First define a function $\bar{f}(T)$ which is given by

$$RT\bar{C}(T) = RTC^a(T) + \bar{f}(T) \quad (29)$$

then substituting equations (13), (14) and (29) into the trace equation (2) gives the following equation for $\bar{f}(T)$[6]

$$\frac{d\bar{f}}{dT} + \frac{F}{T}\bar{f} = G \quad (30)$$

where

$$F = 1 - 2\frac{T}{\beta^a}\frac{d\beta^a}{dT} \quad (31)$$

$$G = \frac{3\beta^a}{RT^2}\left[R/c_V^{Ia}(\beta^a - T\,d\beta^a/dT) - \beta^a\right] \quad (32)$$

$$= -\frac{(\beta^a)^2}{RT^2}\left(1 + 2\frac{T}{\beta^a}\frac{d\beta^a}{dT}\right)$$

where $c_V^{Ia} = 3R/2$. Equations (30) through (32) are of the same form that has already been obtained in Reference 6 for the scalar relativistic trace equation (1) except $\bar{f}(T)$ is a complex number.

The complex number $\bar{f}$ is written as follows

$$\bar{f} = -fe^{j\theta_f} \quad (33)$$

$$f_R = -f\cos\theta_f \quad (34)$$

$$f_I = -f\sin\theta_f \quad (35)$$

with $f > 0$, and where f and θ_f are to be determined. Note that the use of the function f is different from that in Reference 6. In the present paper f is used as a magnitude of a complex number and is always positive. The choice of the negative sign in equation (33) is made so that θ_f is small and does not contain the π associated with negative real and imaginary values of $\bar{f}$ in the re-ions of high temperature. Placing equation (33) into equation (30) gives

$$e^{j\theta_f}\left(\frac{df}{dT} + jf\frac{d\theta_f}{dT} + \frac{F}{T}f\right) = -G \quad (36)$$

Taking the real and imaginary parts of equation (36) yields

$$\cos \theta_f \left(\frac{df}{dT} + \frac{F}{T} f\right) - \sin \theta_f \, f \frac{d\theta_f}{dT} = - G \tag{37}$$

$$\sin \theta_f \left(\frac{df}{dT} + \frac{F}{T} f\right) + \cos \theta_f \, f \frac{d\theta_f}{dT} = 0 \tag{38}$$

Equations (37) and (38) are the two required equations for determining f and θ_f . Equations (37) and (38) can be rewritten as

$$\tan \theta_f = - \frac{f \, d\theta_f / dT}{\frac{df}{dT} + \frac{F}{T} f} \tag{39}$$

$$\left(\frac{df}{dT} + \frac{F}{T} f\right)^2 + (f \, d\theta_f / dT)^2 = G^2 \tag{40}$$

Equations (37) and (38) are difficult to solve without some approximations. In this paper a solution of equations (37) and (38) is obtained that is valid only for high temperatures.

An approximate solution for equations (37) and (38) can be obtained by assuming that θ_f is small so that

$$\sin \theta_f \sim \theta_f \sim 0 \tag{41}$$

$$\cos \theta_f \sim 1 \tag{42}$$

$$f = f_o \tag{43}$$

Substituting equations (41) through (43) into equations (37) and (38) yields

$$\frac{df_o}{dT} + \frac{F}{T} f_o = - G \tag{44}$$

$$- \theta_f G + f_o \frac{d\theta_f}{dT} = 0 \tag{45}$$

where all second order terms of θ_f are dropped. It has already been shown that the solution to equation (44) is[6]

$$f_o = 3RT(B^a)^2 \, \ell n \, \psi^a \tag{46}$$

so that from equations (33) through (35) it follows that

$$\bar{f} \sim - 3RT(B^a)^2 \, \ell n \, \psi^a \, e^{j\theta_f} \tag{47}$$

$$f_R \sim - 3RT(B^a)^2 \, \ell n \, \psi^a \cos \theta_f \tag{48}$$

$$f_I \sim - 3RT(B^a)^2 \, \ell n \, \psi^a \sin \theta_f \tag{49}$$

This solution is valid only when $f_o > 0$ since in fact f is the magnitude of a complex number. Therefore the assumption that θ_f is small and that $f_o > 0$ restricts the validity of the approximate solution to the regions of high temperature. For this case $f_R < 0$ and $f_I \lesseqgtr 0$. Combining equations (29) and (47) gives the following approximate relationships for the renormalized third virial coefficients

$$\bar{C}(T) = C^a(T) - 3(B^a)^2 \, \ell n \, \psi^a \, e^{j\theta_f} \tag{50}$$

$$C_R(T) = C^a(T) - 3(B^a)^2 \, \ell n \, \psi^a \cos \theta_f \tag{51}$$

$$C_I(T) = - 3(B^a)^2 \, \ell n \, \psi^a \sin \theta_f \tag{52}$$

It is the real value of the third virial coefficient C_R that is measured from experimental pressure versus volume curves at constant temperature.

It remains only to obtain θ_f from a solution of equation (45) which can be rewritten as

$$d\theta_f/\theta_f = G/f_o \, dT \tag{53}$$

$$\ell n|\theta_f| = \int G/f_o \, dT \tag{54}$$

Combining equations (24), (32) and (46) gives

$$G = - R(B^a)^2 \left(3 + 2 \frac{T}{B^a} \frac{dB^a}{dT}\right) \tag{55}$$

$$G/f_o = - \left(\frac{1}{T} + \frac{2}{3B^a} \frac{dB^a}{dT}\right) / \, \ell n \, \psi^a \tag{56}$$

But from equation (12) it follows that

$$\frac{d\,\ell n\,\psi^a}{dT} = \frac{1}{T} + \frac{2}{3B^a}\frac{dB^a}{dT} \tag{57}$$

Combining equations (54), (56) and (57) gives

$$\ell n|\theta_f| = -\int \frac{1}{\ell n\,\psi^a}\frac{d\,\ell n\,\psi^a}{dT}\,dT = -\ell n(\ell n\,\psi^a) + \ell n\,b \tag{58}$$

and therefore

$$\theta_f = \pm b/\ell n\,\psi^a \tag{59}$$

where b = constant whose value can only be determined from the full solution of equations (37) and (38). The solution in equation (59) is valid only when θ_f is a small number, and in general this limits the application of equation (59) to the regions of high temperature. Note that equations (37) and (38) are unchanged for $\theta_f \rightarrow -\theta_f$, so that either $\pm\theta_f$ are valid solutions, and therefore these equations exhibit degeneracy.

The relativistic third virial coefficient can be obtained from equations (50) through (52) as

$$\bar{C} = Ce^{j\theta_C} = C_R + jC_I \tag{60}$$

$$C_R = C\cos\theta_C \tag{61}$$

$$C_I = C\sin\theta_C \tag{62}$$

and therefore

$$\tan\theta_C = C_I/C_R \tag{63}$$

$$\sin\theta_C = C_I/C \tag{64}$$

$$\cos\theta_C = C_R/C \tag{65}$$

where C_R and C_I are given by equations (51) and (52) and where the magnitude of the third virial coefficient is given by

$$C^2 = C_R^2 + C_I^2 \tag{66}$$

$$= [C^a - 3(B^a)^2 \, \ell n \, \psi^a]^2 + 6C^a(B^a)^2 \, \ell n \, \psi^a \, (1 - \cos \theta_f)$$

when a single phase θ_f of the gas is present. At high temperatures $C_R < 0$, and either $C_I < 0$ for positive θ_f, or $C_I > 0$ for $\theta_f < 0$. For $C_I < 0$ it follows from equations (63) through (65) that $\theta_C = \pi + \theta_C'$ and θ_C is in the third quadrant, while for $C_I > 0$ it follows that $\theta_C = \pi - \theta_C'$ and θ_C is in the second quadrant. Therefore equations (37) and (38) have degenerate solutions, and the real gases can appear in two states corresponding to $\pm\theta_f$. If a real gas is in a state which is a mixture of fraction α with $\theta_f > 0$ and fraction $(1 - \alpha)$ with $\theta_f < 0$ it follows

$$C_I(T) = -3(2\alpha - 1)\,(B^a)^2 \, \ell n \, \psi^a \sin |\theta_f| \tag{52A}$$

Therefore for equal mixtures of both phases $C_I = 0$ and $\theta_C = \pi$. The function $C_R(T)$ is determined in static pressure versus volume measurements at constant temperature, and is not affected by the sign of the phase function θ_f. Finally it should be noted that the fourth virial coefficient can be written as

$$\bar{D} = De^{j\theta_D} \tag{67}$$

but no calculations to determine D and θ_D have been done.

3. <u>INTERNAL PHASE ANGLES OF THERMODYNAMIC FUNCTIONS.</u> This section considers the calculation of the pressure, internal energy, entropy, enthalpy, and free energy of real gases with broken internal symmetry. The relativistic pressure and internal energy density for a broken symmetry real gas are given in equations (13) and (14). The corresponding expressions for the entropy density, enthalpy density and free energy density are given for asymmetric real gases as a generalization of the standard results in the literature.[8] The basic thermodynamic functions for a broken symmetry real gas are then given as follows

A. <u>Pressure.</u>

The pressure is written as

$$\bar{P} = Pe^{j\theta_P} = nRT(1 + nB + n^2\bar{C} + n^3\bar{D} + \cdots) \tag{68}$$

or in component form

$$P_R = nRT(1 + nB + n^2C\cos\theta_C + n^3D\cos\theta_D + \cdots) \tag{69}$$

$$P_I = nRT(n^2C\sin\theta_C + n^3D\sin\theta_D + \cdots) \tag{70}$$

$$\tan\theta_P = \frac{n^2(C\sin\theta_C + nD\sin\theta_D + \cdots)}{(1 + nB + n^2C\cos\theta_C + n^3D\cos\theta_D + \cdots)} \tag{71}$$

$$\sim n^2C\sin\theta_C \qquad \text{low density} \tag{72}$$

The magnitude of the pressure is given by

$$P = (P_R^2 + P_I^2)^{1/2} \tag{73}$$

It is the real part of the pressure P_R that is measured in laboratory experiments. For $\theta_f > 0$ and $\theta_C = \pi + \theta_C'$ it follows from equations (69) through (72) that $\theta_P < 0$ in the regions of high temperature, but for the case $\theta_f < 0$ and $\theta_C = \pi - \theta_C'$ it follows that $\theta_P > 0$. For an equal mixture of states with $\theta_f > 0$ and $\theta_f < 0$ it follows that $\theta_C = \pi$ and $\theta_P = 0$ neglecting the effects of the fourth and higher virial coefficients.

B. Internal Energy Density.

The energy density for an asymmetric real gas is written as

$$\bar{E} = Ee^{j\theta_E} = nRT\left(\frac{3}{2} - nT\frac{\partial B}{\partial T} - \frac{1}{2}n^2T\frac{\partial \bar{C}}{\partial T} - \frac{1}{3}n^3T\frac{\partial \bar{D}}{\partial T} - \cdots\right)$$

Using equations (60) and (67) in equation (74) gives

$$E_R = nRT\left[\frac{3}{2} - nT\frac{\partial B}{\partial T} - \frac{1}{2}n^2I_C\cos(\theta_C + \beta_{C,T}) - \frac{1}{3}n^3I_D\cos(\theta_D + \beta_{D,T}) - \cdots\right] \tag{75}$$

$$E_I = -n^3RT\left[\frac{1}{2}I_C\sin(\theta_C + \beta_{C,T}) + \frac{1}{3}nI_D\sin(\theta_D + \beta_{D,T}) + \cdots\right] \tag{76}$$

$$\tan\theta_E = E_I/E_R \tag{77}$$

$$\sim -\frac{1}{3}n^2I_C\sin(\theta_C + \beta_{C,T}) \qquad \text{low density} \tag{78}$$

where

$$I_C = \sqrt{(T\,\partial C/\partial T)^2 + (CT\,\partial\theta_C/\partial T)^2} \tag{79}$$

$$I_D = \sqrt{(T\,\partial D/\partial T)^2 + (DT\,\partial\theta_D/\partial T)^2} \tag{80}$$

$$\tan\beta_{C,T} = C\,\frac{\partial\theta_C/\partial T}{\partial C/\partial T} \tag{81}$$

$$\tan\beta_{D,T} = D\,\frac{\partial\theta_D/\partial T}{\partial D/\partial T} \tag{82}$$

The magnitude of the internal energy density is given by

$$E = \sqrt{E_R^2 + E_I^2} \tag{83}$$

It is E_R that is measured in the laboratory

C. Entropy.

The entropy density for an asymmetric real gas is written as

$$\bar{s} = se^{j\theta_s} = -\,nR[\ell n(nRT) + n(B + T\,\frac{\partial B}{\partial T}) + \frac{1}{2}\,n^2(\bar{C} + T\,\frac{\partial\bar{C}}{\partial T}) + \frac{1}{3}\,n^3(\bar{D} + T\,\frac{\partial\bar{D}}{\partial T}) + \cdots\,] \tag{84}$$

Using equations (60) and (67) in equation (84) gives

$$s_R = -\,nR[\ell n(nRT) + n(B + T\,\frac{\partial B}{\partial T}) + \frac{1}{2}\,n^2J_C\cos(\theta_C + \alpha_{C,T}) + \frac{1}{3}\,n^3J_D\cos(\theta_D + \alpha_{D,T}) + \cdots\,] \tag{85A}$$

$$s_I = -\,n^3R[\frac{1}{2}\,J_C\sin(\theta_C + \alpha_{C,T}) + \frac{1}{3}\,nJ_D\sin(\theta_D + \alpha_{D,T}) + \cdots\,] \tag{85B}$$

$$\tan\theta_s = s_I/s_R \tag{86}$$

$$\sim \frac{1}{2}\,n^2J_C\,\frac{\sin(\theta_C + \alpha_{C,T})}{\ell n(nRT)} \qquad \text{low density} \tag{87}$$

where nRT > 1 and where

$$J_C = \sqrt{(C + T\,\partial C/\partial T)^2 + (CT\,\partial\theta_C/\theta T)^2} \tag{88}$$

$$J_D = \sqrt{(D + T\,\partial D/\partial T)^2 + (DT\,\partial\theta_D/\partial T)^2} \tag{89}$$

$$\tan\alpha_{C,T} = \frac{CT\,\partial\theta_C/\partial T}{C + T\,\partial C/\partial T} \tag{90}$$

$$\tan\alpha_{D,T} = \frac{DT\,\partial\theta_D/\partial T}{D + T\,\partial D/\partial T} \tag{91}$$

The magnitude of the entropy per unit volume is given by

$$s = \sqrt{s_R^2 + s_I^2} \tag{92}$$

The real quantity is the entropy density measured in the laboratory.

D. Enthalpy.

The enthalpy density of an asymmetric real gas is given by

$$\bar{h} = he^{j\theta_h} = nRT[\frac{5}{2} + n(B - T\frac{\partial B}{\partial T}) + \frac{1}{2}n^2(2\bar{C} - T\frac{\partial\bar{C}}{\partial T}) + \frac{1}{3}n^3(3\bar{D} - T\frac{\partial\bar{D}}{\partial T}) + \cdots] \tag{93}$$

Placing equations (60) and (67) into equation (93) gives

$$h_R = nRT[\frac{5}{2} + n(B - T\frac{\partial B}{\partial T}) + \frac{1}{2}n^2K_C\cos(\theta_C - \eta_C) + \frac{1}{3}n^3K_D\cos(\theta_D - \eta_D) + \cdots] \tag{94}$$

$$h_I = n^3RT[\frac{1}{2}K_C\sin(\theta_C - \eta_C) + \frac{1}{3}nK_D\sin(\theta_D - \eta_D) + \cdots] \tag{95}$$

$$\tan\theta_h = h_I/h_R \tag{96}$$

$$\sim \frac{1}{5}n^2K_C\sin(\theta_C - \eta_C) \qquad \text{low density} \tag{97}$$

where

$$K_C = \sqrt{(2C - T\,\partial C/\partial T)^2 + (CT\,\partial\theta_C/\partial T)^2} \tag{98}$$

$$K_D = \sqrt{(3D - T\,\partial D/\partial T)^2 + (DT\,\partial\theta_D/\partial T)^2} \tag{99}$$

$$\tan\eta_C = \frac{CT\,\partial\theta_C/\partial T}{2C - T\,\partial C/\partial T} \tag{100}$$

$$\tan\eta_D = \frac{DT\,\partial\theta_D/\partial T}{3D - T\,\partial D/\partial T} \tag{101}$$

The magnitude of the enthalpy density is given by

$$h = \sqrt{h_R^2 + h_I^2} \tag{102}$$

The value of h_R is obtained from laboratory measurements.

E. Free Energy Density.

The complex number free energy density for an asymmetric real gas is given by

$$\bar{a} = ae^{j\theta_a} = nRT[\frac{3}{2} + \ell n(nRT) + nB + \frac{1}{2}n^2\bar{C} + \frac{1}{3}n^3\bar{D} + \cdots] \tag{103}$$

The real and imaginary parts of equation (103) are

$$a_R = nRT[\frac{3}{2} + \ell n(nRT) + nB + \frac{1}{2}n^2C\cos\theta_C + \frac{1}{3}n^3D\cos\theta_D + \cdots] \tag{104}$$

$$a_I = n^3RT(\frac{1}{2}C\sin\theta_C + \frac{1}{3}nD\sin\theta_D + \cdots) \tag{105}$$

$$\tan\theta_a = a_I/a_R \tag{105A}$$

$$\sim \frac{1}{2}n^2C\sin\theta_C/[\frac{3}{2} + \ell n(nRT)] \quad \text{low density} \tag{106}$$

The real part of the free energy density is a physically measurable quantity.

4. HEAT CAPACITY AND GRÜNEISEN PARAMETER. The calculation of the heat capacity and Grüneisen parameter for a real gas with broken internal symmetries

is performed in this section. The complex number molar heat capacity is obtained from equation (79) to be[8]

$$\bar{C}_V = C_V e^{j\theta_{CV}} = C_{VR} + jC_{VI} = R\left(\frac{3}{2} - nC_{V1} - \frac{1}{2}n^2\bar{C}_{V2} - \cdots\right) \quad (108)$$

where

$$C_{V1} = T^2\frac{\partial^2 B}{\partial T^2} + 2T\frac{\partial B}{\partial T} \quad (109)$$

$$\bar{C}_{V2} = T^2\frac{\partial^2 \bar{C}}{\partial T^2} + 2T\frac{\partial \bar{C}}{\partial T} = C_{V2R} + jC_{V2I} \quad (110)$$

Substituting equation (60) into equation (110) gives

$$\bar{C}_{V2} = I_{2C}e^{j\theta_{2C}} + 2I_C e^{j\theta_{1C}} \quad (111)$$

$$C_{V2R} = I_{2C}\cos\theta_{2C} + 2I_C\cos\theta_{1C} \quad (112)$$

$$C_{V2I} = I_{2C}\sin\theta_{2C} + 2I_C\sin\theta_{1C} \quad (113)$$

where I_C is given by equation (79) and

$$I_{2C} = \sqrt{L_{2C}^2 + M_{2C}^2} \quad (114)$$

$$L_{2C} = T^2\frac{\partial^2 C}{\partial T^2} - C\left(T\frac{\partial \theta_C}{\partial T}\right)^2 \quad (114A)$$

$$M_{2C} = CT^2\frac{\partial^2 \theta_C}{\partial T^2} + 2\left(T\frac{\partial C}{\partial T}\right)\left(T\frac{\partial \theta_C}{\partial T}\right) \quad (114B)$$

$$\theta_{2C} = \theta_C + \gamma_{C,T} \quad (115)$$

$$\tan\gamma_{C,T} = M_{2C}/L_{2C} \quad (116)$$

$$\theta_{1C} = \theta_C + \beta_{C,T} \quad (117)$$

where $\beta_{C,T}$ is given by equation (81). Then it follows from equation (108) that

$$C_{VR} = R(\frac{3}{2} - nC_{V1} - \frac{1}{2} n^2 C_{V2R} - \cdots) \quad (118)$$

$$C_{VI} = - \frac{1}{2} n^2 R C_{V2I} \quad (119)$$

$$\tan \theta_{CV} = C_{VI}/C_{VR} \quad (120)$$

$$\sim - \frac{1}{3} n^2 (I_{2C} \sin \theta_{2C} + 2I_C \sin \theta_{1C}) \quad (121)$$

where equation (121) is valid for low densities.

The Grüneisen parameter for an asymmetric real gas can be written as[9]

$$\bar{\gamma} = \gamma e^{j\theta_\gamma} = \gamma_R + j\gamma_I = \frac{2}{3}(1 + n\gamma_1 + n^2\bar{\gamma}_2 + \cdots) \quad (122)$$

where

$$\gamma_1 = f_1 + g_1 \quad (123)$$

$$\bar{\gamma}_2 = \bar{f}_2 + \bar{g}_2 + f_1 g_1 \quad (124)$$

$$f_1 = T \frac{\partial B}{\partial T} + B \quad (125)$$

$$g_1 = \frac{2}{3}(T^2 \frac{\partial^2 B}{\partial T^2} + 2T \frac{\partial B}{\partial T}) \quad (126)$$

$$\bar{f}_2 = T \frac{\partial \bar{C}}{\partial T} + \bar{C}$$

$$\bar{g}_2 = \frac{1}{3}(T^2 \frac{\partial^2 \bar{C}}{\partial T^2} + 2T \frac{\partial \bar{C}}{\partial T}) + \frac{4}{9}(T^2 \frac{\partial^2 B}{\partial T^2} + 2T \frac{\partial B}{\partial T})^2 \quad (128)$$

From equation (122) it follows that

$$\gamma_R = \frac{2}{3}(1 + n\gamma_1 + n^2\gamma_{2R} + \cdots) \quad (129)$$

$$\gamma_I = \frac{2}{3} n^2 \gamma_{2I} + \cdots \quad (130)$$

where

$$\gamma_{2R} = f_{2R} + g_{2R} + f_1 g_1 \tag{131}$$

$$\gamma_{2I} = f_{2I} + g_{2I} \tag{132}$$

$$f_{2R} = J_C \cos(\theta_C + \alpha_{C,T}) \tag{133}$$

$$f_{2I} = J_C \sin(\theta_C + \alpha_{C,T}) \tag{134}$$

$$g_{2R} = \frac{1}{3}(I_{2C} \cos \theta_{2C} + 2I_C \cos \theta_{1C}) + \frac{4}{9}\left(T^2 \frac{\partial^2 B}{\partial T^2} + 2T \frac{\partial B}{\partial T}\right)^2 \tag{135}$$

$$g_{2I} = \frac{1}{3}(I_{2C} \sin \theta_{2C} + 2I_C \sin \theta_{1C}) \tag{136}$$

where J_C and $\alpha_{C,T}$ are given by equations (88) and (90) respectively, I_C and θ_{1C} are given by equations (79) and (117) respectively, and where I_{2C} and θ_{2C} are given by equations (114) and (115) respectively.

5. SUPERPOSITION OF THERMODYNAMIC FUNCTIONS. Consider a mixture of two interacting gases with broken internal symmetries. There will be interactions between the two species of gas as well as self interactions within each species. Therefore the total pressure and internal energy is written as

$$\bar{P} = \bar{P}_1 + \bar{P}_2 + \bar{P}_{12} \tag{137}$$

$$\bar{U} = \bar{U}_1 + \bar{U}_2 + \bar{U}_{12} \tag{138}$$

where $\bar{P}_{12}$ and $\bar{U}_{12}$ are the interspecies interaction pressure and internal energy respectively. For asymmetric real gases the terms in equations (137) and (138) are written as

$$\bar{P} = Pe^{j\theta_P} \qquad \bar{P}_1 = P_1 e^{j\theta_{P1}} \tag{139}$$

$$\bar{P}_2 = P_2 e^{j\theta_{P2}} \qquad \bar{P}_{12} = P_{12} e^{j\theta_{P12}} \tag{140}$$

$$\bar{U} = Ue^{j\theta_U} \qquad \bar{U}_1 = U_1 e^{j\theta_{U1}} \tag{141}$$

$$\bar{U}_2 = U_2 e^{j\theta_{U2}} \qquad \bar{U}_{12} = U_{12} e^{j\theta_{U12}} \tag{142}$$

Equations (137) and (138) can be written in component form as follows

$$P \cos \theta_P = P_1 \cos \theta_{P1} + P_2 \cos \theta_{P2} + P_{12} \cos \theta_{P12} \tag{143}$$

$$P \sin \theta_P = P_1 \sin \theta_{P1} + P_2 \sin \theta_{P2} + P_{12} \sin \theta_{P12} \tag{144}$$

$$U \cos \theta_U = U_1 \cos \theta_{U1} + U_2 \cos \theta_{U2} + U_{12} \cos \theta_{U12} \tag{145}$$

$$U \sin \theta_U = U_1 \sin \theta_{U1} + U_2 \sin \theta_{U2} + u_{12} \sin \theta_{U12} \tag{146}$$

From equations (143) through (146) it follows that

$$\tan \theta_P = \frac{P_1 \sin \theta_{P1} + P_2 \sin \theta_{P2} + P_{12} \sin \theta_{P12}}{P_1 \cos \theta_{P1} + P_2 \cos \theta_{P2} + P_{12} \cos \theta_{P12}} \tag{147}$$

$$\tan \theta_U = \frac{U_1 \sin \theta_{U1} + U_2 \sin \theta_{U2} + U_{12} \sin \theta_{U12}}{U_1 \cos \theta_{U1} + U_2 \cos \theta_{U2} + U_{12} \cos \theta_{U12}} \tag{148}$$

$$P = (\psi_D + \psi_I)^{1/2} \tag{149}$$

$$U = (\phi_D + \phi_I)^{1/2} \tag{150}$$

where

$$\psi_D = P_1^2 + P_2^2 + P_{12}^2 \tag{151}$$

$$\psi_I = 2P_1P_2 \cos (\theta_{P1} - \theta_{P2}) + 2P_1P_{12} \cos (\theta_{P1} - \theta_{P12}) + 2P_2P_{12} \cos (\theta_{P2} - \theta_{P12}) \tag{152}$$

$$\phi_D = U_1^2 + U_2^2 + U_{12}^2 \tag{153}$$

$$\phi_I = 2U_1U_2 \cos (\theta_{U1} - \theta_{U2}) + 2U_1U_{12} \cos (\theta_{U1} - \theta_{U12}) + 2U_2U_{12} \cos (\theta_{U2} - \theta_{U12}) \tag{154}$$

Therefore mixtures of two asymmetric real gases should exhibit interference in regard to the component pressures and internal energies, and the magnitude of the total pressure and internal energy should exhibit a small oscillation as the density of the interacting mixture is increased. This is also true of the measured pressure and internal energy which can respectively be written as $P \cos \theta_P$ and $U \cos \theta_U$.

6. CONCLUSIONS. Real gases are expected to exhibit broken internal symmetries that manifest themselves as internal phase angles associated with the thermodynamic functions such as pressure, internal energy and entropy. These internal phase angles arise from the third and higher virial coefficients and are due to renormalization effects associated with the interaction of matter with spacetime as described by equation (2). The ideal gas and the second virial coefficient of an interacting gas are unaffected by spacetime interactions. The phase angle associated with the third virial coefficient can be determined from the solution of two simultaneous first order differential equations. These equations are generally not easy to solve analytically, but yield a simple solution for the high temperature regions of the real gas where the phase angle of the third virial coefficient is small. The virial form of the state equation for real gases allows the phase angles associated with the pressure, internal energy, entropy, enthalpy, and free energy to be calculated in terms of density and temperature. The existence of internal phase angles for the thermodynamic state functions suggests that mixtures of real gases will produce an interference phenomenon wherein the total pressure and internal energy will oscillate slightly as the density of the system is increased.

It can also be conjectured that parabolic waves of the form[15,16]

$$\frac{\partial \theta_U}{\partial \bar{t}} = f(\theta_U) + D_U \frac{\partial^2 \theta_U}{\partial \bar{x}^2} \tag{155}$$

$$\frac{\partial \theta_P}{\partial \bar{t}} = f(\theta_P) + D_P \frac{\partial^2 \theta_P}{\partial \bar{x}^2} \tag{156}$$

can exist in asymmetric gases and liquids as well as in asymmetric solids and quantum liquids, and these wave motions may have interesting applications to thermodynamics, hydrodynamics and chemical and biological cycles. The internal phase angles of the pressure and other thermodynamic functions of the real gases are also expected to play an important role in the determination of the equilibrium configuration of stars and planets. This is true because the internal phase angles of the radial coordinates in a star are determined by the internal phase angle of the pressure. Therefore the complex number values of the third and higher virial coefficients are intimately involved in stellar equilibrium calculations.

ACKNOWLEDGEMENT

The author wishes to thank Elizabeth K. Klein for typing this paper.

REFERENCES

1. Aitchison, I. J. R. and Hey, A. J. G., Gauge Theories in Particle Physics, Adam Hilger Ltd., Bristol, United Kingdom, 1982.

2. O'Raifeartaigh, L., Group Structure of Gauge Theories, Cambridge University Press, London, 1986.

3. Bailin, D., Weak Interactions, Adam Hilger, Bristol, 1982.

4. Grassie, A. D. C., The Superconducting State, Sussex University Press, London, 1975.

5. Weiss, R. A., "Thermodynamic Gauge Theory of Solids and Quantum Liquids with Internal Phase", Fifth Army Conference on Applied Mathematics and Computing, West Point, New York, ARO 88-1, June 15-18, 1987, p. 649.

6. Weiss, R. A., Relativistic Thermodynamics, Exposition Press, New York, 1976.

7. Hirschfelder, J. O., Curtiss, C. F. and Bird, R. B., Molecular Theory of Gases and Liquids, John Wiley, New York, 1954.

8. Beattie, J. A., "Thermodynamic Properties of Real Gases and Mixtures of Real Gases", article in Thermodynamics and Physics of Matter, edited by F. D. Rossini, Princeton University Press, 1955, p. 240.

9. Weiss, R. A., "Relativistic Wave Equations for Real Gases", Fourth Army Conference on Applied Mathematics and Computing, Cornell University, ARO 87-1, May 27-30, 1986, p. 341.

10. Sears, F. W., Thermodynamics, the Kinetic Theory of Gases, and Statistical Mechanics, Addison-Wesley, Reading, MA, 1953.

11. Planck, M., Theory of Heat, MacMillan, New York, 1949.

12. Pauli, W., Pauli Lectures on Physics: Volume 3. Thermodynamics and the Kinetic Theory of Gases, MIT Press, Cambridge, MA, 1973.

13. Sommerfeld, A., Thermodynamics and Statistical Mechanics, Academic Press, New York, 1955.

14. Michelson, P. F., Price, J. C. and Taber, R. C., "Resonant-Mass Detectors of Gravitational Radiation", Science, Vol. 237, 10 July 1987, p. 150.

15. Winfree, A. T., The Geometry of Biological Time, Springer-Verlag, New York, 1980.

16. Winfree, A. T., When Time Breaks Down, Princeton University Press, 1987.

G

MAXWELL'S EQUATIONS WITH BROKEN INTERNAL SYMMETRIES

Richard A. Weiss
U. S. Army Engineer Waterways Experiment Station
Vicksburg, Mississippi 39180

ABSTRACT. On account of the broken symmetries of the thermodynamic ground state and excited states of bulk matter and the vacuum, the electric and magnetic fields in bulk matter and the vacuum exhibit broken internal symmetries. Maxwell's equations are formulated for an electromagnetic field with broken internal symmetry. Lorentz covariance is expressed in terms of space and time coordinates that have broken symmetries represented by internal phase angles. Special relativity mechanics in bulk matter and the vacuum with broken symmetries is formulated for particles whose kinematic and dynamic variables exhibit internal phases. Electromagnetic wave equations for broken symmetry matter and vacuum are developed and the gauge conditions for the electromagnetic potential are developed. The vacuum state is shown to have properties that are essentially similar to those of a bulk matter system, and in particular both exhibit broken internal symmetry. The description of electromagnetic effects in matter and the vacuum must properly account for the broken symmetry of the fields and space and time coordinates, and the internal phase angles of the electromagnetic field vectors must be determined jointly with the internal phase angles of the space and time coordinates. A better knowledge of electromagnetic interactions in bulk matter will be useful for understanding electromagnetic wave propagation in the atmosphere and for comprehending the complex processes that occur when high energy microwave beams interact with matter.

1. INTRODUCTION. Electrodynamics is a theory that is based on the Lorentz covariant set of Maxwell's equations and on the symmetry of the gauge group U(1).[1-3] This theory has charges and currents as the sources of the electromagnetic field. Maxwell's equations are a set of partial differential equations that determine the space and time variation of the electric and magnetic fields that are associated with the distribution of charges and currents. Classically, the charges and currents are situated in a passive space and time background (the vacuum) which is assumed to be inert and plays no active part in the determination of the fields. In quantum electrodynamics, the vacuum is taken to be a polarizable medium which can affect the energy levels of charged particle configurations. The active vacuum is one of the great discoveries of twentieth century physics, and has been experimentally verified in a number of ways including a measurement of the Lamb shift of energy levels.[4-6]

In this paper an additional vacuum effect on the electromagnetic field is suggested to manifest itself through the fact that space and time coordinates within asymmetric bulk matter or vacuum acquire internal phase angles (broken symmetries). The electric and magnetic field vectors also acquire broken symmetries. The internal phase angles of the space and time coordinates and of the electromagnetic field vectors are due to the interaction of Minkowski spacetime with bulk matter, the electromagnetic field, and the vacuum. The internal phase angles of the electromagnetic field vectors must be determined jointly with the internal phase angles of the space and time coordinates, and it is the joint solution of Maxwell's equations and the equations of motion of charged particles for a system with broken internal symmetries that accomplishes this task.

The relativistic values of the electromagnetic field vectors in asymmetric bulk matter or vacuum must satisfy the relativistic trace equation for radiation.[7-8] This radiation trace equation relates the renormalized radiation pressure to the corresponding nonrelativistic radiation pressure. The trace equation for radiation is derived from the relativistic trace equation for the ground state of bulk matter which is written as[7,8]

$$\bar{U} + T\left(\frac{d\bar{U}}{dT}\right)_{\bar{P}V} - 3V\frac{d}{dV}(\bar{P}V)_{\bar{U}} = U^a + T\left(\frac{dU^a}{dT}\right)_{p^aV} \tag{1}$$

or equivalently as

$$(1 - \bar{b} + T\frac{\partial}{\partial T} - \bar{b}V\frac{\partial}{\partial V})\bar{E} - 3(1 + \bar{\gamma} + V\frac{\partial}{\partial V} - \bar{\gamma}T\frac{\partial}{\partial T})\bar{P} = \psi^a \tag{2}$$

where

$$\psi^a = (T\frac{\partial}{\partial T} - b^aV\frac{\partial}{\partial V} + 1 - b^a)E^a \tag{3}$$

and where $\bar{U}$, $\bar{E}$, $\bar{P}$, $\bar{\gamma}$, and $\bar{b}$ are complex number representations of the internal energy, energy density, pressure, and the gauge parameters, T = absolute temperature, and V = volume of specified number of particles. The complex number Grüneisen parameter is defined as

$$\bar{\gamma} = \frac{V}{\bar{C}_V}\frac{\partial\bar{P}}{\partial T} = \frac{\partial\bar{P}/\partial T}{\partial\bar{E}/\partial T} = \gamma e^{\theta_\gamma} \tag{4}$$

where γ and θ_γ = magnitude and phase of the Grüneisen parameter respectively. The corresponding equation for radiation in matter with internal phases is derived from equation (2) to be[8]

$$(1 - \bar{b} + T\frac{\partial}{\partial T} - \bar{b}V\frac{\partial}{\partial V})\bar{E}_r - \bar{\beta}_r(T\frac{\partial\bar{P}}{\partial T} - \bar{P}) \tag{5}$$

$$- 3[(1 + \bar{\gamma} + V\frac{\partial}{\partial V} - \bar{\gamma}T\frac{\partial}{\partial T})\bar{P}_r - \bar{\delta}_r(T\frac{\partial\bar{P}}{\partial T} - \bar{P})] = \psi_r^a$$

where

$$\psi_r^a = (T\frac{\partial}{\partial T} - b^a V\frac{\partial}{\partial V} + 1 - b^a)E_r^a - \beta_r^a(T\frac{\partial P^a}{\partial T} - P^a) \tag{6}$$

and where $\bar{E}_r$, $\bar{P}_r$, $\bar{\beta}_r$, and $\bar{\delta}_r$ are the complex number radiation energy density, radiation pressure, and two radiation gauge functions respectively.[8] The radiation Grüneisen parameter is defined by

$$\gamma_r = \frac{\partial P_r}{\partial T} / \frac{\partial E_r}{\partial T} \tag{6A}$$

Throughout this paper the index "a" will refer to nonrelativistic (unrenormalized) calculations. Equation (1) with its right hand side set equal to zero represents the asymmetric ground state of the vacuum, while equation (5) with its right hand side equal to zero represents the excited (radiation) states of the asymmetric vacuum.

The relativistic trace equations for the ground and excited states of bulk matter and the vacuum imply that the ground state and excited state pressure fields have broken symmetries.[8] In turn, this implies that all of the descriptive variables of particles and fields located in asymmetric bulk matter or vacuum also exhibit broken symmetries. Therefore the space and time coordinates as well as the electric and magnetic field vectors will exhibit broken symmetries as manifested by internal phase angles. The space and time coordinates are written as

$$\bar{x} = xe^{j\theta_x} \tag{7}$$

$$\bar{y} = ye^{j\theta_y} \tag{8}$$

$$\bar{z} = ze^{j\theta_z} \tag{9}$$

$$\bar{t} = te^{j\theta_t} \tag{10}$$

and the derivatives with respect to the space and time coordinates are written as

$$\partial/\partial\bar{x} = e^{-j\theta dx} \cos \beta_{x,x}\ \partial/\partial x \tag{11}$$

$$\partial/\partial\bar{t} = e^{-j\theta dt} \cos \beta_{t,t}\ \partial/\partial t \tag{12}$$

where

$$\tan \beta_{x,x} = x\,\partial\theta_x/\partial x \tag{13}$$

$$\tan \beta_{t,t} = t\,\partial\theta_t/\partial t \tag{14}$$

$$\cos \beta_{x,x} = [1 + (x\, \partial\theta_x/\partial x)^2]^{-1/2} \tag{15}$$

$$\cos \beta_{t,t} = [1 + (t\, \partial\theta_t/\partial t)^2]^{-1/2} \tag{16}$$

$$\theta_{dx} = \theta_x + \beta_{x,x} \tag{17}$$

$$\theta_{dt} = \theta_t + \beta_{t,t} \tag{17A}$$

In a condensed notation the coordinates and derivatives are written with $\eta = t, x, y, z$ as follows

$$\bar{\eta} = \eta e^{j\theta_\eta} \tag{18}$$

$$\partial/\partial\bar{\eta} = e^{-j\theta_{d\eta}} \cos \beta_{\eta,\eta}\, \partial/\partial\eta \tag{19}$$

$$\theta_{d\eta} = \theta_\eta + \beta_{\eta,\eta} \tag{20}$$

$$\tan \beta_{\eta,\eta} = \eta\, \partial\theta_\eta/\partial\eta \tag{21}$$

$$\cos \beta_{\eta,\eta} = [1 + (\eta\, \partial\theta_\eta/\partial\eta)^2]^{-1/2} \tag{22}$$

Note that it is the real parts of the complex number quantities such as space and time coordinates, electric and magnetic field vectors, pressure and energy that are the measured quantities.

This paper develops Maxwell's equations for electromagnetic fields that have broken internal symmetries and for space and time coordinates that also have broken internal symmetries. Section 2 considers the fields and coordinates with broken symmetry, while Section 3 develops Maxwell's equations and the equations of motion of charged particles in an electromagnetic field in a broken symmetry vacuum or bulk matter system. Section 4 develops the consequences of assuming the validity of Lorentz covariance for coordinate systems with broken internal symmetry. In Section 5 the electromagnetic wave equations and their gauge conditions are written for systems with internal phase angles. Finally, Section 6 develops the equations of the relativistic vacuum from the corresponding bulk matter equations, and a broken symmetry condition for the vacuum state is suggested. Therefore all of the conclusions for the bulk matter state with broken internal symmetries are also valid for the broken symmetry vacuum state. The conventional coordinates t_a , x_a , y_a , z_a are related to the gauge rotated coordinates t , x , y and z by $t = t_a$, $x = x_a$, $y = y_a$ and $z = z_a$.

2. THE BROKEN SYMMETRY OF ELECTROMAGNETIC FIELDS. For electromagnetic waves within asymmetric bulk matter or vacuum, the electric and magnetic fields

are expected to acquire internal phase angles. This is due to the fact that the spacetime coordinates and the kinematical and dynamical variables of particles in asymmetric bulk matter or vacuum exhibit broken internal symmetries. In particular the particle velocity, and therefore the electric current for charged particles, has an internal phase angle. Therefore the cartesian components of the electric and magnetic field vectors in asymmetric bulk matter or vacuum can be written as

$$\bar{E}_\alpha = E_\alpha e^{j\theta_{E\alpha}} = E_{\alpha R} + jE_{\alpha I} \tag{23}$$

$$\bar{D}_\alpha = D_\alpha e^{j\theta_{D\alpha}} = D_{\alpha R} + jD_{\alpha I} \tag{24}$$

$$\bar{H}_\alpha = H_\alpha e^{j\theta_{H\alpha}} = H_{\alpha R} + jH_{\alpha I} \tag{25}$$

$$\bar{B}_\alpha = B_\alpha e^{j\theta_{B\alpha}} = B_{\alpha R} + jB_{\alpha I} \tag{26}$$

where α = x, y, and z. The phase angles $\theta_{E\alpha}$, $\theta_{D\alpha}$, $\theta_{H\alpha}$, and $\theta_{B\alpha}$ are in general functions of space and time of the general form

$$\theta_{E\alpha} = \theta_{E\alpha}(x,y,z,t,\theta_x,\theta_y,\theta_z,\theta_t) \tag{27}$$

The field vector amplitudes are also functions of space and time, as for example

$$E_\alpha = E_\alpha(x,y,z,t,\theta_x,\theta_y,\theta_z,\theta_t) \tag{28}$$

The imaginary number j will be used to refer to internal phase angles that are associated with broken symmetries, while the imaginary number i will refer to external phase angles. For plane waves the magnitudes of the field vectors in equations (23) through (26) may be written as

$$E_\alpha = A_{E\alpha} e^{i\xi} \tag{29}$$

$$D_\alpha = A_{D\alpha} e^{i\xi} \tag{30}$$

$$H_\alpha = A_{H\alpha} e^{i\xi} \tag{31}$$

$$B_\alpha = A_{B\alpha} e^{i\xi} \tag{32}$$

where

$$\xi = k_x x + k_y y + k_z z - \omega t \tag{32A}$$

where $A_{E\alpha}$, $A_{D\alpha}$, $A_{H\alpha}$, and $A_{B\alpha}$ are constants; k_x , k_y , k_z = wavenumber components, and ω = magnitude of the frequency.

In general the internal phase angles $\theta_{E\alpha}$, $\theta_{D\alpha}$, $\theta_{H\alpha}$, and $\theta_{B\alpha}$ are functions of space and time, and it follows from equations (18) through (22) and (23) through (26) that

$$\partial\bar{E}_\alpha/\partial\bar{\eta} = e^{j\Phi_{E\alpha,\eta}} \cos \beta_{\eta,\eta}\, W_{E\alpha,\eta} \tag{33}$$

$$\partial\bar{D}_\alpha/\partial\bar{\eta} = e^{j\Phi_{D\alpha,\eta}} \cos \beta_{\eta,\eta}\, W_{D\alpha,\eta} \tag{34}$$

$$\partial\bar{H}_\alpha/\partial\bar{\eta} = e^{j\Phi_{H\alpha,\eta}} \cos \beta_{\eta,\eta}\, W_{H\alpha,\eta} \tag{35}$$

$$\partial\bar{B}_\alpha/\partial\bar{\eta} = e^{j\Phi_{B\alpha,\eta}} \cos \beta_{\eta,\eta}\, W_{B\alpha,\eta} \tag{36}$$

where η = t, x, y, z and where

$$W_{E\alpha,\eta} = \sqrt{(\partial E_\alpha/\partial\eta)^2 + (E_\alpha\, \partial\theta_{E\alpha}/\partial\eta)^2} \tag{37}$$

$$W_{D\alpha,\eta} = \sqrt{(\partial D_\alpha/\partial\eta)^2 + (D_\alpha\, \partial\theta_{D\alpha}/\partial\eta)^2} \tag{38}$$

$$W_{H\alpha,\eta} = \sqrt{(\partial H_\alpha/\partial\eta)^2 + (H_\alpha\, \partial\theta_{H\alpha}/\partial\eta)^2} \tag{39}$$

$$W_{B\alpha,\eta} = \sqrt{(\partial B_\alpha/\partial\eta)^2 + (B_\alpha\, \partial\theta_{B\alpha}/\partial\eta)^2} \tag{40}$$

$$\Phi_{E\alpha,\eta} = \theta_{E\alpha} + \beta_{E\alpha,\eta} - \theta_{d\eta} \tag{41}$$

$$\Phi_{D\alpha,\eta} = \theta_{D\alpha} + \beta_{D\alpha,\eta} - \theta_{d\eta} \tag{42}$$

$$\Phi_{H\alpha,\eta} = \theta_{H\alpha} + \beta_{H\alpha,\eta} - \theta_{d\eta} \tag{43}$$

$$\Phi_{B\alpha,\eta} = \theta_{B\alpha} + \beta_{B\alpha,\eta} - \theta_{d\eta} \tag{44}$$

$$\tan \beta_{E\alpha,\eta} = E_\alpha \frac{\partial\theta_{E\alpha}/\partial\eta}{\partial E_\alpha/\partial\eta} \tag{45}$$

$$\tan \beta_{D\alpha,\eta} = D_\alpha \frac{\partial\theta_{D\alpha}/\partial\eta}{\partial D_\alpha/\partial\eta} \tag{46}$$

$$\tan \beta_{H\alpha,\eta} = H_\alpha \frac{\partial\theta_{H\alpha}/\partial\eta}{\partial H_\alpha/\partial\eta} \tag{47}$$

$$\tan \beta_{B\alpha,\eta} = B_\alpha \frac{\partial\theta_{B\alpha}/\partial\eta}{\partial B_\alpha/\partial\eta} \tag{48}$$

and where

$$\theta_{d\eta} = \theta_\eta + \beta_{\eta,\eta} \tag{49}$$

If the electric and magnetic fields have an external time dependence given by equations (29) through (32) it follows that

$$\frac{\partial\bar{E}_\alpha}{\partial\bar{t}} = e^{j\Phi_{E\alpha,t}} \cos \beta_{t,t}\, W_{E\alpha,t} \tag{50}$$

and

$$W_{E\alpha,t} = A_{E\alpha} e^{i\xi} \sqrt{-\omega^2 + \left(\frac{\partial\theta_{E\alpha}}{\partial t}\right)^2} = -iA_{E\alpha} e^{i\xi} \sqrt{\omega^2 - \left(\frac{\partial\theta_{E\alpha}}{\partial t}\right)^2} \tag{51}$$

where ξ is given by equation (32A), with similar expressions for the magnetic vector components. Similarly, the derivative with respect to the coordinate $\bar{x}$ is given by

$$\frac{\partial\bar{E}_\alpha}{\partial\bar{x}} = e^{j\Phi_{E\alpha,x}} \cos \beta_{x,x}\, W_{E\alpha,x} \tag{52}$$

where

$$W_{E\alpha,x} = A_{E\alpha} e^{i\xi} \sqrt{-k_x^2 + \left(\frac{\partial\theta_{E\alpha}}{\partial x}\right)^2} = iA_{E\alpha} e^{i\xi} \sqrt{k_x^2 - \left(\frac{\partial\theta_{E\alpha}}{\partial x}\right)^2} \tag{53}$$

with similar expressions for the magnetic field vector components and for the derivatives with respect to $\bar{y}$ and $\bar{z}$.

It is sometimes conventient to write the derivatives in equations (33) and (36) in the following alternative forms

$$\frac{\partial \bar{E}_\alpha}{\partial \bar{\eta}} = \cos \beta_{\eta,\eta} \; e^{-j\theta_{d\eta}} \frac{\partial \bar{E}_\alpha}{\partial \eta} = R_{1\eta}(E_{\alpha R},E_{\alpha I}) + jI_{1\eta}(E_{\alpha R},E_{\alpha I}) \tag{54}$$

$$\frac{\partial \bar{B}_\alpha}{\partial \bar{\eta}} = \cos \beta_{\eta,\eta} \; e^{-j\theta_{d\eta}} \frac{\partial \bar{B}_\alpha}{\partial \eta} = R_{1\eta}(B_{\alpha R},B_{\alpha I}) + jI_{1\eta}(B_{\alpha R},B_{\alpha I}) \tag{55}$$

where

$$\begin{aligned} R_{1\eta}(E_{\alpha R},E_{\alpha I}) &= \cos \beta_{\eta,\eta} \; (\cos \theta_{d\eta} \; \partial E_{\alpha R}/\partial\eta + \sin \theta_{d\eta} \; \partial E_{\alpha I}/\partial\eta) \\ &= \cos \beta_{\eta,\eta} \; \cos \Phi_{E\alpha,\eta} \; W_{E\alpha,\eta} \\ &\sim \cos \beta_{\eta,\eta} \; \cos (\theta_{E\alpha} - \theta_{d\eta}) \; \partial E_\alpha/\partial\eta \end{aligned} \tag{56}$$

$$\begin{aligned} I_{1\eta}(E_{\alpha R},E_{\alpha I}) &= \cos \beta_{\eta,\eta} \; (- \sin \theta_{d\eta} \; \partial E_{\alpha R}/\partial\eta + \cos \theta_{d\eta} \; \partial E_{\alpha I}/\partial\eta) \\ &= \cos \beta_{\eta,\eta} \; \sin \Phi_{E\alpha,\eta} \; W_{E\alpha,\eta} \\ &\sim \cos \beta_{\eta,\eta} \; \sin (\theta_{E\alpha} - \theta_{d\eta}) \; \partial E_\alpha/\partial\eta \end{aligned} \tag{57}$$

$$\begin{aligned} R_{1\eta}(B_{\alpha R},B_{\alpha I}) &= \cos \beta_{\eta,\eta} \; (\cos \theta_{d\eta} \; \partial B_{\alpha R}/\partial\eta + \sin \theta_{d\eta} \; \partial B_{\alpha I}/\partial\eta) \\ &= \cos \beta_{\eta,\eta} \; \cos \Phi_{B\alpha,\eta} \; W_{B\alpha,\eta} \\ &\sim \cos \beta_{\eta,\eta} \; \cos (\theta_{B\alpha} - \theta_{d\eta}) \; \partial B_\alpha/\partial\eta \end{aligned} \tag{58}$$

$$I_{1\eta}(B_{\alpha R}, B_{\alpha I}) = \cos \beta_{\eta,\eta} \; (- \sin \theta_{d\eta} \; \partial B_{\alpha R}/\partial\eta + \cos \theta_{d\eta} \; \partial B_{\alpha I}/\partial\eta) \tag{59}$$

$$= \cos \beta_{\eta,\eta} \; \sin \Phi_{B\alpha,\eta} \; W_{B\alpha,\eta}$$

$$\sim \cos \beta_{\eta,\eta} \; \sin (\theta_{B\alpha} - \theta_{d\eta}) \; \partial B_{\alpha}/\partial\eta$$

and where

$$E_{\alpha R} = E_{\alpha} \cos \theta_{E\alpha} \tag{60}$$

$$E_{\alpha I} = E_{\alpha} \sin \theta_{E\alpha} \tag{61}$$

$$B_{\alpha R} = B_{\alpha} \cos \theta_{B\alpha} \tag{62}$$

$$B_{\alpha I} = B_{\alpha} \sin \theta_{B\alpha} \tag{63}$$

The second derivatives of the field vectors are obtained by consecutively applying the first derivative operators that appear in equation (54) as follows

$$\frac{\partial^2 \bar{E}_{\alpha}}{\partial \bar{\eta}^2} = \cos \beta_{\eta,\eta} \; e^{-j\theta_{d\eta}} \frac{\partial}{\partial\eta} (\cos \beta_{\eta,\eta} \; e^{-j\theta_{d\eta}} \frac{\partial \bar{E}_{\alpha}}{\partial\eta}) \tag{64}$$

where $\alpha = x, y, z$ and $\eta = x, y, z, t$. For simplicity it will be assumed that $\beta_{\eta,\eta}$ and $\theta_{d\eta}$ are slowly varying functions of η . Within the limits of this approximation the second derivatives of the field vectors are written as

$$\frac{\partial^2 \bar{E}_{\alpha}}{\partial \bar{\eta}^2} \sim \cos^2 \beta_{\eta,\eta} \; e^{-2j\theta_{d\eta}} \frac{\partial^2 \bar{E}_{\alpha}}{\partial\eta^2} = R_{2\eta}(E_{\alpha R}, E_{\alpha I}) + jI_{2\eta}(E_{\alpha R}, E_{\alpha I}) \tag{65}$$

$$\sim \cos^2 \beta_{\eta,\eta} \; e^{j\Psi_{E\alpha,\eta}} \; T_{E\alpha,\eta}$$

$$\sim \cos^2 \beta_{\eta,\eta} \; e^{j(\theta_{E\alpha} - 2\theta_{d\eta})} \frac{\partial^2 E_{\alpha}}{\partial\eta^2}$$

$$\frac{\partial^2 \bar{H}_\alpha}{\partial \bar{\eta}^2} \sim \cos^2 \beta_{\eta,\eta} \; e^{-2j\theta_{d\eta}} \frac{\partial^2 \bar{H}_\alpha}{\partial \eta^2} = R_{2\eta}(H_{\alpha R}, H_{\alpha I}) + jI_{2\eta}(H_{\alpha R}, H_{\alpha I}) \tag{66}$$

$$\sim \cos^2 \beta_{\eta,\eta} \; e^{j\Psi_{H\alpha,\eta}} \; T_{H\alpha,\eta}$$

$$\sim \cos^2 \beta_{\eta,\eta} \; e^{j(\theta_{H\alpha} - 2\theta_{d\eta})} \frac{\partial^2 H_\alpha}{\partial \eta^2}$$

where

$$T_{E\alpha,\eta} = \sqrt{\left(\frac{\partial^2 E_\alpha}{\partial \eta^2}\right)^2 + \left(E_\alpha \frac{\partial^2 \theta_{E\alpha}}{\partial \eta^2}\right)^2} \tag{66A}$$

$$T_{H\alpha,\eta} = \sqrt{\left(\frac{\partial^2 H_\alpha}{\partial \eta^2}\right)^2 + \left(H_\alpha \frac{\partial^2 \theta_{H\alpha}}{\partial \eta^2}\right)^2} \tag{66B}$$

$$\Psi_{E\alpha,\eta} = \theta_{E\alpha} + \delta_{E\alpha,\eta} - 2\theta_{d\eta} \tag{66C}$$

$$\Psi_{H\alpha,\eta} = \theta_{H\alpha} + \delta_{H\alpha,\eta} - 2\theta_{d\eta} \tag{66D}$$

and where

$$R_{2\eta}(E_{\alpha R}, E_{\alpha I}) = \cos^2 \beta_{\eta,\eta} \left[\cos(2\theta_{d\eta}) \frac{\partial^2 E_{\alpha R}}{\partial \eta^2} + \sin(2\theta_{d\eta}) \frac{\partial^2 E_{\alpha I}}{\partial \eta^2}\right] \tag{67}$$

$$\sim \cos^2 \beta_{\eta,\eta} \cos \Psi_{E\alpha,\eta} \; T_{E\alpha,\eta}$$

$$\sim \cos^2 \beta_{\eta,\eta} \cos(\theta_{E\alpha} - 2\theta_{d\eta}) \frac{\partial^2 E_\alpha}{\partial \eta^2}$$

$$I_{2\eta}(E_{\alpha R},E_{\alpha I}) = \cos^2 \beta_{\eta,\eta} \left[- \sin(2\theta_{d\eta}) \frac{\partial^2 E_{\alpha R}}{\partial \eta^2} + \cos(2\theta_{d\eta}) \frac{\partial^2 E_{\alpha I}}{\partial \eta^2}\right] \tag{68}$$

$$\sim \cos^2 \beta_{\eta,\eta} \sin \Psi_{E\alpha,\eta} T_{E\alpha,\eta}$$

$$\sim \cos^2 \beta_{\eta,\eta} \sin(\theta_{E\alpha} - 2\theta_{d\eta}) \frac{\partial^2 E_\alpha}{\partial \eta^2}$$

$$R_{2\eta}(H_{\alpha R},H_{\alpha I}) = \cos^2 \beta_{\eta,\eta} \left[\cos(2\theta_{d\eta}) \frac{\partial^2 H_{\alpha R}}{\partial \eta^2} + \sin(2\theta_{d\eta}) \frac{\partial^2 H_{\alpha I}}{\partial \eta^2}\right] \tag{69}$$

$$\sim \cos^2 \beta_{\eta,\eta} \cos \Psi_{H\alpha,\eta} T_{H\alpha,\eta}$$

$$\sim \cos^2 \beta_{\eta,\eta} \cos(\theta_{H\alpha} - 2\theta_{d\eta}) \frac{\partial^2 H_\alpha}{\partial \eta^2}$$

$$I_{2\eta}(H_{\alpha R},H_{\alpha I}) = \cos^2 \beta_{\eta,\eta} \left[- \sin(2\theta_{d\eta}) \frac{\partial^2 H_{\alpha R}}{\partial \eta^2} + \cos(2\theta_{d\eta}) \frac{\partial^2 H_{\alpha I}}{\partial \eta^2}\right] \tag{70}$$

$$\sim \cos^2 \beta_{\eta,\eta} \sin \Psi_{H\alpha,\eta} T_{H\alpha,\eta}$$

$$\sim \cos^2 \beta_{\eta,\eta} \sin(\theta_{H\alpha} - 2\theta_{d\eta}) \frac{\partial^2 H_\alpha}{\partial \eta^2}$$

and

$$\tan \delta_{E\alpha,\eta} = E_\alpha \frac{\partial^2 \theta_{E\alpha}}{\partial \eta^2} \Big/ \frac{\partial^2 E_\alpha}{\partial \eta^2} \tag{70A}$$

$$\tan \delta_{H\alpha,\eta} = H_\alpha \frac{\partial^2 \theta_{H\alpha}}{\partial \eta^2} \Big/ \frac{\partial^2 H_\alpha}{\partial \eta^2} \tag{70B}$$

Similar equations can be written for the D_α and B_α components. In this way the derivatives necessary for evaluating Maxwell's equations in asymmetric bulk matter or vacuum can be evaluated.

3. <u>MAXWELL'S EQUATIONS WITH BROKEN INTERNAL SYMMETRIES.</u> This section develops Maxwell's equations for electromagnetic fields whose field vectors have internal phase angles. The asymmetric bulk matter or vacuum in which the electromagnetic fields exist also have broken internal symmetries in the static pressure field, Grüneisen parameter, and in the space and time coordinates of each point in the system. The internal phase angles of the ambient medium such as θ_P, θ_γ, θ_x, θ_y, θ_z and θ_t must be calculated in conjunction with the internal phase angles of the electromagnetic field vectors. The quantities of θ_P and θ_γ are obtained from the ground state relativistic trace equation (1) and equation (4) that defines the relativistic Grüneisen function.

The unrenormalized Maxwell equations for charges and currents are written as[9-17]

$$\vec{\nabla}_a \cdot \vec{B}^a = 0 \tag{71}$$

$$\vec{\nabla}_a \cdot \vec{D}^a = \rho_q^a \tag{72}$$

$$\vec{\nabla}_a \times \vec{H}^a = \partial\vec{D}^a/\partial t_a + \vec{j}^a \tag{73}$$

$$\vec{\nabla}_a \times \vec{E}^a = -\partial\vec{B}^a/\partial t_a \tag{74}$$

where $\vec{B}^a$ = unrenormalized magnetic induction vector, $\vec{D}^a$ = unrenormalized electric displacement vector, ρ_q^a = unrenormalized charge density, $\vec{H}^a$ = unrenormalized magnetic field vector, $\vec{j}^a$ = unrenormalized current density vector, and $\vec{E}^a$ = unrenormalized electric field vector. Equations (71) through (74) represent eight equations, six of which are independent. The simplest constitutive equations are the following

$$\vec{B}^a = \mu^a\vec{H}^a \tag{75}$$

$$\vec{D}^a = \varepsilon^a\vec{E}^a \tag{76}$$

where μ^a = unrenormalized magnetic permeability, and ε^a = unrenormalized dielectric constant (permittivity). More complicated constitutive equations are often used.[18] In general

$$\mu^a = \mu^a(P^a,\gamma^a) \qquad \varepsilon^a = \varepsilon^a(P^a,\gamma^a) \tag{77}$$

where P^a and γ^a are functions of density and temperature.

Within asymmetric bulk matter or vacuum a similar set of Maxwell's equations must be valid except now the renormalized electric and magnetic field vectors must have internal phases, and the space and time coordinates must also have internal phase angles. Therefore equations (71) through (74) can be written for the electromagnetic field in bulk matter or vacuum with broken internal symmetries as follows

$$\vec{\bar{\nabla}} \cdot \vec{\bar{B}} = 0 \tag{78}$$

$$\vec{\bar{\nabla}} \cdot \vec{\bar{D}} = \bar{\rho}_q \tag{79}$$

$$\vec{\bar{\nabla}} \times \vec{\bar{H}} = \partial\vec{\bar{D}}/\partial\bar{t} + \vec{\bar{j}} \tag{80}$$

$$\vec{\bar{\nabla}} \times \vec{\bar{E}} = -\partial\vec{\bar{B}}/\partial\bar{t} \tag{81}$$

where $\vec{\bar{B}}$ = renormalized complex number magnetic induction vector, $\vec{\bar{D}}$ = renormalized complex number electric displacement vector, ρ_q = renormalized charge density, $\vec{\bar{H}}$ = renormalized complex number magnetic field vector, $\vec{\bar{j}}$ = renormalized complex number current density vector, and $\vec{\bar{E}}$ = renormalized complex number electric field vector. Equations (78) through (81) are complex number vector equations and represent a total of sixteen equations, twelve of which are independent. Note that $\bar{\rho}_q \neq \rho_q^a$ as can be seen from equations (72) and (79) on account of $\vec{\bar{D}} \neq \vec{D}^a$. Since $\bar{\rho}_q = \bar{n}q$ and $\rho_q^a = n^a q^a$ it follows that $\bar{n} = ne^{j\theta_n}$ and $n = n^a \cos\beta_{x,x} \cos\beta_{y,y} \cos\beta_{z,z}$ and $\theta_n = -\theta_x - \theta_y - \theta_z - \beta_{x,x} - \beta_{y,y} - \beta_{z,z}$ where it was assumed that $q = q^a$. The renormalized current density is given by $\vec{\bar{j}} = \bar{n}q\vec{\bar{v}}$, where $\vec{\bar{v}}$ = vector particle velocity with internal phase. The internal phase angle of the particle velocity is a function of θ_x , θ_y , θ_z , θ_t , and clearly $\vec{\bar{j}} \neq \vec{j}^a$.

The simplest renormalized constitutive equations are written as

$$\vec{\bar{B}} = \mu\vec{\bar{H}} \tag{82}$$

$$\vec{\bar{D}} = \varepsilon\vec{\bar{E}} \tag{83}$$

where μ = renormalized magnetic permeability, and ε = renormalized permittivity. Taking account of the fact that equations (82) and (83) are vector equations with real and imaginary parts, it is clear that they represent twelve equations.

The state equations for μ and ε will be assumed to be given by

$$\mu = \mu^a(P,\gamma) \tag{84}$$

$$\varepsilon = \varepsilon^a(P,\gamma) \tag{85}$$

where the functions are evaluated at the values of the renormalized pressure and Grüneisen function as determined from a solution of equation (1).

The component form of the renormalized Maxwell equations (78) through (81) are written as

$$\partial\bar{B}_x/\partial\bar{x} + \partial\bar{B}_y/\partial\bar{y} + \partial\bar{B}_z/\partial\bar{z} = 0 \tag{86}$$

$$\partial\bar{D}_x/\partial\bar{x} + \partial\bar{D}_y/\partial\bar{y} + \partial\bar{D}_z/\partial\bar{z} = \bar{\rho}_q \tag{87}$$

$$\partial\bar{H}_y/\partial\bar{x} - \partial\bar{H}_x/\partial\bar{y} = \partial\bar{D}_z/\partial\bar{t} + \bar{j}_z \tag{88}$$

$$\partial\bar{H}_z/\partial\bar{y} - \partial\bar{H}_y/\partial\bar{z} = \partial\bar{D}_x/\partial\bar{t} + \bar{j}_x \tag{89}$$

$$\partial\bar{H}_x/\partial\bar{z} - \partial\bar{H}_z/\partial\bar{x} = \partial\bar{D}_y/\partial\bar{t} + \bar{j}_y \tag{90}$$

$$\partial\bar{E}_y/\partial\bar{x} - \partial\bar{E}_x/\partial\bar{y} = -\,\partial\bar{B}_z/\partial\bar{t} \tag{91}$$

$$\partial\bar{E}_z/\partial\bar{y} - \partial\bar{E}_y/\partial\bar{z} = -\,\partial\bar{B}_x/\partial\bar{t} \tag{92}$$

$$\partial\bar{E}_x/\partial\bar{z} - \partial\bar{E}_z/\partial\bar{x} = -\,\partial\bar{B}_y/\partial\bar{t} \tag{93}$$

Using equations (33) through (49) to evaluate derivatives allows Maxwell's equations to be rewritten as

$$W_{Bx,x} \cos \Phi_{Bx,x} \cos \beta_{x,x} + W_{By,y} \cos \Phi_{By,y} \cos \beta_{y,y} + W_{Bz,z} \cos \Phi_{Bz,z} \cos \beta_{z,z} = 0 \tag{94}$$

$$W_{Bx,x} \sin \Phi_{Bx,x} \cos \beta_{x,x} + W_{By,y} \sin \Phi_{By,y} \cos \beta_{y,y} + W_{Bz,z} \sin \Phi_{Bz,z} \cos \beta_{z,z} = 0 \tag{95}$$

$$W_{Dx,x} \cos \Phi_{Dx,x} \cos \beta_{x,x} + W_{Dy,y} \cos \Phi_{Dy,y} \cos \beta_{y,y} + W_{Dz,z} \cos \Phi_{Dz,z} \cos \beta_{z,z} = \rho_q \cos \theta_{\rho q} \tag{96}$$

$$W_{Dx,x} \sin \Phi_{Dx,x} \cos \beta_{x,x} + W_{Dy,y} \sin \Phi_{Dy,y} \cos \beta_{y,y} + W_{Dz,z} \sin \Phi_{Dz,z} \cos \beta_{z,z} = \rho_q \sin \theta_{\rho q} \tag{97}$$

$$W_{Hy,x} \cos \Phi_{Hy,x} \cos \beta_{x,x} - W_{Hx,y} \cos \Phi_{Hx,y} \cos \beta_{y,y} = W_{Dz,t} \cos \Phi_{Dz,t} \cos \beta_{t,t} + j_z \cos \theta_{jz} \tag{98}$$

$$W_{Hy,x} \sin \Phi_{Hy,x} \cos \beta_{x,x} - W_{Hx,y} \sin \Phi_{Hx,y} \cos \beta_{y,y} = W_{Dz,t} \sin \Phi_{Dz,t} \cos \beta_{t,t} + j_z \sin \theta_{jz} \tag{99}$$

$$W_{Hz,y} \cos \Phi_{Hz,y} \cos \beta_{y,y} - W_{Hy,z} \cos \Phi_{Hy,z} \cos \beta_{z,z} = W_{Dx,t} \cos \Phi_{Dx,t} \cos \beta_{t,t} + j_x \cos \theta_{jx} \tag{100}$$

$$W_{Hz,y} \sin \Phi_{Hz,y} \cos \beta_{y,y} - W_{Hy,z} \sin \Phi_{Hy,z} \cos \beta_{z,z} = W_{Dx,t} \sin \Phi_{Dx,t} \cos \beta_{t,t} + j_x \sin \theta_{jx} \tag{101}$$

$$W_{Hx,z} \cos \Phi_{Hx,z} \cos \beta_{z,z} - W_{Hz,x} \cos \Phi_{Hz,x} \cos \beta_{x,x} \quad (102)$$

$$= W_{Dy,t} \cos \Phi_{Dy,t} \cos \beta_{t,t} + j_y \cos \theta_{jy}$$

$$W_{Hx,z} \sin \Phi_{Hx,z} \cos \beta_{z,z} - W_{Hz,x} \sin \Phi_{Hz,x} \cos \beta_{x,x} \quad (103)$$

$$= W_{Dy,t} \sin \Phi_{Dy,t} \cos \beta_{t,t} + j_y \sin \theta_{jy}$$

$$W_{Ey,x} \cos \Phi_{Ey,x} \cos \beta_{x,x} - W_{Ex,y} \cos \Phi_{Ex,y} \cos \beta_{y,y} \quad (104)$$

$$= - W_{Bz,t} \cos \Phi_{Bz,t} \cos \beta_{t,t}$$

$$W_{Ey,x} \sin \Phi_{Ey,x} \cos \beta_{x,x} - W_{Ex,y} \sin \Phi_{Ex,y} \cos \beta_{y,y} \quad (105)$$

$$= - W_{Bz,t} \sin \Phi_{Bz,t} \cos \beta_{t,t}$$

$$W_{Ez,y} \cos \Phi_{Ez,y} \cos \beta_{y,y} - W_{Ey,z} \cos \Phi_{Ey,z} \cos \beta_{z,z} \quad (106)$$

$$= - W_{Bx,t} \cos \Phi_{Bx,t} \cos \beta_{t,t}$$

$$W_{Ez,y} \sin \Phi_{Ez,y} \cos \beta_{y,y} - W_{Ey,z} \sin \Phi_{Ey,z} \cos \beta_{z,z} \quad (107)$$

$$= - W_{Bx,t} \sin \Phi_{Bx,t} \cos \beta_{t,t}$$

$$W_{Ex,z} \cos \Phi_{Ex,z} \cos \beta_{z,z} - W_{Ez,x} \cos \Phi_{Ez,x} \cos \beta_{x,x} \quad (108)$$

$$= - W_{By,t} \cos \Phi_{By,t} \cos \beta_{t,t}$$

$$W_{Ex,z} \sin \Phi_{Ex,z} \cos \beta_{z,z} - W_{Ez,x} \sin \Phi_{Ez,x} \cos \beta_{x,x} \tag{109}$$

$$= - W_{By,t} \sin \Phi_{By,t} \cos \beta_{t,t}$$

Maxwell's equations can be written in an alternative but equivalent form by using equations (54) through (59) as follows

$$R_{1x}(B_{xR},B_{xI}) + R_{1y}(B_{yR},B_{yI}) + R_{1z}(B_{zR},B_{zI}) = 0 \tag{110}$$

$$I_{1x}(B_{xR},B_{xI}) + I_{1y}(B_{yR},B_{yI}) + I_{1z}(B_{zR},B_{zI}) = 0 \tag{111}$$

$$R_{1x}(D_{xR},D_{xI}) + R_{1y}(D_{yR},D_{yI}) + R_{1z}(D_{zR},D_{zI}) = \rho_q \cos \theta_{\rho q} \tag{112}$$

$$I_{1x}(D_{xR},D_{xI}) + I_{1y}(D_{yR},D_{yI}) + I_{1z}(D_{zR},D_{zI}) = \rho_q \sin \theta_{\rho q} \tag{113}$$

$$R_{1x}(H_{yR},H_{yI}) - R_{1y}(H_{xR},H_{xI}) = R_{1t}(D_{zR},D_{zI}) + j_{zR} \tag{114}$$

$$I_{1x}(H_{yR},H_{yI}) - I_{1y}(H_{xR},H_{xI}) = I_{1t}(D_{zR},D_{zI}) + j_{zI} \tag{115}$$

$$R_{1y}(H_{zR},H_{zI}) - R_{1z}(H_{yR},H_{yI}) = R_{1t}(D_{xR},D_{xI}) + j_{xR} \tag{116}$$

$$I_{1y}(H_{zR},H_{zI}) - I_{1z}(H_{yR},H_{yI}) = I_{1t}(D_{xR},D_{xI}) + j_{xI} \tag{117}$$

$$R_{1z}(H_{xR},H_{xI}) - R_{1x}(H_{zR},H_{zI}) = R_{1t}(D_{yR},D_{yI}) + j_{yR} \tag{118}$$

$$I_{1z}(H_{xR},H_{xI}) - I_{1x}(H_{zR},H_{zI}) = I_{1t}(D_{yR},D_{yI}) + j_{yI} \tag{119}$$

$$R_{1x}(E_{yR},E_{yI}) - R_{1y}(E_{xR},E_{xI}) = - R_{1t}(B_{zR},B_{zI}) \tag{120}$$

$$I_{1x}(E_{yR},E_{yI}) - I_{1y}(E_{xR},E_{xI}) = - I_{1t}(B_{zR},B_{zI}) \tag{121}$$

$$R_{1y}(E_{zR},E_{zI}) - R_{1z}(E_{yR},E_{yI}) = - R_{1t}(B_{xR},B_{xI}) \tag{122}$$

$$I_{1y}(E_{zR},E_{zI}) - I_{1z}(E_{yR},E_{yI}) = - I_{1t}(B_{xR},B_{xI}) \tag{123}$$

$$R_{1z}(E_{xR},E_{xI}) - R_{1x}(E_{zR},E_{zI}) = - R_{1t}(B_{yR},B_{yI}) \tag{124}$$

$$I_{1z}(E_{xR},E_{xI}) - I_{1x}(E_{zR},E_{zI}) = - I_{1t}(B_{yR},B_{yI}) \tag{125}$$

where

$$j_{zR} = j_z \cos \theta_{jz} \qquad j_{zI} = j_z \sin \theta_{jz} \tag{126}$$

$$j_{xR} = j_x \cos \theta_{jx} \qquad j_{xI} = j_x \sin \theta_{jx} \tag{127}$$

$$j_{yR} = j_y \cos \theta_{jy} \qquad j_{yI} = j_y \sin \theta_{jy} \tag{128}$$

The radiation pressure is related to the radiation energy density for isotropic radiation with broken internal symmetry by the following approximate formula

$$\bar{P}_r = P_r e^{j\theta_{Pr}} = \frac{1}{3} \bar{\mathcal{E}}_r = \frac{1}{3} \mathcal{E}_r e^{j\theta_{\mathcal{E}r}} \tag{129}$$

Equation (129) is exact only for symmetrical isotropic radiation. Equation (129) gives the following approximate equations

$$P_r = \frac{1}{3} \mathcal{E}_r \tag{130}$$

$$\theta_{Pr} = \theta_{\mathcal{E}r} \tag{131}$$

where $\bar{\mathcal{E}}_r$ = radiation energy density with broken internal symmetry, that is related to the electromagnetic field vectors as follows

$$\bar{\mathcal{E}}_r = \frac{\varepsilon}{2} (\bar{E}_x^2 + \bar{E}_y^2 + \bar{E}_z^2) + \frac{\mu}{2} (\bar{H}_x^2 + \bar{H}_y^2 + \bar{H}_z^2) \tag{132}$$

Equation (132) is equivalent to the following two equations

$$\mathcal{E}_r \cos \theta_{\mathcal{E}r} = \frac{\varepsilon}{2}[E_x^2 \cos (2\theta_{Ex}) + E_y^2 \cos (2\theta_{Ey}) + E_z^2 \cos (2\theta_{Ez})] \quad (133)$$

$$+ \frac{\mu}{2}[H_x^2 \cos (2\theta_{Hx}) + H_y^2 \cos (2\theta_{Hy}) + H_z^2 \cos (2\theta_{Hz})]$$

$$\mathcal{E}_r \sin \theta_{\mathcal{E}r} = \frac{\varepsilon}{2}[E_x^2 \sin (2\theta_{Ex}) + E_y^2 \sin (2\theta_{Ey}) + E_z^2 \sin (2\theta_{Ez})] \quad (134)$$

$$+ \frac{\mu}{2}[H_x^2 \sin (2\theta_{Hx}) + H_y^2 \sin (2\theta_{Hy}) + H_z^2 \sin (2\theta_{Hz})]$$

from which $\mathcal{E}_r$ and $\theta_{\mathcal{E}r}$ can be immediately obtained. The unrenormalized radiation density is given by equation (132) with the bars removed and with the superscript "a" inserted on all quantities.

In addition to Maxwell's equations several other equations are required to form a complete set of equations to determine the phase angles of the spacetime coordinates as well as the phase angles of the electromagnetic field vectors. Six of the additional equations required are the equations of motion for charged bulk matter (plasma). These six equations are given by the complex number vector nonrelativistic Euler equations combined with the Lorentz force as follows[19-22]

$$\rho d\bar{v}_\alpha/d\bar{t} = - \cos \beta_{\alpha,\alpha} (\partial\bar{P}/\partial\alpha + \partial\bar{P}_r/\partial\alpha) - \rho\partial\bar{W}/\partial\bar{\alpha} + \rho_q(\bar{\vec{E}} + \bar{\vec{v}} \times \bar{\vec{B}})_\alpha \quad (135)$$

where $\alpha = x , y , z$, ρ = mass density, $\bar{v}_\alpha$ = spatial components of particle velocity with internal phase, $\bar{P}$ = static pressure with internal phase, $\bar{P}_r$ = radiation pressure with internal phase, $\bar{W}$ = external potential (such as gravity) with internal phase, and $\bar{\vec{v}}$ = particle velocity vector with internal phase = $(\bar{v}_x , \bar{v}_y , \bar{v}_z)$. The static pressure and external potential are complex numbers with internal phase angles and are written as

$$\bar{P} = Pe^{j\theta_P} \quad (136)$$

$$\bar{W} = We^{j\theta_W} \quad (137)$$

The time derivatives of the velocity components in equation (135) are given by the following six equations

$$\bar{a}_\alpha = d\bar{v}_\alpha/d\bar{t} = \partial\bar{v}_\alpha/\partial\bar{t} + \sum_\sigma \bar{v}_\sigma \, \partial\bar{v}_\alpha/\partial\bar{\sigma} \quad (138)$$

where the following six equations define the complex number velocity

$$\bar{v}_\alpha = d\bar{\alpha}/d\bar{t} = v_\alpha e^{j\theta_{v\alpha}} \tag{139}$$

$$v_\alpha = \sqrt{\frac{(d\alpha/dt)^2 + (\alpha\, d\theta_\alpha/dt)^2}{1 + (t\, d\theta_t/dt)^2}} \tag{140}$$

$$\theta_{v\alpha} = \theta_\alpha + \beta_{\alpha,t} - \theta_t - \beta_{t,t} \tag{141}$$

where

$$\tan \beta_{\alpha,t} = \alpha \frac{d\theta_\alpha/dt}{d\alpha/dt} \tag{142}$$

$$\frac{d\theta_\alpha}{dt} = \frac{\partial\theta_\alpha}{\partial t} + \frac{\partial\theta_\alpha}{\partial x}\frac{dx}{dt} + \frac{\partial\theta_\alpha}{\partial y}\frac{dy}{dt} + \frac{\partial\theta_\alpha}{\partial z}\frac{dz}{dt} \tag{143}$$

In addition, the continuity equation

$$\partial\bar{\rho}_q/\partial\bar{t} + \bar{\vec{\nabla}} \cdot (\bar{\rho}_q \bar{\vec{v}}) = 0 \tag{144A}$$

is necessary to determine θ_t and $\rho_q(x,y,z,t)$. Equation (144A) has two components because it is a complex number equation.

The Maxwell equations for broken symmetry matter, equations (94) through (109) or equivalently equations (110) through (125), are not sufficient by themselves to determine the internal phase angles of the space and time coordinates. This is because the twelve independent Maxwell equations (88) through (93), the twelve constitutive equations (82) and (83), the two components of the ground state trace equation (1), the two components of the ground state Grüneisen parameter equation (4), the two components of the excited state trace equation (5), the two components of the radiation Grüneisen parameter equation (6A), the two state equations (84) and (85) for the renormalized magnetic permeability and electric permittivity, and the two components of the continuity equation, represent thirty-six equations. However, thus far only thirty-five field and matter variables have been enumerated and these are: E_α, $\theta_{E\alpha}$; H_α, $\theta_{H\alpha}$; B_α, $\theta_{B\alpha}$; $\mathcal{E}$, $\theta_{\mathcal{E}}$; γ, θ_γ; $\mathcal{E}_r$, $\theta_{\mathcal{E}r}$; γ_r, $\theta_{\gamma r}$; ε, μ; and ρ_q. But these thirty-five quantities are related to nineteen kinematic and dynamic variables because of the space and time derivatives of the field vectors in Maxwell's equations (33) through (48) and because of the appearance of the current density (velocity) in Maxwell's equations. The nineteen kinematic and dynamic variables are: x, y, z, v_x, v_y, v_z, a_x, a_y, a_z and the corresponding phase angles θ_x, θ_y, θ_z, θ_{vx}, θ_{vy}, θ_{vz},

θ_{ax}, θ_{ay}, θ_{az} and by itself θ_t. In these calculations the magnitude of the time t is taken to be a free and independent variable. Therefore a total of fifty-four unknown quantities need to be calculated, and thus far only thirty-six equations have been enumerated. The additional necessary eighteen equations are: the six equations of motion (135), the six kinematic acceleration equations (138), and the six kinematic velocity equations (139). Thus there are fifty-four equations and fifty-four unknown variables to be determined. Note that the two components of the complex number scalar continuity equation introduces only one new unknown quantity ρ_q, and this leaves the second component equation to determine θ_t which stands by itself because t is taken to be an independent variable.

The relativistic trace equations (1) and (5) play an important part in the calculation of the renormalized electromagnetic fields in asymmetric bulk matter or vacuum. Starting with the unrenormalized ground state energy density and Grüneisen parameter, E^a and γ^a respectively, equation (1) is used to calculate the renormalized values of the ground state energy density E, θ_E and the ground state Grüneisen parameter γ, θ_γ. The renormalized values of magnetic permeability μ and dielectric constant ε are expressed in terms of P and γ through equations (84) and (85). In addition to Maxwell's equations, the radiation trace equation (5), in conjunction with equations (133) and (134) that relate the radiation energy density to the electromagnetic field vectors, determines the renormalized field vectors in terms of the corresponding unrenormalized values. The solution of the unrenormalized Maxwell equations (71) through (74) gives the unrenormalized field vectors in terms of the unrenormalized charge density ρ_q^a and current density j^a. The unrenormalized energy density is then calculated in terms of the unrenormalized field vectors using equations (133) and (134). Then equation (5) is applied again in conjuction with equations (133) and (134) and Maxwell's equations to obtain the renormalized field vectors. Finally the renormalized charge and current density are obtained from the renormalized field vectors by using equations (79) and (80).

For electromagnetic waves in the vacuum, the total density ρ and the charge density ρ_q that appear in equations (135) refer to test charges placed within the vacuum to measure the electromagnetic field strengths. Therefore the case of electromagnetic waves in the vacuum is formally equivalent to the case of electromagnetic waves in bulk matter. For the vacuum

$$\begin{aligned}
&\theta_x = \theta_x^{(v)} && \theta_y = \theta_y^{(v)} && \theta_z = \theta_z^{(v)} \\
&\theta_{vx} = \theta_{vx}^{(v)} && \theta_{vy} = \theta_{vy}^{(v)} && \theta_{vz} = \theta_{vz}^{(v)} \\
&\theta_{ax} = \theta_{ax}^{(v)} && \theta_{ay} = \theta_{ay}^{(v)} && \theta_{az} = \theta_{az}^{(v)} \\
&\theta_t = \theta_t^{(v)} && \theta_P = \theta_P^{(v)} && \theta_U = \theta_U^{(v)}
\end{aligned} \tag{145}$$

where (v) refers to the vacuum state (see Section 6).

The energy conservation equation is the first integral of equation (135) and in its simplest form is written as[19]

$$\frac{1}{2}\rho\bar{v}^2 + \bar{P} + \bar{P}_r + \bar{W} - \rho_q \int \vec{\bar{E}} \cdot \vec{\bar{v}}\, d\bar{t} = \text{constant} \qquad (146)$$

where $\vec{\bar{v}}$ = complex number vector velocity whose complex number magnitude is given by

$$\bar{v}^2 = \bar{v}_x^2 + \bar{v}_y^2 + \bar{v}_z^2 \qquad (147)$$

where

$$\bar{v} = ve^{j\theta_v} \qquad (147A)$$

$$\bar{v}_\alpha = v_\alpha e^{j\theta_{v\alpha}} \qquad (147B)$$

Note that

$$\vec{\bar{E}} \cdot \vec{\bar{v}} = \sum_{\alpha=1}^{3} \bar{E}_\alpha \bar{v}_\alpha \qquad (148)$$

Were it possible to neglect the charge density term by making ρ_q vanishingly small, equation (146) becomes

$$\frac{1}{2}\rho\bar{v}^2 + \bar{P}_r + \bar{P} + \bar{W} = \text{constant} \qquad (149)$$

where the mass density ρ refers to a test probe. Finally, the dynamical equations for relativistic bulk matter with broken internal symmetry are given by the following generalization of equation (135) for low pressures[23]

$$\rho\bar{\gamma}^2 \frac{d\bar{v}_\alpha}{d\bar{t}} = - \cos\beta_{\alpha,\alpha}\left[\frac{\partial\bar{P}}{\partial\alpha} + \frac{\partial\bar{P}_r}{\partial\alpha} + \frac{dt}{d\alpha}\left(\frac{\partial\bar{P}}{\partial t} + \frac{\partial\bar{P}_r}{\partial t}\right)\right] \qquad (150)$$

$$- \rho\frac{\partial\bar{W}}{\partial\bar{\alpha}} + \rho_q\bar{\gamma}(\vec{\bar{E}} + \vec{\bar{v}} \times \vec{\bar{B}})_\alpha$$

where $\bar{\gamma}$ = complex velocity factor that is defined in Section 4, and where c = light speed in the vacuum.

4. LORENTZ INVARIANCE IN ASYMMETRIC BULK MATTER AND VACUUM. This section considers the Lorentz invariance of Maxwell's equations in bulk matter and vacuum with broken internal symmetries. Maxwell's equations for symmetric systems, equations (71) through (74), are invariant under the Lorentz transformation of coordinate systems[9,10,16,23-34]

$$x'_a = \gamma_a(x_a - v_a t_a) \tag{151}$$

$$t'_a = \gamma_a(t_a - v_a x_a/c^2) \tag{152}$$

where v_a = relative speed of coordinate systems, and the standard velocity factor is given by

$$\gamma_a = (1 - \beta_a^2)^{-1/2} \tag{153}$$

where $\beta_a = v_a/c$, where c = light speed in vacuum. The Lorentz transformation can be obtained by requiring the form invariance of the Minkowski metric as follows

$$x_a^2 - c^2 t_a^2 = x_a'^2 - c^2 t_a'^2 \tag{154}$$

General relativity, which is not considered in this paper, uses a Riemann metric.[23-25]

The form of Maxwell's equations for charges and currents in asymmetric bulk matter or vacuum, equations (78) through (81), is the same as that for symmetric bulk matter or vacuum, equations (71) through (74). The only difference is that in asymmetric systems the field vectors, current density, and spacetime coordinates are complex numbers. Therefore by the same analysis that shows the symmetric Maxwell equations (71) through (74) to be form covariant under the real number Lorentz transformation equations (151) through (153), it follows that the asymmetric Maxwell equations (78) through (81) are form covariant under the following complex number Lorentz transformations

$$\bar{x}' = \bar{\gamma}(\bar{x} - \bar{v}\bar{t}) \tag{155}$$

$$\bar{t}' = \bar{\gamma}(\bar{t} - \bar{v}\bar{x}/c^2) \tag{156}$$

where $\bar{v}$ = complex number relative speed of the two coordinate systems, and the complex number velocity factor for an asymmetric system is given by

$$\bar{\gamma} = (1 - \bar{\beta}^2)^{-1/2} \tag{157}$$

where $\bar{\beta} = \bar{v}/c$. Also, simple algebra shows that equations (155) through (157) satisfy

$$\bar{x}'^2 - c^2\bar{t}'^2 = \bar{x}^2 - c^2\bar{t}^2 \tag{158}$$

where $\bar{x}$, $\bar{t}$, $\bar{x}'$ and $\bar{t}'$ are space and time coordinates that exhibit broken internal symmetry, and in general

$$\bar{x} = xe^{j\theta_x} \qquad \bar{x}' = x'e^{j\theta'_x} \tag{159}$$

$$\bar{t} = te^{j\theta_t} \qquad \bar{t}' = t'e^{j\theta'_t} \tag{160}$$

and the relative speed of the coordinate systems is written as

$$\bar{v} = ve^{j\theta_v} \qquad \bar{\beta} = \beta e^{j\theta_v} \tag{161}$$

where θ_x , θ'_x , θ_t , θ'_t and θ_v are functions of P and θ_P of the ambient asymmetric bulk matter or vacuum.

Combining equations (157) and (161) gives

$$\bar{\gamma} = (f - jb)^{-1/2} = \sqrt{\frac{f + jb}{f^2 + b^2}} = \gamma e^{j\theta_\gamma} \tag{162}$$

where

$$f = 1 - \beta^2 \cos (2\theta_v) \tag{163}$$

$$b = \beta^2 \sin (2\theta_v) \tag{164}$$

From equation (162) it follows that for an asymmetric system the magnitude and internal phase angle of the velocity factor are given by

$$\gamma = (f^2 + b^2)^{-1/4} = [1 - 2\beta^2 \cos (2\theta_v) + \beta^4]^{-1/4} \tag{165}$$

$$\tan (2\theta_\gamma) = \frac{\beta^2 \sin (2\theta_v)}{1 - \beta^2 \cos (2\theta_v)} \tag{166}$$

Note that $\beta = v/c$, where now v = magnitude of the complex number velocity that appears in equation (161). Also, if $\theta_v = 0$ then equation (165) reduces to equation (153).

The Lorentz transformations in equations (155) and (156) can be written in the form of real and imaginary components as follows

$$x' \cos \theta'_x = \gamma[x \cos(\theta_x + \theta_\gamma) - vt \cos(\theta_v + \theta_t + \theta_\gamma)] \tag{167}$$

$$x' \sin \theta'_x = \gamma[x \sin(\theta_x + \theta_\gamma) - vt \sin(\theta_v + \theta_t + \theta_\gamma)] \tag{168}$$

$$t' \cos \theta'_t = \gamma[t \cos(\theta_t + \theta_\gamma) - vx/c^2 \cos(\theta_v + \theta_x + \theta_\gamma)] \tag{169}$$

$$t' \sin \theta'_t = \gamma[t \sin(\theta_t + \theta_\gamma) - vx/c^2 \sin(\theta_v + \theta_x + \theta_\gamma)] \tag{170}$$

where γ is given by equation (165). From equations (167) and (168) x' and θ'_x can be calculated as follows

$$x'^2 = \gamma^2[x^2 + v^2t^2 - 2vtx \cos(\theta_t + \theta_v - \theta_x)] \tag{171}$$

$$\tan \theta'_x = \frac{x \sin(\theta_x + \theta_\gamma) - vt \sin(\theta_v + \theta_t + \theta_\gamma)}{x \cos(\theta_x + \theta_\gamma) - vt \cos(\theta_v + \theta_t + \theta_\gamma)} \tag{172}$$

while from equations (169) and (170) t' and θ'_t can be calculated in the following manner

$$t'^2 = \gamma^2[t^2 + v^2x^2/c^4 - 2vxt/c^2 \cos(\theta_x + \theta_v - \theta_t)] \tag{173}$$

$$\tan \theta'_t = \frac{t \sin(\theta_t + \theta_\gamma) - vx/c^2 \sin(\theta_v + \theta_x + \theta_\gamma)}{t \cos(\theta_t + \theta_\gamma) - vx/c^2 \cos(\theta_v + \theta_x + \theta_\gamma)} \tag{174}$$

From equations (171) and (173) the Minkowski interval can be written as

$$x'^2 - c^2t'^2 = \gamma^2[(1 - \beta^2)(x^2 - c^2t^2) - 4vtx \sin \theta_v \sin(\theta_x - \theta_t)] \tag{175}$$

where γ is given by equation (165). If $\theta_v = 0$ equation (175) reduces to equation (154).

Consider now some properties of the velocity factor γ given by equation (165). The first thing to see is that γ is not singular for real values of β. In fact it is easy to show that the roots of the denominator in equation (165) are given by

$$\beta^4 - 2\beta^2 \cos(2\theta_v) + 1 = 0 \tag{176}$$

or

$$\beta = e^{\pm j\theta_v} \tag{177}$$

Only when $\theta_v = 0$ does equation (176) have a real root $\beta = 1$ which agrees with equation (153). By taking the derivative of γ given by equation (165) and setting the result equal to zero it is easy to show that γ has a maximum value (assuming θ_v to be independent of velocity) given by

$$[\gamma]_{max} = [\sin(2\theta_v)]^{-1/2} \tag{178}$$

and this maximum value of γ occurs at a value of β given by

$$[\beta]_{max\ \gamma} = [\cos(2\theta_v)]^{1/2} < 1 \tag{179}$$

Combining equations (166) and (179) gives the following value of θ_γ at the maximum point of γ

$$[\theta_\gamma]_{max\ \gamma} = \pi/4 - \theta_v \tag{180}$$

The values of γ and θ_γ for $\beta = 1$ are obtained from equations (165) and (166) respectively as

$$[\gamma]_{\beta=1} = [2\sin\theta_v]^{-1/2} \tag{181}$$

$$[\theta_\gamma]_{\beta=1} = \pi/4 - \theta_v/2 \tag{182}$$

The functions γ and θ_γ appear in Figures 1 and 2 respectively. As shown by equations (179) the maximum value of γ occurs for $\beta < 1$, and if θ_v is small the maximum value of γ occurs close to $\beta = 1$. Within asymmetric bulk matter or vacuum γ is nonsingular. For $\beta \sim 0$ it follows from equations (165) and (166) that

$$\gamma \sim 1 + \frac{1}{2}\beta^2 \cos(2\theta_v) \tag{183}$$

$$\theta_\gamma \sim \frac{1}{2}\beta^2 \sin(2\theta_v) \tag{184}$$

For $\beta \to \infty$ it follows from equations (165) and (166) that

$$\gamma \to 1/\beta \to 0 \tag{185}$$

$$\theta_\gamma \to \pi/2 - \theta_v \tag{186}$$

where θ_v is assumed to be independent of particle velocity. It is assumed that θ_v depends on P and θ_P of the ambient medium.

The complex number de Broglie wavelength for a relativistic particle moving at velocity $\bar{v}$ is[35]

$$\bar{\lambda} = h/\bar{p} = h/(m\bar{\gamma}\bar{v}) = \lambda e^{j\theta_\lambda} \tag{187}$$

where $\bar{p}$ = complex number momentum whose magnitude is given by $p = m\gamma v$, and h = Planck's constant. From equation (187) it follows that

$$\lambda = h/(m\gamma v) = \lambda_c/(\gamma\beta) \tag{188}$$

$$\theta_\lambda = -\theta_\gamma - \theta_v \tag{189}$$

where λ_c = Compton wavelength given by[30]

$$\lambda_c = h/(mc) \tag{190}$$

Three special cases can be considered.

<u>Case 1.</u> $\beta = [\beta]_{max\ \gamma} = [\cos(2\theta_v)]^{1/2}$ (191)

It follows from equations (178), (179), (180), (188), and (189) that

$$\lambda = \lambda_c[\tan(2\theta_v)]^{1/2} \tag{192}$$

$$\theta_\lambda = -\pi/4 \tag{193}$$

<u>Case 2.</u> $\beta = 1$ (194)

In this case it follows from equations (181), (182) and (188) that

$$\lambda = \lambda_c\ (2\sin\theta_v)^{1/2} \tag{195}$$

$$\theta_\lambda = -\pi/4 - \theta_v/2 \tag{196}$$

<u>Case 3.</u> $\beta \to \infty$ (197)

In the limit $\beta \to \infty$ equations (185), (186), (188) and (189) give

$$\lambda \to \lambda_c \qquad (198)$$

$$\theta_\lambda \to -\pi/2 \qquad (199)$$

Therefore as β increases without limit the de Broglie wavelength increases to a limiting value of λ_c . It is assumed that θ_v is independent of velocity.

The total energy of a particle located in asymmetric bulk matter or vacuum is given by the following generalization of the standard results of special relativity[24]

$$\bar{\varepsilon} = \varepsilon e^{j\theta_\varepsilon} = \bar{\gamma} mc^2 \qquad (200)$$

where the complex number velocity factor is given by equation (157), and m = proper mass. Therefore the total energy of a particle has the same properties as $\bar{\gamma}$, so that

$$\varepsilon = \gamma mc^2 \qquad (201)$$

$$\theta_\varepsilon = \theta_\gamma \qquad (202)$$

where the magnitude and internal phase angle of the velocity factor is given by equations (165) and (166) respectively. The kinetic energy of a particle in asymmetric bulk matter or vacuum is given by[24]

$$\bar{\varepsilon}_K = \varepsilon_K e^{j\theta_K} = (\bar{\gamma} - 1)mc^2 \qquad (203)$$

The component form of equation (203) is written as

$$\varepsilon_K \cos \theta_K = (\gamma \cos \theta_\gamma - 1)mc^2 \qquad (204)$$

$$\varepsilon_K \sin \theta_K = \gamma \sin \theta_\gamma \; mc^2 \qquad (205)$$

and therefore for a broken symmetry system

$$\tan \theta_K = \frac{\gamma \sin \theta_\gamma}{\gamma \cos \theta_\gamma - 1} \qquad (206)$$

$$\varepsilon_K = mc^2(\gamma^2 - 2\gamma \cos \theta_\gamma + 1)^{1/2} \qquad (207)$$

Placing equations (183) and (184) into equations (206) and (207) shows that for $\beta \sim 0$

$$\theta_K \sim 2\theta_v \tag{208}$$

$$\varepsilon_K \sim \frac{1}{2} mv^2 \tag{209}$$

which agrees with the nonrelativistic limit obtained directly from equations (157) and (203) namely

$$\bar{\varepsilon}_K \sim \frac{1}{2} m\bar{v}^2 \tag{210}$$

Figures 3 and 4 show ε_K and θ_K in terms of β .

Specific values of the total energy, kinetic energy, and momentum will now be considered for some characteristic values of β .

<u>Case 1.</u> $\beta \sim 0$

$$\gamma \sim 1 + \beta^2/2 \cos(2\theta_v) \tag{211}$$

$$\theta_\gamma \sim \beta^2/2 \sin(2\theta_v) \tag{212}$$

$$\varepsilon \sim mc^2 + \frac{1}{2} mv^2 \tag{212A}$$

$$\theta_\varepsilon \sim \beta^2/2 \sin(2\theta_v) \tag{212B}$$

$$\varepsilon_K \sim \frac{1}{2} mv^2 \tag{212C}$$

$$\theta_K \sim 2\theta_v \tag{212D}$$

$$p \sim mv \tag{212E}$$

$$\theta_p \sim \theta_v \tag{212F}$$

<u>Case 2.</u> $\beta = [\beta]_{max\ \gamma} = [\cos(2\theta_v)]^{1/2}$

$$\gamma = [\sin(2\theta_v)]^{-1/2} \tag{213}$$

$$\theta_\gamma = \pi/4 - \theta_v \tag{214}$$

$$\varepsilon = mc^2[\sin(2\theta_v)]^{-1/2} \tag{215}$$

$$\theta_\varepsilon = \pi/4 - \theta_v \tag{216}$$

$$\varepsilon_K = mc^2\left[\frac{1}{\sin(2\theta_v)} - \frac{2\cos(\pi/4 - \theta_v)}{\sqrt{\sin(2\theta_v)}} + 1\right]^{1/2} \tag{217}$$

$$\varepsilon_K \sim \varepsilon \text{ for small } \theta_v \tag{218A}$$

$$\tan\theta_K = \frac{\sin(\pi/4 - \theta_v)}{\cos(\pi/4 - \theta_v) - \sqrt{\sin(2\theta_v)}} \tag{218B}$$

$$\theta_K \sim \theta_\gamma \text{ for small } \theta_v \tag{218C}$$

$$p/mc = \gamma\beta = [\cot(2\theta_v)]^{1/2} \tag{218D}$$

$$\theta_p = \pi/4 \tag{218E}$$

<u>Case 3.</u> $\beta = 1$

$$\gamma = (2\sin\theta_v)^{-1/2} \tag{219}$$

$$\theta_\gamma = \pi/4 - \theta_v/2 \tag{220}$$

$$\varepsilon = mc^2(2\sin\theta_v)^{-1/2} = cp \tag{221}$$

$$\theta_\varepsilon = \pi/4 - \theta_v/2 \tag{222}$$

$$\varepsilon_K = mc^2\left[\frac{1}{2\sin\theta_v} - \frac{2\cos(\pi/4 - \theta_v/2)}{\sqrt{2\sin\theta_v}} + 1\right]^{1/2} \tag{223}$$

$$\varepsilon_K \sim \varepsilon \text{ for small } \theta_v \tag{224}$$

$$\tan \theta_K = \frac{\sin(\pi/4 - \theta_v/2)}{\cos(\pi/4 - \theta_v/2) - \sqrt{2 \sin \theta_v}} \qquad (224A)$$

$$\theta_K \sim \theta_\gamma \text{ for small } \theta_v \qquad (224B)$$

$$p/mc = \gamma\beta = (2 \sin \theta_v)^{-1/2} \qquad (224C)$$

$$\theta_p = \pi/4 + \theta_v/2 \qquad (224D)$$

<u>Case 4.</u> $\beta \to \infty$

$$\gamma \to 1/\beta \to 0 \qquad (225)$$

$$\theta_\gamma \to \pi/2 - \theta_v \qquad (226)$$

$$\varepsilon \to mc^2/\beta \to 0 \qquad (227)$$

$$\theta_\varepsilon \to \pi/2 - \theta_v \qquad (228)$$

$$\varepsilon_K \to mc^2[1 - \frac{1}{\beta} \cos(\pi/2 - \theta_v)] \to mc^2 \qquad (229)$$

$$\theta_K \to \pi \qquad (230A)$$

$$p/mc = \gamma\beta \to 1 \qquad (230B)$$

$$\theta_p \to \pi/2 \qquad (230C)$$

In direct analogy to the standard expression for relativistic momentum, the momentum of a particle located in asymmetric bulk matter or vacuum is written as[24]

$$\bar{p} = pe^{j\theta_p} = m\bar{\gamma}\bar{v} \qquad (231)$$

so that

$$p = m\gamma v \qquad (232)$$

$$\theta_p = \theta_\gamma + \theta_v \qquad (233)$$

where γ and θ_γ are given by equations (165) and (166) respectively. From equations (157), (204) and (231) it follows that the single particle energy is

$$\bar{\varepsilon}^2 = c^2\bar{p}^2 + m^2c^4 \qquad (234)$$

which shows the four-vector status of $\bar{\varepsilon}$ and $\bar{p}$. Equation (234) has two component equations

$$\varepsilon^2 \cos(2\theta_\varepsilon) = c^2p^2 \cos(2\theta_p) + m^2c^4 \qquad (235)$$

$$\varepsilon^2 \sin(2\theta_\varepsilon) = c^2p^2 \sin(2\theta_p) \qquad (236)$$

Equations (231) through (236) are equivalent to equations (165), (166), (205) and (206). From equations (235) and (236) it follows that

$$\tan(2\theta_\varepsilon) = \frac{p^2 \sin(2\theta_p)}{p^2 \cos(2\theta_p) + m^2c^2} \qquad (237)$$

$$\gamma = \frac{\varepsilon}{mc^2} = [(\frac{p}{mc})^4 + 2(\frac{p}{mc})^2 \cos(2\theta_p) + 1]^{1/4} \qquad (238)$$

It should be remembered that for asymmetric matter or vacuum, an interaction potential and a gauge potential needs to be added to obtain the total single particle energy

$$\bar{\varepsilon}_i = \bar{\varepsilon} + \bar{V}_e + \bar{V}_g \qquad (239)$$

Equation (232) can be written as

$$p/mc = \gamma\beta \qquad (240A)$$

Combining equations (165) and (240A) and setting the derivative of the momentum equal to zero gives the following value of β for maximum momentum

$$[\beta]_{max\ p} = [\cos(2\theta_v)]^{-1/2} = 1/[\beta]_{max\ \gamma} > 1 \tag{240B}$$

for which

$$[\gamma]_{max\ p} = [\cot(2\theta_v)]^{1/2} = [p/mc]_{max\ \gamma} \tag{240C}$$

so that the maximum momentum is

$$[p/mc]_{max} = [\sin(2\theta_v)]^{-1/2} = [\gamma]_{max}$$

where equation (178) has been used. Figures 5 and 6 give p and θ_p in terms of β .

The following arguments show why numerical values of θ_v for the asymmetric vacuum cannot be obtained from the experimental results of the Michelson-Morley experiment.[10] The generalization of the standard relativistic velocity addition formula to a system with broken internal symmetry is[24]

$$\bar{w} = \frac{\bar{u} + \bar{v}}{1 + \bar{u}\bar{v}/c^2} \tag{241}$$

where $\bar{u}$ = particle velocity relative to a reference frame that itself is moving at a velocity $\bar{v}$, and $\bar{w}$ = particle velocity relative to a frame of reference from which the moving frame has a velocity $\bar{v}$. Writing the velocities as

$$\bar{w} = we^{j\theta_w} \qquad \bar{u} = ue^{j\theta_u} \qquad \bar{v} = ve^{j\theta_v} \tag{241A}$$

gives the following velocity addition formula for asymmetric bulk matter or vacuum

$$we^{j\theta_w} = \frac{A + jB}{C + jD} \tag{242}$$

$$w = \left[\frac{A^2 + B^2}{C^2 + D^2}\right]^{1/2} \tag{243}$$

$$\theta_w = \theta_N - \theta_D \tag{244}$$

$$\tan\theta_N = B/A \tag{245}$$

$$\tan\theta_D = D/C \tag{246}$$

$$A = u \cos \theta_u + v \cos \theta_v \tag{247}$$

$$B = u \sin \theta_u + v \sin \theta_v \tag{248}$$

$$C = 1 + uv/c^2 \cos (\theta_u + \theta_v) \tag{249}$$

$$D = uv/c^2 \sin (\theta_u + \theta_v) \tag{250}$$

It is easy to show that

$$A^2 + B^2 = u^2 + v^2 + 2uv \cos (\theta_u - \theta_v) \tag{251}$$

$$C^2 + D^2 = 1 + u^2v^2/c^4 + 2uv/c^2 \cos (\theta_u + \theta_v) \tag{252}$$

Equations (251) and (252) are completely general.

It will be assumed that the internal phase of the massive particle velocity is independent of the magnitude of the velocity so that $\theta_u = \theta_v = \theta$ for massive particles and

$$A^2 + B^2 = (u + v)^2 \tag{253}$$

$$C^2 + D^2 = 1 + u^2v^2/c^4 + 2uv/c^2 \cos(2\theta) \tag{254}$$

Consider now the case of two massive particles each travelling at the light speed so that $u = v = c$ and $\theta = \theta_u = \theta_v = \theta_c \neq 0$, then equations (243), (244) and (247) through (254) yield for two massive particles

$$\begin{array}{lll} A = 2c \cos \theta_c & B = 2c \sin \theta_c & \\ C = 1 + \cos(2\theta_c) & D = \sin(2\theta_c) & \\ w = c/\cos \theta_c & \theta_N = \theta_D = \theta_c & \theta_w = 0 \end{array} \tag{255}$$

Because $\theta_w = 0$, it follows that the measured relative velocity is given by $w_m = w$. If $\theta_c = 0$ in equation (255), the standard special relativistic result $w = c$ is regained.

For two photons the situation is $u = v = c$ and $\theta_u = \theta_v = \theta_c = 0$ and equations (243), (244) and (247) through (252) give

$$A = 2c \qquad B = 0 \qquad C = 2$$

$$A^2 + B^2 = 4c^2 \qquad C^2 + D^2 = 4 \qquad D = 0 \tag{256}$$

$$\theta_N = \theta_D = 0 \qquad w = c \qquad \theta_w = 0$$

Because $\theta_w = 0$ it follows that the measured relative speed is $w_m = w = c$.

Consider now the case of the Michelson and Morley experiment.[10,24-34] For this case the photons with $u = c$ and $\theta_c = 0$ move relative to the earth which has speed v and internal phase angle of speed θ_v . From equations (247) through (250) it follows that

$$A = c + v \cos \theta_v \tag{257}$$

$$B = v \sin \theta_v \tag{258}$$

$$C = 1 + v/c \cos \theta_v = A/c \tag{259}$$

$$D = v/c \sin \theta_v = B/c \tag{260}$$

Then equations (257) through (260) and equations (245) and (246) give

$$A^2 + B^2 = (c + v)^2 - 4cv \sin^2(\theta_v/2) \tag{261}$$

$$C^2 + D^2 = (A^2 + B^2)/c^2 \tag{262}$$

$$\theta_N = \theta_D = \tan^{-1}[(v \sin \theta_v)/(c + v \cos \theta_v)] \tag{263}$$

Equations (261) and (262) agree with equations (251) and (252) with $u = c$ and $\theta_u = \theta_c = 0$. Combining equations (243) and (244) with equations (261) through (263) yields

$$w = c \qquad \theta_w = 0 \tag{264}$$

for the relative velocity and its internal phase angle for photons moving relative to the earth. Because $\theta_w = 0$ it follows that the measured relative speed is $w_m = w = c$. Thus the Michelson and Morley experiment cannot be used to determine the magnitude v or the internal phase angle θ_v of the earth's velocity because the analysis of this experiment using the broken symmetry form of the velocity addition formula nevertheless yields the standard special relativity result given in equation (264). This is true even though the v and θ_v appear in equations (257) through (262) for the earth moving in spacetime with broken coordinate symmetries. The speed of the photons relative to the earth does not depend on either v or θ_v .

Alternatively, measurements of the velocity factor γ from particle accelerator experiments may eventually produce a maximum value of γ which will immediately determine the value of θ_v . For this, particles with $\beta > 1$ would have to be observed. If no such particles are ever found, it would show that $\theta_v = 0$ for the vacuum, and that the vacuum is symmetric. Experiment can only resolve this issue. Experiments to determine θ_v for asymmetric bulk matter may be easier because θ_v for bulk matter is expected to be larger than $\theta_v^{(v)}$ for the vacuum. Note that astronomical objects with $\beta > 1$ have apparently already been observed, and their explanation in terms of conventional effects can be given only with much difficulty.[36]

Finally, the laws of motion of a relativistic particle in asymmetric bulk matter or vacuum are considered. Newton's law of motion is modified by special relativity to give the following dynamical equation of motion for a force in the direction of motion[24]

$$F^a = \frac{d}{dt}(m\gamma_a v_a) = m\gamma_a^3\, dv_a/dt_a = m\gamma_a^3\, a_a \tag{265}$$

where a_a = conventionally calculated acceleration, and γ_a is given by equation (153). The generalization of this equation to the case of particle motion in asymmetric bulk matter or vacuum is

$$\bar{F} = \frac{d}{d\bar{t}}(m\bar{\gamma}\bar{v}) = m\bar{\gamma}^3\, d\bar{v}/d\bar{t} = m\bar{\gamma}^3\, \bar{a} \tag{266}$$

where $\bar{\gamma}$ is given by equation (157) and where $\bar{t}$, $\bar{v}$, $\bar{a}$, and $\bar{F}$ are the gauge rotated time, velocity, acceleration and force respectively. Therefore

$$\bar{a} = ae^{j\theta_a} \tag{267}$$

$$\bar{F} = Fe^{j\theta_F} \tag{268}$$

Combining equation (162) with equations (266) through (268) gives the force in the direction of motion as

$$F = m\gamma^3 a \tag{269}$$

$$\theta_F = 3\theta_\gamma + \theta_a \tag{270}$$

where γ and θ_γ are given by equations (165) and (166) respectively.

5. ELECTROMAGNETIC WAVE EQUATIONS. A direct result of Maxwell's equations is a set of wave equations that describe the time and space dependence of the electric and magnetic field vectors in a material body or vacuum.[9,10] This section considers the construction of electromagnetic wave equations for matter and radiation with broken internal symmetries. The standard equation of telegraphy that determines the electric (or magnetic) field in a conducting medium is[9,10]

$$\nabla_a^2 E_\alpha^a = \varepsilon^a \mu^a \frac{\partial^2 E_\alpha^a}{\partial t_a^2} + \mu^a \sigma^a \frac{\partial E_\alpha^a}{\partial t_a} \tag{271}$$

where $\alpha = x$, y and z , and σ^a = unrenormalized conductivity. The Laplacian operator is defined as

$$\nabla_a^2 = \frac{\partial^2}{\partial x_a^2} + \frac{\partial^2}{\partial y_a^2} + \frac{\partial^2}{\partial z_a^2} \tag{272}$$

The prescription introduced in this paper to handle electromagnetism in matter or vacuum with broken internal symmetries is to use gauge rotated field vectors and gauge rotated space and time coordinates. Applying this prescription to equation (271) yields

$$\bar{\nabla}^2 \bar{E}_\alpha = \varepsilon\mu \frac{\partial^2 \bar{E}_\alpha}{\partial \bar{t}^2} + \mu\sigma \frac{\partial \bar{E}_\alpha}{\partial \bar{t}} \tag{273}$$

The complex number Laplacian is given by

$$\bar{\nabla}^2 = \frac{\partial^2}{\partial \bar{x}^2} + \frac{\partial^2}{\partial \bar{y}^2} + \frac{\partial^2}{\partial \bar{z}^2} \tag{274}$$

The first and second derivative terms in equation (273) have already been evaluated in Section 2. Using the notation developed in equations (54) through (59) and (65) through (70) allows equation (273) to be written as six real number relations as follows

$$\sum_\beta R_{2\beta}(E_{xR},E_{xI}) \sim \varepsilon\mu R_{2t}(E_{xR},E_{xI}) + \mu\sigma R_{1t}(E_{xR},E_{xI}) \tag{275}$$

$$\sum_\beta R_{2\beta}(E_{yR},E_{yI}) \sim \varepsilon\mu R_{2t}(E_{yR},E_{yI}) + \mu\sigma R_{1t}(E_{yR},E_{yI}) \tag{276}$$

$$\sum_\beta R_{2\beta}(E_{zR},E_{zI}) \sim \varepsilon\mu R_{2t}(E_{zR},E_{zI}) + \mu\sigma R_{1t}(E_{zR},E_{zI}) \tag{277}$$

$$\sum_{\beta} I_{2\beta}(E_{xR},E_{yR}) \sim \varepsilon\mu I_{2t}(E_{xR},E_{xI}) + \mu\sigma I_{1t}(E_{xR},E_{xI}) \tag{278}$$

$$\sum_{\beta} I_{2\beta}(E_{yR},E_{yI}) \sim \varepsilon\mu I_{2t}(E_{yR},E_{yI}) + \mu\sigma I_{1t}(E_{yR},E_{yI}) \tag{279}$$

$$\sum_{\beta} I_{2\beta}(E_{zR},E_{zI}) \sim \varepsilon\mu I_{2t}(E_{zR},E_{zI}) + \mu\sigma I_{1t}(E_{zR},E_{zI}) \tag{280}$$

where the sum is over $\beta = x$, y and z .

The standard equations that determine the electromagnetic potentials are written as[9,10]

$$\nabla_a^2\phi^a - \varepsilon^a\mu^a \frac{\partial^2\phi^a}{\partial t_a^2} = -\rho_q^a/\varepsilon^a \tag{281}$$

$$\nabla_a^2 A_\alpha^a - \varepsilon^a\mu^a \frac{\partial^2 A_\alpha^a}{\partial t_a^2} = -\mu^a j_\alpha^a \tag{282}$$

The generalization of these equations to electromagnetic fields in asymmetric bulk matter or vacuum is as follows

$$\bar{\nabla}^2\bar{\phi} - \varepsilon\mu \frac{\partial^2\bar{\phi}}{\partial\bar{t}^2} = -\bar{\rho}_q/\varepsilon \tag{283}$$

$$\bar{\nabla}^2\bar{A}_\alpha - \varepsilon\mu \frac{\partial^2\bar{A}_\alpha}{\partial\bar{t}^2} = -\mu\bar{j}_\alpha \tag{284}$$

where the complex number electromagnetic potentials are written as

$$\bar{\phi} = \phi e^{j\theta_\phi} = \phi_R + j\phi_I \tag{285}$$

$$\bar{A}_\alpha = A_\alpha e^{j\theta_{A\alpha}} = A_{\alpha R} + jA_{\alpha I} \tag{286}$$

Using the notation of equations (65) and (66) allows equation (283) to be written as the following two relations

$$\sum_{\beta} R_{2\beta}(\phi_R,\phi_I) - \varepsilon\mu R_{2t}(\phi_R,\phi_I) \sim -\rho_q/\varepsilon \cos\theta_{\rho q} \tag{287}$$

$$\sum_{\beta} I_{2\beta}(\phi_R,\phi_I) - \varepsilon\mu I_{2t}(\phi_R,\phi_I) \sim - \rho_q/\varepsilon \sin \theta_{\rho q} \quad (288)$$

while equation (284) can be written as the following six approximations

$$\sum_{\beta} R_{2\beta}(A_{xR},A_{xI}) - \varepsilon\mu R_{2t}(A_{xR},A_{xI}) \sim - \mu j_x \cos \theta_{jx} \quad (289)$$

$$\sum_{\beta} R_{2\beta}(A_{yR},A_{yI}) - \varepsilon\mu R_{2t}(A_{yR},A_{yI}) \sim - \mu j_y \cos \theta_{jy} \quad (290)$$

$$\sum_{\beta} R_{2\beta}(A_{zR},A_{zI}) - \varepsilon\mu R_{2t}(A_{zR},A_{zI}) \sim - \mu j_z \cos \theta_{jz} \quad (291)$$

$$\sum_{\beta} I_{2\beta}(A_{xR},A_{xI}) - \varepsilon\mu I_{2t}(A_{xR},A_{xI}) \sim - \mu j_x \sin \theta_{jx} \quad (292)$$

$$\sum_{\beta} I_{2\beta}(A_{yR},A_{yI}) - \varepsilon\mu I_{2t}(A_{yR},A_{yI}) \sim - \mu j_y \sin \theta_{jy} \quad (293)$$

$$\sum_{\beta} I_{2\beta}(A_{zR},A_{zI}) - \varepsilon\mu I_{2t}(A_{zR},A_{zI}) \sim - \mu j_z \sin \theta_{jz} \quad (294)$$

Finally the gauge conditions for an electromagnetic field with broken internal symmetry is written as[9,10]

$$\bar{\vec{\nabla}} \cdot \bar{\vec{A}} + \varepsilon\mu \frac{\partial \bar{\phi}}{\partial \bar{t}} = 0 \quad (295)$$

which can be written in terms of real and imaginary components as

$$\sum_{\beta} R_{1\beta}(A_{\beta R},A_{\beta I}) + \varepsilon\mu R_{1t}(\phi_R,\phi_I) = 0 \quad (296)$$

$$\sum_{\beta} I_{1\beta}(A_{\beta R},A_{\beta I}) + \varepsilon\mu I_{1t}(\phi_R,\phi_I) = 0 \quad (297)$$

where the sum is over $\beta = x$, y and z.

6. VACUUM WITH BROKEN INTERNAL SYMMETRIES. Of special importance to the propagation of electromagnetic waves are the properties of the vacuum state. The vacuum state may exhibit the same broken internal symmetries as does bulk

matter. Consider the vacuum state to have a zero temperature state coupled to a thermal state in such a way that the vacuum energy density and pressure for low temperatures are given by

$$\bar{E}^{(v)} = \bar{E}_o^{(v)} + \bar{E}_j^{(v)} T^j + \cdots = E^{(v)} e^{j\theta_E^{(v)}} \tag{298}$$

$$\bar{P}^{(v)} = \bar{P}_o^{(v)} + \bar{P}_j^{(v)} T^j + \cdots = P^{(v)} e^{j\theta_P^{(v)}} \tag{299}$$

where $\bar{E}^{(v)}$ and $\bar{P}^{(v)}$ = vacuum energy density and pressure respectively, $\bar{E}_o^{(v)}$ and $\bar{P}_o^{(v)}$ = zero temperature vacuum energy density and pressure respectively, and $\bar{E}_j^{(v)}$ and $\bar{P}_j^{(v)}$ = thermal coefficients for the vacuum energy density and pressure respectively. The vacuum Grüneisen parameter is defined by

$$\bar{\gamma}_o^{(v)} = \gamma_o^{(v)} e^{j\theta_{\gamma o}^{(v)}} \tag{300}$$

$$= \left[\frac{\partial \bar{P}^{(v)}/\partial T}{\partial \bar{E}^{(v)}/\partial T}\right]_{T=o} = \frac{\bar{P}_j^{(v)}}{\bar{E}_j^{(v)}} = \frac{1}{(j-1)} \frac{V}{\bar{U}_j^{(v)}} \frac{\partial \bar{U}_j^{(v)}}{\partial V}$$

where $\bar{U}_j^{(v)} = V\bar{E}_j^{(v)}$, and where j = index that describes the thermal properties of the vacuum.

The energy density $\bar{E}_o^{(v)}$ and Grüneisen parameter $\bar{\gamma}_o^{(v)}$ for the zero temperature vacuum are calculated from the simultaneous solution of two differential equations

$$\bar{E}_o^{(v)} - 3\left\{[1 + \bar{\gamma}_o^{(v)}]\bar{P}_o^{(v)} - \bar{K}_o^{(v)}\right\} = 0 \tag{301}$$

$$1 + j + \frac{j\bar{\gamma}_o^{(v)}\bar{P}_o^{(v)}}{\bar{P}_o^{(v)} - \bar{K}_o^{(v)}} + 3n\frac{d\bar{\gamma}_o^{(v)}}{dn} = 0 \tag{302}$$

which are just equations (252) and (253) of Reference 8 with their right hand sides set equal to zero. A trivial solution of equations (252) and (253) of Reference 8 with their right hand sides equal to zero is just $\bar{E}_o^{(v)} = 0$ and $\bar{E}_j^{(v)} = 0$ which is equivalent to the unrenormalized vacuum $E_o^a = 0$ and $E_j^a = 0$. A non-trivial solution is obtained by simultaneously solving equations (301) and (302). It is easy to show that equation (301) can be written as

$$3n^2 \frac{d^2\bar{E}_o^{(v)}}{dn^2} - 3[1 + \bar{\gamma}_o^{(v)}]n \frac{d\bar{E}_o^{(v)}}{dn} + [3\bar{\gamma}_o^{(v)} + 4]\bar{E}_o^{(v)} = 0 \qquad (303)$$

The vacuum radiation energy density and pressure are written as[8]

$$\bar{E}_r^{(v)} = \bar{E}_{or}^{(v)} + \bar{E}_{jr}^{(v)}T^j + \cdots \qquad (304)$$

$$\bar{P}_r^{(v)} = \bar{P}_{or}^{(v)} + \bar{P}_{jr}^{(v)}T^j + \cdots \qquad (305)$$

while the zero temperature radiation Gruneisen parameter for the vacuum is given by

$$\bar{\gamma}_{or}^{(v)} = \gamma_{or}^{(v)} e^{j\theta_{\gamma or}^{(v)}} \qquad (306)$$

$$= \left[\frac{\partial\bar{P}_r^{(v)}/\partial T}{\partial\bar{E}_r^{(v)}/\partial T}\right]_{T=o} = \frac{\bar{P}_{jr}^{(v)}}{\bar{E}_{jr}^{(v)}} = \frac{1}{(j-1)} \frac{V}{\bar{U}_{jr}^{(v)}} \frac{\partial\bar{U}_{jr}^{(v)}}{\partial V}$$

The vacuum radiation equations are then written as[8]

$$\bar{E}_{or}^{(v)} - 3\left\{[1 + \bar{\gamma}_o^{(v)}]\bar{P}_{or}^{(v)} - \bar{K}_{or}^{(v)}\right\} - 3 \frac{\bar{E}_{jr}^{(v)}}{\bar{E}_j^{(v)}} \bar{P}_o^{(v)}[\bar{\gamma}_{or}^{(v)} - \bar{\gamma}_o^{(v)}] = 0 \qquad (307)$$

$$j\bar{E}_j^{(v)}[\bar{\alpha}^{(v)}\bar{K}_{or}^{(v)} - \bar{\beta}^{(v)}\bar{P}_{or}^{(v)}] + \bar{E}_{jr}^{(v)}\bar{S}_{jr}^{(v)} = 0 \qquad (308)$$

which are just equations (287) and (288) of Reference 8 with their right hand sides set equal to zero and a superscript (v) added to indicate a vacuum solution.

Therefore in principle asymmetric vacuum state is formally identical to the asymmetric bulk matter state. In fact, the vacuum is simpler than the bulk matter state as can be seen by comparing equations (301), (302), (307) and (308) with equations (252), (253), (287) and (288) respectively of Reference 8. The vacuum is expected to exhibit a broken internal symmetry state that is described by $\theta_P^{(v)}$ and $\theta_\gamma^{(v)}$. The broken symmetry of the vacuum will impress broken symmetries on the kinematic and dynamic variable of particles moving in the vacuum. Similarly, electromagnetic waves in the vacuum are expected to possess electric and magnetic fields and a spacetime coordinate description that exhibit internal phase angles.

7. CONCLUSION. The effects of the broken symmetry of space and time on electromagnetism in matter and the vacuum is considered, and Maxwell's equations with broken internal symmetries are developed. The Lorentz covariance of these equations is assumed to be valid but must now be represented in the form of complex number Lorentz transformations. The results of the Michelson-Morley experiment cannot be used to place a limit on the magnitude or the internal phase angle of the velocity of a particle moving in the vacuum. Experiments conducted in asymmetric bulk matter may be fruitful because the internal symmetries of spacetime are larger in this case than for the vacuum. The wave equations and gauge conditions for electromagnetic waves with broken internal symmetries are easily developed. Finally, the broken symmetry properties of the vacuum are obtained by solving a set of coupled differential equations which are similar in form to the corresponding equations for asymmetric bulk matter.

ACKNOWLEDGEMENT

The author wishes to thank Elizabeth K. Klein for typing this paper.

REFERENCES

1. O'Raifeartaigh, L., Group Structure of Gauge Theories, Cambridge University Press, Cambridge, 1986.

2. Taylor, J., Gauge Theories of Weak Interactions, Cambridge University Press, Cambridge, 1976.

3. Ramond, P., Field Theory: A Modern Primer, Benjamin, New York, 1981.

4. Mandl, F. and Shaw, G., Quantum Field Theory, John Wiley, New York, 1984.

5. Itzykson, C. and Zuber, J., Quantum Field Theory, McGraw-Hill, New York, 1980.

6. de Wit, B. and Smith, J., Field Theory in Particle Physics, North-Holland, New York, 1986.

7. Weiss, R. A., Relativistic Thermodynamics, Vols. 1 and 2, Exposition Press, New York, 1976.

8. Weiss, R. A., "Thermodynamic Gauge Theory of Solids and Quantum Liquids with Internal Phase", Fifth Army Conference on Applied Mathematics and Computing, West Point, New York, June 15-18, 1987, p. 649.

9. Jackson, J., Classical Electrodynamics, John Wiley, New York, 1975.

10. Panofsky, W. and Phillips, M., Classical Electricity and Magnetism, Addison-Wesley, Reading, MA, 1955.

11. Stratton, J., Electromagnetic Theory, McGraw-Hill, New York, 1941.

12. Smythe, W., Static and Dynamic Electricity, McGraw-Hill, New York, 1950.

13. Becker, R., Electromagnetic Fields and Interactions, Dover, New York, 1964.

14. Jeans, J., The Mathematical Theory of Electricity and Magnetism, Cambridge University Press, Cambridge, 1951.

15. Menzel, D., Mathematical Physics, Prentice-Hall, New York, 1953.

16. Whittaker, E., A History of the Theories of Aether and Electricity, Philosophical Library, New York, 1951.

17. Sommerfeld, A., Electrodynamics, Academic, New York, 1952.

18. Kong, J., Electromagnetic Wave Theory, John Wiley, New York, 1986.

19. Chandrasekhar, S., Plasma Physics, University of Chicago Press, Chicago, 1960.

20. Artsimovich, L. A., A Physicists ABC on Plasma, MIR Publishers, Moscow, 1978.

21. Spitzer, L., Physics of Fully Ionized Gases, John Wiley-Interscience, New York, 1964.

22. Thompson, W. B., An Introduction to Plasma Physics, Addison-Wesley, Reading, Massachusetts, 1962.

23. Weinberg, S., Gravitation and Cosmology: Principles and Applications of the General Theory of Relativity, John Wiley, New York, 1972.

24. Pauli, W., Theory of Relativity, Pergamon, New York, 1958.

25. Møller, C., The Theory of Relativity, Oxford, New York, 1952.

26. Synge, J. L., Relativity: The Special Theory, North-Holland, Amsterdam, 1965.

27. Bergmann, P., Introduction to the Theory of Relativity, Prentice-Hall, New York, 1953.

28. Eddington, A., The Mathematical Theory of Relativity, Cambridge University Press, Cambridge, 1952.

29. Weyl, H., Space-Time-Matter, Dover, New York, 1922.

30. Robertson, H. and Noonan, T., Relativity and Cosmology, Saunders, Philadelphia, 1968.

31. Tolman, R., Relativity Thermodynamics and Cosmology, Oxford, New York, 1934.

32. Aharoni, J., The Special Theory of Relativity, Oxford, New York, 1959.

33. Misner, C., Thorne, K. and Wheeler, J., Gravitation, Freeman, San Francisco, 1973.

34. Zeldovich, Ya. and Novikov, I., Relativistic Astrophysics Volume 1 Stars and Relativity, University of Chicago Press, Chicago, 1971.

35. Born, M., Atomic Physics, Hafner, New York, 1953.

36. Zensus, J. A. and Pearson, T. J., Eds., Superluminal Radio Sources, Cambridge University Press, New York, 1987.

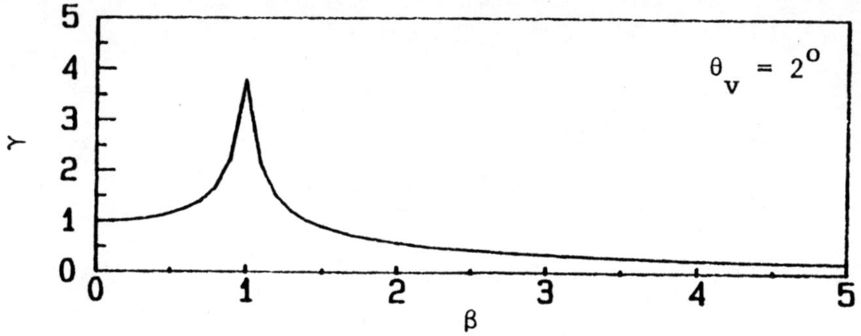

Figure 1. Magnitude of velocity factor versus magnitude of velocity ratio.

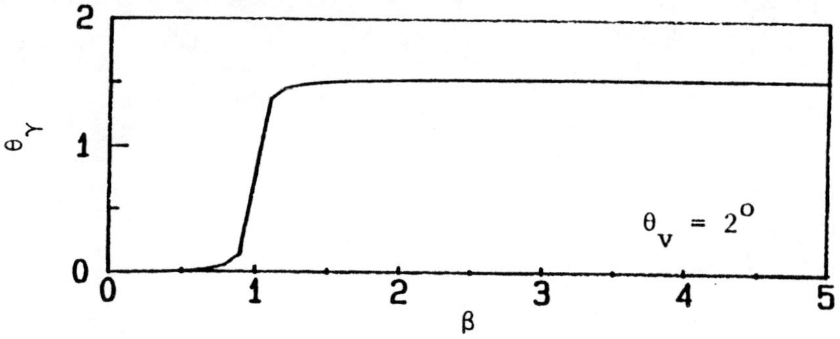

Figure 2. Phase angle of velocity factor versus magnitude of velocity ratio.

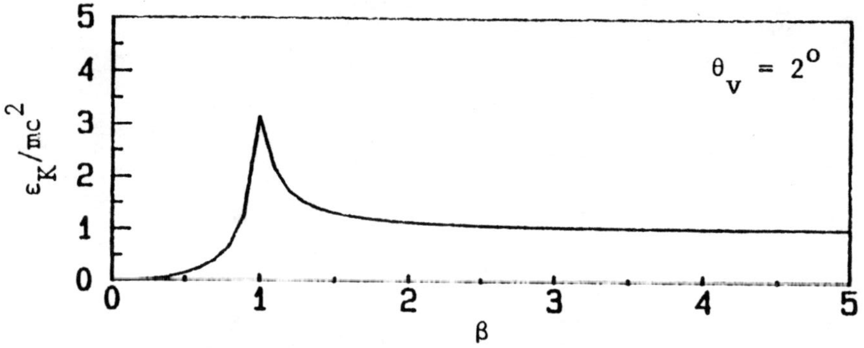

Figure 3. Normalized magnitude of kinetic energy versus magnitude of velocity ratio.

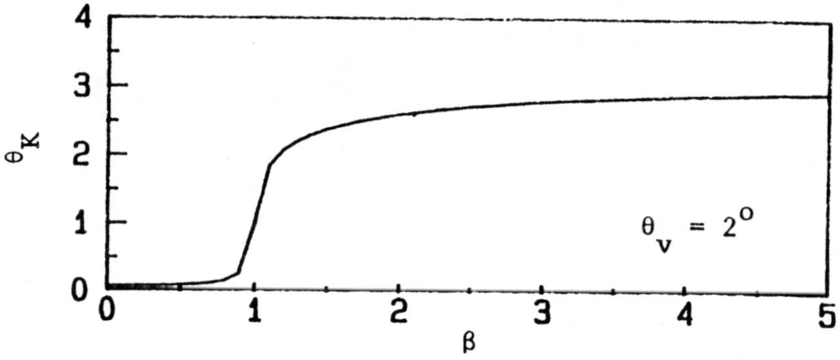

Figure 4. Phase angle of kinetic energy versus magnitude of velocity ratio.

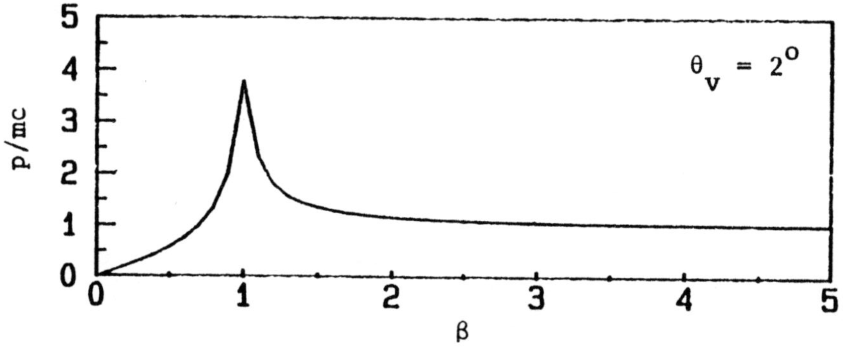

Figure 5. Normalized magnitude of momentum versus magnitude of velocity ratio.

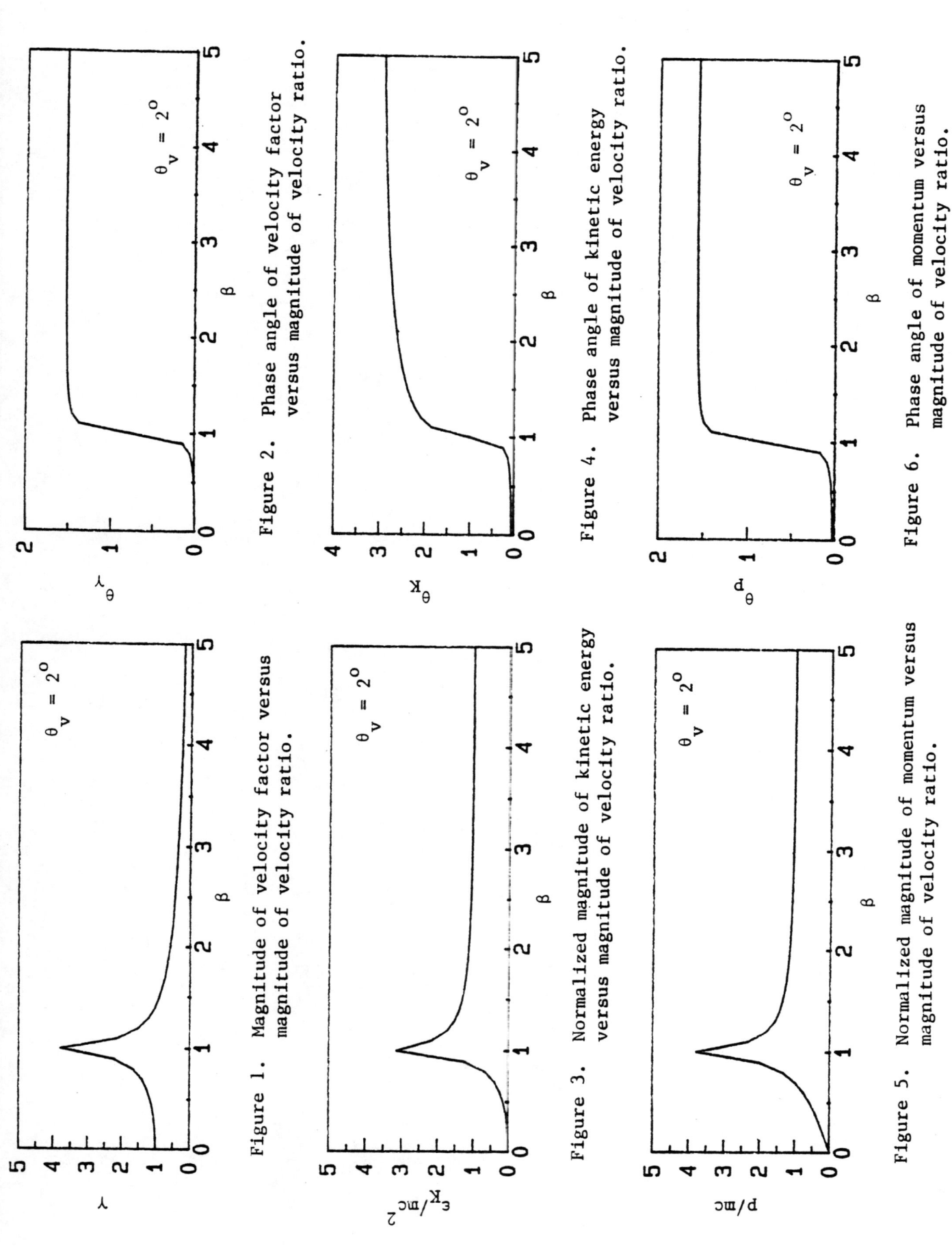

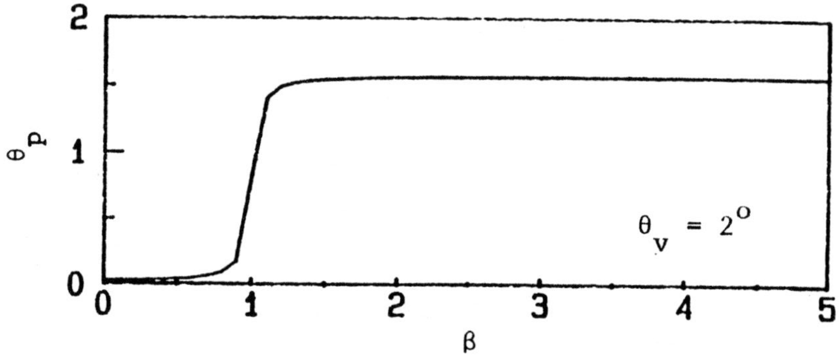

Figure 6. Phase angle of momentum versus magnitude of velocity ratio.

H

GAUGE THEORY OF ATOMIC PROCESSES

Richard A. Weiss
U. S. Army Engineer Waterways Experiment Station
Vicksburg, Mississippi 39180

ABSTRACT. Atomic particle processes that occur within bulk matter or the vacuum are expected to be influenced by the broken symmetries of the thermodynamic ground and excited states of these systems. Internal phase angles are associated with the space and time coordinates and kinematic and dynamic variables of particles and radiation in bulk matter or vacuum with broken internal symmetries. A broken symmetry photon gas in bulk matter or vacuum is considered, and the radiation pressure and energy density is calculated. The geometric angles between kinematic variables and between dynamic variables have internal phase angles. This affects the description of the photoelectric effect, Compton effect, and Coulomb scattering in bulk matter and the vacuum. Thomson, Compton, Rutherford, Mott, Bhabha, and Møller scattering processes in broken symmetry systems are investigated. The Schrödinger and Dirac equations are developed for a particle located in bulk matter or vacuum with broken internal symmetries. This work will have applications to nuclear explosions and the interaction of directed energy beams with matter.

1. INTRODUCTION.

In the past decade, great advances were made in the theory of the elementary forces which bind the universe. These advances developed through the realization that gauge theory is the natural framework for describing the four basic interactions that occur in nature.[1-3] Gauge theory was first formulated many years ago by Hermann Weyl, but only recently has its real importance to physics been understood.[1] In some cases when gauge symmetry is broken spontaneously by some special set of forces, a coherent state of matter can be formed as in the case of superconductivity where the Cooper pairs of electrons break the ground state gauge symmetry through electron-phonon interactions.[4]

It has been suggested that vacuum interactions with bulk matter may produce a coherent ground state which is described by thermodynamic functions that possess internal phase angles.[5] This coherent broken symmetry ground state can possibly influence the microscopic processes that take place in bulk matter. The effect can occur in two ways: first, through the Euler equations by which fluid elements are expected to have space and time coordinates and kinematic and dynamic variables that have internal phase angles. Secondly, a microscopic gauge interaction between material particles can be induced by Minkowski spacetime, and this complex number gauge interaction will impart internal phases to the space and time coordinates and to the kinematic and dynamic variables of individual particles. Therefore individual particles in bulk matter require complex numbers for their kinematic and dynamic descriptions and for their coordinate locations in space and time. The same conclusions are valid for the vacuum with broken internal symmetries, because the vacuum can be considered as a special simplified case of bulk matter.

The coherent state of bulk matter is due to spacetime interactions, and these have been described by a bulk matter relativistic trace equation whose scalar form for symmetrical bulk matter is[6]

$$U_s + T\left(\frac{dU_s}{dT}\right)_{P_sV} - 3V\frac{d}{dV}(P_sV)_{U_s} = U^a + T\left(\frac{dU^a}{dT}\right)_{p^aV} \tag{1}$$

where U_s = renormalized internal energy for symmetric bulk matter, P_s = renormalized pressure for symmetric bulk matter, T = absolute temperature, V = volume of substance, and U^a and P^a = corresponding nonrelativistic internal energy and pressure. Throughout this paper the index "a" will refer to nonrelativistic (unrenormalized) calculations. The complex number form of the relativistic trace equation that describes the coherent broken symmetry state of bulk matter is given by[5]

$$\bar{U} + T\left(\frac{d\bar{U}}{dT}\right)_{\bar{P}V} - 3V\frac{d}{dV}(\bar{P}V)_{\bar{U}} = U^a + T\left(\frac{dU^a}{dT}\right)_{p^aV} \tag{2}$$

or equivalently as

$$(1 - \bar{b} + T\frac{\partial}{\partial T} - \bar{b}V\frac{\partial}{\partial V})\bar{E} - 3(1 + \bar{\gamma} + V\frac{\partial}{\partial V} - \bar{\gamma}T\frac{\partial}{\partial T})\bar{P} = \psi^a \tag{3}$$

where

$$\psi^a = (T\frac{\partial}{\partial T} - b^aV\frac{\partial}{\partial V} + 1 - b^a)E^a \tag{4}$$

and where $\bar{U}$, $\bar{E}$, $\bar{P}$, $\bar{\gamma}$, and $\bar{b}$ are complex number representations of the internal energy, energy density, pressure, and the gauge parameters.[5] With their right hand sides set equal to zero, equations (2) and (3) describe the broken symmetry thermodynamic ground state of the vacuum. Therefore the broken symmetry thermodynamic ground state of the vacuum is a simpler special case of the broken symmetry state of bulk matter.

Due to the spacetime interactions with bulk matter, the single particle energy must contain a gauge potential that produces the difference between U and U^a at the macroscopic level. Corresponding to equation (1) the noninteracting single particle energy is given by[7]

$$\varepsilon_{is}^{free} = \sqrt{c^2p_s^2 + m^2c^4} + V_g^s \tag{5}$$

where p_s = single particle momentum for a symmetrical system, c = light speed, m = proper mass, and V_g^s = scalar gauge potential for symmetrical matter. The gauge potential is zero when PV = αU where α = constant, and $U = U^a$.[6] When it has a non-zero value, the gauge potential breaks the Lorentz symmetry of the system.[6] The condition $V_g^s \neq 0$ is valid for the general case of a noninteracting zero temperature Fermi gas (for which $PV \neq \alpha U$), except for the low density non-relativistic case and the high density ultra-relativistic case for which $PV = \alpha U$ and $V_g^s = 0$.[6] The single particle energy for the non-interacting case corresponding to equation (2) for broken symmetry matter is given by

$$\bar{\varepsilon}_i^{free} = \sqrt{c^2\bar{p}^2 + m^2c^4} + \bar{V}_g \tag{6}$$

$$\bar{V}_g = V_g e^{j\theta_{Vg}} \tag{7}$$

where $\bar{p}$ = complex number single particle momentum, and $\bar{V}_g$ = complex number gauge potential. In a similar fashion, $\bar{V}_g = 0$ when $\bar{P}V = \alpha\bar{U}$. It is just the derivative terms in equations (1) and (2), required for gauge invariance, which produce the spacetime interaction gauge potentials that prevent the single particle energy and momentum from being four vectors in equations (5) and (6) when $PV \neq \alpha U$ and $\bar{P}V \neq \alpha\bar{U}$ respectively. For the interacting case, the single particle energy is written as

$$\varepsilon_i^s = \sqrt{c^2p_s^2 + m^2c^4} + V_g^s + V_e^s \tag{8}$$

corresponding to equation (1) for a symmetrical system, and as

$$\bar{\varepsilon}_i = \sqrt{c^2\bar{p}^2 + m^2c^4} + \bar{V}_g + \bar{V}_e \tag{9}$$

corresponding to the relativistic trace equation (2) for a system with broken internal symmetry. For the broken symmetry case the external potential is written as

$$\bar{V}_e = V_e e^{j\theta_{Ve}} \tag{10}$$

In this paper the "s" refers to a symmetrical renormalized system.

The gauge potentials V_g or $\bar{V}_g$ are determined indirectly from the solution of the trace equations (1) or (2). Consider the trace equation (2). This equation is solved to determine the bulk matter internal energy $\bar{U}$ in terms of the unrenormalized internal energy U^a . The unrenormalized thermodynamic functions are determined from the unrenormalized partition function Z^a which is given by[8,9]

$$Z^a = \int \eta e^{-\beta H^a} dq_a\, dp_a \tag{11}$$

where η = degeneracy, $\beta = 1/(kT)$, q_a and p_a = conventional generalized coordinates and momenta respectively, and the unrenormalized Hamiltonian is given by

$$H^a(q_a, p_a) = \sqrt{c^2p_a^2 + m^2c^4} + V_e^a \tag{12}$$

where V_e^a = unrenormalized external potential, and[8,9]

$$U^a = -\left(\frac{\partial\, \ell n\, Z^a}{\partial\beta}\right)_V \tag{13}$$

$$P^a = \frac{1}{\beta}\left(\frac{\partial\, \ell n\, Z^a}{\partial V}\right)_\beta \tag{14}$$

The trace equation (2) is then used to determine the renormalized complex number internal energy and pressure $\bar{U}$ and $\bar{P}$ respectively. These values of $\bar{U}$ and $\bar{P}$ are then used to determine the complex number renormalized partition function $\bar{Z}$ from

$$\bar{U} = -\left(\frac{\partial\, \ell n\, \bar{Z}}{\partial\beta}\right)_V \tag{15}$$

$$\bar{P} = \frac{1}{\beta}\left(\frac{\partial\, \ell n\, \bar{Z}}{\partial V}\right)_\beta \tag{16}$$

where

$$\bar{Z} = \int \eta e^{-\beta\bar{H}} d\bar{q}\, d\bar{p} = Ze^{j\theta_Z} \tag{17}$$

and $\bar{q}$ and $\bar{p}$ = complex number generalized coordinates and momenta respectively, and where the renormalized complex number Hamiltonian is given by

$$\bar{H}(\bar{q},\bar{p}) = \sqrt{c^2\bar{p}^2 + m^2c^4} + \bar{V}_g + \bar{V}_e \tag{18}$$

From equations (15), (16), and (17) it follows that

$$U \cos\theta_U = -\left(\frac{\partial\, \ell n\, Z}{\partial\beta}\right)_V \tag{19}$$

$$U \sin\theta_U = -\left(\frac{\partial\theta_Z}{\partial\beta}\right)_V \tag{20}$$

$$P \cos\theta_P = \frac{1}{\beta}\left(\frac{\partial\, \ell n\, Z}{\partial V}\right)_\beta \tag{21}$$

$$P \sin\theta_P = \frac{1}{\beta}\left(\frac{\partial\theta_Z}{\partial V}\right)_\beta \tag{22}$$

Therefore equations (19) through (22) can be used to determine Z and θ_Z for a relativistic system with broken internal symmetry. From a knowledge of $\bar{Z}$, equations (17) and (18) are then inverted to determine $\bar{V}_g$ and $\bar{V}_e$. In summary,

$$Z^a \rightarrow P^a, U^a \rightarrow \bar{P}, \bar{U} \rightarrow \bar{Z} \rightarrow \bar{V}_g, \bar{V}_e \tag{23}$$

Because $\bar{V}_g$ and $\bar{V}_e$ are complex numbers, it is expected that the coordinates of space and time must also be complex numbers. Note that it is the real parts of the complex number quantities such as coordinates, momentum, energy, pressure, frequency, angles and scattering cross sections that are the measured quantities.

Therefore, macroscopic local gauge invariance suggests the existence of a symmetry breaking microscopic gauge potential. Also, the macroscopic broken symmetry state required by equation (2) suggests that the space and time coordinates and the kinematic and dynamic quantities such as single particle velocity, acceleration, and force should be represented by complex numbers that include a description of internal phase angles. This paper indicates the effects of microscopic internal phase angles on the photon gas, and on such elementary atomic processes as the photoelectric effect, Compton effect, and Coulomb scattering. The forms of the Dirac and Schrödinger equations for particles in bulk matter or vacuum with broken symmetry are developed.

2. BROKEN SYMMETRY PHOTON GAS IN BULK MATTER. This section describes a photon gas with broken internal symmetry interacting with bulk matter that also has internal phase angles. The spectral energy density of a symmetrical photon gas is given by Planck's law as follows[10-12]

$$E_{\nu s} = \frac{A}{e^{h\nu/kT} - 1} = E_{\nu s}^{(v)} = E_{\nu a} = E_{\nu a}^{(v)} \tag{24}$$

where

$$A = \frac{8\pi h\nu^3}{c^3} \tag{25}$$

and where $E_{\nu s}$ = spectral energy density of radiation in symmetric matter, $E_{\nu s}^{(v)}$ = spectral energy density in the symmetric vacuum, h = Planck's constant, k = Boltzmann's constant, and T = absolute temperature. For this case the total energy density is given by the Stefan-Boltzmann law[10,11]

$$E_{rs} = \int_0^\infty E_{\nu s} d\nu = \sigma T^4 = E_{rs}^{(v)} = E_{ra} = E_{ra}^{(v)} \tag{26}$$

where E_{rs} and $E_{rs}^{(v)}$ = total energy density for a symmetrical photon gas in symmetric matter and the symmetric vacuum respectively, and σ = Stefan-Boltzmann constant. The pressure of the symmetrical photon gas is given for the symmetrical vacuum by

$$P_{\nu s}^{(v)} = \frac{1}{3} E_{\nu s}^{(v)} = P_{\nu a}^{(v)} = \frac{1}{3} E_{\nu a}^{(v)} \tag{27}$$

$$P_{rs}^{(v)} = \frac{1}{3} E_{rs}^{(v)} = \frac{1}{3} \sigma T^4 = \frac{1}{3} E_{ra}^{(v)} \tag{28}$$

where $P_{\nu s}^{(v)}$ and $P_{rs}^{(v)}$ = spectral and total radiation pressure respectively for the symmetrical photon gas in symmetric vacuum.

In order to write Planck's law for radiation in matter or the vacuum with broken internal symmetries, a complex number form of the radiation frequency is adopted and written as

$$\bar{\nu} = \nu e^{j\theta_\nu} = \nu(\cos\theta_\nu + j \sin\theta_\nu) \tag{29}$$

where $\bar{\nu}$ = complex number frequency, ν = magnitude of frequency, and θ_ν = frequency phase angle. For radiation in the asymmetrical vacuum, the radiation frequency is written as

$$\bar{\nu}^{(v)} = \nu e^{j\theta_\nu^{(v)}} \tag{30}$$

where $\theta_\nu^{(v)}$ = internal phase angle of the photon frequency in the asymmetric vacuum. Using equations (24) and (29), the complex number form of Planck's law is written as

$$\bar{E}_\nu = \frac{\bar{A}}{e^{h\bar{\nu}/kT} - 1} \tag{31}$$

where

$$\bar{A} = \frac{8\pi h \bar{\nu}^3}{c^3} \tag{32}$$

Placing equation (29) into equation (31) yields

$$\bar{E}_\nu = \frac{A}{D}(B + jC) = E_\nu e^{j\theta_{E\nu}} \tag{33}$$

$$E_\nu = \frac{A}{D}\sqrt{B^2 + C^2} \tag{34}$$

$$\theta_{E\nu} = \tan^{-1}\left(\frac{C}{B}\right) \tag{35}$$

where A is given by equation (25) and where

$$D = e^{2x} - 2 \cos y \, e^{x} + 1 \tag{36}$$

$$B = \cos(3\theta_{\nu}) \, (\cos y \, e^{x} - 1) + \sin(3\theta_{\nu}) \sin y \, e^{x} \tag{37}$$

$$C = \sin(3\theta_{\nu}) \, (\cos y \, e^{x} - 1) - \cos(3\theta_{\nu}) \sin y \, e^{x} \tag{38}$$

$$x = \frac{h\nu}{kT} \cos \theta_{\nu} \tag{39}$$

$$y = \frac{h\nu}{kT} \sin \theta_{\nu} \tag{40}$$

The same expressions that are valid for radiation in matter with broken internal symmetry are also valid for radiation in the vacuum with broken internal symmetry if the substitution $\theta_\nu \rightarrow \theta_\nu^{(v)}$ is made in equations (36) through (40). This gives the photon energy density for radiation in the asymmetric vacuum as

$$\bar{E}_{\nu}^{(v)} = E_{\nu}^{(v)} e^{j\theta_{E\nu}^{(v)}} \tag{41}$$

where $E_{\nu}^{(v)}$ and $\theta_{E\nu}^{(v)}$ are obtained from equations (34) and (35) respectively.

Matter in general has a spectral index of refraction, but it does not enter into the photon spectral energy density as given by the Planck function.[14] This is because the Planck function is universal and does not include specific properties of matter.[14] The index of refraction does not enter the spectral energy calculation whether or not the radiation is symmetric or not. Later in this paper it will be shown that the index of refraction does enter the calculation of the total energy density of photons in matter due to an increased photon density in matter.

The spectral radiation pressure associated with the complex spectral radiation energy density given in equation (31) is obtained by a generalization of a radiation pressure formula given in the literature for mechanical radiation in matter as[15]

$$P_{\nu a} = \left(\frac{1}{3} + \frac{n}{W_{\nu}^{a}} \frac{dW_{\nu}^{a}}{dn}\right) E_{\nu a} = \left(\frac{1}{3} - \frac{n}{\mu_{\nu a}} \frac{d\mu_{\nu a}}{dn}\right) E_{\nu a} \tag{42}$$

where $P_{\nu a}$ = spectral radiation pressure, n = particle number density for matter, and W_ν^a = speed of waves of frequency ν. The generalization of equation (42) to the case of broken symmetry electromagnetic radiation is given by

$$\bar{P}_\nu = \left(\frac{1}{3} - \frac{n}{\bar{\mu}_\nu}\frac{d\bar{\mu}_\nu}{dn}\right)\bar{E}_\nu \tag{43}$$

where $\bar{P}_\nu$ = complex number spectral radiation pressure in matter, and $\bar{\mu}_\nu$ = complex number spectral index of refraction for electromagnetic waves in matter. For the vacuum with broken internal symmetry, equation (43) reduces to

$$\bar{P}_\nu^{(v)} = \frac{1}{3}\bar{E}_\nu^{(v)} \tag{44}$$

because $\bar{\mu}_\nu$ = 1 for the vacuum. If the complex number spectral index of refraction is written as

$$\bar{\mu}_\nu = \mu_\nu e^{j\theta_{\mu\nu}} \tag{45}$$

then

$$\frac{n}{\bar{\mu}_\nu}\frac{d\bar{\mu}_\nu}{dn} = \frac{n}{\mu_\nu}\frac{d\mu_\nu}{dn} + jn\frac{d\theta_{\mu\nu}}{dn} = H_\nu e^{j\beta_{\mu\nu,n}}$$

and then equation (43) can be rewritten as

$$\bar{P}_\nu = E_\nu\left[\frac{1}{3}e^{j\theta_{E\nu}} - H_\nu e^{j(\theta_{E\nu} + \beta_{\mu\nu,n})}\right] \tag{47}$$

where

$$H_\nu = \sqrt{\left(\frac{n}{\mu_\nu}\frac{d\mu_\nu}{dn}\right)^2 + \left(n\frac{d\theta_{\mu\nu}}{dn}\right)^2} \tag{48}$$

and where

$$\tan\beta_{\mu\nu,n} = n\frac{d\theta_{\mu\nu}}{dn}\Bigg/\frac{n}{\mu_\nu}\frac{d\mu_\nu}{dn} \tag{49}$$

and finally where $\theta_{E\nu}$ is given by equation (35). For the vacuum μ_ν = 1 and $\theta_{\mu\nu}$ = 0 so that

$$\bar{P}_\nu^{(v)} = \frac{1}{3} E_\nu^{(v)} e^{j\theta_{E\nu}^{(v)}} \tag{50}$$

where $\theta_{E\nu}^{(v)}$ is given by equation (35) with the substitution $\theta_\nu \rightarrow \theta_\nu^{(v)}$. For the symmetric vacuum $\theta_\nu^{(v)} = 0$ and $\theta_{E\nu}^{(v)} = 0$ so that

$$P_{\nu s}^{(v)} = \frac{1}{3} E_{\nu s}^{(v)} = \frac{1}{3} E_{\nu s} = P_{\nu a}^{(v)} \tag{51}$$

where $E_{\nu s} = E_{\nu s}^{(v)}$ for the symmetric vacuum or symmetric matter is given by equation (24). For symmetric radiation in symmetric matter equation (43) becomes

$$P_{\nu s} = \left(\frac{1}{3} - \frac{n}{\mu_{\nu s}} \frac{d\mu_{\nu s}}{dn}\right) E_{\nu s} \neq P_{\nu a} \tag{52}$$

where $P_{\nu s}$ = spectral radiation pressure for symmetric matter. Note that although $E_{\nu s}^{(v)} = E_{\nu s}$ one has $P_{\nu s} \neq P_{\nu s}^{(v)}$ on account of the spectral index of refraction that appears in equation (52).

Writing the complex spectral radiation pressure for asymmetric radiation as

$$\bar{P}_\nu = P_\nu e^{j\theta_{P\nu}} \tag{53}$$

and using equation (47) gives

$$P_\nu \cos \theta_{P\nu} = E_\nu \left[\frac{1}{3} \cos \theta_{E\nu} - H_\nu \cos (\theta_{E\nu} + \beta_{\mu\nu,n})\right] \tag{54}$$

$$P_\nu \sin \theta_{P\nu} = E_\nu \left[\frac{1}{3} \sin \theta_{E\nu} - H_\nu \sin (\theta_{E\nu} + \beta_{\mu\nu,n})\right] \tag{55}$$

Equations (54) and (55) give for the asymmetric vacuum

$$P_\nu^{(v)} \cos \theta_{P\nu}^{(v)} = \frac{1}{3} E_\nu^{(v)} \cos \theta_{E\nu}^{(v)} \tag{56}$$

$$P_\nu^{(v)} \sin \theta_{P\nu}^{(v)} = \frac{1}{3} E_\nu^{(v)} \sin \theta_{E\nu}^{(v)} \tag{57}$$

which can also be obtained directly from equation (50). From equations (56) and (57) it follows for the asymmetric vacuum that

$$P_\nu^{(v)} = \frac{1}{3} E_\nu^{(v)} \tag{58}$$

$$\theta_{P\nu}^{(v)} = \theta_{E\nu}^{(v)} \tag{59}$$

A comparison of equations (51) and (58) shows that this equation holds for both the symmetric and asymmetric vacuum. From equations (54) and (55) it follows that

$$\tan \theta_{P\nu} = \frac{\frac{1}{3} \sin \theta_{E\nu} - H_\nu \sin (\theta_{E\nu} + \beta_{\mu\nu,n})}{\frac{1}{3} \cos \theta_{E\nu} - H_\nu \cos (\theta_{E\nu} + \beta_{\mu\nu,n})} \tag{60}$$

and

$$P_\nu = E_\nu \sqrt{\frac{1}{9} + H_\nu^2 - \frac{2}{3} H_\nu \cos \beta_{\mu\nu,n}} \tag{61}$$

When $\beta_{\mu\nu,n} = 0$ and $\theta_{\mu\nu} = 0$ equation (61) reduces to the case of symmetric radiation in symmetric matter

$$P_{\nu s} = E_{\nu s}\left(\frac{1}{3} - H_{\nu s}\right) \tag{62}$$

where

$$H_{\nu s} = \frac{n}{\mu_{\nu s}} \frac{d\mu_{\nu s}}{dn} \tag{63}$$

for symmetric matter.

In order to determine the phase angles θ_ν and $\theta_\nu^{(v)}$ in terms of the pressure internal phase angle θ_P that is associated with the state equation of bulk matter, the mechanical equilibrium of matter and radiation with broken internal symmetries must be considered. Consider a piece of matter bathed in the surrounding radiation of a vacuum with broken internal symmetry. Were no radiation present, the matter would have a zero pressure $P = 0$ because the situation corresponds to a minimum value of the binding energy at the equilibrium density. When radiation is present inside the matter and outside in the vacuum, the density of matter shifts from its $P = 0$ equilibrium value to a new value for which $P \neq 0$. The new equilibrium density depends on the internal and external radiation densities (pressures) which in turn depends on the temperature and frequency for monochromatic radiation, or solely on temperature for thermal radiation.[16] The relationship between the induced matter pressure and the radiation pressure of a monochromatic radiation field will now be developed for matter and radiation with broken internal symmetries.

For matter in mechanical equilibrium with internal and external monochromatic radiation fields, the equilibrium condition at the surface boundary is

$$\bar{P} = \bar{P}_\nu^{(v)} - \bar{P}_\nu \tag{64}$$

where $\bar{P}_\nu^{(v)}$ = spectral radiation pressure in vacuum, $\bar{P}_\nu$ = spectral radiation pressure in matter, and $\bar{P}$ = complex matter mechanical pressure induced by the radiation fields. In component form equation (64) can be rewritten as

$$P \cos \theta_P = P_\nu^{(v)} \cos \theta_{P\nu}^{(v)} - P_\nu \cos \theta_{P\nu} \tag{65}$$

$$P \sin \theta_P = P_\nu^{(v)} \sin \theta_{P\nu}^{(v)} - P_\nu \sin \theta_{P\nu} \tag{66}$$

Combining equations (65) and (66) with equations (54) through (57) gives

$$P \cos \theta_P = \frac{1}{3} E_\nu^{(v)} \cos \theta_{E\nu}^{(v)} - E_\nu \left[\frac{1}{3} \cos \theta_{E\nu} - H_\nu \cos (\theta_{E\nu} + \beta_{\mu\nu,n})\right] \tag{67}$$

$$P \sin \theta_P = \frac{1}{3} E_\nu^{(v)} \sin \theta_{E\nu}^{(v)} - E_\nu \left[\frac{1}{3} \sin \theta_{E\nu} - H_\nu \sin (\theta_{E\nu} + \beta_{\mu\nu,n})\right] \tag{68}$$

For the vacuum the following conditions have been used

$$\mu_\nu^{(v)} = 1 \qquad \theta_{\mu\nu}^{(v)} = 0 \qquad \beta_{\mu\nu,n}^{(v)} = 0 \qquad H_\nu^{(v)} = 0 \tag{69}$$

Note that for the asymmetric vacuum $\theta_\nu^{(v)} \neq 0$ just as for the case of radiation in matter one has $\theta_\nu \neq 0$. From equations (65) and (66) it follows that

$$\tan \theta_P = \frac{P_\nu^{(v)} \sin \theta_{P\nu}^{(v)} - P_\nu \sin \theta_{P\nu}}{P_\nu^{(v)} \cos \theta_{P\nu}^{(v)} - P_\nu \cos \theta_{P\nu}} \tag{70}$$

$$P^2 = [P_\nu^{(v)}]^2 + P_\nu^2 - 2P_\nu P_\nu^{(v)} \cos [\theta_{P\nu}^{(v)} - \theta_{P\nu}] \tag{71}$$

For the case $\theta_{P\nu} = 0$ and $\theta_{P\nu}^{(v)} = 0$ equations (65) through (71) reduce to their proper scalar forms, for instance equation (71) gives the induced pressure in symmetrical matter as

$$P_s = P_{\nu s}^{(v)} - P_{\nu s} = \frac{1}{3} E_{\nu s}^{(v)} - (\frac{1}{3} - H_{\nu s})E_{\nu s} \tag{72}$$

But for symmetrical matter and symmetrical vacuum $E_{\nu s}^{(v)} = E_{\nu s}$, so that equation (72) can be written as

$$P_s = H_{\nu s} E_{\nu s} \tag{73}$$

In general P and θ_P are functions of matter density, and equations (70) and (71) can be satisfied only if the equilibrium density of matter is altered by the radiation fields. Therefore equilibrium at a surface requires that the internal phase of matter and radiation are related by equations (70) and (71).

Now the total (integrated) energy density and associated pressure needs to be determined for radiation in matter and the vacuum with broken internal symmetries. The integrated radiation energy density is obtained from equations (31) through (40) by the following integral

$$\bar{E}_r = E_r e^{j\theta_{Er}} = \int \bar{E}_\nu \, d\bar{\nu} = \int_0^\infty E_\nu \sqrt{1 + (\nu \, d\theta_\nu/d\nu)^2} \; e^{j(\theta_{E\nu} + \theta_\nu + \beta_{\nu,\nu})} \, d\nu \tag{74}$$

where

$$\tan \beta_{\nu,\nu} = \nu \frac{d\theta_\nu}{d\nu} \tag{75}$$

and where the following result was used

$$d\bar{\nu} = e^{j\theta_\nu} (d\nu + j\nu d\theta_\nu) = e^{j(\theta_\nu + \beta_{\nu,\nu})} \sqrt{1 + (\nu \, d\theta_\nu/d\nu)^2} \, d\nu \tag{76}$$

Therefore the gauge rotated frequency must be used to evaluate the integrated radiation energy density. The radiation energy density has the following real and imaginary parts

$$E_r^R = E_r \cos \theta_{Er} = \int_0^\infty E_\nu \sqrt{1 + (\nu \, d\theta_\nu/d\nu)^2} \cos (\theta_{E\nu} + \theta_\nu + \beta_{\nu,\nu}) \, d\nu \tag{77}$$

$$E_r^I = E_r \sin \theta_{Er} = \int_0^\infty E_\nu \sqrt{1 + (\nu \, d\theta_\nu/d\nu)^2} \sin (\theta_{E\nu} + \theta_\nu + \beta_{\nu,\nu}) \, d\nu \tag{78}$$

The magnitude and phase angle of the radiation energy is given in terms of these integrals as follows

$$E_r = \sqrt{(E_r^R)^2 + (E_r^I)^2} \qquad (79)$$

$$\tan \theta_{Er} = E_r^I / E_r^R \qquad (80)$$

The evaluation of the integrals in equations (74), (77), and (78) is not simple due to the complicated form of the spectral energy density given by equations (34) and (35). It follows from equations (34), (35), (77), and (78) that the energy density for asymmetric radiation does not have the T^4 temperature behaviour that is valid for symmetric radiation according to equation (26). The radiation pressure is given by

$$\bar{P}_r = \frac{1}{3} \bar{E}_r + \Delta\bar{P}_r \qquad (81)$$

where ΔP_r = small difference in radiation pressure due to internal phase angles, and considering only the complex number values of the Planck function. Material properties, such as the index of refraction, do not enter the Planck analysis.[17] Later in this paper it will be shown that the index of refraction enters the expressions for radiation energy density and pressure due to an increased photon density in matter. For the asymmetric vacuum the integrated radiation density is given by

$$\bar{E}_r^{(v)} = \int_0^\infty E_\nu^{(v)} \sqrt{1 + (\nu\, d\theta_\nu^{(v)}/d\nu)^2}\; e^{j[\theta_{E\nu}^{(v)} + \theta_\nu^{(v)} + \beta_{\nu,\nu}^{(v)}]} d\nu \qquad (82)$$

$$= E_r^{(v)} e^{j\theta_{Er}^{(v)}} = E_r^{R(v)} + jE_r^{I(v)}$$

$$\tan \theta_{Er}^{(v)} = E_r^{I(v)} / E_r^{R(v)} \qquad (83)$$

which is formally identical in structure to equations (74) through (80) for asymmetric radiation in matter. Asymmetric radiation in the vacuum also does not have a T^4 dependence. The radiation pressure for the asymmetric vacuum is given by

$$\bar{P}_r^{(v)} = \frac{1}{3} \bar{E}_r^{(v)} + \Delta\bar{P}_r^{(v)} \qquad (84)$$

where $\Delta\bar{P}_r^{(v)}$ = small difference in vacuum radiation pressure due to internal phase angles.

Considering the effects of increased photon number density due to the index of refraction the energy density and pressure of symmetrical thermal radiation in symmetrical matter is given by[16]

$$E_{rMs} = \sigma\mu_s^3 T^4 \tag{85}$$

$$P_{rMs} = \sigma\mu_s^3 T^4 \left(\frac{1}{3} - \frac{n}{\mu_s}\frac{d\mu_s}{dn}\right) \tag{86}$$

where E_{rMs} and P_{rMs} = measured radiation energy density and pressure for symmetrical matter and radiation, and where μ_s = density dependent index of refraction averaged over frequency. The term μ_s^3 arises from the general thermodynamic relations between pressure and energy denstiy.[16] A comparison of equations (26), (28), (85), and (86) shows that $E_{rMs} \neq E_{rs}$ and $P_{rMs} \neq P_{rs}$. The expressions in (26) and (28) are totally independent of any reference to material parameters (such as μ_s) and are the results of local thermodynamic equilibrium.[17] This is why the T^4 law is universal in the sense that it applies to all symmetric thermal radiation in symmetric matter or vacuum. The presence of the μ_s^3 term in equation (85) represents a diffusion effect where the photon number density is increased due to their slower speed in matter as compared to the vacuum.[16] The important point is that the μ_s^3 term (or any other dependence on material properties) does not originate from the Planck distribution. The subscript M (for measured value) is added to all expressions that include a μ_s^3 dependence.

In analogy to equations (85) and (86) for symmetric thermal radiation in symmetric matter, the measured radiation energy density and pressure for an asymmetric system is written as

$$\bar{E}_{rM} = \bar{W}(\bar{\mu})\bar{E}_r = E_{rM}e^{j\theta_{ErM}} \tag{87}$$

$$\bar{P}_{rM} = \bar{E}_{rM}\left(\frac{1}{3} - \frac{n}{\bar{\mu}}\frac{d\bar{\mu}}{dn}\right) = P_{rM}e^{j\theta_{PrM}} \tag{88}$$

where $\bar{E}_r$ = energy density for asymmetric radiation in matter as calculated from the complex number Planck function given in equation (74), $\bar{E}_{rM}$ = measured thermal radiation energy density in an asymmetric system, $\bar{\mu}$ = density dependent complex number index of refraction for asymmetric matter and averaged over frequency, and where $\bar{W}(\bar{\mu})$ = yet to be determined function of the complex number refraction index. The density dependent, frequency averaged, index of refraction for asymmetric matter is written as

$$\bar{\mu} = \mu e^{j\theta_\mu} \tag{89}$$

Note that equation (88) is already an approximation because the factor 1/3 holds only for symmetric radiation as shown in equation (81). For asymmetric matter and radiation the integrals in equations (77) and (78) have not been evaluated due to their complexity and so values of $\bar{\bar{E}}_r$ in equation (87) have not been found, but it is clear from equation (74) that the leading term of $\bar{\bar{E}}_r$ is the scalar σT^4 corresponding to symmetric radiation

$$\bar{\bar{E}}_r = \sigma T^4 + \Delta\bar{\bar{E}}_r \tag{90}$$

where $\Delta\bar{\bar{E}}_r$ is small if θ_ν is small. From equations (81) and (90) it follows that

$$\bar{\bar{P}}_r = \frac{1}{3}\sigma T^4 + \frac{\Delta\bar{\bar{E}}_r}{3} + \Delta\bar{\bar{P}}_r \tag{90A}$$

The determination of the exact value of $\bar{W}(\bar{\mu})$ is not possible since the value of $\bar{\bar{E}}_r$ has not been determined. However, an approximate value of the factor $\bar{W}(\bar{\mu})$ can be determined by using the first order term in equation (90). This is done by first considering the Gibbs-Helmholtz relation (also called Maxwell's relation) applied to $\bar{\bar{E}}_{rM}$ and $\bar{\bar{P}}_{rM}$ as follows[16]

$$-n\frac{\partial\bar{\bar{E}}_{rM}}{\partial n} + \bar{\bar{E}}_{rM} = T\frac{\partial\bar{\bar{P}}_{rM}}{\partial T} - \bar{\bar{P}}_{rM} \tag{91}$$

Combining equations (87) and (88) with equation (91) yields

$$-n\frac{\partial}{\partial n}(\bar{W}\bar{\bar{E}}_r) + \bar{W}\bar{\bar{E}}_r = \left(\frac{1}{3} - \frac{n}{\bar{\mu}}\frac{d\bar{\mu}}{dn}\right)\left[T\frac{\partial}{\partial T}(\bar{W}\bar{\bar{E}}_r) - \bar{W}\bar{\bar{E}}_r\right] \tag{92}$$

Then assuming $\bar{W}$ is a function of density alone, and $\bar{\bar{E}}_r$ is given by the scalar term in equation (90) so that $\bar{\bar{E}}_r \sim \sigma T^4$, it follows from equation (92) that

$$-n\frac{d\bar{W}}{dn} + \bar{W} \sim 3\bar{W}\left(\frac{1}{3} - \frac{n}{\bar{\mu}}\frac{d\bar{\mu}}{dn}\right) \tag{93}$$

or

$$\frac{n}{\bar{W}}\frac{d\bar{W}}{dn} \sim 3\frac{n}{\bar{\mu}}\frac{d\bar{\mu}}{dn} \tag{94}$$

from which

$$\bar{W} \sim \bar{\mu}^3 \tag{95}$$

From equations (87), (88), and (95) the following approximations are obtained

$$\bar{E}_{rM} = \bar{\mu}^3 \bar{E}_r \quad (96)$$

$$\bar{P}_{rM} = \bar{\mu}^3 \bar{E}_r \left(\frac{1}{3} - \frac{n}{\bar{\mu}} \frac{d\bar{\mu}}{dn} \right) \quad (97)$$

From equations (87), (89), and (96) it follows that

$$E_{rM} = \mu^3 E_r \quad (98)$$

$$\theta_{ErM} = \theta_{Er} + 3\theta_\mu \quad (99)$$

where E_r and θ_{Er} are given by equations(79) and (80) respectively. All further analysis based on equations (96) and (97) is limited to the same approximations that went into the derivation of equations (96) and (97) namely, that all asymmetries are small.

The detailed calculation of the radiation pressure proceeds from equation (97). From equation (89) it follows that

$$\frac{n}{\bar{\mu}} \frac{d\bar{\mu}}{dn} = He^{j\beta_{\mu,n}} \quad (100)$$

where

$$H = \sqrt{\left(\frac{n}{\mu}\frac{d\mu}{dn}\right)^2 + \left(n\frac{d\theta_\mu}{dn}\right)^2} \quad (101)$$

$$\tan \beta_{\mu,n} = n \frac{d\theta_\mu}{dn} \Big/ \frac{n}{\mu}\frac{d\mu}{dn} \quad (102)$$

Then the measured thermal radiation pressure in asymmetric bulk matter is obtained from equations (97) through (99) as the following approximation

$$\bar{P}_{rM} = \mu^3 E_r \left[\frac{1}{3} e^{j\theta_{ErM}} - He^{j(\theta_{ErM} + \beta_{\mu,n})} \right] \quad (103)$$

The component forms of equation (103) are

$$P_{rM} \cos \theta_{PrM} = \mu^3 E_r \left[\frac{1}{3} \cos \theta_{ErM} - H \cos (\theta_{ErM} + \beta_{\mu,n}) \right] \quad (104)$$

$$P_{rM} \sin \theta_{PrM} = \mu^3 E_r \left[\frac{1}{3} \sin \theta_{ErM} - H \sin (\theta_{ErM} + \beta_{\mu,n}) \right] \quad (105)$$

From equations (104) and (105) it follows that

$$\tan \theta_{PrM} = \frac{\frac{1}{3} \sin \theta_{ErM} - H \sin (\theta_{ErM} + \beta_{\mu,n})}{\frac{1}{3} \cos \theta_{ErM} - H \cos (\theta_{ErM} + \beta_{\mu,n})} \tag{106}$$

$$P_{rM} = \mu^3 E_r \sqrt{\frac{1}{9} - \frac{2}{3} H \cos \beta_{\mu,n} + H^2} \tag{107}$$

For the case $\beta_{\mu,n} = 0$ and $\theta_\mu = 0$ for symmetric radiation, equation (107) becomes

$$P_{rMs} = \mu_s^3 E_{rs} \left(\frac{1}{3} - H_s \right) \tag{108}$$

which is just equation (86) because

$$H_s = \frac{n}{\mu_s} \frac{d\mu_s}{dn} \tag{109}$$

and E_{rs} is given by equation (26).

For the vacuum $\mu = 1$, $\theta_\mu = 0$, $\beta_{\mu,n} = 0$, and $H = 0$, so that from equations (98) and (99) it follows that

$$E_{rM}^{(v)} = E_r^{(v)} \tag{110}$$

$$\theta_{ErM}^{(v)} = \theta_{Er}^{(v)} \tag{111}$$

For radiation in the asymmetric vacuum it follows from equations (104) and (105) that the following approximate equations are valid

$$P_r^{(v)} \cos \theta_{Pr}^{(v)} = \frac{1}{3} E_r^{(v)} \cos \theta_{Er}^{(v)} \tag{112}$$

$$P_r^{(v)} \sin \theta_{Pr}^{(v)} = \frac{1}{3} E_r^{(v)} \sin \theta_{Er}^{(v)} \tag{113}$$

For the vacuum it follows from equations (112) and (113) that approximately

$$\theta_{Pr}^{(v)} = \theta_{Er}^{(v)} \tag{114}$$

$$P_r^{(v)} = \frac{1}{3} E_r^{(v)} \tag{115}$$

where $\theta_{Er}^{(v)}$ and $E_r^{(v)}$ are obtained from the evaluation of the integral in equation (82).

Consider now the equilibrium equations at the surface of asymmetric matter that is bathed in asymmetric thermal radiation of the vacuum. The condition for mechanical equilibrium at the surface of the body is that the induced mechanical pressure is

$$\bar{P} = \bar{P}_r^{(v)} - \bar{P}_{rM} \tag{116}$$

or equivalently

$$P \cos \theta_P = P_r^{(v)} \cos \theta_{Pr}^{(v)} - P_{rM} \cos \theta_{PrM} \tag{117}$$

$$P \sin \theta_P = P_r^{(v)} \sin \theta_{Pr}^{(v)} - P_{rM} \sin \theta_{PrM} \tag{118}$$

Combining equations (117), (118) and equations (104), (105), (112), and (113) gives the following approximations

$$P \cos \theta_P = \frac{1}{3} E_r^{(v)} \cos \theta_{Er}^{(v)} - \mu^3 E_r \left[\frac{1}{3} \cos \theta_{ErM} - H \cos (\theta_{ErM} + \beta_{\mu,n})\right] \tag{119}$$

$$P \sin \theta_P = \frac{1}{3} E_r^{(v)} \sin \theta_{Er}^{(v)} - \mu^3 E_r \left[\frac{1}{3} \sin \theta_{ErM} - H \sin (\theta_{ErM} + \beta_{\mu,n})\right] \tag{120}$$

From equations (117) and (118) it follows that

$$P^2 = [P_r^{(v)}]^2 + P_{rM}^2 - 2P_r^{(v)} P_{rM} \cos [\theta_{Pr}^{(v)} - \theta_{PrM}] \tag{121}$$

$$\tan \theta_P = \frac{P_r^{(v)} \sin \theta_{Pr}^{(v)} - P_{rM} \sin \theta_{PrM}}{P_r^{(v)} \cos \theta_{Pr}^{(v)} - P_{rM} \cos \theta_{PrM}} \tag{122}$$

For the case of symmetric matter and symmetric radiation in matter and the vacuum, equation (121) becomes[16]

$$P_s = P_{rs}^{(v)} - P_{rMs} \tag{123}$$

$$= P_{rs}^{(v)} - \mu_s^3 E_{rs} \left(\frac{1}{3} - H_s\right)$$

$$= P_{rs}^{(v)}\left[1 - 3\mu_s^3 \left(\frac{1}{3} - H_s\right)\right] = P_{rs}^{(v)}(1 - \mu_s^3 + 3\mu_s^3 H_s)$$

where $P_{rs}^{(v)}$ is given by equation (28). Equation (123) can also be obtained directly from equation (119).

The functions θ_ν and E_ν have not been obtained explicitly, and therefore the radiation energy E_r is also unknown. For the purposes of the rest of this paper it is sufficient to understand that the frequency of photons in asymmetric bulk matter or vacuum has an internal phase angle that will manifest itself in the interactions of photons with other atomic particles.

3. BROKEN SYMMETRY OF ANGLES IN ASYMMETRIC BULK MATTER AND VACUUM. Within asymmetric bulk matter or the asymmetric vacuum, the internal phase angles of the coordinates produce a broken symmetry in various geometrical quantities such as, for example, angles. These broken symmetry angles enter the basic calculations of atomic processes that are treated in this paper. Consider first the fact that angles have internal phases, a result which can be deduced from the law of cosines which, for the complex number lengths that appear in asymmetric bulk matter or vacuum, can be written as

$$\cos\bar{\phi} = \frac{\bar{a}^2 + \bar{b}^2 - \bar{c}^2}{2\bar{a}\bar{b}} \qquad (124)$$

where $\bar{a}$, $\bar{b}$, and $\bar{c}$ are the sides of a plane triangle and $\bar{\phi}$ is the angle opposite side $\bar{c}$. Therefore it is clear that $\bar{\phi}$ and $\cos\bar{\phi}$ are complex numbers and can be written as

$$\bar{\phi} = \phi e^{j\theta_\phi} \qquad (125)$$

$$\cos\bar{\phi} = C_\phi e^{-j\theta_{c\phi}}$$

where C_ϕ = magnitude of $\cos\bar{\phi}$, and $\theta_{c\phi}$ = phase angle associated with $\cos\bar{\phi}$. It also follows that

$$\sin\bar{\phi} = S_\phi e^{j\theta_{s\phi}} \qquad (127)$$

where S_ϕ = magnitude of $\sin\bar{\phi}$, and $\theta_{s\phi}$ = phase angle of $\sin\bar{\phi}$. From the following expression

$$\cos\bar{\phi} = \frac{1}{2}\left(e^{j\bar{\phi}} + e^{-j\bar{\phi}}\right) \qquad (128)$$

it follows by elementary algebra that

$$\cos\bar{\phi} = \cos\phi_R \cosh\phi_I - j\sin\phi_R \sinh\phi_I \qquad (129)$$

where

$$\bar{\phi} = \phi_R + j\phi_I = \phi(\cos\theta_\phi + j\sin\theta_\phi) \tag{130}$$

Then the combining equations (126), (129), and (130) gives

$$C_\phi = \sqrt{\cos^2(\phi\cos\theta_\phi) + \sinh^2(\phi\sin\theta_\phi)} \tag{131}$$

$$\tan\theta_{c\phi} = \tan(\phi\cos\theta_\phi)\tanh(\phi\sin\theta_\phi) \tag{132}$$

In a similar fashion from

$$\sin\bar{\phi} = \frac{1}{2j}(e^{j\bar{\phi}} - e^{-j\bar{\phi}}) \tag{133}$$

it follows that

$$\sin\bar{\phi} = \sin\phi_R\cosh\phi_I + j\cos\phi_R\sinh\phi_I \tag{134}$$

and combining equations (127), (130), and (134) gives

$$S_\phi = \sqrt{\sin^2(\phi\cos\theta_\phi) + \sinh^2(\phi\sin\theta_\phi)} \tag{135}$$

$$\tan\theta_{s\phi} = \cot(\phi\cos\theta_\phi)\tanh(\phi\sin\theta_\phi) \tag{136}$$

These results will be used in Sections 5 and 6 where particle scattering in asymmetric bulk matter or vacuum is considered. The measured angle is given by $\phi_m = \phi\cos\theta_\phi = \phi_a$ where ϕ_a = conventional angle between two lines.

4. PHOTOELECTRIC EFFECT IN ASYMMETRIC BULK MATTER OR VACUUM. A very simple atomic process is the photoelectric effect wherein a photon collides with an electron that is bound in matter. If the photon has sufficient energy it will overcome the binding energy of the electron, and the electron will leave its site in the matter lattice with an excess kinetic energy.[17,18] The description of the process that occurs in bulk matter or vacuum with broken symmetries is similar to that of the standard analysis for the case where the electrons and photons move in symmetrical bulk matter or vacuum, except that now the kinematic variables for the photons and electrons become complex numbers. This is related to the broken symmetry of space and time in bulk matter and the vacuum.

Within asymmetric bulk matter the binding energy of an electron is described by a complex number potential $\bar{W}$, so that the binding energy is $e\bar{W}$, where e = electric charge. The conservation of energy then requires[17]

$$\frac{1}{2}m\bar{v}^2 = h\bar{\nu} - e\bar{W} \tag{137}$$

where m = electron mass, $\bar{v}$ = complex number electron velocity, and as before $\bar{\nu}$ = complex number frequency of the photon. Within asymmetric bulk matter or vacuum the electron velocity has a broken internal symmetry and is written as

$$\bar{v} = ve^{j\theta_v} \tag{138}$$

where v and θ_v = magnitude and internal phase angle respectively of the electron velocity. The complex number binding potential is written as

$$\bar{W} = We^{j\theta_W} \tag{139}$$

where W and θ_W = magnitude and internal phase angle of the binding potential. As described in Section 2 the photon frequency is a complex number for a photon propagating in asymmetric bulk matter or vacuum, and is written as in equation (29). It is assumed that ν , θ_ν , W , and θ_W are known quantities and that equation (137) can be used to determine the unknown complex number speed of the ejected electron.

The two scalar equations equivalent to equation (137) are

$$\frac{1}{2} mv^2 \cos (2\theta_v) = h\nu \cos \theta_\nu - eW \cos \theta_W \tag{140}$$

$$\frac{1}{2} mv^2 \sin (2\theta_v) = h\nu \sin \theta_\nu - eW \sin \theta_W \tag{141}$$

These two equations can be used to determine the unknown quantities v and θ_v as follows

$$\tan (2\theta_v) = \frac{h\nu \sin \theta_\nu - eW \sin \theta_W}{h\nu \cos \theta_\nu - eW \cos \theta_W} \tag{142}$$

$$\frac{1}{4} m^2v^4 = h^2\nu^2 + e^2W^2 - 2h\nu eW \cos (\theta_\nu - \theta_W) \tag{143}$$

A plot of the kinetic energy of the electron versus frequency is shown in Figure 1, while a plot of the internal phase angle of the electron kinetic energy $2\theta_v$ versus frequency is shown in Figure 2. These two figures show that there is a discontinuity in the kinetic energy magnitude and phase angle at a threshold frequency which is obtained from equation (140) by taking $2\theta_v = \pi/2$ or

$$h\nu_t \cos \theta_{\nu t} = eW \cos \theta_W \tag{144}$$

where ν_t = threshold frequency, and $\theta_{\nu t}$ = internal phase angle of the frequency at the threshold frequency. The electron kinetic energy at the threshold frequency is obtained from equation (141) to be

$$\frac{1}{2} mv_t^2 = h\nu_t \sin \theta_{\nu t} - eW \sin \theta_W \tag{145}$$

$$= eW (\cos \theta_W \tan \theta_{\nu t} - \sin \theta_W)$$

The threshold kinetic energy given by equation (145) is the minimum kinetic energy that the ejected electron can have in asymmetric bulk matter or vacuum. Below the threshold frequency the photoelectric process will not occur. If all phase angles are set equal to zero, the standard results are regained that the threshold frequency is given by $h\nu_t = eW$, and the minimum kinetic energy of the ejected electron is zero. Note that the measured electron kinetic energy is given by equation (140) which is linear in the photon frequency ν. The measured frequency is equal to $\nu \cos \theta_\nu$.

5. THOMSON SCATTERING AND THE COMPTON EFFECT IN ASYMMETRIC BULK MATTER AND VACUUM. The elastic scattering of photons by electrons is called Thomson scattering. For this case the photon energy is much smaller than the electron mass energy mc^2 and, if the electron is bound in an atom, the photon energy is larger than the binding energy so that the electrons can be considered to be free. For this case the differential cross section is given by[13]

$$I^a(\phi_a) = \frac{r_o^2}{2}(1 + \cos^2 \phi_a) \tag{146}$$

where r_o = classical electron radius, and where ϕ_a = conventional scattering angle. The corresponding differential cross section for Thomson scattering of photons by electrons in asymmetric bulk matter or vacuum is given by

$$\bar{I}(\bar{\phi}) = Ie^{j\theta_I} = \frac{r_o^2}{2}(1 + \cos^2 \bar{\phi}) \tag{147}$$

Combining equations (126) and (147) gives

$$I \cos \theta_I = \frac{r_o^2}{2}[1 + C_\phi^2 \cos (2\theta_{c\phi})] \tag{148}$$

$$I \sin \theta_I = - \frac{r_o^2}{2} C_\phi^2 \sin (2\theta_{c\phi}) \tag{149}$$

or

$$\tan \theta_I = - \frac{C_\phi^2 \sin (2\theta_{c\phi})}{1 + C_\phi^2 \cos (2\theta_{c\phi})} \tag{150}$$

$$I^2 = \frac{r_o^4}{4}[1 + C_\phi^4 + 2C_\phi^2 \cos (2\theta_{c\phi})] \tag{151}$$

The Compton effect is the name associated with the quantum scattering of photons by electrons with a transfer of momentum and energy from the photons to the electrons.[11,17] The description of this process using quanta of light was one of the early successes of quantum theory. This process is conventionally described by assuming that the photon and electron propagate in the symmetric vacuum, and applying the laws of conservation of energy and momentum to the colliding particles. When this process occurs within asymmetric bulk matter or vacuum, the same conservation laws are expected to be valid except now the kinematical parameters of the photon and the electron have broken symmetries and are represented by complex numbers.

Within bulk matter or vacuum with broken internal symmetries, a photon of initial frequency $\bar{\nu}$ collides with a stationary electron, and a new photon of frequency $\bar{\nu}'$ is emitted at an angle $\bar{\phi}$ with respect to the initial photon direction, and the electron recoils with speed $\bar{v}$ in a direction $\bar{\psi}$ with respect to the initial photon direction. Then the conservation of energy for the nonrelativistic case gives[11]

$$h\bar{\nu} = \frac{1}{2} m\bar{v}^2 + h\bar{\nu}' \tag{152}$$

while the conservation of momentum yields two equations[11]

$$\frac{h\bar{\nu}}{c} = \frac{h\bar{\nu}'}{c} \cos \bar{\phi} + m\bar{v} \cos \bar{\psi} \tag{153}$$

$$\frac{h\bar{\nu}'}{c} \sin \bar{\phi} = m\bar{v} \sin \bar{\psi} \tag{154}$$

These equations can be used to determine the three unknown complex number quantities $\bar{\nu}'$, $\bar{v}$, and $\bar{\phi}$ in terms of the known quantities $\bar{\nu}$ and $\bar{\psi}$. These three conservation equations are expressed in terms of complex numbers and are therefore equivalent to six scalar equations. The two components of the nonrelativistic energy conservation equation (152) are

$$h\nu \cos \theta_\nu = \frac{1}{2} mv^2 \cos (2\theta_v) + h\nu' \cos \theta'_\nu \tag{155}$$

$$h\nu \sin \theta_\nu = \frac{1}{2} mv^2 \sin (2\theta_v) + h\nu' \sin \theta'_\nu \tag{156}$$

The four momentum conservation equations obtained from equations (153) and (154) are respectively

$$\frac{h\nu}{c} \cos \theta_\nu = \frac{h\nu'}{c} C_\phi \cos (\theta'_\nu - \theta_{c\phi}) + mvC_\psi \cos (\theta_v - \theta_{c\psi}) \tag{157}$$

$$\frac{h\nu}{c} \sin \theta_\nu = \frac{h\nu'}{c} C_\phi \sin (\theta'_\nu - \theta_{c\phi}) + mvC_\psi \sin (\theta_v - \theta_{c\psi}) \tag{158}$$

$$\frac{h\nu'}{c} S_\phi = mvS_\psi \tag{159}$$

$$\theta_\nu' + \theta_{s\phi} = \theta_v + \theta_{s\psi} \tag{160}$$

where C_ϕ and S_ϕ are given by equations (131) and (135) respectively, and $\theta_{c\phi}$ and $\theta_{s\phi}$ are given by equations (132) and (136) respectively, and similarly

$$C_\psi = \sqrt{\cos^2(\psi \cos \theta_\psi) + \sinh^2(\psi \sin \theta_\psi)} \tag{161}$$

$$S_\psi = \sqrt{\sin^2(\psi \cos \theta_\psi) + \sinh^2(\psi \sin \theta_\psi)} \tag{162}$$

$$\tan \theta_{c\psi} = \tan(\psi \cos \theta_\psi) \tanh(\psi \sin \theta_\psi) \tag{163}$$

$$\tan \theta_{s\psi} = \cot(\psi \cos \theta_\psi) \tanh(\psi \sin \theta_\psi) \tag{164}$$

where

$$\bar{\psi} = \psi e^{j\theta_\psi} = \psi_R + j\psi_I \tag{165}$$

$$\cos \bar{\psi} = C_\psi e^{-j\theta_{c\psi}} \tag{166}$$

$$\sin \bar{\psi} = S_\psi e^{j\theta_{s\psi}} \tag{167}$$

The six equations (155) through (160) can be solved simultaneously for the six unknowns ν', θ_ν', v, θ_v, ϕ, and θ_ϕ in terms of the four known quantities ν, θ_ν, ψ, and θ_ψ.

For a bulk matter system or vacuum with broken internal symmetries, the relativistic analogs of the energy and momentum conservation equations (152) through (154) are[12,17]

$$h\bar{\nu} = mc^2(\bar{\gamma} - 1) + h\bar{\nu}' \tag{168}$$

$$\frac{h\bar{\nu}}{c} = \frac{h\bar{\nu}'}{c} \cos \bar{\phi} + m\bar{\gamma}\bar{v} \cos \bar{\psi} \tag{169}$$

$$\frac{h\bar{\nu}'}{c} \sin \bar{\phi} = m\bar{\gamma}\bar{v} \sin \bar{\psi} \tag{170}$$

where the complex number velocity factor for a particle with a velocity that has a broken symmetry is

$$\bar{\gamma} = \gamma e^{j\theta_\gamma} = (1 - \bar{v}^2/c^2)^{-1/2} \tag{171}$$

and where the magnitude and internal phase angle of the complex number boost is

$$\gamma = (f^2 + b^2)^{-1/4} \tag{172}$$

$$\tan (2\theta_\gamma) = b/f \tag{173}$$

where

$$b = v^2/c^2 \sin (2\theta_v) \tag{174}$$

$$f = 1 - v^2/c^2 \cos (2\theta_v) \tag{175}$$

The six scalar component equations corresponding to equations (168) through (170) are

$$h\nu \cos \theta_\nu = mc^2 (\gamma \cos \theta_\gamma - 1) + h\nu' \cos \theta'_\nu \tag{176}$$

$$h\nu \sin \theta_\nu = mc^2\gamma \sin \theta_\gamma + h\nu' \sin \theta'_\nu \tag{177}$$

$$\frac{h\nu}{c} \cos \theta_\nu = \frac{h\nu'}{c} C_\phi \cos (\theta'_\nu - \theta_{c\phi}) + m\gamma v C_\psi \cos (\theta_\gamma + \theta_v - \theta_{c\psi}) \tag{178}$$

$$\frac{h\nu}{c} \sin \theta_\nu = \frac{h\nu'}{c} C_\phi \sin (\theta'_\nu - \theta_{c\phi}) + m\gamma v C_\psi \sin (\theta_\gamma + \theta_v - \theta_{c\psi}) \tag{179}$$

$$\frac{h\nu'}{c} S_\phi = m\gamma v S_\psi \tag{180}$$

$$\theta'_\nu + \theta_{s\phi} = \theta_\gamma + \theta_v + \theta_{s\psi} \tag{181}$$

In the limit $v/c \rightarrow 0$ equations (176) through (184) reduce to equations (155) through (160) by noting that

$$\gamma \to 1 + \frac{1}{2} v^2/c^2 \cos(2\theta_v) \to 1 \tag{182}$$

$$\theta_\gamma \to \frac{1}{2} v^2/c^2 \sin(2\theta_v) \to 0 \tag{183}$$

The six equations (155) through (160) or (176) through (181) can be solved numerically using Brown's algorithm for the solution of simultaneous nonlinear equations. This algorithm is a modification of Newton's method and requires no derivative evaluations.[19]

The solution of equations (168) through (170) can be obtained by direct analogy to the solution for the standard Compton effect as follows[17]

$$\bar{\lambda}' - \bar{\lambda} = \lambda_o(1 - \cos\bar{\phi}) \tag{184}$$

where

$$\bar{\lambda}' = \lambda' e^{j\theta'_\lambda} = c/\bar{\nu}' = c/\nu' e^{-j\theta'_\nu} \tag{185}$$

$$\bar{\lambda} = \lambda e^{j\theta_\lambda} = c/\bar{\nu} = c/\nu e^{-j\theta_\nu} \tag{186}$$

and where λ_o = Compton wavelength = $h/(mc)$. The scalar equivalents for equation (184) are

$$\lambda' \cos\theta'_\lambda = \lambda\cos\theta_\lambda + \lambda_o(1 - C_\phi \cos\theta_{c\phi}) \tag{187}$$

$$\lambda' \sin\theta'_\lambda = \lambda\sin\theta_\lambda + \lambda_o C_\phi \sin\theta_{c\phi} \tag{188}$$

From equations (187) and (188) it follows that

$$\tan\theta'_\lambda = \frac{\lambda\sin\theta_\lambda + \lambda_o C_\phi \sin\theta_{c\phi}}{\lambda\cos\theta_\lambda + \lambda_o(1 - C_\phi\cos\theta_{c\phi})} \tag{189}$$

$$(\lambda')^2 = \lambda^2 + 2\lambda\lambda_o[\cos\theta_\lambda - C_\phi\cos(\theta_\lambda + \theta_{c\phi})] + \lambda_o^2(1 - 2C_\phi\cos\theta_{c\phi} + C_\phi^2) \tag{190}$$

Equations (189) and (190) give the wavelength internal phase angle and wavelength magnitude respectively of the scattered photon in asymmetric bulk matter or vacuum. The corresponding frequency equations can be obtained from equations (189) and (190) by noting that $\lambda' = c/\nu'$, $\lambda = c/\nu$, $\theta'_\lambda = -\theta'_\nu$, and $\theta_\lambda = -\theta_\nu$. Note that equation (187) gives the change in measured wavelengths, and this wavelength difference is independent of the wavelength itself.

Consider now the differential cross section for Compton scattering in asymmetric bulk matter or vacuum. The standard Compton scattering differential cross section is given by the Klein-Nishina formula[13,20]

$$I^a(\phi_a) = \frac{r_o^2}{2}\left(\frac{\nu_a'}{\nu_a}\right)^2\left(\frac{\nu_a}{\nu_a'} + \frac{\nu_a'}{\nu_a} - \sin^2 \phi_a\right) \tag{191}$$

where ν_a = conventionally determined initial photon frequency, and ν_a' = conventionally determined scattered photon frequency. The generalization to the differential scattering cross section for Compton scattering within bulk matter or the vacuum with broken internal symmetries follows from equation (191) as

$$\bar{I}(\bar{\phi}) = \frac{r_o^2}{2}\left(\frac{\bar{\nu}'}{\bar{\nu}}\right)^2\left(\frac{\bar{\nu}}{\bar{\nu}'} + \frac{\bar{\nu}'}{\bar{\nu}} - \sin^2 \bar{\phi}\right) \tag{192}$$

or equivalently as

$$\bar{I}(\bar{\phi}) = \frac{r_o^2}{2}\left(\frac{\nu'}{\nu}\right)^2\left(\frac{\nu}{\nu'} e^{j\Gamma_1} + \frac{\nu'}{\nu} e^{j\Gamma_2} - S_\phi^2 e^{j\Gamma_3}\right) \tag{193}$$

where ν and ν' = magnitudes of the complex number initial and scattered photon frequencies respectively, and where

$$\Gamma_1 = \theta_\nu' - \theta_\nu \tag{194}$$

$$\Gamma_2 = 3(\theta_\nu' - \theta_\nu) \tag{195}$$

$$\Gamma_3 = 2(\theta_\nu' - \theta_\nu + \theta_{s\phi}) \tag{196}$$

Therefore from equation (193) it follows that

$$I \cos \theta_I = \frac{r_o^2}{2}\left(\frac{\nu'}{\nu}\right)^2\left(\frac{\nu}{\nu'} \cos \Gamma_1 + \frac{\nu'}{\nu} \cos \Gamma_2 - S_\phi^2 \cos \Gamma_3\right) \tag{197}$$

$$I \sin \theta_I = \frac{r_o^2}{2}\left(\frac{\nu'}{\nu}\right)^2\left(\frac{\nu}{\nu'} \sin \Gamma_1 + \frac{\nu'}{\nu} \sin \Gamma_2 - S_\phi^2 \sin \Gamma_3\right) \tag{198}$$

from which I and θ_I can easily be obtained. The measured differential cross section = $I \cos \theta_I$.

6. COULOMB SCATTERING IN BULK MATTER AND THE VACUUM WITH BROKEN INTERNAL SYMMETRIES. This section considers Rutherford, Mott, Bhabha, and Møller scat-

tering in asymmetric bulk matter and vacuum.

A. Rutherford Scattering

The α-particle scattering experiments of Rutherford are one of the cornerstones of knowledge about atomic structure. These experiments measured the scattering angles of α-particles interacting with atomic nuclei of charge Ze. The basic formulas for Rutherford scattering give the differential scattering cross section as[17]

$$I^a(\phi_a) = \frac{A}{v_a^4} \csc^4 \frac{\phi_a}{2} \tag{199}$$

where

$$A = \left(\frac{Z'Ze^2}{2m}\right)^2 \tag{200}$$

ϕ_a = measured scattering angle, v_a = conventionally calculated initial relative speed of the α-particle and the atomic nucleus, Z' = atomic number of incident particle (Z' = 2 for α-particle), Z = atomic number of the atomic nucleus, and m = reduced mass of the incident particle and the atomic nucleus. In addition to the differential cross section, the other quantity that is often calculated is the number of particles deviated through and angle between ϕ_a and $\phi_a + d\phi_a$ which is given by[17]

$$\frac{dN^a}{d\phi_a} = \frac{4\pi A}{v_a^4} \cot \frac{\phi_a}{2} \csc^2 \frac{\phi_a}{2} \tag{201}$$

These formulas were deduced by considering the scattering of an α-particle by an isolated atomic nucleus situated in a symmetrical vacuum.

For Rutherford scattering within asymmetric bulk matter or vacuum equations (199) and (201) need to be modified because the indicent α-particle speed $\bar{v}$ is now a complex number, and because the deflection angle $\bar{\phi}$ is also a complex number. Therefore equations (199) and (201) must now be written as

$$\bar{I}(\bar{\phi}) = \frac{A}{\bar{v}^4} \csc^4 \frac{\bar{\phi}}{2} \tag{202}$$

$$\frac{d\bar{N}}{d\bar{\phi}} = \frac{4\pi A}{\bar{v}^4} \cot \frac{\bar{\phi}}{2} \csc^2 \frac{\bar{\phi}}{2} \tag{203}$$

where $\bar{I}(\bar{\phi})$ = complex number differential scattering cross section, $\bar{\phi}$ = complex number deflection angle, $d\bar{N}/d\bar{\phi}$ = complex number of particles deviated through $\bar{\phi}$ and $\bar{\phi} + d\bar{\phi}$, and $\bar{v}$ = complex number initial α-particle speed. Because $\bar{v}$ and $\bar{\phi}$ are phase rotated, the number of α-particles scattered will also include a phase rotated part, so that

$$\bar{N} = Ne^{j\theta_N} \tag{204}$$

where N and θ_N = magnitude and internal phase angle respectively of the number of scattered particles.

Using the following standard trigonometric formulas

$$\sin^2 \frac{\bar{\phi}}{2} = \frac{1}{2} (1 - \cos \bar{\phi}) \tag{205}$$

$$\tan \frac{\bar{\phi}}{2} = \frac{\sin \bar{\phi}}{1 + \cos \bar{\phi}} \tag{206}$$

$$\cos^2 \frac{\bar{\phi}}{2} = \frac{1}{2} (1 + \cos \bar{\phi}) \tag{207}$$

and combining them with equations (126) and (127) gives

$$\csc^2 \frac{\bar{\phi}}{2} = K_s e^{-jx_\phi} \tag{208}$$

$$\csc^4 \frac{\bar{\phi}}{2} = K_s^2 e^{-2jx_\phi} \tag{209}$$

$$\cot \frac{\bar{\phi}}{2} = K_t e^{-jz_\phi} \tag{211}$$

$$\sec^2 \frac{\bar{\phi}}{2} = K_{se} e^{jy_\phi} \tag{212}$$

where

$$K_s = 2(1 - 2C_\phi \cos \theta_{c\phi} + C_\phi^2)^{-1/2} \tag{213}$$

$$K_t = \frac{1}{S_\phi} (1 + 2C_\phi \cos \theta_{c\phi} + C_\phi^2)^{1/2} \tag{214}$$

$$K_{se} = 2(1 + 2C_\phi \cos \theta_{c\phi} + C_\phi^2)^{-1/2} \tag{215}$$

$$\tan x_\phi = \frac{C_\phi \sin \theta_{c\phi}}{1 - C_\phi \cos \theta_{c\phi}} \tag{216}$$

$$\tan z_\phi = \frac{\sin \theta_{s\phi} + C_\phi \sin (\theta_{c\phi} + \theta_{s\phi})}{\cos \theta_{s\phi} + C_\phi \cos (\theta_{c\phi} + \theta_{s\phi})} \tag{217}$$

$$\tan y_\phi = \frac{C_\phi \sin \theta_{c\phi}}{1 + C_\phi \cos \theta_{c\phi}} \tag{218}$$

where C_ϕ , S_ϕ , $\theta_{c\phi}$, and $\theta_{s\phi}$ are given by equations (131), (135), (132), and (136) respectively.

Combining equations (208) and (209) with equation (202) gives

$$\bar{I}(\bar{\phi}) = Ie^{j\theta_I} = \frac{AK_s^2}{v^4} e^{-j(4\theta_v + 2x_\phi)} \tag{219}$$

or

$$I = \frac{AK_s^2}{v^4} \tag{220}$$

$$\theta_I = - 4\theta_v - 2x_\phi \tag{221}$$

which are the equations for the magnitude and internal phase of the complex number differential cross section for Rutherford scattering in asymmetric matter or vacuum. Combining equations (208) and (211) with equation (203) gives

$$\frac{d\bar{N}}{d\bar{\phi}} = \left|\frac{d\bar{N}}{d\bar{\phi}}\right| e^{j\theta_{N\phi}} = \frac{4\pi AK_sK_t}{v^4} e^{-j(4\theta_v + x_\phi + z_\phi)} \tag{222}$$

and therefore

$$\left|\frac{d\bar{N}}{d\bar{\phi}}\right| = \sqrt{\frac{\left(\frac{dN}{d\phi}\right)^2 + N^2\left(\frac{d\theta_N}{d\phi}\right)^2}{1 + \phi^2\left(\frac{d\theta_\phi}{d\phi}\right)^2}} = 4\pi AK_sK_t/v^4 \tag{223}$$

$$\theta_{N\phi} = \theta_N + \beta_{N,\phi} - \theta_\phi - \beta_{\phi,\phi} = - 4\theta_v - x_\phi - z_\phi \tag{224}$$

where

$$\tan \beta_{N,\phi} = N \frac{d\theta_N/d\phi}{dN/d\phi} \tag{225}$$

$$\tan \beta_{\phi,\phi} = \phi \frac{d\theta_\phi}{d\phi} \tag{226}$$

which gives the magnitude and phase angle of the number of scattered particles. The measured scattering cross section is given by $I \cos \theta_I$.

B. Mott Scattering

Mott scattering describes the Coulomb scattering of two identical fermions such as, for example, two protons. The differential scattering cross section for two protons scattering in the symmetric vacuum is described in the center of mass coordinates by the following equation[21-26]

$$I^a(\phi_a) = \frac{A}{v_a^4} [\csc^4 \frac{\phi_a}{2} + \sec^4 \frac{\phi_a}{2} - \csc^2 \frac{\phi_a}{2} \sec^2 \frac{\phi_a}{2} \cos (2\xi_a \, \ell n \tan \frac{\phi_a}{2})] \tag{227}$$

where

$$A = (\frac{e^2}{2m})^2 \tag{228}$$

$$\xi_a = \frac{e^2}{\hbar v_a} \tag{229}$$

and m = reduced mass = $m_p/2$ where m_p = proton mass, and v_a = conventionally determined relative speed of the two protons. For the scattering of two protons within asymmetric bulk matter or vacuum, the differential scattering cross section is written as a complex number as follows

$$\bar{I}(\bar{\phi}) = \frac{A}{\bar{v}^4} [\csc^4 \frac{\bar{\phi}}{2} + \sec^4 \frac{\bar{\phi}}{2} - \csc^2 \frac{\bar{\phi}}{2} \sec^2 \frac{\bar{\phi}}{2} \cos (2\bar{\xi} \, \ell n \tan \frac{\bar{\phi}}{2})] \tag{230}$$

where

$$\bar{\xi} = \frac{e^2}{\hbar \bar{v}} = \frac{e^2}{\hbar v} e^{-j\theta_v} = \xi e^{-j\theta_v} \tag{231}$$

Equation (230) can be rewritten as

$$\bar{I}(\bar{\phi}) = \frac{A}{v^4} (J_R e^{-j\theta_{JR}} + J_E e^{-j\theta_{JE}} + J_I e^{-j\theta_{JI}}) \tag{232}$$

where the Rutherford term and the exchange term are written as

$$J_R = K_s^2 \tag{233}$$

$$J_E = K_{se}^2 \tag{234}$$

$$\theta_{JR} = 4\theta_v + 2x_\phi \tag{235}$$

$$\theta_{JE} = 4\theta_v - 2y_\phi \tag{236}$$

The interaction term is written as

$$\bar{J}_I = - K_s K_{se} \cos \bar{G}\, e^{-j(4\theta_v + x_\phi - y_\phi)} \tag{237}$$

where from equations (211) and (230) it follows that

$$\bar{G} = Ge^{j\theta_G} = 2\,\bar{\xi}\,\ell n \left(\frac{e^{jz_\phi}}{K_t}\right) \tag{238}$$

$$= 2\xi(\cos \theta_v - j \sin \theta_v)\,(jz_\phi - \ell n\, K_t)$$

$$= 2\xi[z_\phi \sin \theta_v - \ell n\, K_t \cos \theta_v + j(z_\phi \cos \theta_v + \ell n\, K_t \sin \theta_v)]$$

so that

$$G = 2\xi \sqrt{z_\phi^2 + (\ell n\, K_t)^2} \tag{239}$$

$$\tan \theta_G = \frac{z_\phi \cos \theta_v + \ell n\, K_t \sin \theta_v}{z_\phi \sin \theta_v - \ell n\, K_t \cos \theta_v} \tag{240}$$

The interaction term in equation (237) can be rewritten as

$$\bar{J}_I = J_I e^{-j\theta_{JI}} \tag{241}$$

where

$$J_I = - K_s K_{se} C_G \tag{242}$$

$$\theta_{JI} = 4\theta_v + x_\phi - y_\phi + \theta_{cG} \tag{243}$$

$$C_G = \sqrt{\cos^2 (G \cos \theta_G) + \sinh^2 (G \sin \theta_G)} \tag{244}$$

$$\tan \theta_{cG} = \tan (G \cos \theta_G) \tanh (G \sin \theta_G) \tag{245}$$

The two equations for determining I and θ_I are obtained from equations (232) through (245) as

$$I \cos \theta_I = \frac{A}{v^4} (J_R \cos \theta_{JR} + J_E \cos \theta_{JE} + J_I \cos \theta_{JI}) \tag{246}$$

$$I \sin \theta_I = - \frac{A}{v^4} (J_R \sin \theta_{JR} + J_E \sin \theta_{JE} + J_I \sin \theta_{JI}) \tag{247}$$

In this manner a theory of Mott scattering in an asymmetric medium is developed which is consistent with the gauge theory of the asymmetric background medium. The measured cross section is = $I \cos \theta_I$.

C. Bhabha Scattering

Bhabha scattering is electron-positron scattering $e^+ + e^- \rightarrow e^+ + e^-$ by photon exchange and pair annihilation. The differential scattering cross section in the center of mass system and in the high energy limit ($E_a >> m$) is given for the symmetrical vacuum by[27]

$$I^a = \frac{B}{\gamma_a^2 v_a^2} \left[\frac{1 + \cos^4 \frac{\phi_a}{2}}{\sin^4 \frac{\phi_a}{2}} + \frac{1}{2} (1 + \cos^2 \phi_a) - 2 \frac{\cos^4 \frac{\phi_a}{2}}{\sin^2 \frac{\phi_a}{2}} \right] \tag{248}$$

where

$$B = \frac{1}{2} \left(\frac{\alpha}{mc}\right)^2 \tag{249}$$

$$\gamma_a = (1 - v_a^2/c^2)^{-1/2} \tag{250}$$

α = fine structure constant, m = reduced mass of electron, and v_a = conventionally determined speed in center of mass system. The first term in equation (248) is the photon exchange term, the second term is the pair annihilation contribution, while the third term represents the interference between the first two terms.[22]

The corresponding cross section for Bhabha scattering in asymmetric bulk matter or vacuum is written as

$$\bar{I}(\bar{\phi}) = \frac{B}{\bar{\gamma}^2 \bar{v}^2} \left[\frac{1 + \cos^4 \frac{\bar{\phi}}{2}}{\sin^4 \frac{\bar{\phi}}{2}} + \frac{1}{2} (1 + \cos^2 \bar{\phi}) - 2 \frac{\cos^4 \frac{\bar{\phi}}{2}}{\sin^2 \frac{\bar{\phi}}{2}} \right] \qquad (251\text{-}260)$$

where $\bar{v}$ and $\bar{\gamma}$ are given by equations (138) and (171) respectively. Combining equations (208) through (215) with equation (260) gives

$$\bar{I}(\bar{\phi}) = \frac{B}{\gamma^2 v^2} (L_1 e^{-j\Phi_1} + L_2 e^{-j\Phi_2} + L_3 e^{-j\Phi_3} + L_4 e^{-j\Phi_4} + L_5 e^{-j\Phi_5}) \qquad (261)$$

where γ is now the magnitude of the boost for a broken symmetry system and is given by equation (172), and where

$$L_1 = K_s^2 \qquad L_4 = \frac{1}{2} C_\phi^2 \qquad (262)$$

$$L_2 = K_s^2 / K_{se}^2 \qquad L_5 = - 2K_s / K_{se}^2 \qquad (263)$$

$$L_3 = 1/2$$

$$\Phi_1 = 2(\theta_\gamma + \theta_v + x_\phi) \qquad \Phi_4 = 2(\theta_\gamma + \theta_v + \theta_{c\phi}) \qquad (264)$$

$$\Phi_2 = 2(\theta_\gamma + \theta_v + x_\phi + y_\phi) \qquad \Phi_5 = 2(\theta_\gamma + \theta_v + x_\phi/2 + y_\phi) \qquad (265)$$

$$\Phi_3 = 2(\theta_\gamma + \theta_v) \qquad (266)$$

where θ_γ is given by equation (173). From equation (261) it follows that

$$I \cos \theta_I = \frac{B}{\gamma^2 v^2} (L_1 \cos \Phi_1 + L_2 \cos \Phi_2 + L_3 \cos \Phi_3 + L_4 \cos \Phi_4 + L_5 \cos \Phi_5) \qquad (267)$$

$$I \sin \theta_I = - \frac{B}{\gamma^2 v^2} (L_1 \sin \Phi_1 + L_2 \sin \Phi_2 + L_3 \sin \Phi_3 + L_4 \sin \Phi_4 + L_5 \sin \Phi_5) \qquad (268)$$

from which I and θ_I can be easily determined. In equations (261), (267), and (268) the first two terms are due to photon exchange, terms three and four are due to pair annihilation, and term five is the interference term.

D. Møller Scattering

Møller scattering is electron-electron scattering $e^- + e^- \rightarrow e^- + e^-$ by photon exchange. The differential scattering cross section for this process in the center of mass system and for high energy ($E_a >> m$) is given for the symmetrical vacuum by[19,23,27]

$$I^a = \frac{B}{\gamma_a^2 v_a^2}\left[\frac{1 + \cos^4\frac{\phi_a}{2}}{\sin^4\frac{\phi_a}{2}} + \frac{2}{\sin^2\frac{\phi_a}{2}\cos^2\frac{\phi_a}{2}} + \frac{1 + \sin^4\frac{\phi_a}{2}}{\cos^4\frac{\phi_a}{2}}\right] \tag{269}$$

where ϕ_a = scattering angle, and γ_a = ordinary relativistic boost given by equation (250). The first term in equation (269) is due to direct scattering, the second is due to interference, and the third term is the result of exchange scattering.

The corresponding differential scattering cross section for Møller scattering in bulk matter or vacuum with broken internal symmetries is given by

$$\bar{I} = \frac{B}{\bar{\gamma}^2\bar{v}^2}\left[\frac{1 + \cos^4\frac{\bar{\phi}}{2}}{\sin^4\frac{\bar{\phi}}{2}} + \frac{2}{\sin^2\frac{\bar{\phi}}{2}\cos^2\frac{\bar{\phi}}{2}} + \frac{1 + \sin^4\frac{\bar{\phi}}{2}}{\cos^4\frac{\bar{\phi}}{2}}\right] \tag{270}$$

where the particle speed $\bar{v}$ and boost $\bar{\gamma}$ for a broken symmetry system are given by equations (138) and (171) respectively. Combining equation (270) with equations (208) through (215) gives

$$\bar{I} = \frac{B}{\gamma^2 v^2}\left(T_1 e^{-j\psi_1} + T_2 e^{-j\psi_2} + T_3 e^{-j\psi_3} + T_4 e^{-j\psi_4} + T_5 e^{-j\psi_5}\right) \tag{271}$$

where the boost γ for a broken symmetry system is given by equation (172), and where

$$T_1 = K_s^2 \qquad T_4 = K_{se}^2 \tag{272}$$

$$T_2 = K_s^2/K_{se}^2 \qquad T_5 = K_{se}^2/K_s^2 \tag{273}$$

$$T_3 = 2K_sK_{se} \tag{274}$$

where K_s and K_{se} are given by equations (213) and (215) respectively, and where

$$\psi_1 = 2(\theta_\gamma + \theta_v + x_\phi) \qquad \psi_4 = 2(\theta_\gamma + \theta_v - y_\phi) \tag{275}$$

$$\psi_2 = 2(\theta_\gamma + \theta_v + x_\phi + y_\phi) \qquad \psi_5 = 2(\theta_\gamma + \theta_v - x_\phi - y_\phi) \tag{276}$$

$$\psi_3 = 2(\theta_\gamma + \theta_v + x_\phi/2 - y_\phi/2) \tag{277}$$

where x_ϕ and y_ϕ are given by equations (216) and (218) respectively, and where θ_γ is expressed in terms of θ_v by equation (173). From equation (271) it follows that the magnitude and internal phase of the differential cross section for Møller scattering in a broken symmetry system is given by

$$I \cos\theta_I = \frac{B}{\gamma^2 v^2}(T_1 \cos\psi_1 + T_2 \cos\psi_2 + T_3 \cos\psi_3 + T_4 \cos\psi_4 + T_5 \cos\psi_5) \tag{278}$$

$$I \sin\theta_I = -\frac{B}{\gamma^2 v^2}(T_1 \sin\psi_1 + T_2 \sin\psi_2 + T_3 \sin\psi_3 + T_4 \sin\psi_4 + T_5 \sin\psi_5) \tag{279}$$

which can be solved for I and θ_I immediately.

7. DIRAC EQUATION FOR FERMIONS IN ASYMMETRIC BULK MATTER OR VACUUM. The Dirac equation determines the spectrum and eigenfunctions of half-integral spin particles moving in an external potential.[18] The eigenfunctions take the form of four-component spinors, and therefore the Dirac equation for a particle moving in the symmetric vacuum under the influence of an external potential must be equivalent to four equations. In fact the Dirac equation is a matrix equation involving 4 x 4 matrices and is written as[18,19,27-37]

$$(-i\gamma^\mu \frac{\partial}{\partial x_{\mu a}} + m + V_e^a)\psi^a = 0 \tag{280}$$

where $x_{\mu a} = t_a$, x_a , y_a , z_a , and where $\gamma_\mu = \gamma_o$, γ_1 , γ_2 , and γ_3 are the four Dirac matrices. Within asymmetric bulk matter or vacuum the Dirac equation is expected to be written as

$$(-i\gamma^\mu \frac{\partial}{\partial \bar{x}_\mu} + m + \bar{W})\bar{\psi} = 0 \tag{281}$$

where $\bar{\psi}$ = spinor with internal phase given by

$$\bar{\psi} = \psi e^{j\theta_\psi} = \psi_R + j\psi_I \tag{282}$$

$$\psi_R = \psi \cos \theta_\psi \tag{283}$$

$$\psi_I = \psi \sin \theta_\psi \tag{284}$$

and where the complex number potential is written as

$$\bar{W} = \bar{V}_e + \bar{V}_g \tag{285}$$

The gauge rotated time and space coordinates $\bar{x}_\mu = \bar{t}\,,\bar{x}\,,\bar{y}\,,$ and $\bar{z}$ of a particle in bulk matter or vacuum with broken internal symmetries are written as

$$\bar{t} = te^{j\theta_t} \qquad \bar{x} = xe^{j\theta_x} \tag{286}$$

$$\bar{y} = ye^{j\theta_y} \qquad \bar{z} = ze^{j\theta_z} \tag{287}$$

The combined effects of gauge rotated coordinates, gauge rotated external potential (which is a function of the gauge rotated coordinates), and the gauge potential itself $\bar{V}_g$, will manifest themselves in the eigenvalues and eigenfunctions of the Dirac equation for a fermion located in an asymmetric system.

The space and time derivatives that appear in equation (281) are written as

$$\partial/\partial\bar{t} = e^{-j\theta dt}\left[1 + \left(t\,\frac{\partial\theta_t}{\partial t}\right)^2\right]^{-1/2} \partial/\partial t = e^{-j\theta dt} \cos\beta_{t,t}\ \partial/\partial t \tag{288}$$

$$\partial/\partial\bar{x} = e^{-j\theta dx}\left[1 + \left(x\,\frac{\partial\theta_x}{\partial x}\right)^2\right]^{-1/2} \partial/\partial x = e^{-j\theta dx} \cos\beta_{x,x}\ \partial/\partial x \tag{289}$$

$$\partial/\partial\bar{y} = e^{-j\theta dy}\left[1 + \left(y\,\frac{\partial\theta_y}{\partial y}\right)^2\right]^{-1/2} \partial/\partial y = e^{-j\theta dy} \cos\beta_{y,y}\ \partial/\partial y \tag{290}$$

$$\partial/\partial\bar{z} = e^{-j\theta dz}\left[1 + \left(z\,\frac{\partial\theta_z}{\partial z}\right)^2\right]^{-1/2} \partial/\partial z = e^{-j\theta dz} \cos\beta_{z,z}\ \partial/\partial z \tag{291}$$

where

$$\theta_{dt} = \theta_t + \beta_{t,t} = \theta_o + \beta_{o,o} = \theta_{do} \tag{292}$$

$$\theta_{dx} = \theta_x + \beta_{x,x} = \theta_1 + \beta_{1,1} = \theta_{d1} \tag{293}$$

$$\theta_{dy} = \theta_y + \beta_{y,y} = \theta_2 + \beta_{2,2} = \theta_{d2} \tag{294}$$

$$\theta_{dz} = \theta_z + \beta_{z,z} = \theta_3 + \beta_{3,3} = \theta_{d3} \tag{295}$$

and where

$$\tan \beta_{o,o} = \tan \beta_{t,t} = t\, \partial\theta_t/\partial t \tag{296}$$

$$\tan \beta_{1,1} = \tan \beta_{x,x} = x\, \partial\theta_x/\partial x \tag{297}$$

$$\tan \beta_{2,2} = \tan \beta_{y,y} = y\, \partial\theta_y/\partial y \tag{298}$$

$$\tan \beta_{3,3} = \tan \beta_{z,z} = z\, \partial\theta_z/\partial z \tag{299}$$

In this way the necessary space and time derivatives in Dirac's equation for broken symmetry systems are evaluated.

Equation (281) can then be rewritten as

$$(-ie^{-j\theta_{d\mu}} \cos \beta_{\mu,\mu}\, \gamma^\mu \frac{\partial}{\partial x_\mu} + m + \bar{W})\, \bar{\psi} = 0 \tag{300}$$

The two matrix equations corresponding to equation (300) are obtained by taking the real and imaginary parts in the internal space as follows

$$(-i \cos \theta_{d\mu} \cos \beta_{\mu,\mu}\, \gamma^\mu\, \partial/\partial x_\mu + m + W \cos \theta_W)\, \psi_R \tag{301}$$

$$- (i \sin \theta_{d\mu} \cos \beta_{\mu,\mu}\, \gamma^\mu\, \partial/\partial x_\mu + W \sin \theta_W)\, \psi_I = 0$$

$$(i \sin \theta_{d\mu} \cos \beta_{\mu,\mu}\, \gamma^\mu\, \partial/\partial x_\mu + W \sin \theta_W)\, \psi_R \tag{302}$$

$$+ (-i \cos \theta_{d\mu} \cos \beta_{\mu,\mu}\, \gamma^\mu\, \partial/\partial x_\mu + m + W \cos \theta_W)\, \psi_I = 0$$

where it is assumed that the mass is a real number that is not affected by the gauge rotations due to the asymmetric background. Note that from equation (285) it follows that

$$W \cos \theta_W = V_e \cos \theta_{Ve} + V_g \cos \theta_{Vg} \tag{303}$$

$$W \sin \theta_W = V_e \sin \theta_{Ve} + V_g \sin \theta_{Vg} \tag{304}$$

The equations (301) and (302) are equivalent to eight equations for the eight spinor components ψ_R^o , ψ_R^1 , ψ_R^2 , ψ_R^3 , ψ_I^o , ψ_I^1 , ψ_I^2 , and ψ_I^3 , or equivalently ψ^o , ψ^1 , ψ^2 , ψ^3 , $\theta_{\psi o}$, $\theta_{\psi 1}$, $\theta_{\psi 2}$, and $\theta_{\psi 3}$. Therefore Dirac's equation for a fermion located in a background with broken internal symmetry is equivalent to eight independent equations. An approximate solution ignores the imaginary wavefunction components, which gives

$$(- i \cos \theta_{d\mu} \cos \beta_{\mu,\mu} \gamma^\mu \partial/\partial x_\mu + m + W \cos \theta_W) \psi_R = 0 \tag{301A}$$

$$(i \sin \theta_{d\mu} \cos \beta_{\mu,\mu} \gamma^\mu \partial/\partial x_\mu + W \sin \theta_W) \psi_R = 0 \tag{302A}$$

as the Dirac equations with four spinor components.

Alternatively, equation (300) can be combined with equation (282) to give the following set of Dirac equations

$$[- i \cos \theta_{d\mu} \cos \beta_{\mu,\mu} \gamma^\mu(\partial/\partial x_\mu + \partial\theta_\psi/\partial x_\mu) + m + W \cos \theta_W]\psi = 0 \tag{304A}$$

$$[i \sin \theta_{d\mu} \cos \beta_{\mu,\mu} \gamma^\mu(\partial/\partial x_\mu + \partial\theta_\psi/\partial x_\mu) + W \sin \theta_W]\psi = 0 \tag{304B}$$

If the space and time derivatives of θ_ψ can be neglected these equations become

$$(- i \cos \theta_{d\mu} \cos \beta_{\mu,\mu} \gamma^\mu \partial/\partial x_\mu + m + W \cos \theta_W)\psi = 0 \tag{304C}$$

$$(i \sin \theta_{d\mu} \cos \beta_{\mu,\mu} \gamma^\mu \partial/\partial x_\mu + W \sin \theta_W)\psi = 0 \tag{304D}$$

8. SCHRÖDINGER'S EQUATION FOR A PARTICLE WITHIN ASYMMETRIC BULK MATTER OR VACUUM. This section considers the effects of bulk matter and vacuum with broken internal symmetries on Schrödinger's equation for a particle moving in a potential field. The time dependent Schrödinger equation for a particle moving in a potential field in a symmetric vacuum is written as[38-47]

$$[\frac{1}{2m} (p_{xa}^2 + p_{ya}^2 + p_{za}^2) + V_e^a] \psi^a = i\hbar \frac{\partial \psi^a}{\partial t_a} \qquad (305)$$

where the single particle momentum and energy operators are given by

$$P_{\alpha a} = - i\hbar \partial / \partial \alpha_a \qquad E_a = i\hbar \partial / \partial t_a \qquad (306)$$

with $\alpha = x , y , z$. Within asymmetric bulk matter or vacuum it is assumed that space, time, momentum operators, energy operator, potential, and wave functions exhibit broken symmetries and must be represented by complex numbers in internal space. For this case the time dependent Schrödinger equation is written as

$$[\frac{1}{2m} (\bar{p}_x^2 + \bar{p}_y^2 + \bar{p}_z^2) + \bar{W}] \bar{\psi} = i\hbar \frac{\partial \bar{\psi}}{\partial \bar{t}} \qquad (307)$$

where $\bar{W} = \bar{V}_e + \bar{V}_g$ and where

$$\bar{p}_\alpha = p_\alpha e^{j\theta_{p\alpha}} = - i\hbar \partial / \partial \bar{\alpha} = - i\hbar \cos \beta_{\alpha,\alpha} \, e^{-j\theta_{d\alpha}} \, \partial / \partial \alpha \qquad (308)$$

$$\bar{E} = E e^{j\theta_E} = i\hbar \partial / \partial \bar{t} = i\hbar \cos \beta_{t,t} \, e^{-j\theta_{dt}} \, \partial / \partial t \qquad (309)$$

where

$$\cos \beta_{\alpha,\alpha} = \left[1 + \left(\alpha \frac{\partial \theta_\alpha}{\partial \alpha} \right)^2 \right]^{-1/2} \qquad (310)$$

$$\theta_{d\alpha} = \theta_\alpha + \beta_{\alpha,\alpha} \qquad (311)$$

$$\cos \beta_{t,t} = \left[1 + \left(t \frac{\partial \theta_t}{\partial t} \right)^2 \right]^{-1/2} \qquad (312)$$

$$\theta_{dt} = \theta_t + \beta_{t,t} \qquad (313)$$

where $\beta_{t,t}$ and $\beta_{\alpha,\alpha}$ are given in equations (296) through (299). From equations (308) and (309) it follows that

$$p_\alpha = - i\hbar \cos \beta_{\alpha,\alpha} \, \partial / \partial \alpha \qquad (314)$$

$$\theta_{p\alpha} = - \theta_{d\alpha} \qquad (315)$$

$$E = i\hbar \cos \beta_{t,t} \, \partial / \partial t \qquad (316)$$

$$\theta_E = -\theta_{dt} \tag{317}$$

It is easy to show from the Heisenberg uncertainty principle applied to $\bar{p}_\alpha$ and $\bar{\alpha}$ and to $\bar{E}$ and $\bar{t}$ that $\beta_{\alpha,\alpha} < 0$ and $\beta_{t,t} < 0$, so that from equations (296) through (299) it follows that θ_α is a decreasing function of α, and θ_t is a decreasing function of t.

The kinetic energy operator in equation (307) is written as

$$\sum_{\alpha=1}^{3} \frac{\bar{p}_\alpha^2}{2m} \bar{\psi} = -\frac{\hbar^2}{2m} \sum_{\alpha=1}^{3} \cos \beta_{\alpha,\alpha}\, e^{-j\theta_{d\alpha}} \frac{\partial}{\partial\alpha} (\cos \beta_{\alpha,\alpha}\, e^{-j\theta_{d\alpha}} \frac{\partial}{\partial\alpha}) \bar{\psi} \tag{318}$$

For simplicity it is assumed that β_α and θ_α are slowly varying functions of position so that equation (318) can be rewritten as

$$\sum_{\alpha=1}^{3} \frac{\bar{p}_\alpha^2}{2m} \bar{\psi} = -\frac{\hbar^2}{2m} \sum_{\alpha=1}^{3} \cos^2 \beta_{\alpha,\alpha}\, e^{-j2\theta_{d\alpha}} \frac{\partial^2\bar{\psi}}{\partial\alpha^2} \tag{319}$$

Writing the wavefunction as $\bar{\psi} = \psi_R + j\psi_I$ allows equation (307) to be written as two component equations as follows

$$-\frac{\hbar^2}{2m} \sum_{\alpha=1}^{3} \cos^2 \beta_{\alpha,\alpha} \left[\cos (2\theta_{d\alpha}) \frac{\partial^2\psi_R}{\partial\alpha^2} + \sin (2\theta_{d\alpha}) \frac{\partial^2\psi_I}{\partial\alpha^2}\right] \tag{320}$$

$$+ W(\cos \theta_W\, \psi_R - \sin \theta_W\, \psi_I)$$

$$= i\hbar \cos \beta_{t,t} \left(\cos \theta_{dt} \frac{\partial\psi_R}{\partial t} + \sin \theta_{dt} \frac{\partial\psi_I}{\partial t}\right)$$

$$-\frac{\hbar^2}{2m} \sum_{\alpha=1}^{3} \cos^2 \beta_{\alpha,\alpha} \left[-\sin (2\theta_{d\alpha}) \frac{\partial^2\psi_R}{\partial\alpha^2} + \cos (2\theta_{d\alpha}) \frac{\partial^2\psi_I}{\partial\alpha^2}\right] \tag{321}$$

$$+ W(\sin \theta_W\, \psi_R + \cos \theta_W\, \psi_I)$$

$$= i\hbar \cos \beta_{t,t} \left(-\sin \theta_{dt} \frac{\partial\psi_R}{\partial t} + \cos \theta_{dt} \frac{\partial\psi_I}{\partial t}\right)$$

Equations (320) and (321) can be used to determine ψ_R and ψ_I. For the case of a stationary state the wave function components are written as

$$\psi_R = \phi_R e^{-i\varepsilon t/\hbar} \qquad \psi_I = \phi_I e^{-i\varepsilon t/\hbar} \tag{322}$$

and equations (320) and (321) become

$$-\frac{\hbar^2}{2m}\sum_{\alpha=1}^{3}\cos^2\beta_{\alpha,\alpha}\left[\cos(2\theta_{d\alpha})\frac{\partial^2\phi_R}{\partial\alpha^2} + \sin(2\theta_{d\alpha})\frac{\partial^2\phi_I}{\partial\alpha^2}\right] \tag{323}$$

$$+ W(\cos\theta_W\,\phi_R - \sin\theta_W\,\phi_I)$$

$$= \varepsilon\cos\beta_{t,t}(\cos\theta_{dt}\,\phi_R + \sin\theta_{dt}\,\phi_I)$$

$$-\frac{\hbar^2}{2m}\sum_{\alpha=1}^{3}\cos^2\beta_{\alpha,\alpha}\left[-\sin(2\theta_{d\alpha})\frac{\partial^2\phi_R}{\partial\alpha^2} + \cos(2\theta_{d\alpha})\frac{\partial^2\phi_I}{\partial\alpha^2}\right] \tag{324}$$

$$+ W(\sin\theta_W\,\phi_R + \cos\theta_W\,\phi_I)$$

$$= \varepsilon\cos\beta_{t,t}(-\sin\theta_{dt}\,\phi_R + \cos\theta_{dt}\,\phi_I)$$

It is generally quite difficult to determine ϕ_R and ϕ_I (and ε) from equations (323) and (324). The form and magnitude of the functions $\beta_{\alpha,\alpha}$, $\beta_{t,t}$, $\theta_{d\alpha}$, and θ_{dt} depend on the nature, density, and temperature of the asymmetric bulk matter or vacuum surrounding a particle.

Consider the case where the asymmetries are sufficiently small that the imaginary part of the wavefunction can be neglected in equation (323), so that this equation becomes

$$-\frac{\hbar^2}{2m}\sum_{\alpha=1}^{3}\cos^2\beta_{\alpha,\alpha}\cos(2\theta_{d\alpha})\frac{d^2\phi_R}{d\alpha^2} + W\cos\theta_W\,\phi_R = \varepsilon\cos\beta_{t,t}\cos\theta_{dt}\,\phi_R \tag{325}$$

For an isotropic system equation (325) becomes

$$\frac{d^2\phi_R}{dx^2} + \frac{2m}{3\hbar^2 \cos^2 \beta_{x,x} \cos(2\theta_{dx})} (\varepsilon \cos \beta_{t,t} \cos \theta_{dt} - W \cos \theta_W)\, \phi_R = 0 \tag{326}$$

This can be rewritten as

$$\frac{d^2\phi_R}{dx^2} + \frac{2m^*}{3\hbar^2} (\varepsilon - W^*)\, \phi_R = 0 \tag{327}$$

where m^* is an effective given by

$$m^* = \frac{m}{\cos^2 \beta_{x,x} \cos(2\theta_{dx})} \tag{328}$$

and W^* is an energy dependent effective potential given by

$$W^* = \varepsilon(1 - \cos \beta_{t,t} \cos \theta_{dt}) + W \cos \theta_W \tag{329}$$

Equations (323) and (324) can also be written in terms of the magnitude and phase angle of the wavefunction by writing $\phi_R = \phi \cos \theta_\phi$ and $\phi_I = \phi \sin \theta_\phi$. If the derivatives of the phase angle θ_ϕ are sufficiently small and can be neglected then equations (323) and (324) can be rewritten as

$$-\frac{\hbar^2}{2m} \sum_{\alpha=1}^{3} \cos^2 \beta_{\alpha,\alpha} \cos(2\theta_{d\alpha}) \frac{d^2\phi}{d\alpha^2} + W \cos \theta_W\, \phi \tag{323A}$$

$$= \varepsilon \cos \beta_{t,t} \cos \theta_{dt}\, \phi$$

$$+\frac{\hbar^2}{2m} \sum_{\alpha=1}^{3} \cos^2 \beta_{\alpha,\alpha} \sin(2\theta_{d\alpha}) \frac{d^2\phi}{d\alpha^2} + W \sin \theta_W\, \phi \tag{324A}$$

$$= -\, \varepsilon \cos \beta_{t,t} \sin \theta_{dt}\, \phi$$

For a one dimensional system the factor 3 that appears in equations (326) and (327) should be replaced by unity. Therefore in asymmetric bulk matter or vacuum the particle acquires an effective mass, due to spacetime interactions, which is larger than the bare mass. In addition an energy dependent effective potential arises whose value depends on the degree of asymmetry that exists in the background of the particle.

At this point it is easy to treat the Klein-Gordon equation for a particle that is located in an asymmetric background. For a particle in a symmetric vacuum, the Klein-Gordon equation is written as[34-37]

$$\frac{\partial^2 \psi^a}{\partial t_a^2} = c^2 \nabla_a^2 \psi_a - \frac{m^2 c^4}{\hbar^2} \psi_a \tag{330}$$

Within asymmetric bulk matter or vacuum the spacetime interactions induce a broken symmetry in the wave function and in the space and time coordinates, so that the Klein-Gordon equation becomes

$$\frac{\partial^2 \bar{\psi}}{\partial \bar{t}^2} = c^2 \left(\frac{\partial^2 \bar{\psi}}{\partial \bar{x}^2} + \frac{\partial^2 \bar{\psi}}{\partial \bar{y}^2} + \frac{\partial^2 \bar{\psi}}{\partial \bar{z}^2} \right) - \frac{m^2 c^4}{\hbar^2} \bar{\psi} \tag{331}$$

Taking the real and imaginary components of equation (331) gives

$$R_{2t}(\psi_R, \psi_I) \sim c^2 [R_{2x}(\psi_R, \psi_I) + R_{2y}(\psi_R, \psi_I) + R_{2z}(\psi_R, \psi_I)] - \frac{m^2 c^4}{\hbar^2} \psi_R \tag{332}$$

$$I_{2t}(\psi_R, \psi_I) \sim c^2 [I_{2x}(\psi_R, \psi_I) + I_{2y}(\psi_R, \psi_I) + I_{2z}(\psi_R, \psi_I)] - \frac{m^2 c^4}{\hbar^2} \psi_I \tag{333}$$

where

$$R_{2\eta}(\psi_R, \psi_I) = \cos^2 \beta_{\eta,\eta} \left[\cos (2\theta_{d\eta}) \frac{\partial^2 \psi_R}{\partial \eta^2} + \sin (2\theta_{d\eta}) \frac{\partial^2 \psi_I}{\partial \eta^2} \right] \tag{334}$$

$$I_{2\eta}(\psi_R, \psi_I) = \cos^2 \beta_{\eta,\eta} \left[- \sin (2\theta_{d\eta}) \frac{\partial^2 \psi_R}{\partial \eta^2} + \cos (2\theta_{d\eta}) \frac{\partial^2 \psi_I}{\partial \eta^2} \right] \tag{335}$$

where $\eta = t, x, y, z$, and where

$$\theta_{d\eta} = \theta_\eta + \beta_{\eta,\eta} \tag{336}$$

<u>9. CONCLUSION.</u> On account of spacetime interactions with bulk matter and the vacuum, these systems exhibit broken internal symmetries. In the case of black body radiation in asymmetric bulk matter or vacuum, the photons have complex number frequencies which produce a radiation pressure and energy density that have broken internal symmetries. The space and time coordinates within a

broken symmetry system are also gauge rotated and are described by internal phase angles. From this it follows that geometrical angles are described by complex numbers and have internal phase angles. The skewed nature of space and time affects the fundamental scattering processes of atomic particles. All atomic processes that occur in asymmetric bulk matter or vacuum should also have broken symmetries that are manifested in the measured differential cross sections. For broken symmetry quantum systems, the asymmetry produces an effective mass in the Schrödinger equation that is larger than the bare mass of a particle.

ACKNOWLEDGEMENT

The author wishes to thank Elizabeth K. Klein for typing this paper.

REFERENCES

1. Chaichian, M. and Nelipa, N., Introduction to Gauge Field Theories, Springer-Verlag, New York, 1984.

2. Ryder, L. H., Quantum Field Theory, Cambridge University Press, New York, 1985.

3. Dodd, J., The Ideas of Particle Physics, Cambridge University Press, New York, 1984.

4. Aitchison, I. and Hey, A., Gauge Theories in Particle Physics, Adam Hilger, Bristol, 1982.

5. Weiss, R. A., "Thermodynamic Gauge Theory of Solids and Quantum Liquids with Internal Phase", Fifth Army Conference on Applied Mathematics and Computing, West Point, New York, ARO 88-1, June 15-18, 1987, p. 649.

6. Weiss, R. A., Relativistic Thermodynamics, Vols. 1 and 2, Exposition Press, New York, 1976.

7. Hagedorn, R., Relativistic Kinematics, Benjamin, New York, 1964.

8. Huang, K., Statistical Mechanics, John Wiley, New York, 1963.

9. Hill, T. L., An Introduction to Statistical Thermodynamics, Addison-Wesley, New York, 1960.

10. Planck, M., The Theory of Heat Radiation, Dover, New York, 1959.

11. Joos, G., Theoretical Physics, Hafner, New York, 1950.

12. Page, L., Introduction to Theoretical Physics, Van Nostrand, New York, 1952.

13. Rybicki, G. and Lightman, A., Radiative Processes in Astrophysics, John Wiley, New York, 1979.

14. Sommerfeld, A., Thermodynamics and Statistical Mechanics, Academic Press, New York, 1964, p. 148.

15. Brillouin, L., Tensors in Mechanics and Elasticity, Academic Press, New York, 1964.

16. Weiss, R. A., "Radiation Pressure of Light in a Refractive Medium", Journal of Applied Physics, Vol. 47, No. 1, Jan. 1976.

17. Born, M., Atomic Physics, Hafner, New York, 1953.

18. Bethe, H. and Salpeter, E., Quantum Mechanics of One- and Two-Electron Systems, Springer-Verlag, New York, 1957.

19. Brown, K. M., "Solutions of Simultaneous Non-Linear Equations", Communications of the ACM, Vol. 10, No. 11, Nov. 1967, p. 728.

20. DeBenedetti, S., Nuclear Interactions, John Wiley, New York, 1964.

21. Eder, G., Nuclear Forces, MIT Press, Cambridge, MA, 1968.

22. Kursunoglu, B., Modern Quantum Theory, Freeman, San Francisco, 1962.

23. Mott, N. and Massey, H., The Theory of Atomic Collisions, Oxford University Press, Oxford, 1965.

24. Sachs, R., Nuclear Theory, Addison-Wesley, New York, 1953.

25. Bethe, H. and Morrison, P., Elementary Nuclear Theory, John Wiley, New York, 1961.

26. Elton, L., Introductory Nuclear Theory, John Wiley-Interscience, New York, 1959.

27. Mandl, F. and Shaw, G., Quantum Field Theory, John Wiley, New York, 1984.

28. Bjorken, J. and Drell, S., Relativistic Quantum Mechanics, McGraw-Hill, New York, 1964.

29. Bogoliubov, N. and Shirkov, D., Introduction to the Theory of Quantized Fields, John Wiley-Interscience, New York, 1959.

30. Gasiorowicz, S., Elementary Particle Physics, John Wiley, New York, 1966.

31. Bethe, H. A. and Jackiw, R., Intermediate Quantum Mechanics, Benjamin, New York, 1968.

32. Heitler, W., The Quantum Theory of Radiation, Dover, New York, 1984.

33. Dirac, P., The Principles of Quantum Mechanics, Oxford University Press, New York, 1947.

34. Schweber, S., An Introduction to Relativistic Quantum Field Theory, Harper and Row, New York, 1962.

35. Akhiezer, A. and Berestetskii, V., Quantum Electrodynamics, John Wiley-Interscience, New York, 1965.

36. Schweber, S., Bethe, H. and de Hoffmann, F., Mesons and Fields, Vol. 1, Row-Peterson, Evanston, 1956.

37. Feynman, R., Quantum Electrodynamics, Benjamin, New York, 1962.

38. Rojansky, V., Introductory Quantum Mechanics, Prentice-Hall, New York, 1938.

39. Persico, E., Fundamentals of Quantum Mechanics, Prentice-Hall, New York, 1950.

40. Landau, L. and Lifshitz, E., Quantum Mechanics, Addison-Wesley, New York, 1958.

41. Pauling, L. and Wilson, E., Introduction to Quantum Mechanics, McGraw-Hill, New York, 1935.

42. Bohm, D., Quantum Theory, Prentice-Hall, New York, 1951.

43. Kemble, E., The Fundamental Principles of Quantum Mechanics, Dover, New York, 1958.

44. Powell, J. and Crasemann, B., Quantum Mechanics, Addison-Wesley, New York, 1961.

45. Messiah, A., Quantum Mechanics, John Wiley, New York, 1961.

46. Merzbacher, E., Quantum Mechanics, John Wiley, New York, 1961.

47. Mandl, F., Quantum Mechanics, Academic, New York, 1954.

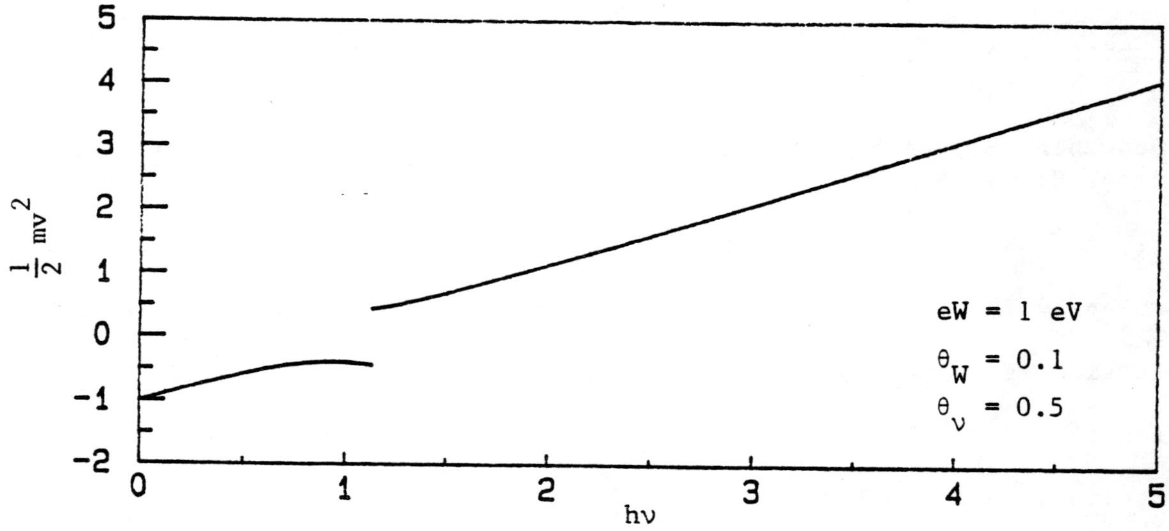

Figure 1. Magnitude of electron kinetic energy versus the magnitude of the photon energy.

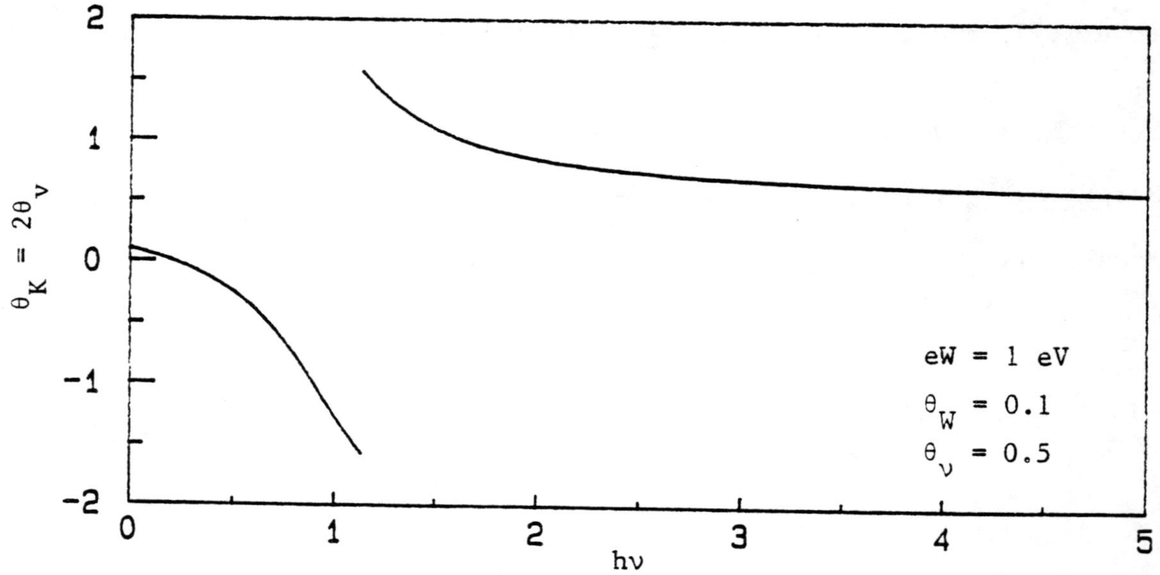

Figure 2. Phase angle of the electron kinetic energy versus the magnitude of the photon energy.

THE BROKEN SYMMETRY OF SPACE AND TIME IN BULK MATTER AND THE VACUUM

Richard A. Weiss
U. S. Army Engineer Waterways Experiment Station
Vicksburg, Mississippi 39180

ABSTRACT. Because the pressure and internal energy of bulk matter and the vacuum are associated with internal phase angles, the space and time coordinates and the kinematic and dynamic variables of an interacting system of particles also exhibit broken internal symmetries. Specifically, in bulk matter or the vacuum with broken internal symmetries, the internal phase angles of the particle velocity, acceleration, and space and time coordinates are related to the internal phase angles of the pressure and internal energy. A procedure is developed for determining the internal phase angles of the kinematic and dynamic variables and of the space and time coordinates in terms of Euler's equations of motion. Continuum mechanics and elasticity solutions for bulk matter require the joint determination of phase angles for the space and time coordinates and the magnitude and internal phase angle of the pressure. Rotating matter with broken space and time symmetries is treated, and it is shown that the conservation of angular momentum is valid for such a system. The gravitational equilibrium configurations of stars and planets are treated for state equations that have broken internal symmetries, and equations are developed that relate the internal phase angles of the space and time coordinates to the internal phase angle of the pressure. Newtonian gravity in matter with broken internal symmetry is considered and applications to the earth's gravity field are suggested. These results will also affect the predicted trajectories of ballistic missiles.

1. INTRODUCTION. The fundamental interactions in nature are formulated as gauge theories. For instance, the theory of gravity is formulated as a gauge theory based on the Lorentz group SO(3,1), while electromagnetism is based on the gauge group U(1).[1] The nongravitational forces are thought to be described by the gauge group SU(3) x SU(2) x U(1).[2,3] In fact the Lie group U(1) and its real value analog $e^{\pm\phi}$ have been shown to be the gauge groups of relativistic thermodynamics.[4] The pressure and energy density of matter described by relativistic thermodynamics are associated with broken symmetries. This is related to the fact that the pressure and energy density can be gauge rotated in such a way as to leave the terms of the basic trace equation of relativistic thermodynamics gauge invariant.[4]

For an interacting bulk matter system the broken symmetry of the state equation is vacuum induced and results from the solution of a complex number trace equation that relates the renormalized (relativistic) state equation to the corresponding ordinary state equation. This trace equation is given by[5,6]

$$\overline{U} + T\left(\frac{d\overline{U}}{dT}\right)_{\overline{P}V} - 3V\frac{d}{dV}(\overline{P}V)_{\overline{U}} = U^a + T\left(\frac{dU^a}{dT}\right)_{P^aV} \tag{1}$$

or equivalently as

$$\left(1 - \overline{b} + T\frac{\partial}{\partial T} - \overline{b}V\frac{\partial}{\partial V}\right)\overline{E} - 3\left(1 + \overline{\gamma} + V\frac{\partial}{\partial V} - \overline{\gamma}T\frac{\partial}{\partial T}\right)\overline{P} = \psi^a \tag{2}$$

where $\overline{U}$, $\overline{E}$, $\overline{P}$, $\overline{\gamma}$, and $\overline{b}$ are complex number representations of the renormalized internal energy, energy density, pressure, and the gauge parameters, and where

$$\overline{b} = \frac{T\dfrac{\partial\overline{P}}{\partial T}}{\overline{P} - \overline{K}_T} \tag{3}$$

$$\overline{K}_T = -V\left(\frac{\partial\overline{P}}{\partial V}\right)_T \tag{4}$$

$$\overline{\gamma} = \frac{V}{\overline{C}_V}\frac{\partial\overline{P}}{\partial T} = \frac{\partial\overline{P}/\partial T}{\partial\overline{E}/\partial T} \tag{5}$$

$$\psi^a = \left(T\frac{\partial}{\partial T} - b^aV\frac{\partial}{\partial V} + 1 - b^a\right)E^a \tag{6}$$

$$b^a = \frac{T(\partial P^a/\partial T)_V}{(P^a - K_T^a)} \tag{7}$$

The quantities E^a, P^a, and K_T^a = unrenormalized values of the energy density, pressure, and bulk modulus respectively. Throughout this paper the index "a" will refer to nonrelativistic (unrenormalized) calculations. The complex number thermodynamic state functions that appear in equations (1) and (2) will be written in terms of their internal phase angles as follows

$$\overline{U} = Ue^{j\theta_U} \tag{8}$$

$$\overline{E} = \overline{U}/V = Ee^{j\theta_U} \tag{9}$$

$$\overline{P} = Pe^{j\theta_P} \tag{10}$$

$$\overline{\gamma} = \gamma e^{j\theta_\gamma} \tag{11}$$

$$\overline{b} = be^{j\theta_b} \tag{12}$$

where θ_U, θ_P, θ_γ, and θ_b = internal phase angles of the internal energy, pressure, Grüneisen parameter, and b gauge parameter respectively. The relativistic ground state of the vacuum is described by equation (1) or (2) with their right hand sides set equal to zero. The vacuum state also has a broken symmetry and in fact the bulk matter state is essentially mathematically equivalent to the vacuum state.

On account of the broken symmetry of the pressure and energy density of bulk matter or the vacuum, time may not unfold in a purely linear fashion but may also rotate in an internal space. Spatial coordinates in bulk matter or the vacuum may also have broken internal symmetries that are associated with internal phase angles. The broken symmetries of space and time in bulk matter or the vacuum are related to the broken symmetries of the state equations for these systems. Thermodynamic and continuum mechanics theories will require the joint determination of the internal phase angles of space and time coordinates along with the pressure and internal energy and their internal phase angles. The gauge rotated space and time coordinates have an effect on the equations of motion of a system of particles and will affect the equilibrium configurations of atomic nuclei, planets and the stars. Note that it is the real parts of the complex number quantities such as space and time coordinates, pressure, energy, velocity and acceleration that are the measured quantities.

The broken symmetry of space and time is related to the broken symmetry of the pressure and internal energy of bulk matter or the vacuum as determined from solutions of equation (1). The right hand side of equation (1) is equal to zero for the case of the vacuum. When matter is present the broken symmetry of space and time can be calculated in two ways: 1, at the macroscopic level through Euler's equations and the complex pressure field for interacting matter (Section 6), and 2, at the single particle level through the action of a complex gauge potential that is induced by vacuum effects. For the vacuum only the second method is possible because the matter density is zero, and the complex gauge potential for the vacuum must be determined.

The complex gauge potential is calculated from the relativistic internal energy and pressure that are obtained from equation (1). This is done by calculating the renormalized complex valued partition function which is defined as[7,8]

$$\bar{Z} = \int \eta e^{-\beta \bar{H}} d\bar{q}\, d\bar{p} \tag{13}$$

where η = degeneracy, $\beta = 1/(kT)$, and where the complex number Hamiltonian is given by

$$\bar{H} = \frac{\bar{p}^2}{2m} + \bar{W} \tag{14}$$

where $\bar{W} = V^a + \bar{V}_g$, where V^a = ordinary external potential, $\bar{V}_g$ = complex number gauge potential that is responsible for the difference between $\bar{U}$ and U^a given in equation (1). The connection between the internal energy and pressure and the partition function is given by[7,8]

$$\bar{U} = -\left(\frac{\partial\, \ell n\, \bar{Z}}{\partial \beta}\right)_V \qquad \bar{P} = \frac{1}{\beta}\left(\frac{\partial\, \ell n\, \bar{Z}}{\partial V}\right)_\beta \qquad (15)$$

where $\bar{U}$ is given by equation (1), so that equations (13) through (15) can be used to determine the complex gauge potential $\bar{V}_g$ in terms of P and θ_P of the complex matter fields. These equations relate the macroscopic pressure field given by equation (1) to the microscopic gauge potential $\bar{V}_g$. For the broken symmetry vacuum the partition function is

$$\bar{Z}^{(v)} = \int \eta e^{-\beta \bar{V}_g}\, d\bar{q}^{(v)}\, d\bar{p}^{(v)} \qquad (13A)$$

from which $\bar{U}^{(v)}$ and $\bar{P}^{(v)}$ can be obtained using equation (15). These values of $\bar{U}^{(v)}$ and $\bar{P}^{(v)}$ must agree with the vacuum solutions of equation (1), and this determines $\bar{V}_g$.

The broken symmetry of the state functions of interacting bulk matter and the vacuum impart a broken symmetry to the velocity, acceleration and space and time coordinates of particles located in bulk matter or the vacuum. Forces exerted in bulk matter or the vacuum will also exhibit broken internal symmetries. The aim of this paper is to relate the broken symmetries of space, time and the kinematic and dynamical variables, to the broken symmetry of the state equations for interacting bulk matter or the vacuum. The paper is organized as follows: Section 2. introduces gauge rotated coordinates, Section 3. treats the geometry of broken internal symmetry, Section 4. considers the kinematics and dynamics of broken symmetry particle systems, Section 5. studies rotating systems with broken internal symmetry, Section 6. introduces the Euler equations for bulk matter with broken symmetry, and Section 7. considers the equilibrium equations of stars and planets whose matter has internal phase.

2. GAUGE ROTATED SPACE AND TIME. In bulk matter or the vacuum the thermodynamic functions such as pressure and internal energy exhibit internal phases (broken symmetry).[6] This suggests that space and time coordinates in bulk matter or the vacuum may also possess broken symmetries. Accordingly the space and time coordinates of particles in bulk matter are written as

$$\bar{x} = xe^{j\theta_x} \qquad (16)$$

$$\bar{y} = ye^{j\theta_y} \qquad (17)$$

$$\bar{z} = ze^{j\theta_z} \qquad (18)$$

$$\bar{t} = te^{j\theta_t} \qquad (19)$$

where the phase angles θ_x, θ_y, θ_z, and θ_t manifest the broken symmetry. It will be assumed that in bulk matter the phase angles can be represented as

$$\theta_x = \theta_x(x,y,z,t) \tag{20}$$

$$\theta_y = \theta_y(x,y,z,t) \tag{21}$$

$$\theta_z = \theta_z(x,y,z,t) \tag{22}$$

$$\theta_t = \theta_t(x,y,z,t) \tag{23}$$

For the vacuum, coordinates will be written as

$$\bar{x}^{(v)} = x^{(v)} e^{j\theta_x^{(v)}} \tag{23A}$$

$$\bar{y}^{(v)} = y^{(v)} e^{j\theta_y^{(v)}} \tag{23B}$$

$$\bar{z}^{(v)} = z^{(v)} e^{j\theta_z^{(v)}} \tag{23C}$$

$$\bar{t}^{(v)} = t^{(v)} e^{j\theta_t^{(v)}} \tag{23D}$$

The differentials of the space and time coordinates can be written as

$$d\bar{x} = e^{j\theta_x}(dx + jx\,d\theta_x) = e^{j\theta_x}[(1 + jx\frac{\partial\theta_x}{\partial x})dx + jx\frac{\partial\theta_x}{\partial y}dy + jx\frac{\partial\theta_x}{\partial z}dz + jx\frac{\partial\theta_x}{\partial t}dt] \tag{24}$$

$$d\bar{y} = e^{j\theta_y}(dy + jy\,d\theta_y) = e^{j\theta_y}[jy\frac{\partial\theta_y}{\partial x}dx + (1 + jy\frac{\partial\theta_y}{\partial y})dy + jy\frac{\partial\theta_y}{\partial z}dz + jy\frac{\partial\theta_y}{\partial t}dt] \tag{25}$$

$$d\bar{z} = e^{j\theta_z}(dz + jz\,d\theta_z) = e^{j\theta_z}[jz\frac{\partial\theta_z}{\partial x}dx + jz\frac{\partial\theta_z}{\partial y}dy + (1 + jz\frac{\partial\theta_z}{\partial z})dz + jz\frac{\partial\theta_z}{\partial t}dt] \tag{26}$$

$$d\bar{t} = e^{j\theta_t}(dt + jt\,d\theta_t) = e^{j\theta_t}[jt\frac{\partial\theta_t}{\partial x}dx + jt\frac{\partial\theta_t}{\partial y}dy + jt\frac{\partial\theta_t}{\partial z}dz + (1 + jt\frac{\partial\theta_t}{\partial t})dt] \tag{27}$$

From equations (24) through (27) it follows that

$$\partial\bar{x}/\partial x = \sqrt{1 + x^2(\partial\theta_x/\partial x)^2}\; e^{j(\theta_x+\beta_{x,x})} \tag{28}$$

$$\partial\bar{x}/\partial y = x\partial\theta_x/\partial y\; e^{j(\theta_x+\pi/2)} \tag{29}$$

$$\partial\bar{x}/\partial z = x\partial\theta_x/\partial z\; e^{j(\theta_x+\pi/2)} \tag{30}$$

$$\partial\bar{x}/\partial t = x\partial\theta_x/\partial t\; e^{j(\theta_x+\pi/2)} \tag{30A}$$

$$\partial\bar{y}/\partial x = y\partial\theta_y/\partial x \, e^{j(\theta_y+\pi/2)} \quad (31)$$

$$\partial\bar{y}/\partial y = \sqrt{1 + y^2(\partial\theta_y/\partial y)^2} \, e^{j(\theta_y+\beta_{y,y})} \quad (32)$$

$$\partial\bar{y}/\partial z = y\partial\theta_y/\partial z \, e^{j(\theta_y+\pi/2)} \quad (33)$$

$$\partial\bar{y}/\partial t = y\partial\theta_y/\partial t \, e^{j(\theta_y+\pi/2)} \quad (33A)$$

$$\partial\bar{z}/\partial x = z\partial\theta_z/\partial x \, e^{j(\theta_z+\pi/2)} \quad (34)$$

$$\partial\bar{z}/\partial y = z\partial\theta_z/\partial y \, e^{j(\theta_z+\pi/2)} \quad (35)$$

$$\partial\bar{z}/\partial z = \sqrt{1 + z^2(\partial\theta_z/\partial z)^2} \, e^{j(\theta_z+\beta_{z,z})} \quad (36)$$

$$\partial\bar{z}/\partial t = z\partial\theta_z/\partial t \, e^{j(\theta_z+\pi/2)} \quad (36A)$$

$$\partial\bar{t}/\partial x = t\partial\theta_t/\partial x \, e^{j(\theta_t+\pi/2)} \quad (37A)$$

$$\partial\bar{t}/\partial y = t\partial\theta_t/\partial y \, e^{j(\theta_t+\pi/2)} \quad (37B)$$

$$\partial\bar{t}/\partial z = t\partial\theta_t/\partial z \, e^{j(\theta_t+\pi/2)} \quad (37C)$$

$$\partial\bar{t}/\partial t = \sqrt{1 + t^2(\partial\theta_t/\partial t)^2} \, e^{j(\theta_t+\beta_{t,t})} \quad (37D)$$

where in equations (28) through (37) the following notation is used

$$\tan \beta_{x,x} = x \frac{\partial\theta_x}{\partial x} \quad (38)$$

$$\tan \beta_{y,y} = y \frac{\partial\theta_y}{\partial y} \quad (39)$$

$$\tan \beta_{z,z} = z \frac{\partial\theta_z}{\partial z} \quad (40)$$

$$\tan \beta_{t,t} = t \frac{\partial\theta_t}{\partial t} \quad (41)$$

The following angles are also useful

$$\tan\beta_{x,y} = x\frac{\partial\theta_x}{\partial y} \qquad \tan\beta_{x,z} = x\frac{\partial\theta_x}{\partial z} \qquad \tan\beta_{x,t} = x\frac{\partial\theta_x}{\partial t} \tag{42}$$

$$\tan\beta_{y,x} = y\frac{\partial\theta_y}{\partial x} \qquad \tan\beta_{y,z} = y\frac{\partial\theta_y}{\partial z} \qquad \tan\beta_{y,t} = y\frac{\partial\theta_y}{\partial t} \tag{43}$$

$$\tan\beta_{z,x} = z\frac{\partial\theta_z}{\partial x} \qquad \tan\beta_{z,y} = z\frac{\partial\theta_z}{\partial y} \qquad \tan\beta_{z,t} = z\frac{\partial\theta_z}{\partial t} \tag{44}$$

$$\tan\beta_{t,x} = t\frac{\partial\theta_t}{\partial x} \qquad \tan\beta_{t,y} = t\frac{\partial\theta_t}{\partial y} \qquad \tan\beta_{t,z} = t\frac{\partial\theta_t}{\partial z} \tag{44A}$$

From equations (16) through (19) it also follows that

$$\frac{\partial}{\partial\bar{\eta}} = e^{-j\theta_{d\eta}}\cos\beta_{\eta,\eta}\frac{\partial}{\partial\eta} \tag{45}$$

where η = x, y, z, and t, and

$$\theta_{d\eta} = \theta_\eta + \beta_{\eta,\eta} \tag{46}$$

$$\tan\beta_{\eta,\eta} = \eta\frac{\partial\theta_\eta}{\partial\eta} \tag{47}$$

$$\cos\beta_{\eta,\eta} = \frac{1}{\sqrt{1 + (\eta\,\partial\theta_\eta/\partial\eta)^2}} \tag{48}$$

The result in equation (45) follows from the fact that if $\bar{y}$, $\bar{z}$, and $\bar{t}$ are constant, then their respective magnitudes y, z, and t are also constant. The measured space and time coordinates are $x_m = x\cos\theta_x$, $y_m = y\cos\theta_y$, $z_m = z\cos\theta_z$ and $t_m = t\cos\theta_t$ respectively. Space and time can be represented by helices whose spiral lengths are $L_x = x\sec\beta_{x,x}$; $L_y = y\sec\beta_{y,y}$; $L_z = z\sec\beta_{z,z}$ and $L_t = t\sec\beta_{t,t}$. The conventional coordinates t_a , x_a , y_a , z_a are related to the gauge rotated coordinates by $t = t_a$, $x = x_a$, $y = y_a$ and $z = z_a$.

The following relationships hold for spherical polar coordinates

$$\bar{r} = r\,e^{j\theta_r} \tag{49}$$

$$\bar{\psi} = \psi\,e^{j\theta_\psi} \tag{50}$$

$$\bar{\phi} = \phi\,e^{j\theta_\phi} \tag{51}$$

where ψ = zenith angle, ϕ = azimuth angle, and where $\theta_r = \theta_r(r,\psi,\phi,t)$, $\theta_\psi = \theta_\psi(r,\psi,\phi,t)$, and $\theta_\phi = \theta_\phi(r,\psi,\phi,t)$ which gives

$$d\bar{r} = e^{j\theta_r}(dr + jrd\theta_r) = e^{j\theta_r}[(1 + jr\frac{\partial\theta_r}{\partial r})dr + jr\frac{\partial\theta_r}{\partial\psi}d\psi + jr\frac{\partial\theta_r}{\partial\phi}d\phi + jr\frac{\partial\theta_r}{\partial t}dt] \quad (52)$$

$$d\bar{\psi} = e^{j\theta_\psi}(d\psi + j\psi d\theta_\psi) = e^{j\theta_\psi}[j\psi\frac{\partial\theta_\psi}{\partial r}dr + (1 + j\psi\frac{\partial\theta_\psi}{\partial\psi})d\psi + j\psi\frac{\partial\theta_\psi}{\partial\phi}d\phi + j\psi\frac{\partial\theta_\psi}{\partial t}dt] \quad (53)$$

$$d\bar{\phi} = e^{j\theta_\phi}(d\phi + j\phi d\theta_\phi) = e^{j\theta_\phi}[j\phi\frac{\partial\theta_\phi}{\partial r}dr + j\phi\frac{\partial\theta_\phi}{\partial\psi}d\psi + (1 + j\phi\frac{\partial\theta_\phi}{\partial\phi})d\phi + j\phi\frac{\partial\theta_\phi}{\partial t}dt] \quad (54)$$

$$d\bar{t} = e^{j\theta_t}(dt + jtd\theta_t) = e^{j\theta_t}[jt\frac{\partial\theta_t}{\partial r}dr + jt\frac{\partial\theta_t}{\partial\psi}d\psi + jt\frac{\partial\theta_t}{\partial\phi}d\phi + (1 + jt\frac{\partial\theta t}{\partial t})dt] \quad (54A)$$

and

$$\partial\bar{r}/\partial r = \sqrt{1 + (r\partial\theta_r/\partial r)^2}\ e^{j(\theta_r+\beta_{r,r})} \quad (55)$$

$$\partial\bar{r}/\partial\psi = r\partial\theta_r/\partial\psi\, e^{j(\theta_r+\pi/2)} \quad (56)$$

$$\partial\bar{r}/\partial\phi = r\partial\theta_r/\partial\phi\, e^{j(\theta_r+\pi/2)} \quad (57)$$

$$\partial\bar{r}/\partial t = r\partial\theta_r/\partial t\, e^{j(\theta_r+\pi/2)} \quad (57A)$$

$$\partial\bar{\psi}/\partial r = \psi\partial\theta_\psi/\partial r\, e^{j(\theta_\psi+\pi/2)} \quad (58)$$

$$\partial\bar{\psi}/\partial\psi = \sqrt{1 + (\psi\partial\theta_\psi/\partial\psi)^2}\ e^{j(\theta_\psi+\beta_{\psi,\psi})} \quad (59)$$

$$\partial\bar{\psi}/\partial\phi = \psi\partial\theta_\psi/\partial\phi\, e^{j(\theta_\psi+\pi/2)} \quad (60)$$

$$\partial\bar{\psi}/\partial t = \psi\partial\theta_\psi/\partial t\, e^{j(\theta_\psi+\pi/2)} \quad (60A)$$

$$\partial\bar{\phi}/\partial r = \phi\partial\theta_\phi/\partial r\, e^{j(\theta_\phi+\pi/2)} \quad (61)$$

$$\partial\bar{\phi}/\partial\psi = \phi\partial\theta_\phi/\partial\psi\, e^{j(\theta_\phi+\pi/2)} \quad (62)$$

$$\partial\bar{\phi}/\partial\phi = \sqrt{1 + (\phi\partial\theta_\phi/\partial\phi)^2}\ e^{j(\theta_\phi+\beta_{\phi,\phi})} \quad (63)$$

$$\partial\bar{\phi}/\partial t = \phi\partial\theta_\phi/\partial t\, e^{j(\theta_\phi+\pi/2)} \quad (63A)$$

$$\partial \bar{t}/\partial r = t\partial\theta_t/\partial r\, e^{j(\theta_t+\pi/2)} \tag{63B}$$

$$\partial \bar{t}/\partial \psi = t\partial\theta_t/\partial \psi\, e^{j(\theta_t+\pi/2)} \tag{63C}$$

$$\partial \bar{t}/\partial \phi = t\partial\theta_t/\partial \phi\, e^{j(\theta_t+\pi/2)} \tag{63D}$$

where

$$\tan \beta_{r,r} = r \frac{\partial\theta_r}{\partial r} \tag{64}$$

$$\tan \beta_{\psi,\psi} = \psi \frac{\partial\theta_\psi}{\partial \psi} \tag{65}$$

$$\tan \beta_{\phi,\phi} = \phi \frac{\partial\theta_\phi}{\partial \phi} \tag{66}$$

where the connection between r, θ_r, ψ, θ_ψ, ϕ, θ_ϕ and x, θ_x, y, θ_y and z, θ_z is given in Section 3. One can also define the following angles

$$\tan \beta_{r,\psi} = r \frac{\partial\theta_r}{\partial \psi} \qquad \tan \beta_{r,\phi} = r \frac{\partial\theta_r}{\partial \phi} \qquad \tan \beta_{r,t} = r \frac{\partial\theta_r}{\partial t} \tag{67}$$

$$\tan \beta_{\psi,r} = \psi \frac{\partial\theta_\psi}{\partial r} \qquad \tan \beta_{\psi,\phi} = \psi \frac{\partial\theta_\psi}{\partial \phi} \qquad \tan \beta_{\psi,t} = \psi \frac{\partial\theta_\psi}{\partial t} \tag{68}$$

$$\tan \beta_{\phi,r} = \phi \frac{\partial\theta_\phi}{\partial r} \qquad \tan \beta_{\phi,\psi} = \phi \frac{\partial\theta_\phi}{\partial \psi} \qquad \tan \beta_{\phi,t} = \phi \frac{\partial\theta_\phi}{\partial t} \tag{69}$$

$$\tan \beta_{t,r} = t \frac{\partial\theta_t}{\partial r} \qquad \tan \beta_{t,\psi} = t \frac{\partial\theta_t}{\partial \psi} \qquad \tan \beta_{t,\phi} = t \frac{\partial\theta_t}{\partial \phi} \tag{69A}$$

The derivatives with respect to the complex spherical polar coordinates are now written in the same form as in equation (45) where now $\eta = t, r, \psi, \phi$. The measured space and time coordinates are $r_m = r \cos \theta_r$, $\psi_m = \psi \cos \theta_\psi$, $\phi_m = \phi \cos \theta_\phi$ and $t_m = t \cos \theta_t$ respectively.

The effects of the different types of forces on the gauge rotation of space and time depend on the relative magnitude and ranges of the forces. Over small distances $< 10^{-18}$ cm the color force dominates, $< 10^{-13}$ cm the strong nuclear force between nucleons dominates, $< 10^{-8}$ cm the electric and magnetic forces of electrons and nuclei dominate.[9,10] For ranges $> 10^{-8}$ cm the long range gravitational force dominates. Therefore when equations (24) through (69) are written, the origin of the coordinates is associated with the origin of the forces involved. Thus for gravity the origin is taken to be the center of the planet or star in question, and the range of r is throughout the gravitating body and beyond because gravity has an infinite range. For nuclear forces the range of r is $r < 10^{-13}$ cm, while for electric forces in an atom $r < 10^{-8}$ cm. The values of θ_r and θ_t depend on the scale at which the dominant forces act. It is the real part of a complex number coordinate that is the quantity measured when a space or time coordinate measurement is made.

3. GEOMETRY OF SPACE IN BULK MATTER AND THE VACUUM. The broken symmetry of coordinates of particles located in bulk matter or the vacuum will influence the calculation of the effects of the basic forces that operate in these media, such as for example pressure and gravity. This section considers the effects of the broken symmetry of coordinates on basic geometrical quantities such as angles, areas, and path lengths. For example, the simple law of cosines for a plane triangle located in a medium with broken symmetry is written as

$$\cos \bar{\phi} = \frac{\bar{a}^2 + \bar{b}^2 - \bar{c}^2}{2\bar{a}\bar{b}} \tag{70}$$

where $\bar{a}$, $\bar{b}$ and $\bar{c}$ are the complex number sides of a plane triangle, and $\bar{\phi}$ is the complex angle opposite side $\bar{c}$. The complex number sides of the triangle can be written as

$$\bar{a} = a e^{j\theta_a} \tag{71}$$

$$\bar{b} = b e^{j\theta_b} \tag{72}$$

$$\bar{c} = c e^{j\theta_c} \tag{73}$$

then

$$\cos \bar{\phi} = \frac{1}{2}\frac{a}{b} e^{j(\theta_a - \theta_b)} + \frac{1}{2}\frac{b}{a} e^{j(\theta_b - \theta_a)} - \frac{c^2}{2ab} e^{j(2\theta_c - \theta_a - \theta_b)} \tag{74}$$

From equation (74) it is clear that $\bar{\phi}$ and $\cos \bar{\phi}$ are complex numbers so that

$$\bar{\phi} = \phi e^{j\theta_\phi} \tag{75}$$

$$\cos \bar{\phi} = C_\phi e^{-j\theta_{c\phi}} \tag{76}$$

where C_ϕ = magnitude of $\cos \bar{\phi}$, and $\theta_{c\phi}$= phase angle associated with $\cos \bar{\phi}$. In the same manner it follows that

$$\sin \bar{\phi} = S_\phi e^{j\theta_{s\phi}} \tag{77}$$

where S_ϕ = magnitude of $\sin \bar{\phi}$, and $\theta_{s\phi}$ = phase angle associated with $\sin \bar{\phi}$. From the well known relation

$$\cos \bar{\phi} = \frac{1}{2} [e^{j\bar{\phi}} + e^{-j\bar{\phi}}] \tag{78}$$

it follows that

$$\cos \bar{\phi} = \cos \phi_R \cosh \phi_I - j \sin \phi_R \sinh \phi_I \tag{79}$$

where from equation (38)

$$\bar{\phi} = \phi_R + j\phi_I = \phi(\cos \theta_\phi + j \sin \theta_\phi) \tag{80}$$

Combining equations (76), (79), and (80) gives

$$C_\phi = \sqrt{\cos^2 (\phi \cos \theta_\phi) + \sinh^2 (\phi \sin \theta_\phi)} \tag{81}$$

$$\tan \theta_{c\phi} = \tan (\phi \cos \theta_\phi) \tanh (\phi \sin \theta_\phi) \tag{82}$$

In a similar manner from

$$\sin \bar{\phi} = \frac{1}{2j} [e^{j\bar{\phi}} - e^{-j\bar{\phi}}] \tag{83}$$

it follows that

$$\sin \bar{\phi} = \sin \phi_R \cosh \phi_I + j \cos \phi_R \sinh \phi_I \tag{84}$$

and combining equations (77), (80) and (84) that

$$S_\phi = \sqrt{\sin^2 (\phi \cos \theta_\phi) + \sinh^2 (\phi \sin \theta_\phi)} \tag{85}$$

$$\tan \theta_{s\phi} = \cot (\phi \cos \theta_\phi) \tanh (\phi \sin \theta_\phi) \tag{86}$$

The law of sines for a plane triangle is given by

$$\frac{\bar{a}}{\sin \bar{A}} = \frac{\bar{b}}{\sin \bar{B}} = \frac{\bar{c}}{\sin \bar{C}} \tag{87}$$

where

$$\bar{A} = Ae^{j\theta_A} \tag{88}$$

with similar expressions for $\bar{B}$ and $\bar{C}$. It follows from equation (87) that

$$\frac{a}{S_A} = \frac{b}{S_B} = \frac{c}{S_C} \tag{89}$$

and

$$\theta_a - \theta_{sA} = \theta_b - \theta_{sB} = \theta_c - \theta_{sC} \tag{90}$$

and where

$$S_A = \sqrt{\sin^2 (A \cos \theta_A) + \sinh^2 (A \sin \theta_A)} \tag{91}$$

$$\tan \theta_{sA} = \cot (A \cos \theta_A) \tanh (A \sin \theta_A) \tag{92}$$

with similar expressions for S_B, S_C, θ_{sB}, and θ_{sC} . It should be noted that for spherical triangles equations (89) and (90) become respectively

$$\frac{S_a}{S_A} = \frac{S_b}{S_B} = \frac{S_c}{S_C} \tag{93}$$

and

$$\theta_{sa} - \theta_{sA} = \theta_{sb} - \theta_{sB} = \theta_{sc} - \theta_{sC} \tag{94}$$

Consider now simple plane areas located within a medium with broken internal symmetry. For example, the area of a triangle of sides $\bar{a}$, $\bar{b}$ and $\bar{c}$ with $\bar{\phi}$ = angle between sides $\bar{a}$ and $\bar{b}$ is given by

$$\bar{A} = \frac{1}{2} \bar{a}\bar{b} \sin \bar{\phi} = Ae^{j\theta_A} \tag{95}$$

where A = magnitude of area, and θ_A = phase angle of area. Combining equations (71), (72), (77), and (88) gives

$$A = \frac{1}{2} abS_\phi \tag{96}$$

$$\theta_A = \theta_a + \theta_b + \theta_{s\phi} \tag{97}$$

where S_ϕ and $\theta_{s\phi}$ are given by equations (85) and (86) respectively. Now consider the area of a circular sector of angle $\bar{\phi}$ which is

$$\bar{A} = \frac{1}{2}\,\bar{r}^2(\bar{\phi} - \sin\bar{\phi}) \tag{98}$$

$$= \frac{1}{2}\,r^2[\phi e^{j(2\theta_r+\theta_\phi)} - S_\phi e^{j(2\theta_r+\theta_{s\phi})}]$$

then it follows that

$$A\cos\theta_A = \frac{1}{2}\,r^2[\phi\cos(2\theta_r + \theta_\phi) - S_\phi\cos(2\theta_r + \theta_{s\phi})] \tag{99}$$

$$A\sin\theta_A = \frac{1}{2}\,r^2[\phi\sin(2\theta_r + \theta_\phi) - S_\phi\sin(2\theta_r + \theta_{s\phi})] \tag{100}$$

From equations (99) and (100) it follows that

$$\tan\theta_A = \frac{\phi\sin(2\theta_r + \theta_\phi) - S_\phi\sin(2\theta_r + \theta_{s\phi})}{\phi\cos(2\theta_r + \theta_\phi) - S_\phi\cos(2\theta_r + \theta_{s\phi})} \tag{101}$$

$$A^2 = \frac{1}{4}\,r^4\left\{\phi^2 + S_\phi^2 - 2\phi S_\phi\cos(\theta_\phi - \theta_{s\phi})\right\} \tag{102}$$

For a full circle obviously

$$A = \pi r^2 \tag{103}$$

$$\theta_A = 2\theta_r \tag{104}$$

For a rectangle of sides $\bar{x}$ and $\bar{y}$ one has

$$A = xy \tag{105}$$

$$\theta_A = \theta_x + \theta_y \tag{106}$$

For these cases, measured area = $A\cos\theta_A$.

Now consider various coordinate systems located in bulk matter or vacuum with broken internal symmetries. For example, for plane polar coordinates

$$\bar{x} = \bar{r}\cos\bar{\phi} = xe^{j\theta_x} \tag{107}$$

$$\bar{y} = \bar{r}\sin\bar{\phi} = ye^{j\theta_y} \tag{108}$$

and

$$\bar{x}^2 + \bar{y}^2 = \bar{r}^2 = r^2e^{2j\theta_r} \tag{109}$$

The scalar equivalents of equations (107) and (108) are

$$x = rC_{\phi} \tag{110}$$

$$y = rS_{\phi} \tag{111}$$

$$\theta_x = \theta_r - \theta_{c\phi} \tag{112}$$

$$\theta_y = \theta_r + \theta_{s\phi} \tag{113}$$

The scalar equivalents of equation (109) are

$$x^2 \cos (2\theta_x) + y^2 \cos (2\theta_y) = r^2 \cos (2\theta_r) \tag{114}$$

$$x^2 \sin (2\theta_x) + y^2 \sin (2\theta_y) = r^2 \sin (2\theta_r) \tag{115}$$

or equivalently

$$r^4 = x^4 + y^4 + 2x^2y^2 \cos [2(\theta_x - \theta_y)] \tag{116}$$

and

$$\tan (2\theta_r) = \frac{x^2 \sin (2\theta_x) + y^2 \sin (2\theta_y)}{x^2 \cos (2\theta_x) + y^2 \cos (2\theta_y)} \tag{117}$$

Finally, substituting equations (110) through (113) into equation (116) gives

$$1 = C_{\phi}^4 + S_{\phi}^4 + 2C_{\phi}^2S_{\phi}^2 \cos [2(\theta_{c\phi} + \theta_{s\phi})] \tag{118}$$

Consider now spherical coordinates located within bulk matter. For this system

$$\bar{x} = \bar{r} \sin \bar{\psi} \cos \bar{\phi} \tag{119}$$

$$\bar{y} = \bar{r} \sin \bar{\psi} \sin \bar{\phi} \tag{120}$$

$$\bar{z} = \bar{r} \cos \bar{\psi} \tag{121}$$

$$\bar{x}^2 + \bar{y}^2 + \bar{z}^2 = \bar{r}^2 \tag{122}$$

The scalar equivalent equations for equations (119) through (121) are

$$x = rS_\psi C_\phi \tag{123}$$

$$y = rS_\psi S_\phi \tag{124}$$

$$z = rC_\psi \tag{125}$$

and

$$\theta_x = \theta_r + \theta_{s\psi} - \theta_{c\phi} \tag{126}$$

$$\theta_y = \theta_r + \theta_{s\psi} + \theta_{s\phi} \tag{127}$$

$$\theta_z = \theta_r - \theta_{c\psi} \tag{128}$$

where C_ϕ and S_ψ are defined in equations (81) and (85) respectively, and $\theta_{c\phi}$ and $\theta_{s\psi}$ be equations (82) and (86) respectively. From equation (122) it follows that

$$x^2 \cos(2\theta_x) + y^2 \cos(2\theta_y) + z^2 \cos(2\theta_z) = r^2 \cos(2\theta_r) \tag{129}$$

$$x^2 \sin(2\theta_x) + y^2 \sin(2\theta_y) + z^2 \sin(2\theta_z) = r^2 \sin(2\theta_r) \tag{130}$$

Equations (129) and (130) give

$$r^4 = x^4 + y^4 + z^4 + 2x^2y^2\cos[2(\theta_x - \theta_y)] + 2y^2z^2 \cos[2(\theta_y - \theta_z)] + 2x^2z^2 \cos[2(\theta_x - \theta_z)] \tag{131}$$

$$\tan(2\theta_r) = \frac{x^2 \sin(2\theta_x) + y^2 \sin(2\theta_y) + z^2 \sin(2\theta_z)}{x^2 \cos(2\theta_x) + y^2 \cos(2\theta_y) + z^2 \cos(2\theta_z)} \tag{132}$$

From equations (123) through (125) and equation (131) it follows that

$$1 = S_\psi^4 C_\phi^4 + S_\psi^4 S_\phi^4 + C_\psi^4 + 2S_\psi^4 C_\phi^2 S_\phi^2 \cos[2(\theta_{c\phi} + \theta_{s\phi})] + 2S_\psi^2 S_\phi^2 C_\psi^2 \cos[2(\theta_{s\psi} + \theta_{s\phi} + \theta_{c\psi})] + 2S_\psi^2 C_\phi^2 C_\psi^2 \cos[2(\theta_{s\psi} - \theta_{c\phi} + \theta_{c\psi})] \tag{133}$$

The last type of coordinate system that will be considered is the polar space coordinates which utilizes direction cosines as follows

$$\bar{x} = \bar{r} \cos \bar{\alpha} \tag{134}$$

$$\bar{y} = \bar{r} \cos \bar{\beta} \tag{135}$$

$$\bar{z} = \bar{r} \cos \bar{\gamma} \tag{136}$$

$$\bar{r}^2 = \bar{x}^2 + \bar{y}^2 + \bar{z}^2 \tag{137}$$

It follows from equations (134) through (136) that

$$x = rC_\alpha \tag{138}$$

$$y = rC_\beta \tag{139}$$

$$z = rC_\gamma \tag{140}$$

$$\theta_x = \theta_r - \theta_{c\alpha} \tag{141}$$

$$\theta_y = \theta_r - \theta_{c\beta} \tag{142}$$

$$\theta_z = \theta_r - \theta_{c\gamma} \tag{143}$$

where

$$C_\alpha = \sqrt{\cos^2 (\alpha \cos \theta_\alpha) + \sinh^2 (\alpha \sin \theta_\alpha)} \tag{144}$$

$$\tan \theta_{c\alpha} = \tan (\alpha \cos \theta_\alpha) \tanh (\alpha \sin \theta_\alpha) \tag{145}$$

with similar expressions for C_β, C_γ, $\theta_{c\beta}$, and $\theta_{c\gamma}$. Equations (129) through (132) also hold for polar space coordinates. The equivalent of equation (133) for polar space coordinates is

$$1 = C_\alpha^4 + C_\beta^4 + C_\gamma^4 + 2C_\alpha^2 C_\beta^2 \cos [2(\theta_{c\beta} - \theta_{c\alpha})] \tag{146}$$

$$+ 2C_\beta^2 C_\gamma^2 \cos [2(\theta_{c\gamma} - \theta_{c\beta})] + 2C_\alpha^2 C_\gamma^2 \cos [2(\theta_{c\gamma} - \theta_{c\alpha})]$$

Consider now the case of rotation of coordinates in a plane that is located within bulk matter or vacuum with broken symmetry. The values of the coordinates in the cartesian system that is rotated through an angle $\bar{\phi}$ are

$$\bar{x}' = \bar{x} \cos \bar{\phi} + \bar{y} \sin \bar{\phi} \tag{147}$$

$$\bar{y}' = \bar{y} \cos \bar{\phi} - \bar{x} \sin \bar{\phi} \tag{148}$$

The component equations for equation (147) are

$$x' \cos \theta_x' = xC_\phi \cos (\theta_x - \theta_{c\phi}) + yS_\phi \cos (\theta_y + \theta_{s\phi}) \tag{149}$$

$$x' \sin \theta_x' = xC_\phi \sin (\theta_x - \theta_{c\phi}) + yS_\phi \sin (\theta_y + \theta_{s\phi}) \tag{150}$$

while the component equations for equation (148) are

$$y' \cos \theta_y' = yC_\phi \cos (\theta_y - \theta_{c\phi}) - xS_\phi \cos (\theta_x + \theta_{\bar{s}\phi}) \tag{151}$$

$$y' \sin \theta_y' = yC_\phi \sin (\theta_y - \theta_{c\phi}) - xS_\phi \sin (\theta_x + \theta_{s\phi}) \tag{152}$$

From equations (149) through (152) it follows that

$$(x')^2 = x^2C_\phi^2 + y^2S_\phi^2 + 2xyC_\phi S_\phi \cos [\theta_x - \theta_y - \theta_{c\phi} - \theta_{s\phi}] \tag{153}$$

$$(y')^2 = y^2C_\phi^2 + x^2S_\phi^2 - 2xyC_\phi S_\phi \cos [\theta_x - \theta_y + \theta_{c\phi} + \theta_{s\phi}] \tag{154}$$

The coordinate internal phase angles in the rotated system are given by

$$\tan \theta_x' = \frac{xC_\phi \sin (\theta_x - \theta_{c\phi}) + yS_\phi \sin (\theta_y + \theta_{s\phi})}{xC_\phi \cos (\theta_x - \theta_{c\phi}) + yS_\phi \cos (\theta_y + \theta_{s\phi})} \tag{155}$$

$$\tan \theta_y' = \frac{yC_\phi \sin (\theta_y - \theta_{c\phi}) - xS_\phi \sin (\theta_x + \theta_{s\phi})}{yC_\phi \cos (\theta_y - \theta_{c\phi}) - xS_\phi \cos (\theta_x + \theta_{s\phi})} \tag{156}$$

From equations (153) and (154) it follows that

$$(x')^2 + (y')^2 = (x^2 + y^2)(C_\phi^2 + S_\phi^2) + 4xyC_\phi S_\phi \sin (\theta_x - \theta_y) \sin (\theta_{c\phi} + \theta_{s\phi}) \tag{156A}$$

which reduces to the standard cartesian result when the internal phase angles are set equal to zero. The Lorentz group of rotations in spacetime are considered in an accompanying paper where Maxwell's equations with broken internal symmetry are considered.

4. BROKEN SYMMETRY OF KINEMATICAL AND DYNAMICAL VARIABLES. This section considers the effects of gauge rotated space and time on kinematics and dynamics. The gauge rotated space and time coordinates that were introduced in Section 2 can be used to define gauge rotated velocity and acceleration of particles located within bulk matter or the vacuum. For instance the components of the velocity of a particle are given by

$$\bar{v}_x = \frac{d\bar{x}}{d\bar{t}} = v_x e^{j\theta_{vx}} \tag{157}$$

$$\bar{v}_y = \frac{d\bar{y}}{d\bar{t}} = v_y e^{j\theta_{vy}} \tag{158}$$

$$\bar{v}_z = \frac{d\bar{z}}{d\bar{t}} = v_z e^{j\theta_{vz}} \tag{159}$$

where

$$v_x = \sqrt{\frac{\left(\frac{dx}{dt}\right)^2 + x^2\omega_{\theta x}^2}{1 + t^2\omega_{\theta t}^2}} \tag{160}$$

$$v_y = \sqrt{\frac{\left(\frac{dy}{dt}\right)^2 + y^2\omega_{\theta y}^2}{1 + t^2\omega_{\theta t}^2}} \tag{161}$$

$$v_z = \sqrt{\frac{\left(\frac{dz}{dt}\right)^2 + z^2\omega_{\theta z}^2}{1 + t^2\omega_{\theta t}^2}} \tag{162}$$

$$\theta_{vx} = \theta_x - \theta_t + \beta_{x,t} - \beta_{t,t} \tag{163}$$

$$\theta_{vy} = \theta_y - \theta_t + \beta_{y,t} - \beta_{t,t} \tag{164}$$

$$\theta_{vz} = \theta_z - \theta_t + \beta_{z,t} - \beta_{t,t} \tag{165}$$

where the internal angular velocities are given by

$$\omega_{\theta t} = d\theta_t/dt \qquad \omega_{\theta x} = d\theta_x/dt \tag{166}$$

$$\omega_{\theta y} = d\theta_y/dt \qquad \omega_{\theta z} = d\theta_z/dt \tag{167}$$

and where

$$\tan \beta_{x,t} = x \frac{d\theta_x/dt}{dx/dt} \tag{168}$$

$$\tan \beta_{y,t} = y \frac{d\theta_y/dt}{dy/dt} \tag{169}$$

$$\tan \beta_{z,t} = z \frac{d\theta_z/dt}{dz/dt} \tag{170}$$

and where $\beta_{t,t}$ is given by equation (41). The internal angular velocities can also be written as

$$\omega_{\theta x} = \frac{d\theta_x}{dt} = \frac{\partial\theta_x}{\partial t} + \frac{\partial\theta_x}{\partial x}\frac{dx}{dt} + \frac{\partial\theta_x}{\partial y}\frac{dy}{dt} + \frac{\partial\theta_x}{\partial z}\frac{dz}{dt} \tag{171}$$

$$\omega_{\theta y} = \frac{d\theta_y}{dt} = \frac{\partial\theta_y}{\partial t} + \frac{\partial\theta_y}{\partial x}\frac{dx}{dt} + \frac{\partial\theta_y}{\partial y}\frac{dy}{dt} + \frac{\partial\theta_y}{\partial z}\frac{dz}{dt} \tag{172}$$

$$\omega_{\theta z} = \frac{d\theta_z}{dt} = \frac{\partial\theta_z}{\partial t} + \frac{\partial\theta_z}{\partial x}\frac{dx}{dt} + \frac{\partial\theta_z}{\partial y}\frac{dy}{dt} + \frac{\partial\theta_z}{\partial z}\frac{dz}{dt} \tag{173}$$

$$\omega_{\theta t} = \frac{d\theta_t}{dt} = \frac{\partial\theta_t}{\partial t} + \frac{\partial\theta_t}{\partial x}\frac{dx}{dt} + \frac{\partial\theta_t}{\partial y}\frac{dy}{dt} + \frac{\partial\theta_t}{\partial z}\frac{dz}{dt} \tag{173A}$$

The conventional special relativistic momentum of a particle moving with a velocity v_x^a is given for a conventional dynamical system by the following standard formula[11]

$$p_x^a = m\gamma_x^a v_x^a \tag{174}$$

where m = mass, $v_x^a = dx_a/dt_a = dx_m/dt_m$ = conventionally calculated velocity, and γ_x^a = ordinary velocity factor (boost) given by[11]

$$\gamma_x^a = [1 - (v_x^a/c)^2]^{-1/2} \tag{175}$$

where c = light speed in the vacuum. These standard formulas are developed by considering the particle to be attached to a coordinate system moving with velocity $v = v_x^a$.[11] In this paper the generalization to bulk matter or vacuum with broken internal symmetries is made by considering the particle to be attached to a coordinate system moving with complex velocity $\bar{v} = \bar{v}_x$, so that the single particle momentum is written as

$$\bar{p}_x = m\bar{\gamma}_x\bar{v}_x = m\gamma_x v_x e^{j(\theta_{vx}+\theta_{\gamma x})} \tag{176}$$

where

$$\bar{\gamma}_x = \gamma_x e^{j\theta_{\gamma x}} = (1 - \bar{v}_x^2/c^2)^{-1/2} \tag{177}$$

gives the complex number velocity factor. The magnitude and phase angle of the complex number velocity factor is given by

$$\gamma_x = [1 - 2(v_x/c)^2 \cos(2\theta_{vx}) + (v_x/c)^4]^{-1/4} \tag{178}$$

$$\tan(2\theta_{\gamma x}) = \frac{(v_x/c)^2 \sin(2\theta_{vx})}{1 - (v_x/c)^2 \cos(2\theta_{vx})} \tag{179}$$

The results in equations (178) and (179) are obtained as a simple generalization of standard special relativity results to the case where space and time have intrinsic broken symmetry, and reduce to the standard result in equation (175) if the internal phase angles are set equal to zero. Note that the measured velocity is $v_{xm} = v_x \cos\theta_{vx} \neq v_x^a$.

The magnitude of the particle velocity is obtained by noting that the complex number particle velocity is written as

$$\bar{v} = ve^{j\theta_v} \tag{180}$$

and from equations (157) through (159) and equation (180) it follows that

$$\bar{v}^2 = \bar{v}_x^2 + \bar{v}_y^2 + \bar{v}_z^2 \tag{181}$$

or

$$v^2e^{2j\theta_v} = v_x^2e^{2j\theta_{vx}} + v_y^2e^{2j\theta_{vy}} + v_z^2e^{2j\theta_{vz}} \tag{182}$$

The component equations corresponding to equation (182) are

$$v^2 \cos(2\theta_v) = v_x^2 \cos(2\theta_{vx}) + v_y^2 \cos(2\theta_{vy}) + v_z^2 \cos(2\theta_{vz}) \qquad (183)$$

$$v^2 \sin(2\theta_v) = v_x^2 \sin(2\theta_{vx}) + v_y^2 \sin(2\theta_{vy}) + v_z^2 \sin(2\theta_{vz}) \qquad (184)$$

From equations (183) and (184) it follows that

$$v^4 = v_x^4 + v_y^4 + v_z^4 + 2v_x^2 v_y^2 \cos[2(\theta_{vx} - \theta_{vy})] \qquad (185)$$

$$+ 2v_x^2 v_z^2 \cos[2(\theta_{vx} - \theta_{vz})] + 2v_y^2 v_z^2 \cos[2(\theta_{vy} - \theta_{vz})]$$

and

$$\tan(2\theta_v) = \frac{v_x^2 \sin(2\theta_{vx}) + v_y^2 \sin(2\theta_{vy}) + v_z^2 \sin(2\theta_{vz})}{v_x^2 \cos(2\theta_{vx}) + v_y^2 \cos(2\theta_{vy}) + v_z^2 \cos(2\theta_{vz})} \qquad (186)$$

where v_x, v_y, v_z, θ_{vx}, θ_{vy}, and θ_{vz} are given by equations (160) through (165) respectively. The measured velocity = $v \cos \theta_v$.

The acceleration components are written as

$$\bar{a}_x = \frac{d\bar{v}_x}{d\bar{t}} = a_x e^{j\theta_{ax}} \qquad (187)$$

$$\bar{a}_y = \frac{d\bar{v}_y}{d\bar{t}} = a_y e^{j\theta_{ay}} \qquad (188)$$

$$\bar{a}_z = \frac{d\bar{v}_z}{d\bar{t}} = a_z e^{j\theta_{az}} \qquad (189)$$

where using the Eulerian derivative gives

$$\bar{a}_x = \frac{d\bar{v}_x}{d\bar{t}} = \frac{\partial \bar{v}_x}{\partial \bar{t}} + \bar{v}_x \frac{\partial \bar{v}_x}{\partial \bar{x}} + \bar{v}_y \frac{\partial \bar{v}_x}{\partial \bar{y}} + \bar{v}_z \frac{\partial \bar{v}_x}{\partial \bar{z}} \qquad (190)$$

$$= \bar{a}_x^{(o)} + \bar{a}_x^{(1)} + \bar{a}_x^{(2)} + \bar{a}_x^{(3)}$$

$$= a_x^{(o)} e^{j\psi_{xo}} + a_x^{(1)} e^{j\psi_{x1}} + a_x^{(2)} e^{j\psi_{x2}} + a_x^{(3)} e^{j\psi_{x3}}$$

$$\bar{a}_y = \frac{d\bar{v}_y}{d\bar{t}} = \frac{\partial \bar{v}_y}{\partial \bar{t}} + \bar{v}_x \frac{\partial \bar{v}_y}{\partial \bar{x}} + \bar{v}_y \frac{\partial \bar{v}_y}{\partial \bar{y}} + \bar{v}_z \frac{\partial \bar{v}_y}{\partial \bar{z}} \tag{191}$$

$$= \bar{a}_y^{(o)} + \bar{a}_y^{(1)} + \bar{a}_y^{(2)} + \bar{a}_y^{(3)}$$

$$= a_y^{(o)} e^{j\psi yo} + a_y^{(1)} e^{j\psi y1} + a_y^{(2)} e^{j\psi y2} + a_y^{(3)} e^{j\psi y3}$$

$$\bar{a}_z = \frac{d\bar{v}_z}{d\bar{t}} = \frac{\partial \bar{v}_z}{\partial \bar{t}} + \bar{v}_x \frac{\partial \bar{v}_z}{\partial \bar{x}} + \bar{v}_y \frac{\partial \bar{v}_z}{\partial \bar{y}} + \bar{v}_z \frac{\partial \bar{v}_z}{\partial \bar{z}} \tag{192}$$

$$= \bar{a}_z^{(o)} + \bar{a}_z^{(1)} + \bar{a}_z^{(2)} + \bar{a}_z^{(3)}$$

$$= a_z^{(o)} e^{j\psi zo} + a_z^{(1)} e^{j\psi z1} + a_z^{(2)} e^{j\psi z2} + a_z^{(3)} e^{j\psi z3}$$

where

$$a_x^{(o)} = \sqrt{\frac{\left(\frac{\partial v_x}{\partial t}\right)^2 + v_x^2\left(\frac{\partial \theta_{vx}}{\partial t}\right)^2}{1 + t^2\left(\frac{\partial \theta_t}{\partial t}\right)^2}} \qquad a_x^{(1)} = v_x\sqrt{\frac{\left(\frac{\partial v_x}{\partial x}\right)^2 + v_x^2\left(\frac{\partial \theta_{vx}}{\partial x}\right)^2}{1 + x^2\left(\frac{\partial \theta_x}{\partial x}\right)^2}} \tag{193}$$

$$a_x^{(2)} = v_y\sqrt{\frac{\left(\frac{\partial v_x}{\partial y}\right)^2 + v_x^2\left(\frac{\partial \theta_{vx}}{\partial y}\right)^2}{1 + y^2\left(\frac{\partial \theta_y}{\partial y}\right)^2}} \qquad a_x^{(3)} = v_z\sqrt{\frac{\left(\frac{\partial v_x}{\partial z}\right)^2 + v_x^2\left(\frac{\partial \theta_{vx}}{\partial z}\right)^2}{1 + z^2\left(\frac{\partial \theta_z}{\partial z}\right)^2}} \tag{194}$$

$$a_y^{(o)} = \sqrt{\frac{\left(\frac{\partial v_y}{\partial t}\right)^2 + v_y^2\left(\frac{\partial \theta_{vy}}{\partial t}\right)^2}{1 + t^2\left(\frac{\partial \theta_t}{\partial t}\right)^2}} \qquad a_y^{(1)} = v_x\sqrt{\frac{\left(\frac{\partial v_y}{\partial x}\right)^2 + v_y^2\left(\frac{\partial \theta_{vy}}{\partial x}\right)^2}{1 + x^2\left(\frac{\partial \theta_x}{\partial x}\right)^2}} \tag{195}$$

$$a_y^{(2)} = v_y\sqrt{\frac{\left(\frac{\partial v_y}{\partial y}\right)^2 + v_y^2\left(\frac{\partial \theta_{vy}}{\partial y}\right)^2}{1 + y^2\left(\frac{\partial \theta_y}{\partial y}\right)^2}} \qquad a_y^{(3)} = v_z\sqrt{\frac{\left(\frac{\partial v_y}{\partial z}\right)^2 + v_y^2\left(\frac{\partial \theta_{vy}}{\partial z}\right)^2}{1 + z^2\left(\frac{\partial \theta_z}{\partial z}\right)^2}} \tag{196}$$

$$\dot{a}_z^{(o)} = \sqrt{\frac{\left(\frac{\partial v_z}{\partial t}\right)^2 + v_z^2\left(\frac{\partial \theta_{vz}}{\partial t}\right)^2}{1 + t^2\left(\frac{\partial \theta_t}{\partial t}\right)^2}} \qquad a_z^{(1)} = v_x\sqrt{\frac{\left(\frac{\partial v_z}{\partial x}\right)^2 + v_z^2\left(\frac{\partial \theta_{vz}}{\partial x}\right)^2}{1 + x^2\left(\frac{\partial \theta_x}{\partial x}\right)^2}} \tag{197}$$

$$a_z^{(2)} = v_y\sqrt{\frac{\left(\frac{\partial v_z}{\partial y}\right)^2 + v_z^2\left(\frac{\partial \theta_{vz}}{\partial y}\right)^2}{1 + y^2\left(\frac{\partial \theta_y}{\partial y}\right)^2}} \qquad a_z^{(3)} = v_z\sqrt{\frac{\left(\frac{\partial v_z}{\partial z}\right)^2 + v_z^2\left(\frac{\partial \theta_{vz}}{\partial z}\right)^2}{1 + z^2\left(\frac{\partial \theta_z}{\partial z}\right)^2}} \tag{198}$$

$$\psi_{xo} = \theta_{vx} - \theta_t + \beta_{vx,t} - \beta_{t,t} \tag{199}$$

$$\psi_{x1} = 2\theta_{vx} - \theta_x + \beta_{vx,x} - \beta_{x,x} \tag{200}$$

$$\psi_{x2} = \theta_{vy} + \theta_{vx} - \theta_y + \beta_{vx,y} - \beta_{y,y} \tag{201}$$

$$\psi_{x3} = \theta_{vz} + \theta_{vx} - \theta_z + \beta_{vx,z} - \beta_{z,z} \tag{202}$$

$$\psi_{yo} = \theta_{vy} - \theta_t + \beta_{vy,t} - \beta_{t,t} \tag{203}$$

$$\psi_{y1} = \theta_{vx} + \theta_{vy} - \theta_x + \beta_{vy,x} - \beta_{x,x} \tag{204}$$

$$\psi_{y2} = 2\theta_{vy} - \theta_y + \beta_{vy,y} - \beta_{y,y} \tag{205}$$

$$\psi_{y3} = \theta_{vz} + \theta_{vy} - \theta_z + \beta_{vy,z} - \beta_{z,z} \tag{206}$$

$$\psi_{zo} = \theta_{vz} - \theta_t + \beta_{vz,t} - \beta_{t,t} \tag{207}$$

$$\psi_{z1} = \theta_{vx} + \theta_{vz} - \theta_x + \beta_{vz,x} - \beta_{x,x} \tag{208}$$

$$\psi_{z2} = \theta_{vy} + \theta_{vz} - \theta_y + \beta_{vz,y} - \beta_{y,y} \tag{209}$$

$$\psi_{z3} = 2\theta_{vz} - \theta_z + \beta_{vz,z} - \beta_{z,z} \tag{210}$$

where

$$\tan \beta_{vx,t} = v_x \frac{\partial\theta_{vx}/\partial t}{\partial v_x/\partial t} \qquad \tan \beta_{vx,x} = v_x \frac{\partial\theta_{vx}/\partial x}{\partial v_x/\partial x} \tag{211}$$

$$\tan \beta_{vx,y} = v_x \frac{\partial\theta_{vx}/\partial y}{\partial v_x/\partial y} \qquad \tan \beta_{vx,z} = v_x \frac{\partial\theta_{vx}/\partial z}{\partial v_x/\partial z} \tag{212}$$

$$\tan \beta_{vy,t} = v_y \frac{\partial\theta_{vy}/\partial t}{\partial v_y/\partial t} \qquad \tan \beta_{vy,x} = v_y \frac{\partial\theta_{vy}/\partial x}{\partial v_y/\partial x} \tag{213}$$

$$\tan \beta_{vy,y} = v_y \frac{\partial\theta_{vy}/\partial y}{\partial v_y/\partial y} \qquad \tan \beta_{vy,z} = v_y \frac{\partial\theta_{vy}/\partial z}{\partial v_y/\partial z} \tag{214}$$

$$\tan \beta_{vz,t} = v_z \frac{\partial\theta_{vz}/\partial t}{\partial v_z/\partial t} \qquad \tan \beta_{vz,x} = v_z \frac{\partial\theta_{vz}/\partial x}{\partial v_z/\partial x} \tag{215}$$

$$\tan \beta_{vz,y} = v_z \frac{\partial\theta_{vz}/\partial y}{\partial v_z/\partial y} \qquad \tan \beta_{vz,z} = v_z \frac{\partial\theta_{vz}/\partial z}{\partial v_z/\partial z} \tag{216}$$

Equations (199) through (210) can be further reduced by using equations (163) through (165).

Combining equations (187) through (189) with (190) through (192) gives

$$a_x \cos \theta_{ax} = a_x^{(o)} \cos \psi_{xo} + a_x^{(1)} \cos \psi_{x1} + a_x^{(2)} \cos \psi_{x2} + a_x^{(3)} \cos \psi_{x3} \tag{217}$$

$$a_x \sin \theta_{ax} = a_x^{(o)} \sin \psi_{xo} + a_x^{(1)} \sin \psi_{x1} + a_x^{(2)} \sin \psi_{x2} + a_x^{(3)} \sin \psi_{x3} \tag{218}$$

$$a_y \cos \theta_{ay} = a_y^{(o)} \cos \psi_{yo} + a_y^{(1)} \cos \psi_{y1} + a_y^{(2)} \cos \psi_{y2} + a_y^{(3)} \cos \psi_{y3} \tag{219}$$

$$a_y \sin \theta_{ay} = a_y^{(o)} \sin \psi_{yo} + a_y^{(1)} \sin \psi_{y1} + a_y^{(2)} \sin \psi_{y2} + a_y^{(3)} \sin \psi_{y3} \tag{220}$$

$$a_z \cos \theta_{az} = a_z^{(o)} \cos \psi_{zo} + a_z^{(1)} \cos \psi_{z1} + a_z^{(2)} \cos \psi_{z2} + a_z^{(3)} \cos \psi_{z3} \quad (221)$$

$$a_z \sin \theta_{az} = a_z^{(o)} \sin \psi_{zo} + a_z^{(1)} \sin \psi_{z1} + a_z^{(2)} \sin \psi_{z2} + a_z^{(3)} \sin \psi_{z3} \quad (222)$$

These equations can be used to determine a_x, a_y, a_z, θ_{ax}, θ_{ay}, and θ_{az}. For the special case when there is no spatial variation of the velocity field it follows that

$$a_x = a_x^{(o)} \quad (223)$$

$$a_y = a_y^{(o)} \quad (224)$$

$$a_z = a_z^{(o)} \quad (225)$$

$$\theta_{ax} = \psi_{xo} = \theta_x - 2\theta_t - 2\beta_{t,t} + \beta_{vx,t} + \beta_{x,t} \quad (226)$$

$$\theta_{ay} = \psi_{yo} = \theta_y - 2\theta_t - 2\beta_{t,t} + \beta_{vy,t} + \beta_{y,t} \quad (227)$$

$$\theta_{az} = \psi_{zo} = \theta_z - 2\theta_t - 2\beta_{t,t} + \beta_{vz,t} + \beta_{z,t} \quad (228)$$

In general for $\alpha = x, y, z$

$$\theta_{a\alpha} = \theta_\alpha - 2\theta_t + 2(\beta_{\alpha,\alpha} - \beta_{t,t}) \quad (228A)$$

The complex magnitude of the particle acceleration is written as

$$\bar{a} = ae^{j\theta_a} \quad (229)$$

and from equations (187) through (189) it follows that

$$\bar{a}^2 = \bar{a}_x^2 + \bar{a}_y^2 + \bar{a}_z^2 \quad (230)$$

The component equations corresponding to equation (230) are

$$a^2 \cos (2\theta_a) = a_x^2 \cos (2\theta_{ax}) + a_y^2 \cos (2\theta_{ay}) + a_z^2 \cos (2\theta_{az}) \quad (231)$$

$$a^2 \sin (2\theta_a) = a_x^2 \sin (2\theta_{ax}) + a_y^2 \sin (2\theta_{ay}) + a_z^2 \sin (2\theta_{az}) \quad (232)$$

It follows from equations (231) and (232) that

$$a^4 = a_x^4 + a_y^4 + a_z^4 + 2a_x^2a_y^2 \cos [2(\theta_{ax} - \theta_{ay})] \quad (233)$$

$$+ 2a_x^2a_z^2 \cos [2(\theta_{ax} - \theta_{az})] + 2a_y^2a_z^2 \cos [2(\theta_{ay} - \theta_{az})]$$

and

$$\tan(2\theta_a) = \frac{a_x^2 \sin(2\theta_{ax}) + a_y^2 \sin(2\theta_{ay}) + a_z^2 \sin(2\theta_{az})}{a_x^2 \cos(2\theta_{ax}) + a_y^2 \cos(2\theta_{ay}) + a_z^2 \cos(2\theta_{az})} \tag{234}$$

where a_x, a_y, a_z, and θ_{ax}, θ_{ay}, θ_{az} are given by equations (217) through (222). The measured acceleration $= a \cos \theta_a$.

For a particle moving in bulk matter or vacuum with broken symmetry and not acted upon by forces, the momentum is constant and equation (176) gives

$$m\gamma_x v_x = C_{vx} \tag{235}$$

$$\theta_{vx} + \theta_{\gamma x} = C'_{vx} \tag{236}$$

where C_{vx} and C'_{vx} are constants of the motion. Equations (160), (163), (235), and (236) give

$$\frac{m^2\gamma_x^2\left[\left(\frac{dx}{dt}\right)^2 + x^2\omega_{\theta x}^2\right]}{1 + t^2\omega_{\theta t}^2} = C_{vx}^2 \tag{237}$$

$$\theta_{\gamma x} + \theta_x - \theta_t + \beta_{x,t} - \beta_{t,t} = C'_{vx} \tag{238}$$

Equation (237) shows that there is a transfer of energy between the linear motion and the internal phase motion. Equation (238) shows that there is also a transfer between θ_x and θ_t because equation (238) can be rewritten as

$$\theta_{\gamma x} + \theta_x + \tan^{-1}\left(x\,\frac{d\theta_x/dt}{dx/dt}\right) - \theta_t - \tan^{-1}(t\, d\theta_t/dt) = C'_{vx} \tag{239}$$

The nonrelativistic equations of motion of a particle moving in a potential field W^a are given by[11]

$$m\ddot{x}^a = m\dot{v}_x^a = ma_x^a = -\partial W^a/\partial x^a \tag{240}$$

$$m\ddot{y}^a = m\dot{v}_y^a = ma_y^a = -\partial W^a/\partial y^a \tag{241}$$

$$m\ddot{z}^a = m\dot{v}_z^a = ma_z^a = -\partial W^a/\partial z^a \tag{242}$$

The corresponding relativistic equations of motion for particles in a medium not having broken internal symmetry are[12]

$$m(\gamma_x^a)^3 a_x^a = -\partial W^a/\partial x^a \qquad (243)$$

$$m\gamma_x^a a_y^a = -\partial W^a/\partial y^a \qquad (244)$$

$$m\gamma_x^a a_z^a = -\partial W^a/\partial z^a \qquad (245)$$

where the standard velocity factor is given by equation (175). Consider now a conservative force acting on a particle located in bulk matter or vacuum with broken internal symmetries in the space and time coordinates. If the complex number potential is written as

$$\bar{W} = We^{j\theta_W} \qquad (246)$$

then the nonrelativistic equations of motion are written as

$$m\bar{a}_x = ma_x e^{j\theta_{ax}} = -\partial\bar{W}/\partial\bar{x} \qquad (247)$$

$$m\bar{a}_y = ma_y e^{j\theta_{ay}} = -\partial\bar{W}/\partial\bar{y} \qquad (248)$$

$$m\bar{a}_z = ma_z e^{j\theta_{az}} = -\partial\bar{W}/\partial\bar{z} \qquad (249)$$

where $\bar{a}_x$, $\bar{a}_y$, and $\bar{a}_z$ are given by equations (190) through (192); and a_x, a_y, a_z, θ_{ax}, θ_{ay}, and θ_{az} are obtained from equations (217) through (222). If the theory of special relativity is considered in conjunction with broken internal symmetry, equations (247) through (249) become

$$m\bar{\gamma}_x^3\bar{a}_x = m\gamma_x^3 a_x e^{j(\theta_{ax}+3\theta_{\gamma x})} = -\partial\bar{W}/\partial\bar{x} \qquad (250)$$

$$m\bar{\gamma}_x\bar{a}_y = m\gamma_x a_y e^{j(\theta_{ay}+\theta_{\gamma x})} = -\partial\bar{W}/\partial\bar{y} \qquad (251)$$

$$m\bar{\gamma}_x\bar{a}_z = m\gamma_x a_z e^{j(\theta_{az}+\theta_{\gamma x})} = -\partial\bar{W}/\partial\bar{z} \qquad (252)$$

where $\bar{\gamma}_x$ is given by equation (177), γ_x and $\theta_{\gamma x}$ are given by equations (178) and (179) respectively, and where the particle is moving instantaneously along the x axis with velocity $\bar{v}_x$. Equations (250) through (252) are simple generalizations of the standard special relativistic inertia terms to the case of particle motion in media with broken internal symmetries.

The derivatives of the broken symmetry potential can be written as

$$- \frac{\partial \bar{W}}{\partial \bar{x}} = - \frac{(dW + jWd\theta_W)}{(dx + jxd\theta_x)} e^{j(\theta_W - \theta_x)} \tag{253}$$

$$= \sqrt{\frac{\left(\frac{\partial W}{\partial x}\right)^2 + W^2\left(\frac{\partial \theta_W}{\partial x}\right)^2}{1 + x^2\left(\frac{\partial \theta_x}{\partial x}\right)^2}} \; e^{j(\pi + \theta_W + \beta_{W,x} - \theta_x - \beta_{x,x})}$$

$$- \frac{\partial \bar{W}}{\partial \bar{y}} = \sqrt{\frac{\left(\frac{\partial W}{\partial y}\right)^2 + W^2\left(\frac{\partial \theta_W}{\partial y}\right)^2}{1 + y^2\left(\frac{\partial \theta_y}{\partial y}\right)^2}} \; e^{j(\pi + \theta_W + \beta_{W,y} - \theta_y - \beta_{y,y})} \tag{254}$$

$$- \frac{\partial \bar{W}}{\partial \bar{z}} = \sqrt{\frac{\left(\frac{\partial W}{\partial z}\right)^2 + W^2\left(\frac{\partial \theta_W}{\partial z}\right)^2}{1 + z^2\left(\frac{\partial \theta_z}{\partial z}\right)^2}} \; e^{j(\pi + \theta_W + \beta_{W,z} - \theta_z - \beta_{z,z})} \tag{255}$$

where

$$\tan \beta_{W,x} = W \frac{\partial \theta_W / \partial x}{\partial W / \partial x} \tag{256}$$

$$\tan \beta_{W,y} = W \frac{\partial \theta_W / \partial y}{\partial W / \partial y} \tag{257}$$

$$\tan \beta_{W,z} = W \frac{\partial \theta_W / \partial z}{\partial W / \partial z} \tag{258}$$

and where $\beta_{x,x}$; $\beta_{y,y}$ and $\beta_{z,z}$ are given by equations (38) through (40) respectively. Combining equations (187) through (198) with equations (250) through (255) gives the following relativistic equations of motion for a particle located in bulk matter or vacuum with broken internal symmetries.

$$m\gamma_x^3 a_x = \sqrt{\frac{\left(\frac{\partial W}{\partial x}\right)^2 + W^2\left(\frac{\partial \theta_W}{\partial x}\right)^2}{1 + x^2\left(\frac{\partial \theta_x}{\partial x}\right)^2}} \tag{259}$$

$$m\gamma_x a_y = \sqrt{\frac{\left(\frac{\partial W}{\partial y}\right)^2 + W^2\left(\frac{\partial \theta_W}{\partial y}\right)^2}{1 + y^2\left(\frac{\partial \theta_y}{\partial y}\right)^2}} \tag{260}$$

$$m\gamma_x a_z = \sqrt{\frac{\left(\frac{\partial W}{\partial z}\right)^2 + W^2\left(\frac{\partial \theta_W}{\partial z}\right)^2}{1 + z^2\left(\frac{\partial \theta_z}{\partial z}\right)^2}} \tag{261}$$

$$\theta_{ax} + 3\theta_{\gamma x} = \pi + \theta_W + \beta_{W,x} - \theta_x - \beta_{x,x} \tag{262}$$

$$\theta_{ay} + \theta_{\gamma x} = \pi + \theta_W + \beta_{W,y} - \theta_y - \beta_{y,y} \tag{263}$$

$$\theta_{az} + \theta_{\gamma x} = \pi + \theta_W + \beta_{W,z} - \theta_z - \beta_{z,z} \tag{264}$$

where γ_x and $\theta_{\gamma x}$ are given by equations (178) and (179) respectively, and where a_x, a_y, a_z and θ_{ax}, θ_{ay}, and θ_{az} are obtained from equations (217) through (222).

A useful form of the nonrelativistic equations of motion for a particle located in asymmetric matter is obtained from equation (247) as follows

$$m \frac{d^2\bar{x}}{d\bar{t}^2} = - \partial\bar{W}/\partial\bar{x} \tag{264A}$$

where $\bar{x}$ is a complex number given by equation (16) whose real and imaginary components are written as $x_R = x \cos \theta_x$ and $x_I = x \sin \theta_x$. Then it follows that

$$\frac{d\bar{x}}{d\bar{t}} = \cos \beta_{t,t}\, e^{-j\theta_{dt}} \frac{d\bar{x}}{dt} = R_{1t}(x_R,x_I) + jI_{1t}(x_R,x_I) \tag{264B}$$

where

$$\theta_{dt} = \theta_t + \beta_{t,t} \tag{264C}$$

$$R_{1t}(x_R,x_I) = \cos \beta_{t,t} \left(\cos \theta_{dt} \frac{dx_R}{dt} + \sin \theta_{dt} \frac{dx_I}{dt}\right) \tag{264D}$$

$$\sim \cos \beta_{t,t} \cos (\theta_x - \theta_{dt}) \frac{dx}{dt}$$

$$I_{1t}(x_R,x_I) = \cos \beta_{t,t} \left(- \sin \theta_{dt} \frac{dx_R}{dt} + \cos \theta_{dt} \frac{dx_I}{dt}\right) \tag{264E}$$

$$\sim \cos \beta_{t,t} \sin (\theta_x - \theta_{dt}) \frac{dx}{dt}$$

Then it follows that

$$\frac{d^2\bar{x}}{d\bar{t}^2} = \cos \beta_{t,t} e^{-j\theta_{dt}} \frac{d}{dt}\left(\cos \beta_{t,t} e^{-j\theta_{dt}} \frac{d\bar{x}}{dt}\right) \tag{264F}$$

$$\sim \cos^2 \beta_{t,t} e^{-2j\theta_{dt}} \frac{d^2\bar{x}}{dt^2} = R_{2t}(x_R,x_I) + jI_{2t}(x_R,x_I)$$

where

$$R_{2t}(x_R,x_I) = \cos^2 \beta_{t,t} \left[\cos (2\theta_{dt}) \frac{d^2x_R}{dt^2} + \sin (2\theta_{dt}) \frac{d^2x_I}{dt^2}\right] \tag{264G}$$

$$\sim \cos^2 \beta_{t,t} \cos (\theta_x - 2\theta_{dt}) \frac{d^2x}{dt^2}$$

$$I_{2t}(x_R,x_I) = \cos^2 \beta_{t,t} \left[- \sin (2\theta_{dt}) \frac{d^2x_R}{dt^2} + \cos (2\theta_{dt}) \frac{d^2x_I}{dt^2}\right] \tag{264H}$$

$$\sim \cos^2 \beta_{t,t} \sin (\theta_x - 2\theta_{dt}) \frac{d^2x}{dt^2}$$

The derivative of the complex potential can be written by noting that $W_R = W \cos \theta_W$ and $W_I = W \sin \theta_W$ so that

$$\frac{\partial \bar{W}}{\partial \bar{x}} = \cos \beta_{x,x}\, e^{-j\theta_{dx}} \frac{\partial \bar{W}}{\partial x} = R_{1x}(W_R, W_I) + jI_{1x}(W_R, W_I) \tag{264I}$$

where

$$\theta_{dx} = \theta_x + \beta_{x,x} \tag{264J}$$

$$R_{1x}(W_R, W_I) = \cos \beta_{x,x} \left(\cos \theta_{dx} \frac{\partial W_R}{\partial x} + \sin \theta_{dx} \frac{\partial W_I}{\partial x}\right) \tag{264K}$$

$$\sim \cos \beta_{x,x} \cos (\theta_W - \theta_{dx}) \frac{\partial W}{\partial x}$$

$$I_{1x}(W_R, W_I) = \cos \beta_{x,x} \left(-\sin \theta_{dx} \frac{\partial W_R}{\partial x} + \cos \theta_{dx} \frac{\partial W_I}{\partial x}\right) \tag{264L}$$

$$\sim \cos \beta_{x,x} \sin (\theta_W - \theta_{dx}) \frac{\partial W}{\partial x}$$

Newton's dynamical equation (264A) can now be written in the following approximate forms

$$mR_{2t}(x_R, x_I) \sim - R_{1x}(W_R, W_I) \tag{264M}$$

$$mI_{2t}(x_R, x_I) \sim - I_{1x}(W_R, W_I) \tag{264N}$$

where R_{2t} and I_{2t} are given by equations (264G) and (264H) respectively, and R_{1x} and I_{1x} are given by equations (264K) and (264L) respectively. A further approximation for relations (264M) and (264N) yields

$$m \cos^2 \beta_{t,t} \cos (\theta_x - 2\theta_{dt}) \frac{d^2x}{dt^2} \sim - \cos \beta_{x,x} \cos (\theta_W - \theta_{dx}) \frac{\partial W}{\partial x} \tag{264O}$$

$$m \cos^2 \beta_{t,t} \sin (\theta_x - 2\theta_{dt}) \frac{d^2x}{dt^2} \sim - \cos \beta_{x,x} \sin (\theta_W - \theta_{dx}) \frac{\partial W}{\partial x} \tag{264P}$$

which gives

$$\theta_x - 2\theta_{dt} \sim \theta_W - \theta_{dx} \tag{264Q}$$

$$m\cos^2 \beta_{t,t} \frac{d^2x}{dt^2} \sim -\cos \beta_{x,x} \frac{\partial W}{\partial x} \tag{264R}$$

which are the approximate equations of motion for a particle in a potential field that is located in asymmetric bulk matter or vacuum. For a nonrelativistic system the measured acceleration is given by

$$a_{xm} = a_x \cos \theta_{ax} \sim \cos^2 \beta_{t,t} \cos \theta_{ax} \frac{d^2x}{dt^2} \tag{264S}$$

while the conventionally calculated acceleration is given by

$$a_x^a = \frac{d^2x_a}{dt_a^2} = \frac{d^2x_m}{dt_m^2} \sim \frac{\cos \theta_x}{\cos^2 \theta_t} \frac{d^2x}{dt^2} \tag{264T}$$

and therefore $a_{xm} \neq a_x^a$. Relations (264Q) and (264R) can be applied to many specific dynamical systems that are located in an asymmetric medium. For instance the vibration of molecules and atoms located in matter are expected to be described by these equations.

There are twenty three unknown variables that are needed to describe a particle in bulk matter or vacuum with broken internal symmetries: x, θ_x, y, θ_y, z, θ_z, v_x, θ_{vx}, v_y, θ_{vy}, v_z, θ_{vz}, a_x, θ_{ax}, a_y, θ_{ay}, a_z, θ_{az}, P, θ_P, γ, θ_γ, and θ_t. The magnitude of the time t is taken to be a totally independent parameter. Twenty two equations have been derived thus far in an attempt to determine the twenty three unknowns: two ground state relativistic equations (1), two equations for the ground state Grüneisen parameter (5), the six kinematic velocity equations (160) through (165), the six kinematic acceleration equations (217) through (222), and the six dynamical equations (259) through (264). By means of the twenty two equations the kinematical and dynamical variables have been expressed in terms of the potential components W and θ_W. But the single particle potential parameters W and θ_W are related through a gauge potential to the complex macroscopic state equation variables P, θ_P, γ, and θ_γ that are determined from equations (1) and (5). This connection is made through a partition function as shown in equations (13) through (15) which determine the gauge potential. But through equation (1), P, θ_P, γ, and θ_γ are related to the unrenormalized pressure and Grüneisen function P^a and γ^a respectively. Therefore it should be possible to express all of the kinematical and dynamical variables in terms of t, P^a, γ^a and the unrenormalized potential V^a.

Clearly one additional equation is necessary in order to have a total of twenty three equations that are needed to determine the twenty three unknown variables. The needed equation is given by the following complex number continuity equation for broken symmetry matter

$$\frac{\partial\bar{\rho}}{\partial\bar{t}} + \vec{\bar{\nabla}}\cdot(\bar{\rho}\vec{\bar{v}}) = 0 \qquad (265)$$

where $\bar{\rho}$ = mass density. Equation (265) can be rewritten as two real number equations as follows

$$G_t \cos\phi_t + G_x \cos\phi_x + G_y \cos\phi_y + G_z \cos\phi_z = 0 \qquad (266)$$

$$G_t \sin\phi_t + G_x \sin\phi_x + G_y \sin\phi_y + G_z \sin\phi_z = 0 \qquad (267)$$

where

$$G_t = \sqrt{\frac{\left(\frac{\partial\rho}{\partial t}\right)^2 + \left(\rho\,\frac{\partial\theta_\rho}{\partial t}\right)^2}{1 + t^2\left(\frac{\partial\theta_t}{\partial t}\right)^2}} \qquad (268A)$$

$$G_x = \sqrt{\frac{\left[\frac{\partial}{\partial x}(\rho v_x)\right]^2 + \left[\rho v_x \frac{\partial}{\partial x}(\theta_\rho + \theta_{vx})\right]^2}{1 + x^2\left(\frac{\partial\theta_x}{\partial x}\right)^2}} \qquad (268B)$$

$$G_y = \sqrt{\frac{\left[\frac{\partial}{\partial y}(\rho v_y)\right]^2 + \left[\rho v_y \frac{\partial}{\partial y}(\theta_\rho + \theta_{vy})\right]^2}{1 + y^2\left(\frac{\partial\theta_y}{\partial y}\right)^2}} \qquad (268C)$$

$$G_z = \sqrt{\frac{\left[\frac{\partial}{\partial z}(\rho v_z)\right]^2 + \left[\rho v_z \frac{\partial}{\partial z}(\theta_\rho + \theta_{vz})\right]^2}{1 + z^2\left(\frac{\partial\theta_z}{\partial z}\right)^2}} \qquad (268D)$$

$$\phi_t = \theta_\rho + \beta_{\rho,t} - \theta_t - \beta_{t,t} \qquad (268E)$$

$$\phi_x = \theta_\rho + \theta_{vx} + \beta_{\rho vx,x} - \theta_x - \beta_{x,x} \qquad (268F)$$

$$\phi_y = \theta_\rho + \theta_{vy} + \beta_{\rho vy,y} - \theta_y - \beta_{y,y} \tag{268G}$$

$$\phi_z = \theta_\rho + \theta_{vz} + \beta_{\rho vz,z} - \theta_z - \beta_{z,z} \tag{268H}$$

where

$$\tan \beta_{\rho v\alpha,\alpha} = \frac{\rho v_\alpha \partial/\partial\alpha(\theta_\rho + \theta_{v\alpha})}{\partial/\partial\alpha(\rho v_\alpha)} \tag{269}$$

$$\tan \beta_{\rho,t} = \frac{\rho\partial\theta_\rho/\partial t}{\partial\rho/\partial t} \tag{269A}$$

where $\alpha = x, y, z$. Therefore the complex number equation (265) has two real components that can be used along with the previously elaborated twenty two equations to obtain ρ and θ_t. There are now a total of twenty four equations and twenty four unknown quantities which can be determined to give

$$x = x(t,P^a,\gamma^a,V^a) \qquad \theta_x = \theta_x(t,P^a,\gamma^a,V^a) \tag{270A}$$

$$y = y(t,P^a,\gamma^a,V^a) \qquad \theta_y = \theta_y(t,P^a,\gamma^a,V^a) \tag{270B}$$

$$z = z(t,P^a,\gamma^a,V^a) \qquad \theta_z = \theta_z(t,P^a,\gamma^a,V^a) \tag{270C}$$

$$v_x = v_x(t,P^a,\gamma^a,V^a) \qquad \theta_{vx} = \theta_{vx}(t,P^a,\gamma^a,V^a) \tag{271A}$$

$$v_y = v_y(t,P^a,\gamma^a,V^a) \qquad \theta_{vy} = \theta_{vy}(t,P^a,\gamma^a,V^a) \tag{271B}$$

$$v_z = v_z(t,P^a,\gamma^a,V^a) \qquad \theta_{vz} = \theta_{vz}(t,P^a,\gamma^a,V^a) \tag{271C}$$

$$a_x = a_x(t,P^a,\gamma^a,V^a) \qquad \theta_{ax} = \theta_{ax}(t,P^a,\gamma^a,V^a) \tag{272A}$$

$$a_y = a_y(t,P^a,\gamma^a,V^a) \qquad \theta_{ay} = \theta_{ay}(t,P^a,\gamma^a,V^a) \tag{272B}$$

$$a_z = a_z(t,P^a,\gamma^a,V^a) \qquad \theta_{az} = \theta_{az}(t,P^a,\gamma^a,V^a) \tag{272C}$$

$$P = P(P^a,\gamma^a,V^a) \qquad \theta_P = \theta_P(P^a,\gamma^a,V^a) \tag{273A}$$

$$\gamma = \gamma(P^a,\gamma^a,V^a) \qquad \theta_\gamma = \theta_\gamma(P^a,\gamma^a,V^a) \tag{273B}$$

$$\rho = \rho(t, P^a, \gamma^a, V^a) \tag{274A}$$

$$\theta_t = \theta_t(t, P^a, \gamma^a, V^a) \tag{274B}$$

where time is treated as an independent variable, and where V^a = unrenormalized potential.

The first integral of equations (247) through (249) is

$$\frac{1}{2}m\bar{v}^2 + \bar{W} = \bar{E} \tag{275}$$

where $\bar{E}$ = complex number total energy, and where $\bar{v}^2$ is given by equation (181). The two scalar component equations corresponding to equation (275) are

$$\frac{1}{2}mv^2 \cos (2\theta_v) + W \cos \theta_W = E \cos \theta_E \tag{276}$$

$$\frac{1}{2}mv^2 \sin (2\theta_v) + W \sin \theta_W = E \sin \theta_E \tag{277}$$

where v and θ_v are given by equations (185) and (186) respectively. The corresponding first integral of the relativistic equations of motion (259) through (261) is[12]

$$(\bar{\gamma}_x - 1)mc^2 + \bar{W} = \bar{E} \tag{278}$$

where the particle is instantaneously moving along the x axis and $\bar{\gamma}_x$ is given by equation (177) The component form of equation (178) is written as

$$(\gamma_x \cos \theta_{\gamma x} - 1)mc^2 + W \cos \theta_W = E \cos \theta_E \tag{279}$$

$$\gamma_x \sin \theta_{\gamma x} mc^2 + W \sin \theta_W = E \sin \theta_E \tag{280}$$

The energy equations determine the magnitudes of coordinates and velocities.

5. ROTATING SYSTEMS IN BULK MATTER AND THE VACUUM. In Section 3 it was shown that the angle between two lines located in bulk matter or the vacuum with broken internal symmetry is expected to have an internal phase angle. This suggests that angular velocity also has a broken internal symmetry. Accordingly the angular speed associated with a complex number geometrical angle given by

$$\bar{\phi} = \phi e^{j\theta_\phi} \tag{281}$$

is written as

$$\bar{\omega} = \omega e^{j\theta_\omega} = \frac{d\bar{\phi}}{d\bar{t}} = e^{j(\theta_\phi - \theta_t)} \left(\frac{d\phi + j\phi d\theta_\phi}{dt + jt d\theta_t} \right) \tag{282}$$

$$= \sqrt{\frac{\left(\frac{d\phi}{dt}\right)^2 + \phi^2 \left(\frac{d\theta_\phi}{dt}\right)^2}{1 + t^2 \left(\frac{d\theta_t}{dt}\right)^2}} \; e^{j(\theta_\phi + \beta_{\phi,t} - \theta_t - \beta_{t,t})}$$

so that

$$\omega = \sqrt{\frac{\left(\frac{d\phi}{dt}\right)^2 + \phi^2 \left(\frac{d\theta_\phi}{dt}\right)^2}{1 + t^2 \left(\frac{d\theta_t}{dt}\right)^2}} = \sqrt{\frac{\omega_\phi^2 + \phi^2 \omega_{\theta\phi}^2}{1 + t^2 \omega_{\theta t}^2}} \tag{283}$$

and

$$\theta_\omega = \theta_\phi + \beta_{\phi,t} - \theta_t - \beta_{t,t} \tag{284}$$

where

$$\tan \beta_{\phi,t} = \phi \frac{d\theta_\phi/dt}{d\phi/dt} \tag{285}$$

The angular speed associated with the internal phase angle of the geometrical phase angle is written as

$$\omega_{\theta\phi} = \frac{d\theta_\phi}{dt} = \frac{\partial\theta_\phi}{\partial t} + \frac{\partial\theta_\phi}{\partial r}\frac{dr}{dt} + \frac{\partial\theta_\phi}{\partial\phi}\frac{d\phi}{dt} \tag{286}$$

and the angular speed of the internal phase angle of the time coordinate $\omega_{\theta t}$ is given in equation (173A) or equivalently by

$$\omega_{\theta t} = \frac{d\theta_t}{dt} = \frac{\partial\theta_t}{\partial t} + \frac{\partial\theta_t}{\partial r}\frac{dr}{dt} + \frac{\partial\theta_t}{\partial\phi}\frac{d\phi}{dt} \tag{286A}$$

and finally where

$$\omega_\phi = \frac{d\phi}{dt} \qquad (287)$$

is the ordinary angular speed associated with the magnitude ϕ of the geometrical angle. Equation (283) is the general expression for angular speed within bulk matter or vacuum with broken internal symmetries. The measured angular speed is given by $\omega \cos \theta_\omega$.

For short periods of time equation (283) shows that

$$\omega \sim \omega_\phi \qquad (288)$$

while for long periods of time

$$\omega \sim \langle \omega_\phi \rangle \frac{\omega_{\theta\phi}}{\omega_{\theta t}} \qquad (289)$$

where

$$\langle \omega_\phi \rangle = \frac{1}{t} \int_0^t \omega_\phi dt = \phi/t \qquad (290)$$

In fact equation (283) shows that for a small t

$$\omega = \omega_\phi \left[1 + \frac{1}{2} \left(\langle \omega_\phi \rangle^2 \frac{\omega_{\theta\phi}^2}{\omega_\phi^2} - \omega_{\theta t}^2 \right) t^2 - \cdots \right] \qquad (291)$$

for $\phi^2 \omega_{\theta\phi}^2 / \omega_\phi^2 << 1$ and $t^2 \omega_{\theta t}^2 << 1$, while for large t it follows that

$$\omega = \frac{\omega_{\theta\phi}}{\omega_{\theta t}} \langle \omega_\phi \rangle \left[1 + \frac{1}{2} \left(\frac{\omega_\phi^2}{\langle \omega_\phi \rangle^2 \omega_{\theta\phi}^2} - \frac{1}{\omega_{\theta t}^2} \right) t^{-2} + \cdots \right] \qquad (292)$$

Consider now the velocity of a particle in bulk matter or vacuum that has a radial and a transverse component. The radial component is given by

$$\bar{v}_r = v_r e^{j\theta_{vr}} = \frac{d\bar{r}}{d\bar{t}} = e^{j(\theta_r - \theta_t)} \left(\frac{dr + jrd\theta_r}{dt + jtd\theta_t} \right) \qquad (293)$$

$$= \sqrt{\frac{\left(\frac{dr}{dt}\right)^2 + r^2 \left(\frac{d\theta_r}{dt}\right)^2}{1 + t^2 \left(\frac{d\theta_t}{dt}\right)^2}} \; e^{j(\theta_r + \beta_{r,t} - \theta_t - \beta_{t,t})}$$

so that

$$v_r = \sqrt{\frac{\left(\frac{dr}{dt}\right)^2 + r^2\omega_{\theta r}^2}{1 + t^2\omega_{\theta t}^2}} \tag{294}$$

$$\theta_{vr} = \theta_r + \beta_{r,t} - \theta_t - \beta_{t,t} \tag{295}$$

where

$$\tan \beta_{r,t} = r\,\frac{d\theta_r/dt}{dr/dt} \tag{296}$$

$$\omega_{\theta r} = d\theta_r/dt = \frac{\partial\theta_r}{\partial t} + \frac{\partial\theta_r}{\partial r}\frac{dr}{dt} + \frac{\partial\theta_r}{\partial\phi}\frac{d\phi}{dt} \tag{297}$$

The transverse component of velocity is given by

$$\bar{v}_\phi = v_\phi e^{j\theta_{v\phi}} = \bar{r}\,\frac{d\bar{\phi}}{d\bar{t}} = \bar{r}\bar{\omega} \tag{298}$$

Combining equations (282) and (298) gives

$$v_\phi = r\omega \tag{299}$$

$$\theta_{v\phi} = \theta_r + \theta_\phi + \beta_{\phi,t} - \theta_t - \beta_{t,t} = \theta_r + \theta_\omega \tag{300}$$

where ω is given by equation (283). The magnitude of the vector sum of the radial and transverse velocities is given by

$$\bar{v}^2 = \bar{v}_r^2 + \bar{v}_\phi^2 = v^2 e^{2j\theta_v} \tag{301}$$

which has the following scalar components

$$v^2 \cos(2\theta_v) = v_r^2 \cos(2\theta_{vr}) + v_\phi^2 \cos(2\theta_{v\phi}) \tag{302}$$

$$v^2 \sin(2\theta_v) = v_r^2 \sin(2\theta_{vr}) + v_\phi^2 \sin(2\theta_{v\phi}) \tag{303}$$

From equation (302) and (303) it follows that

$$v^4 = v_r^4 + v_\phi^4 + 2v_r^2 v_\phi^2 \cos\left[2(\theta_{vr} - \theta_{v\phi})\right] \tag{304}$$

$$\tan(2\theta_v) = \frac{v_r^2 \sin(2\theta_{vr}) + v_\phi^2 \sin(2\theta_{v\phi})}{v_r^2 \cos(2\theta_{vr}) + v_\phi^2 \cos(2\theta_{v\phi})} \tag{305}$$

where

$$\theta_{vr} - \theta_{v\phi} = \beta_{r,t} - \theta_\phi - \beta_{\phi,t} \tag{306}$$

The measured speed = $v \cos \theta_v$.

For ordinary matter rotating about a center of force, the radial and transverse accelerations are written as[11]

$$a_r^a = \frac{d^2 r_a}{dt_a^2} - \omega_a^2 r_a \tag{307}$$

$$a_\phi^a = r_a \frac{d^2\phi_a}{dt_a^2} + 2\frac{dr_a}{dt_a}\omega_a \tag{308}$$

The acceleration of a particle that is orbiting about a center of force located within bulk matter or the vacuum with broken internal symmetry will have the following radial and transverse components

$$\bar{a}_r = \frac{d^2\bar{r}}{d\bar{t}^2} - \bar{\omega}^2\bar{r} = a_r e^{j\theta_{ar}} \tag{309}$$

$$\bar{a}_\phi = \bar{r}\frac{d^2\bar{\phi}}{d\bar{t}^2} + 2\frac{d\bar{r}}{d\bar{t}}\bar{\omega} = a_\phi e^{j\theta_{a\phi}} \tag{310}$$

Each of the four terms in equations (309) and (310) can be evaluated in terms of previously calculated quantities.

The linear radial acceleration term in equation (309) is given by

$$\frac{d^2\bar{r}}{d\bar{t}^2} = \frac{d\bar{v}_r}{d\bar{t}} = \bar{a}_{rr} = a_{rr}e^{j\theta_{arr}} \tag{311}$$

where the time derivative of the radial velocity is given by the Eulerian derivative as follows

$$\bar{a}_{rr} = \frac{d\bar{v}_r}{d\bar{t}} = \frac{\partial \bar{v}_r}{\partial \bar{t}} + \bar{v}_r \frac{\partial \bar{v}_r}{\partial \bar{r}} + \frac{\bar{v}_\phi}{\bar{r}} \frac{\partial \bar{v}_r}{\partial \bar{\phi}} \tag{312}$$

$$= \bar{a}_{rr}^{(o)} + \bar{a}_{rr}^{(1)} + \bar{a}_{rr}^{(2)}$$

$$= a_{rr}^{(o)} e^{j\psi_{ro}} + a_{rr}^{(1)} e^{j\psi_{r1}} + a_{rr}^{(2)} e^{j\psi_{r2}}$$

where

$$a_{rr}^{(o)} = \sqrt{\frac{\left(\frac{\partial v_r}{\partial t}\right)^2 + v_r^2\left(\frac{\partial \theta_{vr}}{\partial t}\right)^2}{1 + t^2\left(\frac{\partial \theta_t}{\partial t}\right)^2}} \tag{313A}$$

$$a_{rr}^{(1)} = v_r \sqrt{\frac{\left(\frac{\partial v_r}{\partial r}\right)^2 + v_r^2\left(\frac{\partial \theta_{vr}}{\partial r}\right)^2}{1 + r^2\left(\frac{\partial \theta_r}{\partial r}\right)^2}} \tag{313B}$$

$$a_{rr}^{(2)} = \frac{v_\phi}{r} \sqrt{\frac{\left(\frac{\partial v_r}{\partial \phi}\right)^2 + v_r^2\left(\frac{\partial \theta_{vr}}{\partial \phi}\right)^2}{1 + \phi^2\left(\frac{\partial \theta_\phi}{\partial \phi}\right)^2}} \tag{313C}$$

$$\psi_{ro} = \theta_{vr} + \beta_{vr,t} - \theta_t - \beta_{t,t} \tag{313D}$$

$$\psi_{r1} = 2\theta_{vr} - \theta_r + \beta_{vr,r} - \beta_{r,r} \tag{313E}$$

$$\psi_{r2} = \theta_{v\phi} + \theta_{vr} - \theta_r - \theta_\phi + \beta_{vr,\phi} - \beta_{\phi,\phi} \tag{313F}$$

where $\beta_{t,t}$, $\beta_{r,r}$, and $\beta_{\phi,\phi}$ are given by equations (41), (64), and (66) respectively, and v_r and θ_{vr} are given by equations (294) and (295) respectively, and v_ϕ and $\theta_{v\phi}$ are given by equations (299) and (300) respectively, and where

$$\tan \beta_{vr,t} = v_r \frac{\partial \theta_{vr}/\partial t}{\partial v_r/\partial t} \tag{314A}$$

$$\tan \beta_{vr,r} = v_r \frac{\partial \theta_{vr}/\partial r}{\partial v_r/\partial r} \tag{314B}$$

$$\tan \beta_{vr,\phi} = v_r \frac{\partial \theta_{vr}/\partial \phi}{\partial v_r/\partial \phi} \tag{314C}$$

From equations (331) and (312) it follows that

$$a_{rr} \cos \theta_{arr} = a_{rr}^{(o)} \cos \psi_{ro} + a_{rr}^{(1)} \cos \psi_{r1} + a_{rr}^{(2)} \cos \psi_{r2} \tag{315}$$

$$a_{rr} \sin \theta_{arr} = a_{rr}^{(o)} \sin \psi_{ro} + a_{rr}^{(1)} \sin \psi_{r1} + a_{rr}^{(2)} \sin \psi_{r2} \tag{316}$$

from which a_{rr} and θ_{arr} can be obtained immediately. For the special case where there is no spatial variation of the velocity field the acceleration equation become

$$a_{rr} = a_{rr}^{(o)} \tag{317}$$

$$\theta_{arr} = \psi_{ro} \tag{318}$$

The centrifugal radial acceleration term in equation (309) is written as

$$\overline{r\omega}^2 = r\omega^2 e^{j(\theta_r + 2\theta_\omega)} = a_{cen} e^{j\theta_{acen}} \tag{319}$$

so that

$$a_{cen} = r\omega^2 \tag{320}$$

$$\theta_{acen} = \theta_r + 2\theta_\omega = \theta_r + 2(\theta_\phi + \beta_{\phi,t} - \theta_t - \beta_{t,t}) \tag{321}$$

where ω and θ_ω are given by equations (283) and (284) respectively.

The first term in the angular acceleration given by equation (310) is written as

$$\bar{r}\,\frac{d^2\bar{\phi}}{d\bar{t}^2} = \bar{r}\,\frac{d\bar{\omega}}{d\bar{t}} = \bar{a}_{\phi\phi} = a_{\phi\phi}e^{j\theta_{a\phi\phi}} \tag{322}$$

where $\bar{\omega}$ is given by equation (282). The time derivative is taken to be an Eulerian derivative (which accounts for differential rotation) as follows

$$\bar{a}_{\phi\phi} = \bar{r}\,\frac{d\bar{\omega}}{d\bar{t}} = \bar{r}\left(\frac{\partial\bar{\omega}}{\partial\bar{t}} + \bar{v}_r\,\frac{\partial\bar{\omega}}{\partial\bar{r}} + \frac{\bar{v}_\phi}{\bar{r}}\,\frac{\partial\bar{\omega}}{\partial\bar{\phi}}\right) \tag{323}$$

$$= \bar{a}^{(o)}_{\phi\phi} = \bar{a}^{(1)}_{\phi\phi} + \bar{a}^{(2)}_{\phi\phi}$$

$$= a^{(o)}_{\phi\phi}e^{j\psi_{\phi o}} + a^{(1)}_{\phi\phi}e^{j\psi_{\phi 1}} + a^{(2)}_{\phi\phi}e^{j\psi_{\phi 2}}$$

where

$$a^{(o)}_{\phi\phi} = r\sqrt{\frac{\left(\frac{\partial\omega}{\partial t}\right)^2 + \omega^2\left(\frac{\partial\theta_\omega}{\partial t}\right)^2}{1 + t^2\left(\frac{\partial\theta_t}{\partial t}\right)^2}} \tag{324A}$$

$$a^{(1)}_{\phi\phi} = rv_r\sqrt{\frac{\left(\frac{\partial\omega}{\partial r}\right)^2 + \omega^2\left(\frac{\partial\theta_\omega}{\partial r}\right)^2}{1 + r^2\left(\frac{\partial\theta_r}{\partial r}\right)^2}} \tag{324B}$$

$$a^{(2)}_{\phi\phi} = v_\phi\sqrt{\frac{\left(\frac{\partial\omega}{\partial\phi}\right)^2 + \omega^2\left(\frac{\partial\theta_\omega}{\partial\phi}\right)^2}{1 + \phi^2\left(\frac{\partial\theta_\phi}{\partial\phi}\right)^2}} \tag{324C}$$

$$\psi_{\phi o} = \theta_r + \theta_\omega + \beta_{\omega,t} - \theta_t - \beta_{t,t} \tag{324D}$$

$$\psi_{\phi 1} = \theta_{vr} + \theta_\omega + \beta_{\omega,r} - \beta_{r,r} \tag{324E}$$

$$\psi_{\phi 2} = \theta_{v\phi} + \theta_\omega + \beta_{\omega,\phi} - \theta_\phi - \beta_{\phi,\phi} \tag{324F}$$

where

$$\tan \beta_{\omega,t} = \omega \frac{\partial\theta_\omega/\partial t}{\partial\omega/\partial t} \tag{325A}$$

$$\tan \beta_{\omega,r} = \omega \frac{\partial\theta_\omega/\partial r}{\partial\omega/\partial r} \tag{325B}$$

$$\tan \beta_{\omega,\phi} = \omega \frac{\partial\theta_\omega/\partial \phi}{\partial\omega/\partial \phi} \tag{325C}$$

From equations (322) and (323) it follows that

$$a_{\phi\phi} \cos \theta_{a\phi\phi} = a^{(o)}_{\phi\phi} \cos \psi_{\phi o} + a^{(1)}_{\phi\phi} \cos \psi_{\phi 1} + a^{(2)}_{\phi\phi} \cos \psi_{\phi 2} \tag{326}$$

$$a_{\phi\phi} \sin \theta_{a\phi\phi} = a^{(o)}_{\phi\phi} \sin \psi_{\phi o} + a^{(1)}_{\phi\phi} \sin \psi_{\phi 1} + a^{(2)}_{\phi\phi} \sin \phi_{\phi 2} \tag{327}$$

from which $a_{\phi\phi}$ and $\theta_{a\phi\phi}$ can be immediately obtained. For the special case where there is no spatial variation of the angular velocity (uniform rotation) it follows that

$$a_{\phi\phi} = a^{(o)}_{\phi\phi} \tag{328}$$

$$\theta_{a\phi\phi} = \psi_{\phi o} \tag{329}$$

Finally the Coriolis term in equation (310) is written as

$$2\bar{\omega}\bar{v}_r = a_c e^{j\theta_{ac}} = 2\omega v_r e^{j(\theta_\omega + \theta_{vr})} \tag{330}$$

and therefore for the Coriolis acceleration

$$a_c = 2\omega v_r \tag{331}$$

$$\theta_{ac} = \theta_\omega + \theta_{vr} = \theta_r + \beta_{r,t} + \theta_\phi + \beta_{\phi,t} - 2(\theta_t + \beta_{t,t}) \tag{332}$$

where v_r is given by equation (294), and θ_ω and θ_{vr} by equations (284) and (295) respectively. Combining equations (284), (324D), (329), and (332) shows that for uniform rotation

$$\theta_{a\phi\phi} = \theta_{ac} - \beta_{r,t} + \beta_{\omega,t} \tag{333}$$

All of the terms in equations (309) and (310) have been evaluated and these equations can be written as

$$\bar{a}_r = a_r e^{j\theta_{ar}} = a_{rr} e^{j\theta_{arr}} - a_{cen} e^{j\theta_{acen}} \tag{334}$$

$$\bar{a}_\phi = a_\phi e^{j\theta_{a\phi}} = a_{\phi\phi} e^{j\theta_{a\phi\phi}} + a_c e^{j\theta_{ac}} \tag{335}$$

The magnitudes and internal phase angles of the radial and transverse components of the acceleration a_r, θ_{ar}, a_ϕ, and $\theta_{a\phi}$ have yet to be calculated. This is done using equations (334) and (335). From equation (334) it follows that

$$a_r \cos \theta_{ar} = a_{rr} \cos \theta_{arr} - a_{cen} \cos \theta_{acen} \tag{336}$$

$$a_r \sin \theta_{ar} = a_{rr} \sin \theta_{arr} - a_{cen} \sin \theta_{acen} \tag{337}$$

and

$$\tan \theta_{ar} = \frac{a_{rr} \sin \theta_{arr} - a_{cen} \sin \theta_{acen}}{a_{rr} \cos \theta_{arr} - a_{cen} \cos \theta_{acen}} \tag{338}$$

$$a_r^2 = a_{rr}^2 + a_{cen}^2 - 2a_{rr}a_{cen} \cos (\theta_{arr} - \theta_{acen}) \tag{239}$$

From equation (335) it follows that

$$a_\phi \cos \theta_{a\phi} = a_{\phi\phi} \cos \theta_{a\phi\phi} + a_c \cos \theta_{ac} \tag{340}$$

$$a_\phi \sin \theta_{a\phi} = a_{\phi\phi} \sin \theta_{a\phi\phi} + a_c \sin \theta_{ac} \tag{341}$$

and

$$\tan \theta_{a\phi} = \frac{a_{\phi\phi} \sin \theta_{a\phi\phi} + a_c \sin \theta_{ac}}{a_{\phi\phi} \cos \theta_{a\phi\phi} + a_c \cos \theta_{ac}} \tag{342}$$

$$a_\phi^2 = a_{\phi\phi}^2 + a_c^2 + 2a_c a_{\phi\phi} \cos (\theta_{a\phi\phi} - \theta_{ac}) \tag{343}$$

In order to complete the calculation of the acceleration, the magnitude and phase angle of the vector sum of the radial and transverse components of acceleration need to be calculated. The complex number magnitude of the vector sum will be written as

$$\bar{a} = a e^{j\theta_a} \tag{344}$$

so that

$$\bar{a}^2 = \bar{a}_\phi^2 + \bar{a}_r^2 \tag{345}$$

from which it follows that

$$a^2 \cos(2\theta_a) = a_\phi^2 \cos(2\theta_{a\phi}) + a_r^2 \cos(2\theta_{ar}) \tag{346}$$

$$a^2 \sin(2\theta_a) = a_\phi^2 \sin(2\theta_{a\phi}) + a_r^2 \sin(2\theta_{ar}) \tag{347}$$

where a_ϕ and $\theta_{a\phi}$ are given by equations (343) and (342) respectively, and a_r and θ_{ar} are given by equations (339) and (338) respectively. From equations (346) and (347) it follows that

$$a^4 = a_\phi^4 + a_r^4 + 2a_r^2 a_\phi^2 \cos[2(\theta_{a\phi} - \theta_{ar})] \tag{348}$$

$$\tan(2\theta_a) = \frac{a_\phi^2 \sin(2\theta_{a\phi}) + a_r^2 \sin(2\theta_{ar})}{a_\phi^2 \cos(2\theta_{a\phi}) + a_r^2 \cos(2\theta_{ar})} \tag{349}$$

The measured acceleration is equal to $a \cos \theta_a$.

The relativistic force equations for a particle moving in bulk matter or vacuum with broken internal symmetry are best written in terms of normal and tangential components. The equations of motion of a particle under the action of normal and tangential forces are written as[11]

$$\bar{F}_N = m\bar{\gamma}_T \bar{a}_N = m\gamma_T a_N e^{j(\theta_{\gamma T} + \theta_{aN})} \tag{350A}$$

$$\bar{F}_T = m\bar{\gamma}_T^3 \bar{a}_T = m\gamma_T^3 a_T e^{j(3\theta_{\gamma T} + \theta_{aT})} \tag{350B}$$

where $\bar{F}_N$ amd $\bar{F}_T$ = normal and transverse complex number forces, $\bar{a}_N$ and $\bar{a}_T$ = complex number normal and transverse accelerations written as

$$\bar{a}_N = a_N e^{j\theta_{aN}} \tag{351A}$$

$$\bar{a}_T = a_T e^{j\theta_{aT}} \tag{351B}$$

and where the transverse velocity boost is written as

$$\bar{\gamma}_T = \gamma_T e^{j\theta_{\gamma T}} = (1 - \bar{v}_T^2/c^2)^{-1/2} \tag{352A}$$

$$\bar{v}_T = v_T e^{j\theta_{vT}} \tag{352B}$$

with

$$\gamma_T = [1 - 2(v_T/c)^2 \cos(2\theta_{vT}) + (v_T/c)^4]^{-1/4} \tag{353}$$

$$\tan(2\theta_{\gamma T}) = \frac{(v_T/c)^2 \sin(2\theta_{vT})}{1 - (v_T/c)^2 \cos(2\theta_{vT})} \tag{354}$$

Consider now the question of the conservation of angular momentum of a body under the action of a radial force field in uniformly rotating bulk matter or vacuum with broken internal symmetry. For a radial force field in a broken symmetry system, equations (340) and (341) become

$$a_{\phi\phi} \cos\theta_{a\phi\phi} + a_c \cos\theta_{ac} = 0 \tag{355}$$

$$a_{\phi\phi} \sin\theta_{a\phi\phi} + a_c \sin\theta_{ac} = 0 \tag{356}$$

In order for equations (355) and (356) to be satisfied, remembering that $a_{\phi\phi} > 0$ and $a_c > 0$, the following conditions must hold

$$a_{\phi\phi} - a_c = 0 \tag{357}$$

$$\tan\theta_{a\phi\phi} = \tan\theta_{ac} \tag{358}$$

or

$$\theta_{a\phi\phi} = \theta_{ac} - \pi \tag{359}$$

For uniform rotation and a radial force, the combination of equations (323), (324A), (331), (294), and (357) gives the following equation

$$r\sqrt{\left(\frac{d\omega}{dt}\right)^2 + \omega^2\left(\frac{d\theta_\omega}{dt}\right)^2} - 2\omega\sqrt{\left(\frac{dr}{dt}\right)^2 + r^2\left(\frac{d\theta_r}{dt}\right)^2} = 0 \tag{360}$$

Because $d\omega/dr < 0$ equation (360) can be written as

$$r \frac{d\omega}{dr} \sqrt{1 + \left(\omega \frac{d\theta_\omega}{d\omega}\right)^2} + 2\omega \sqrt{1 + \left(r \frac{d\theta_r}{dr}\right)^2} = 0 \qquad (361)$$

Combining equations (333) and (359) gives for uniform rotation and a central force

$$\beta_{\omega,t} = \beta_{r,t} - \pi \qquad (362)$$

From equation (362) it follows that for a radial force and uniform rotation

$$\tan \beta_{\omega,t} = \tan \beta_{r,t} \qquad (363)$$

Combining equations (296) and (325A) with equation (363) gives

$$r \frac{d\theta_r}{dr} = \omega \frac{d\theta_\omega}{d\omega} \qquad (364)$$

Substituting equation (364) into equation (361) gives

$$r \frac{d\omega}{dr} + 2\omega = 0 \qquad (365)$$

a differential equation whose solution is

$$\omega r^2 = \text{constant} \qquad (366)$$

where ω is given by equation (283). Dividing equations (364) and (365) gives also

$$2\theta_r + \theta_\omega = 2\theta_r + \theta_\phi + \beta_{\phi,t} - \theta_t - \beta_{t,t} = \text{constant} \qquad (367)$$

so that in fact combining equations (366) and (367) gives

$$\overline{\omega r}^2 = \text{constant} \qquad (368)$$

which is the expression for the conservation of angular momentum for a particle of unit mass uniformly rotating in a central force field that is located in bulk matter or vacuum wherein the space and time coordinates exhibit a broken symmetry.

Equations (283) and (366) show that

$$r^2 \sqrt{\frac{\omega_\phi^2 + \phi^2\omega_{\theta\phi}^2}{1 + t^2\omega_{\theta t}^2}} = \text{constant} \tag{369}$$

Equation (369) allows a connection to be made between the $t = 0$ and $t = \infty$ rotational states of a central force system located in bulk matter or vacuum with broken internal symmetries namely

$$r_o^2\omega_\phi(0) = r_\infty^2 < \omega_\phi(\infty) > \frac{\omega_{\theta\phi}(\infty)}{\omega_{\theta t}(\infty)} \tag{370}$$

where

$$< \omega_\phi(\infty) > = \lim_{t\to\infty} \frac{\phi}{t} \tag{371}$$

In a similar way equation (367) allows a connection to be made between the $t = 0$ and $t = \infty$ values of the internal phase angles of the coordinates of a particle in a central force system located in bulk matter or vacuum

$$2\theta_r(0) + \theta_\phi(0) + \beta_{\phi,t}(0) - \theta_t(0) - \beta_{t,t}(0) \tag{372}$$

$$= 2\theta_r(\infty) + \theta_\phi(\infty) + \beta_{\phi,t}(\infty) - \theta_t(\infty) - \beta_{t,t}(\infty)$$

Equation (369) shows that rotational motion is shared between external and internal angular motions, and this equation may perhaps be of value for describing the rotation of galaxies, neutron stars, molecules, atoms, and atomic nuclei where internal angular motions may exist.

A special case of interest, expecially for gravitationally bound systems such as stars or planets, is the situation where $\theta_r \neq 0$ but $d\theta_r/dt = 0$, $\theta_t = 0$ and $\theta_\phi = 0$. This gives the following results

$$v_r = \frac{dr}{dt} \qquad v_\phi = r\frac{d\phi}{dt} \qquad \omega = \omega_\phi = \frac{d\phi}{dt} \tag{373}$$

$$a_{rr} = \frac{dv_r}{dt} = \frac{d^2r}{dt^2} \qquad a_r = a_{rr} - r\omega_\phi^2 \tag{374}$$

$$a_{\phi\phi} = r\frac{d\omega_\phi}{dt} = r\frac{d^2\phi}{dt^2} \qquad a_c = 2\omega_\phi\frac{dr}{dr} \qquad a_\phi = a_{\phi\phi} + a_c \tag{375}$$

$$a^2 = a_r^2 + a_\phi^2 \qquad v^2 = v_r^2 + v_\phi^2 \tag{376}$$

$$\theta_{vr} = \theta_r \qquad \theta_{v\phi} = \theta_r \qquad \theta_\omega = 0 \tag{377}$$

$$\theta_{arr} = \theta_r \qquad \theta_{ar} = \theta_r \qquad \theta_{a\phi\phi} = \theta_r - \pi \tag{378}$$

$$\theta_{ac} = \theta_r \qquad \theta_{acen} = \theta_r \qquad \theta_{a\phi} = \theta_r \tag{379}$$

$$\theta_a = \theta_r \qquad \beta_{r,t} = 0 \qquad \beta_{\omega,t} = -\pi \tag{380}$$

Therefore the case of a time independent θ_r combined with $\theta_t = 0$ and $\theta_\phi = 0$ gives the standard kinematic and dynamic equations (373) through (376). Thus the effects of a time dependent θ_r with $\theta_t \neq 0$ and $\theta_\phi \neq 0$ can be discerned from anomalies in the rotational motion of stars, molecules, atoms, and atomic nuclei. However, the effects of a time independent θ_r with $\theta_t = 0$ and $\theta_\phi = 0$ can be discovered in non-rotating systems through its effect on the gravity and pressure of non-rotating (or slowly rotating) stars and planets. Section 7 shows the effects of θ_r on the equilibrium configurations of stars and planets.

6. EULER EQUATIONS FOR BROKEN SYMMETRY MATTER. This section considers Euler's equations of motion for a broken symmetry fluid, and is a prelude to the study of stellar and planetary equilibrium which is considered in Section 7. The standard special relativistic Euler equations for the radial and transverse directions are written for low pressures as[12,13]

$$\rho\gamma_a^2 a_r^a = -\left(\frac{\partial P^a}{\partial r^a} + \frac{1}{v_r^a}\frac{\partial P^a}{\partial t^a}\right) - \rho\frac{\partial W^a}{\partial r^a} \tag{381}$$

$$\rho\gamma_a^2 a_\phi^a = -\left(\frac{1}{r^a}\frac{\partial P^a}{\partial \phi^a} + \frac{1}{v_\phi^a}\frac{\partial P^a}{\partial t^a}\right) - \frac{\rho}{r^a}\frac{\partial W^a}{\partial \phi^a} \tag{382}$$

where a_r^a and a_ϕ^a are the conventional radial and transverse components of acceleration, ρ = proper mass density, P^a = pressure, W^a = macroscopic external force potential, and where

$$\gamma_a = (1 - \beta_a^2)^{-1/2} \qquad \beta_a = v_a/c \qquad v_a^2 = v_{ra}^2 + v_{\phi a}^2 \tag{383}$$

In section 7 W^a will be taken to be the gravitational potential.

It has been shown that in bulk matter the pressure has an internal phase angle as represented by equation (10), and that the coordinates within bulk matter also have internal phases such as, for example, is represented by equation (49) for the radial coordinate. Therefore the generalization of the special relativistic Euler equations to the case of bulk matter with broken internal symmetries is written for low pressures as

$$\rho\bar{\gamma}^2\bar{a}_r = \rho\gamma^2 a_r e^{j(\theta_{ar}+2\theta_\gamma)} \tag{384}$$

$$= -\cos\beta_{r,r}\left(\frac{\partial\bar{P}}{\partial r} + \frac{dt}{dr}\frac{\partial\bar{P}}{\partial t}\right) - \rho\,\frac{\partial\bar{W}}{\partial\bar{r}}$$

$$\rho\bar{\gamma}^2\bar{a}_\phi = \rho\gamma^2 a_\phi e^{j(\theta_{a\phi}+2\theta_\gamma)} \tag{385}$$

$$= -\cos\beta_{\phi,\phi}\left(\frac{1}{r}\frac{\partial\bar{P}}{\partial\phi} + \frac{1}{r}\frac{dt}{d\phi}\frac{\partial\bar{P}}{\partial t}\right) - \frac{\rho}{\bar{r}}\frac{\partial\bar{W}}{\partial\bar{\phi}}$$

where $\bar{a}_r$ and $\bar{a}_\phi$ are given by equations (334) and (335) respectively, and $\bar{W}$ can be written as in equation (246). The complex boost is written as

$$\bar{\gamma} = (1 - \bar{\beta}^2)^{-1/2} = \gamma e^{j\theta_\gamma} \tag{386}$$

where

$$\bar{\beta} = \bar{v}/c \qquad \bar{v} = v e^{j\theta_v} \qquad \bar{v}^2 = \bar{v}_r^2 + \bar{v}_\phi^2 \tag{387}$$

and where the boost magnitude and internal phase angle are given as

$$\gamma = [1 - 2\beta^2 \cos(2\theta_v) + \beta^4]^{-1/4} \tag{388}$$

$$\tan(2\theta_\gamma) = \frac{\beta^2 \sin(2\theta_v)}{1 - \beta^2 \cos(2\theta_v)} \tag{388A}$$

The generalization of the relativistic Euler equations for bulk matter with broken internal symmetries can also be written for the $\bar{x}$, $\bar{y}$, and $\bar{z}$ coordinates as[12]

$$\rho\bar{\gamma}^2\bar{a}_x = -\cos\beta_{x,x}\left(\frac{\partial\bar{P}}{\partial x} + \frac{dt}{dx}\frac{\partial\bar{P}}{\partial t}\right) - \rho\,\frac{\partial\bar{W}}{\partial\bar{x}} \tag{389}$$

$$\rho\bar{\gamma}^2\bar{a}_y = -\cos\beta_{y,y}\left(\frac{\partial\bar{P}}{\partial y} + \frac{dt}{dy}\frac{\partial\bar{P}}{\partial t}\right) - \rho\,\frac{\partial\bar{W}}{\partial\bar{y}} \tag{390}$$

$$\rho\bar{\gamma}^2\bar{a}_z = -\cos\beta_{z,z}\left(\frac{\partial\bar{P}}{\partial z} + \frac{dt}{dz}\frac{\partial\bar{P}}{\partial t}\right) - \rho\,\frac{\partial\bar{W}}{\partial\bar{z}} \tag{391}$$

where $\bar{a}_x$, $\bar{a}_y$, and $\bar{a}_z$ are given by equations (190) through (192), $\bar{\gamma}$ is given in terms of $\bar{v}/c$ by equation (386), and where

$$\bar{v}^2 = \bar{v}_x^2 + \bar{v}_y^2 + \bar{v}_x^2 \tag{391A}$$

Equations (384) and (385) or equations (389) through (391) are simple generalizations of the standard special relativistic Euler equations to the case of bulk matter with broken internal symmetry.

Euler's equations will be used to relate the internal phase angles of the coordinates to the internal phase angle of the pressure. From the radial acceleration equation (384) it follows for $\partial\bar{P}/\partial t = 0$ and $\bar{P}/c^2 \sim 0$ that

$$\rho\gamma^2 a_r e^{j(\theta_{ar}+2\theta_\gamma)} = D_P e^{j(\Phi_P+\pi)} + \rho D_W e^{j(\Phi_W+\pi)} \tag{392}$$

where

$$\cos\beta_{r,r}\frac{\partial\bar{P}}{\partial r} = D_P e^{j\Phi_P} \qquad \frac{\partial\bar{W}}{\partial\bar{r}} = D_W e^{j\Phi_W} \tag{393}$$

with

$$D_P = \sqrt{\frac{\left(\frac{\partial P}{\partial r}\right)^2 + P^2\left(\frac{\partial\theta_P}{\partial r}\right)^2}{1 + r^2\left(\frac{\partial\theta_r}{\partial r}\right)^2}} \tag{394}$$

$$D_W = \sqrt{\frac{\left(\frac{\partial W}{\partial r}\right)^2 + W^2\left(\frac{\partial\theta_W}{\partial r}\right)^2}{1 + r^2\left(\frac{\partial\theta_r}{\partial r}\right)^2}} \tag{395}$$

$$\Phi_P = \theta_P + \beta_{P,r} \tag{396}$$

$$\Phi_W = \theta_W + \beta_{W,r} - \theta_r - \beta_{r,r} \tag{397}$$

and where

$$\tan \beta_{P,r} = P \frac{\partial\theta_P/\partial r}{\partial P/\partial r} \tag{398}$$

$$\tan \beta_{W,r} = W \frac{\partial\theta_W/\partial r}{\partial W/\partial r} \tag{399}$$

and where $\beta_{r,r}$ is given by equation (64). For the case of an external potential it follows from the radial equation of motion (392) that

$$\rho\gamma^2 a_r \cos(\theta_{ar} + 2\theta\gamma) \tag{400}$$

$$= D_P \cos(\Phi_P + \pi) + \rho D_W \cos(\Phi_W + \pi)$$

$$= - D_P \cos\Phi_P - \rho D_W \cos\Phi_W$$

$$\rho\gamma^2 a_r \sin(\theta_{ar} + 2\theta_\gamma) \tag{401}$$

$$= D_P \sin(\Phi_P + \pi) + \rho D_W \sin(\Phi_W + \pi)$$

$$= - D_P \sin\Phi_P - \rho D_W \sin\Phi_W$$

From equation (400) and (401) it follows that

$$\rho^2\gamma^4 a_r^2 = D_P^2 + \rho^2 D_W^2 + 2\rho D_P D_W \cos(\Phi_W - \Phi_P) \tag{402}$$

Equations (400) and (401) determine a_r and θ_{ar} . Note that a_r and θ_{ar} are related to the component acceleration terms through equations (338) and (339). Expressions similar to equations (400) and (401) can be derived for the transverse acceleration from equation (385).

Consider now the case of static equilibrium. In this case the acceleration terms in equations (400) and (401) are equal to zero, with result

$$D_P = \rho D_W \tag{403}$$

$$\tan(\Phi_P + \pi) = \tan(\Phi_W + \pi) \quad \text{or} \quad \tan\Phi_P = \tan\Phi_W \tag{404}$$

$$\Phi_P = \Phi_W + \pi \qquad (405)$$

Because $D_P > 0$ and $D_W > 0$, the only way equations (400) and (401) can have their left hand sides equal to zero is to have $D_P = \rho D_W$ and

$$\cos \Phi_P = - \cos \Phi_W \qquad (406)$$

$$\sin \Phi_P = - \sin \Phi_W \qquad (407)$$

which requires equation (405) to be valid while at the same time satisfying equation (404). Combining equations (396), (397) and (405) gives

$$\theta_P + \beta_{P,r} = \theta_W + \beta_{W,r} - \theta_r - \beta_{r,r} + \pi \qquad (408)$$

Equations (403) and (408) are the equations for static equilibrium for the Euler equations describing bulk matter with broken internal symmetries under the action of an external potential (which also has a broken symmetry). The phase angle θ_P is determined by the relativistic state equation as shown in Reference 6 for solids and quantum liquids, and in an accompanying paper for the real gases. Therefore since θ_W and $\beta_{W,r}$ are related to the coordinates r and θ_r, it is equations (403) and (408) that relate the phase angle θ_r of the radial coordinate to P and θ_P of the equation of state. This will be made explicitly clear in Section 7 where gravitational equilibrium in stars and planets is considered.

Strictly speaking, only for a bulk matter system in which an external potential acts can one define a variation of θ_r with spatial coordinates, because only in this case can a physical choice or origin of coordinates be made (such as the center of a star or planet) from which to measure the coordinate r and thereby evaluate the denominators in equations (394) and (395). Only then is there a fixed reference point from which to calculate the variation of the phase angles such as θ_r, θ_v, and θ_a over macroscopic distances. However, θ_r, θ_v, and θ_a are determined by the broken symmetry of the local pressure θ_P and the broken symmetry of the local potential θ_W through equations (384) and (385).

7. EQUILIBRIUM OF STARS AND PLANETS. The equilibrium of stars and planets that are composed of matter with broken internal symmetries can be obtained from the complex number form of Euler's equation (384) or the equivalent equations (403) and (408). The gravitational potential energy that includes the effects of the broken symmetry of the space coordinates is written as

$$\bar{W} = We^{j\theta_W} = - \frac{GM}{\bar{r}} = \frac{GM}{r} e^{j(\pi-\theta_r)} \qquad (409)$$

corresponding to a gravitation force

$$\bar{F} = Fe^{j\theta_F} = -\frac{GM}{\bar{r}^2} = \frac{GM}{r^2}\, e^{j(\pi - 2\theta_r)} \tag{410}$$

so that

$$W = \frac{GM}{r} \tag{411}$$

$$\theta_W = \pi - \theta_r \tag{412}$$

where M = M(r) = mass at radius r. Newtonian gravity is assumed to be valid in this paper, so that the force is dependent only on $\bar{r}$ (through $\bar{r}^{-2}$). No explicit dependence on the angular coordinates $\bar{\psi}$ or $\bar{\phi}$ is assumed. However, the radial coordinate phase angle θ_r can depend on angles, $\theta_r = \theta_r(r,\psi,\phi)$.

The first equilibrium condition that is derived from the Euler equation is given by equation (403). Substituting equations (411) and (412) into equation (395) gives

$$D_W = \frac{GM}{r^2} \tag{413}$$

and therefore substituting equations (394) and (413) into equation (403) gives the first equilibrium equation for a gravitating star as

$$\sqrt{\left(\frac{\partial P}{\partial r}\right)^2 + P^2\left(\frac{\partial \theta_P}{\partial r}\right)^2} = \frac{GM\rho}{r^2}\sqrt{1 + \left(r\,\frac{\partial \theta_r}{\partial r}\right)^2} \tag{414}$$

Considering the fact that in a gravitating star or planet $\partial P/\partial r < 0$, equation (414) can be rewritten as

$$\frac{\partial P}{\partial r}\sqrt{1 + P^2\left(\frac{\partial \theta_P/\partial r}{\partial P/\partial r}\right)^2} = -\frac{GM\rho}{r^2}\sqrt{1 + \left(r\,\frac{\partial \theta_r}{\partial r}\right)^2} \tag{415}$$

which reduces to the standard stellar equilibrium equation for $\theta_P = 0$ and $\theta_r = 0$, namely[14]

$$\frac{\partial P}{\partial r} = -\frac{GM\rho}{r^2} \tag{416}$$

where the mass is related to the density and radial coordinates by

$$\cos\beta_{r,r}\,\frac{\partial M}{\partial r} = 4\pi r^2 \rho \tag{417}$$

Note that equation (415) can also be rewritten as

$$\frac{\cos \beta_{r,r}}{\cos \beta_{P,r}} \frac{\partial P}{\partial r} = - \frac{GM\rho}{r^2} \qquad (418)$$

If the terms involving the internal phase angles in equation (415) are assumed to be small it follows from this equation by expanding the radicals and solving a quadratic equation for $\partial P/\partial r$ that to a first approximation

$$\frac{\partial P}{\partial r} = - \frac{GM\rho}{r^2} \Psi \qquad (419)$$

where

$$\Psi = 1 + \frac{1}{2}\left(r \frac{\partial \theta_r}{\partial r}\right)^2 - \frac{1}{2}\left(\frac{rP}{GM\rho}\right)^2 \left(r \frac{\partial \theta_P}{\partial r}\right)^2 \qquad (420)$$

Therefore to first order the pressure gradient in equation (419) for stellar and planetary interiors with broken internal symmetry differs from the conventional result given in equation (416) by two opposing terms that are related to θ_r and θ_P respectively. Solving for the mass M from equation (414) and placing the expression in equation (417) gives the following combined equilibrium equation

$$\cos \beta_{r,r} \frac{1}{r^2} \frac{\partial}{\partial r} \left[\frac{r^2}{\rho} \sqrt{\frac{\left(\frac{\partial P}{\partial r}\right)^2 + P^2\left(\frac{\partial \theta_P}{\partial r}\right)^2}{1 + r^2\left(\frac{\partial \theta_r}{\partial r}\right)^2}} \right] = 4\pi G\rho \qquad (421)$$

or equivalently as

$$\cos \beta_{r,r} \frac{1}{r^2} \frac{\partial}{\partial r} \left[\frac{r^2}{\rho} \frac{\partial P}{\partial r} \cos \beta_{r,r} \sqrt{1 + P^2\left(\frac{\partial \theta_P/\partial r}{\partial P/\partial r}\right)^2} \right] = - 4\pi G\rho \qquad (422)$$

Similarly, using equation (419) for this purpose gives

$$\cos \beta_{r,r} \frac{1}{r^2} \frac{\partial}{\partial r} \left(\frac{r^2}{\rho \Psi} \frac{\partial P}{\partial r} \right) = - 4\pi G\rho \qquad (423)$$

where Ψ is given by equation (420).

The second gravitational equilibrium equation can be obtained by noting that equations (399), (412) and (64) yield

$$\beta_{W,r} = \beta_{r,r} - \pi \tag{424}$$

where $\beta_{r,r}$ is given by equation (64), so that it follows from equations (408), (412), and (424) that the second gravitational equilibrium equation is

$$\theta_P + \beta_{P,r} = -2\theta_r + \pi \tag{425}$$

where $\beta_{P,r}$ is given by equation (398). Equation (425) can be used to solve θ_r in terms of θ_P because this equation can be written as

$$\theta_P + \tan^{-1}\left(P\,\frac{\partial\theta_P/\partial r}{\partial P/\partial r}\right) = -2\theta_r + \pi \tag{426}$$

Equation (427) can be simplified by writing

$$\beta_{P,r} = \pi + \beta'_{P,r} \tag{427}$$

where $\beta'_{P,r}$ is a small quantity which can be positive or negative. Combining equations (425) and (427) gives the second gravitational equilibrium condition as

$$\theta_P + \beta'_{P,r} = -2\theta_r \tag{428}$$

From equation (398) it follows that the case of $\theta_P > 0$ and $\partial\theta_P/\partial r < 0$ (corresponding to planets and degenerate stars such as neutron stars and white dwarfs) gives $\beta_{P,r} > \pi$ or $\beta'_{P,r} > 0$, and from equations (428) and (64) it follows that $\theta_r < 0$. For gaseous stars it may be possible to have $\theta_P > 0$ or $\theta_P < 0$ because of a degeneracy in the state equation of the relativistic real gas (see accompanying paper on real gases). For gaseous stars with $\theta_P < 0$ and $\partial\theta_P/\partial r > 0$ it follows from equation (398) that $\beta_{P,r} < \pi$ or $\beta'_{P,r} < 0$ and therefore from equation (428) it follows that $\theta_r > 0$. This analysis assumes that $\partial P/\partial r < 0$ for all stars and planets. Combining equations (396) and (397) with equations (412), (424), and (425) gives

$$\Phi_P = \pi - 2\theta_r \tag{429}$$

$$\Phi_W = -2\theta_r \tag{430}$$

Equation (429) follows from the fact that

$$\cos\beta_{r,r}\,\frac{\partial\bar{P}}{\partial r} = -\frac{GM\rho}{\bar{r}^2} \tag{431}$$

Equation (428) is the second equilibrium equation derived from the general Euler equilibrium equation (408).

Equation (422), or the approximation equation (423), along with the equilibrium equation for the internal phases given in equation (428) are the two equilibrium equations for a gravitationally bound star or planet. These equations involve P, ρ, θ_P, and θ_r, so that clearly two additional equations are required for a complete solution of the equilibrium configuration (actually an energy generation equation is also required). The two additional equations that are required are the state equations which specify

$$P = P(\rho,T) \tag{432}$$

$$\theta_P = \theta_P(\rho,T) \tag{433}$$

the magnitude and internal phase angle of the complex number pressure. Equations (432) and (433) can be used to develop the following relationships

$$\frac{\partial P}{\partial r} = \frac{\partial P}{\partial \rho}\frac{\partial \rho}{\partial r} + \frac{\partial P}{\partial T}\frac{\partial T}{\partial r} \tag{434A}$$

$$\frac{\partial P}{\partial \psi} = \frac{\partial P}{\partial \rho}\frac{\partial \rho}{\partial \psi} + \frac{\partial P}{\partial T}\frac{\partial T}{\partial \psi} \tag{434B}$$

$$\frac{\partial P}{\partial \phi} = \frac{\partial P}{\partial \rho}\frac{\partial \rho}{\partial \phi} + \frac{\partial P}{\partial T}\frac{\partial T}{\partial \phi} \tag{434C}$$

$$\frac{\partial \theta_P}{\partial r} = \frac{\partial \theta_P}{\partial \rho}\frac{\partial \rho}{\partial r} + \frac{\partial \theta_P}{\partial T}\frac{\partial T}{\partial r} \tag{435A}$$

$$\frac{\partial \theta_P}{\partial \psi} = \frac{\partial \theta_P}{\partial \rho}\frac{\partial \rho}{\partial \psi} + \frac{\partial \theta_P}{\partial T}\frac{\partial T}{\partial \psi} \tag{435B}$$

$$\frac{\partial \theta_P}{\partial \phi} = \frac{\partial \theta_P}{\partial \rho}\frac{\partial \rho}{\partial \phi} + \frac{\partial \theta_P}{\partial T}\frac{\partial T}{\partial \phi} \tag{435C}$$

where r, ψ, and ϕ are the spherical coordinates whose origin is at the center of the star. Defining the following quantities

$$\tan \beta_{P,\rho} = P\,\frac{\partial \theta_P/\partial \rho}{\partial P/\partial \rho} \tag{436}$$

$$\tan \beta_{P,T} = P\,\frac{\partial \theta_P/\partial T}{\partial P/\partial T} \tag{437}$$

allows equations (435A) through (435C) to be written as

$$\frac{\partial\theta_P}{\partial r} = \frac{1}{P}\left(\tan\beta_{P,\rho}\frac{\partial P}{\partial\rho}\frac{\partial\rho}{\partial r} + \tan\beta_{P,T}\frac{\partial P}{\partial T}\frac{\partial T}{\partial r}\right) \tag{438A}$$

$$\frac{\partial\theta_P}{\partial\psi} = \frac{1}{P}\left(\tan\beta_{P,\rho}\frac{\partial P}{\partial\rho}\frac{\partial\rho}{\partial\psi} + \tan\beta_{P,T}\frac{\partial P}{\partial T}\frac{\partial T}{\partial\psi}\right) \tag{438B}$$

$$\frac{\partial\theta_P}{\partial\phi} = \frac{1}{P}\left(\tan\beta_{P,\rho}\frac{\partial P}{\partial\rho}\frac{\partial\rho}{\partial\phi} + \tan\beta_{P,T}\frac{\partial P}{\partial T}\frac{\partial T}{\partial\phi}\right) \tag{438C}$$

also

$$\tan\beta_{P,r} = \frac{\tan\beta_{P,\rho}\frac{\partial P}{\partial\rho}\frac{\partial\rho}{\partial r} + \tan\beta_{P,T}\frac{\partial P}{\partial T}\frac{\partial T}{\partial r}}{\frac{\partial P}{\partial\rho}\frac{\partial\rho}{\partial r} + \frac{\partial P}{\partial T}\frac{\partial T}{\partial r}} \tag{439A}$$

$$\tan\beta_{P,\psi} = \frac{\tan\beta_{P,\rho}\frac{\partial P}{\partial\rho}\frac{\partial\rho}{\partial\psi} + \tan\beta_{P,T}\frac{\partial P}{\partial T}\frac{\partial T}{\partial\psi}}{\frac{\partial P}{\partial\rho}\frac{\partial\rho}{\partial\psi} + \frac{\partial P}{\partial T}\frac{\partial T}{\partial\psi}} \tag{439B}$$

$$\tan\beta_{P,\phi} = \frac{\tan\beta_{P,\rho}\frac{\partial P}{\partial\rho}\frac{\partial\rho}{\partial\phi} + \tan\beta_{P,T}\frac{\partial P}{\partial T}\frac{\partial T}{\partial\phi}}{\frac{\partial P}{\partial\rho}\frac{\partial\rho}{\partial\phi} + \frac{\partial P}{\partial T}\frac{\partial T}{\partial\phi}} \tag{439C}$$

Similarly for the internal phase angle of the radial coordinate

$$\frac{\partial\theta_r}{\partial r} = \frac{\partial\theta_r}{\partial\rho}\frac{\partial\rho}{\partial r} + \frac{\partial\theta_r}{\partial T}\frac{\partial T}{\partial r} \tag{440A}$$

$$\frac{\partial\theta_r}{\partial\psi} = \frac{\partial\theta_r}{\partial\rho}\frac{\partial\rho}{\partial\psi} + \frac{\partial\theta_r}{\partial T}\frac{\partial T}{\partial\psi} \tag{440B}$$

$$\frac{\partial\theta_r}{\partial\phi} = \frac{\partial\theta_r}{\partial\rho}\frac{\partial\rho}{\partial\phi} + \frac{\partial\theta_r}{\partial T}\frac{\partial T}{\partial\phi} \tag{440C}$$

which can be used to evaluate equations (64) and (67). Equations similar to equations (440) hold for θ_ψ and θ_ϕ , but these internal phase angles are taken to be zero in the simplest theory of gravitational equilibrium. In any case, it is clear that the determination of the equilibrium configuration of stars and planets require the determination of $\theta_r(r)$ and $\theta_P(r)$ as part of the solution. Both of these phase angles must approach their vacuum values, $\theta_r^{(v)}$ and $\theta_P^{(v)}$, at the surface of the star or planet.

The magnitude $P(\rho,T)$ and internal phase angle $\theta_P(\rho,T)$ of the relativistic pressure are obtained from a solution of the relativistic trace equation (1) along with the magnitudes and internal phase angles of the other thermodynamic functions.[6] A $T = 0$ degenerate neutron gas state equation with a pressure described by $P^o(\rho)$ and $\theta_P^o(\rho)$, which is obtained from the solution of the $T = 0$ form of the relativistic trace equation (1), can serve as an adequate description of a neutron star.[6] The radial variation of the internal phase angle of the radial coordinates of a neutron star can be determined from $P^o(\rho)$ and $\theta_P^o(\rho)$ using equations (422) and (428). For the interacting classical or quantum gases that occur in ordinary stars, the internal phase angle $\theta_P(\rho,T)$ can be evaluated from the relativistic third and higher virial coefficients of a real classical or quantum gas at high temperatures. The relativistic third and higher virial coefficients are obtained from a solution of the relativistic trace equation (1) for the real gases. Therefore the relativistic third and higher virial coefficients of the state equation of real gases will play an important role in the determination of the equilibrium conditions of ordinary gaseous stars.

The equilibrium of gravitating planets is treated in a slightly different manner than for stars, but the two basic equilibrium equations (422) and (428) are also valid for gravitating planets. Equation (422) will be written in a slightly different form for planets. As in the case for stars, the complex number equilibrium equation is written as

$$\cos \beta_{r,r} \frac{\partial \bar{P}}{\partial r} = - \frac{GM\rho}{\bar{r}^2} \tag{441}$$

or

$$D_P e^{j\Phi_P} = \frac{G\rho M}{r^2} e^{j(\pi - 2\theta_r)} \tag{442}$$

where D_P is given by equation (394) and Φ_P is given by equation (396). Equation (441) can be rewritten in terms of a density derivative by introducing the bulk modulus at constant entropy $\bar{K}_S$. In order to determine $\bar{K}_S$, the bulk modulus at constant temperature $\bar{K}_T$ must first be introduced. The constant temperature bulk modulus is given by[6]

$$\bar{K}_T = \rho \frac{\partial \bar{P}}{\partial \rho} = K_T e^{j(\theta_P + \beta_{P,\rho})} \tag{443}$$

where

$$K_T = \sqrt{\left(\rho \frac{\partial P}{\partial \rho}\right)^2 + P^2\left(\rho \frac{\partial \theta_P}{\partial \rho}\right)^2} \tag{444}$$

and where $\beta_{P,\rho}$ is given by equation (436). The bulk modulus at constant entropy is easily found to be given by

$$\bar{K}_S = \bar{K}_T + \bar{\gamma}\left(T \frac{\partial \bar{P}}{\partial T}\right) = K_S e^{j\theta_{KS}} \tag{445}$$

where the complex number Grüneisen function $\bar{\gamma}$ is given in equation (5). Equation (445) can be written in component form as

$$K_S \cos \theta_{KS} = K_T \cos (\theta_P + \beta_{P,\rho}) + \gamma N \cos (\theta_\gamma + \theta_P + \beta_{P,T}) \tag{446}$$

$$K_S \sin \theta_{KS} = K_T \sin (\theta_P + \beta_{P,\rho}) + \gamma N \sin (\theta_\gamma + \theta_P + \beta_{P,T}) \tag{447}$$

where expressions for the magnitude γ and internal phase θ_γ of the Grüneisen function are given in Reference 6, $\beta_{P,\rho}$ and $\beta_{P,T}$ are given by equations (436) and (437) respectively, and where[6]

$$N = \sqrt{\left(T \frac{\partial P}{\partial T}\right)^2 + P^2\left(T \frac{\partial \theta_P}{\partial T}\right)^2} \tag{448}$$

Equations (446) and (447) give immediately

$$\tan \theta_{KS} = \frac{K_T \sin (\theta_P + \beta_{P,\rho}) + \gamma N \sin (\theta_\gamma + \theta_P + \beta_{P,T})}{K_T \cos (\theta_P + \beta_{P,\rho}) + \gamma N \cos (\theta_\gamma + \theta_P + \beta_{P,T})} \tag{449}$$

$$K_S^2 = K_T^2 + \gamma^2 N^2 + 2\gamma N K_T \cos (\theta_\gamma + \beta_{P,T} - \beta_{P,\rho}) \tag{450}$$

which allow the calculation of the phase angle and magnitude of the bulk modulus at constant entropy.

Combining equations (434A), (441) and (445) gives the following approximation for a planet with broken symmetry matter[17]

$$\cos \beta_{r,r} \frac{\partial \rho}{\partial r} = - \frac{G\rho^2 M}{\bar{K}_S \bar{r}^2} = - \frac{G\rho M}{\bar{v}_S^2 \bar{r}^2} \tag{451}$$

where the adiabatic velocity of elastic waves in a material with broken internal symmetry is given as

$$\bar{v}_S^2 = v_S^2 e^{2j\theta_{vS}} = \frac{\bar{K}_S}{\rho} \qquad (452)$$

or

$$v_S^2 = \frac{K_S}{\rho} \qquad (453)$$

$$2\theta_{vS} = \theta_{KS} \qquad (454)$$

Equation (451) can also be rewritten as

$$\cos\beta_{r,r}\frac{\partial\rho}{\partial r} = -\frac{G\rho M}{r^2 v_S^2} \qquad (455)$$

$$\theta_{vS} = -\theta_r \qquad (456)$$

where

$$\cos\beta_{r,r} = [1 + (r\,\partial\theta_r/\partial r)^2]^{-1/2} \qquad (457)$$

Substituting the expression for the mass in equation (455) into equation (417) gives

$$\cos\beta_{r,r}\frac{1}{r^2}\frac{\partial}{\partial r}\left(\frac{r^2}{\rho}\frac{\partial\rho}{\partial r}v_S^2\cos\beta_{r,r}\right) = -4\pi G\rho \qquad (458)$$

where v_S is obtained from equations (450) and (453). Equation (458) is the first equilibrium equation for gravitating planets and is the analog of equation (422) for stars, while equation (456) is the second equilibrium equation for gravitating planets with broken internal symmetry and is the analog of equation (428) for stars with broken internal symmetry. Finally it should be pointed out that for matter with broken internal symmetries the adiabatic wave velocity is given by a simple formula, analogous to the conventional formula for symmetric matter, as follows[17]

$$\bar{v}_S^2 = \bar{\alpha}^2 - \frac{4}{3}\bar{\beta}^2 \qquad (459)$$

where $\bar{\alpha}$ and $\bar{\beta}$ = compression and shear wave velocities respectively for matter with broken internal symmetries. Writing

$$\bar{\alpha} = \alpha e^{j\theta_\alpha} \qquad \bar{\beta} = \beta e^{j\theta_\beta} \tag{460}$$

gives

$$v_S^2 \cos(2\theta_{vS}) = \alpha^2 \cos(2\theta_\alpha) - \frac{4}{3}\beta^2 \cos(2\theta_\beta) \tag{461}$$

$$v_S^2 \sin(2\theta_{vS}) = \alpha^2 \sin(2\theta_\alpha) - \frac{4}{3}\beta^2 \sin(2\theta_\beta) \tag{462}$$

which are equivalent to the following equations

$$\tan(2\theta_{vS}) = \frac{\alpha^2 \sin(2\theta_\alpha) - \frac{4}{3}\beta^2 \sin(2\theta_\beta)}{\alpha^2 \cos(2\theta_\alpha) - \frac{4}{3}\beta^2 \cos(2\theta_\beta)} \tag{463}$$

$$v_S^2 = \alpha^4 + \frac{16}{9}\beta^4 - \frac{8}{3}\alpha^2\beta^2 \cos[2(\theta_\alpha - \theta_\beta)] \tag{464}$$

The measured adiabatic wave velocity = $v_S \cos\theta_{vS}$, while the measured compression and shear wave velocities = $\alpha \cos\theta_\alpha$ and $\beta \cos\theta_\beta$.

A knowledge of P and θ_P as a function of density and temperature can be obtained experimentally from high pressure measurements on earth materials such as olivine and gabbro. Alternatively P and θ_P can be obtained from the solution of the relativistic trace equation (1) if the unrenormalized pressure P^a and Grüneisen function γ^a can be estimated from atomic structure.[6] The seismic wave velocity v_S and its internal phase angle θ_{vS} can then be obtained from equations (453) and (454) respectively. Finally, equations (463) and (464) can be inverted to find the relativistic values of the compression wave velocity α and the shear wave velocity β . It may be possible to reverse the arguments and measure α and β which gives v_S and θ_{vS} by equations (463) and (464) and then obtain P and θ_P from equations (449), (450), (453), and (454). Equations (456) and (458) are the equilibrium equations for a planet whose solution gives $\rho(r)$ and $\theta_r(r)$ in terms of P and θ_P . As in the case of the equilibrium calculation for stars, two auxiliary state equations of the form given in equations (432) and (433) are required. In any case, it is clear that P, θ_P, and θ_r are required for an understanding of the equilibrium configuration and seismic properties of a planet.

From the previous analysis it is clear that the Newtonian force of gravity acting on a unit mass at a distance r from the center of a spherical body of mass M(r) with broken internal symmetry is written as

$$\bar{F} = -\frac{GM}{\bar{r}^2} = -\frac{GM}{r^2} e^{j2\theta_r} \tag{465}$$

$$F_R = -\frac{GM}{r^2}\cos(2\theta_r) = -GM/r_m^2 \cos^2\theta_r \cos(2\theta_r) \tag{466}$$

where $r_m = r \cos \theta_r$ = measured value of the radial distance between two points, $\bar{F}$ = complex Newtonian gravity force with internal phase, and F_R = real part of the gravity force in the radial direction which is the measured gravity force. The force F_R must be compared to the force F_a = conventional Newtonian gravity force for asymmetric matter which is given by

$$F_a = - \frac{GM}{r_m^2} = - \frac{GM}{r^2 \cos^2 \theta_r} \tag{467}$$

The difference $F_R - F_a$ is given by

$$F_D = F_R - F_a \tag{468}$$

$$= \frac{GM}{r^2} [\cos^{-2} \theta_r - \cos (2\theta_r)]$$

$$= \frac{GM}{r_m^2} [1 - \cos^2 \theta_r \cos (2\theta_r)]$$

$$\sim + 3\theta_r^2 GM/r^2$$

$$\sim + 3\theta_r^2 GM/r_m^2$$

where the last two approximations are valid for small θ_r , and where $\theta_r = \theta_r(r,\psi,\phi)$ is a function of the spherical polar coordinates of the unit mass. Therefore the effect of broken symmetry matter on Newtonian gravity is to imply that there is a new additional repulsive gravity force F_D in operation which does not have a strictly r^{-2} dependence on radial coordinates. But in fact gravity in the planets is Newtonian in form (neglecting general relativity effects) and has a $\bar{r}^{-2}$ dependence as given in equation (465) for broken symmetry matter. The apparent deviation from Newtonian gravity is due to the internal phase angle $\theta_r(r,\psi,\phi)$ of the radial coordinate which can have a complicated coordinate dependence because of the inhomogeneous nature of the earth's core, mantle and crust. Equation (466) shows that F_R does not have an r^{-2} (or r_m^{-2}) dependence on coordinates.

The rate of change of the force of gravity with radial distance is obtained for broken symmetry matter from equation (466) to be

$$\frac{\partial F_R}{\partial r} = \frac{2GM}{r^3} [\cos (2\theta_r) + r \frac{\partial \theta_r}{\partial r} \sin (2\theta_r)] - 4\pi G\rho \cos (2\theta_r) \tag{469}$$

and for radial variations only

$$\frac{\partial F_R}{\partial r_m} = \frac{\partial F_R}{\partial r}\frac{\partial r}{\partial r_m} = \frac{\partial F_R}{\partial r} / (\cos\theta_r - \sin\theta_r \, r\frac{\partial\theta_r}{\partial r}) \tag{470}$$

The corresponding variation for the conventionally calculated Newtonian gravity force given by equation (467) is

$$\frac{\partial F_a}{\partial r} = \frac{2GM}{r^3 \cos^2\theta_r}(1 - \tan\theta_r \, r\frac{\partial\theta_r}{\partial r}) - \frac{4\pi G\rho}{\cos^2\theta_r} \tag{471}$$

$$\frac{\partial F_a}{\partial r_m} = \frac{\partial F_a}{\partial r} / (\cos\theta_r - \sin\theta_r \, r\frac{\partial\theta_r}{\partial r}) \tag{472}$$

Introduce the parameter

$$D = \frac{\partial F_R/\partial r_m - \partial F_a/\partial r_m}{\partial F_a/\partial r_m} = \frac{\partial F_R/\partial r - \partial F_a/\partial r}{\partial F_a/\partial r} \tag{473}$$

then a simple calculation shows that to second order in θ_r (there are no first order terms)

$$D = -3\theta_r^2(1 - \eta) \tag{474}$$

where

$$\eta = \frac{\frac{r}{\theta_r}\frac{\partial\theta_r}{\partial r}}{1 - \frac{2\pi\rho r^3}{M}} \tag{475}$$

For planets $\theta_r < 0$ and $\partial\theta_r/\partial r > 0$, so that in general η should be small and negative.

Therefore experimental measurements of the variation of the force of gravity with height should indicate $D < 0$, while measurements of the gravity force itself should yield $F_D > 0$. The net result of the internal phase of the radial coordinate is that the measured gravity force given by equation (466) should be slightly weaker than that predicted by the conventional Newtonian force given by equation (467). A weaker than Newtonian gravity force has been experimentally observed in geophysical measurements and in new Eötvös experiments.[18-23] These results have been interpreted to be due to a new finite range repulsive force associated with gravity.[19,20] Reference 21 contains many citations to the literature in this field. However the results of the present paper show that in fact the weaker attractive force that is observed may be due to ordinary Newtonian gravity operating in matter with broken internal symmetries as in equation (465). A complete understanding of the earth's gravity field will require a detailed knowledge of the internal phase of the radial coordinate $\theta_r(r,\psi,\phi)$ and its variation with location. The orbits of satellites and ballistic missiles will be affected by the internal phase function $\theta_r(r,\psi,\phi)$, and perturbations in these orbits that are not explained totally by shape and density variations in the earth may lead to techniques for determining local values of $\theta_r(r,\psi,\phi)$.

8. CONCLUSION. By means of a relativistic trace equation, the Minkowski metric of spacetime impresses a broken symmetry on the matter and vacuum that are embedded in spacetime. The broken internal symmetries of matter and the vacuum are manifested at the microscopic level through the internal phase angles that are associated with the coordinates and the kinematic and dynamic variables for single particles. At the macroscopic level the broken symmetries appear in the thermodynamic functions such as pressure and internal energy of interacting systems. Within bulk matter and the vacuum, space and time exhibit broken symmetries that are manifested by internal phase angles that produce the broken symmetries of the kinematic and dynamic parameters and the broken symmetry of geometrical constructs such as angles, lengths and areas. The physical rotation of matter must also be associated with the rotation of the internal phase angles of the space and time coordinates. The internal phase angles of the kinematic and dynamic parameters of bulk matter fluid elements are determined by the Euler equations for broken symmetry matter. The calculation of the equilibrium configurations of stars, planets and other gravitationally bound systems such as galaxies must include the determination of the spatial dependence of the internal phase angle of the radial coordinate along with the spatial variation of the pressure and density. This can only be done if the state equation of broken symmetry matter is known from solutions of the basic complex number relativistic trace equation. The fact that time and space are gauge rotated quantities should affect the basic calculations of astrophysics, geophysics, and the engineering disciplines.

ACKNOWLEDGEMENT

The author wishes to thank Elizabeth K. Klein for typing this paper.

REFERENCES

1. O'Raifeartaigh, L. O., Group Structure of Gauge Theories, Cambridge University Press, London, 1986.

2. Becher, P., Böhm, M. and Joos, H., Gauge Theories, John Wiley, New York, 1984.

3. Konopleva, N. and Popov, V., Gauge Fields, Harwood, New York, 1981.

4. Weiss, R. A., "Scale Invariant Equations for Relativistic Waves", Fourth Army Conference on Applied Mathematics and Computing, Cornell University, ARO 87-1, May 27-30, 1986, p. 307.

5. Weiss, R. A., Relativistic Thermodynamics, Vols. 1 and 2, Exposition Press, New York, 1976.

6. Weiss, R. A., "Thermodynamic Gauge Theory of Solids and Quantum Liquids with Internal Phase", Fifth Army Conference on Applied Mathematics and Computing, West Point, ARO 88-1, June 15-18, 1987, p. 649.

7. Huang, K., Statistical Mechanics, John Wiley, New York, 1963.

8. Yourgrau, W., van der Merwe, A., and Raw, G., Treatise on Irreversible and Statistical Thermophysics, Dover Publications, Inc., New York, 1982.

9. Zee, A., Unity of Forces in the Universe, Vols. 1 and 2, World Scientific, Singapore, 1982.

10. Sakurai, J., Invariance Principles and Elementary Particles, Princeton University Press, Princeton, 1964.

11. Goldstein, H., Classical Mechanics, Addison-Wesley, New York, 1980.

12. Pauli, W., Theory of Relativity, Pergamon, New York, 1958.

13. Weinberg, S., Gravitation and Cosmology, John Wiley, New York, 1972.

14. Chandrasekhar, S., An Introduction to the Study of Stellar Structure, Dover, New York, 1957.

15. Jeffreys, H., The Earth, Cambridge University Press, New York, 1962.

16. Gutenberg, B., Physics of the Earth's Interior, Academic Press, New York, 1959.

17. Magnitskiy, V. A., The Internal Structure and Physics of the Earth, NASA Technical Translation TT F-395, NASA, Wash., D. C., April 1967.

18. Stacey, F. D. in Science Underground, edited by Nieto, M. M., Haxton, W.C., Hoffman, C. M., Kolb, E. W., Sandberg, V. D., and Toevs, J. W., AIP Conf. Proc. No. 96, AIP, New York, 1983.

19. Fischbach, E., Sudarsky, D., Szafer, A., and Talmadge, C., "Reanalysis of the Eötvös Experiment," Phys. Rev. Lett., Vol. 56, Jan. 6, 1986, p. 3.

20. Nieto, M. M., Goldman, T. and Hughes, R. J., "Phenomenological Aspects of New Gravitational Forces, I. Rapidly Rotating Compact Objects, II. Static Planetary Potentials, III. Slowly Rotating Astronomical Bodies" (with Macrae, K. I.), Phys. Rev. D, Vol. 36, No. 12, 15 Dec. 1987, p. 3684.

21. Stacey, F. D., Tuck, G. J., Moore, G. I., Holding, S. C., Goodwin, B. D., and Zhou, R., "Geophysics and the Law of Gravity", Rev. Mod. Phys., Vol. 59, No. 1, Jan 1987, p. 157.

22. Stacey, F. D., Tuck, G. J. and Moore, G. I., Phys. Rev. D. 36, 2374, 1987.

23. Fitch, V. L., Isaila, M. V. and Palmer, M. A., "Limits on the Existence of a Material-Dependent Intermediate-Range Force", Phys. Rev. Lett., Vol. 60, May 2, 1988, p. 1801.

J

NEWTONIAN GRAVITY IN MATTER WITH BROKEN INTERNAL SYMMETRY

Richard A. Weiss
U. S. Army Engineer Waterways Experiment Station
Vicksburg, Mississippi 39180

ABSTRACT. The pressure field in matter is associated with a broken internal symmetry which manisfests itself through the broken internal symmetry of space and time coordinates. This introduces an apparent non-Newtonian behaviour of gravity in matter. The effective Newtonian gravitational constant for a spherical body composed of matter with broken internal symmetry is calculated and determined to be a function of radial distance from the center of a planet or star. The gravity field of a rotating geometrically asymmetric planet composed of matter with broken internal symmetries is investigated. A theoretical analysis of Eötvös, mine shaft, borehole and tower gravity variation experiments is presented in terms of Newtonian gravity in matter with broken internal symmetry. It is found that the discrepancies from Newtonian gravity can be described by ordinary Newtonian gravity in matter with broken internal symmetry combined with the variation of atmospheric pressure down a mine shaft or borehole and up a tower. This research will affect the calculation of trajectories of missiles and projectiles in the earth's atmosphere, and will have applications to geophysics and astrophysics.

1. INTRODUCTION. Discrepancies between Newton's law of gravitation and the measured variation of gravity with distance and composition of the attracting bodies have been observed. These discrepancies appeared first in the measurements of the variation of the gravity force with depth in mine shafts.[1-4] These measurements indicate a larger value of the gravitational constant than is found from laboratory Eötvös experiments.[1-4] On the other hand, recent experiments on the variation of gravity up the length of a tower suggest a value of the gravitational constant which is less than that measured in the laboratory by Eötvös experiments.[5] Differences in behaviour from Newtonian gravity have also been reported for the Eötvös type of experiments and with beam balance experiments.[6-15] Other evidence for non-Newtonian behaviour has been presented from solar system and stellar system measurements.[16-18]

Attempts to explain these measured results by the introduction of new types of gravitational forces (the "fifth" and "sixth" forces) that have finite ranges of the order of hundreds or thousands of meters have been suggested.[9-23] These new forces would represent the effects of massive spin 0 and spin 1 supersymmetric partners to the ordinary massless spin 2 graviton that mediates Newtonian gravitation with its infinite range.[19-23] Much criticism of the reality of these finite range forces has been presented.[24,25] This is due in part to the difficulties of separating extraneous effects due to geological structure from the possible intrinsic non-Newtonian behaviour of gravity. In fact recent data from a borehole in the ice of a glacier in Greenland suggests that the gravitation constant is less than that measured by laboratory Eötvös experiments, and this disagrees with the results given in Reference 1-4 but agrees with the observations in Reference 5. The state of both the experimental and theoretical situation is therefore uncertain.

This paper suggests an alternative explanation for the apparent non-Newtonian behaviour of gravity in the earth which is based on ordinary Newtonian gravitation and the broken symmetry of the thermodynamic and mechanical parameters of bulk matter such as pressure and internal energy.[26,27] Some results have already been obtained toward describing the apparent non-Newtonian behaviour of gravity in terms of the ordinary Newtonian gravity field in matter with broken internal symmetries.[28] This was done by showing that the space and time coordinates exhibit broken symmetries in matter where the pressure has a broken internal symmetry.[28] Section 2 introduces the relationship between Newtonian gravity and the broken internal symmetries of space and time, Section 3 deals with complex number coordinates and the measurement of space and time, Section 4 considers Newtonian gravity for rotating non-spherical masses composed of matter that induces broken symmetries in the pressure and coordinates, Section 5 presents a theory for the description of the Eötvös, mine shaft, borehole and tower experiments, and finally Section 6 gives a numerical calculation of the expected values of the internal phase angles of the radial and angular coordinates due to the earth's gravity field.

2. NEWTONIAN GRAVITY AND BROKEN INTERNAL SYMMETRIES. A gauge theory of relativistic thermodynamics has been developed which is based on a trace equation which for completely symmetrical matter or radiation is given by[29]

$$U + T\left(\frac{dU}{dT}\right)_{PV} - 3V\,\frac{d}{dV}(PV)_{U} = U^{a} + T\left(\frac{dU^{a}}{dT}\right)_{P^{a}V} \tag{1}$$

where U = relativistic internal energy, P = relativistic pressure, T = absolute temperature, V = volume of substance, and U^a and P^a = corresponding nonrelativistic internal energy and pressure. Throughout this paper the index "a" will refer to nonrelativistic calculations. The temperature and volume are parameters for both the renormalized and unrenormalized systems. The trace equation for matter whose thermodynamic functions have broken internal symmetries is given by[27]

$$\bar{U} + T\left(\frac{d\bar{U}}{dT}\right)_{\bar{P}V} - 3V\,\frac{d}{dV}(\bar{P}V)_{\bar{U}} = U^{a} + T\left(\frac{dU^{a}}{dT}\right)_{P^{a}V} \tag{2}$$

where $\bar{U}$ and $\bar{P}$ are complex number representations of the renormalized internal energy and pressure respectively, and where T and V are the magnitudes of the complex number temperature and volume respectively. Equation (2) can be further simplified by using the following complex number form of the Gibbs-Helmholtz-Maxwell equation[27]

$$\frac{\partial\bar{U}}{\partial V} = T\,\frac{\partial\bar{P}}{\partial T} - \bar{P} \tag{3}$$

The complex numbers $\bar{U}$ and $\bar{P}$ that appear in equation (2) are written as[27]

$$\bar{U} = Ue^{j\theta_U} \tag{4}$$

$$\bar{P} = Pe^{j\theta_P} \tag{5}$$

where U, P, θ_U and θ_P can be obtained from a solution of equations (2) and (3). The temperature and volume parameters that appear in equation (2) are real numbers. However the temperature and volume themselves are complex numbers that are written as

$$\bar{T} = Te^{j\theta_T} \tag{6}$$

$$\bar{V} = Ve^{j\theta_V} \tag{7}$$

where T and V are the magnitudes of the temperature and volume and it is these quantities that appear in the trace equation (2). The measured thermodynamic quantities are given by[28]

$$U_m = U \cos \theta_U \tag{8}$$

$$P_m = P \cos \theta_P \tag{9}$$

$$V_m = V \cos \theta_V \tag{10}$$

$$T_m = T \cos \theta_T \tag{11}$$

The phase angles θ_U and θ_P are obtained from equations (2) and (3), while θ_V and θ_T are related to coordinate and velocity internal phase angles as will be shown later.

The determination of the space and time coordinate internal phase angles follows from the complex number Euler equations[28]

$$\rho d\bar{v}/d\bar{t} = - \cos \beta_{r,r} \; \partial\bar{P}/\partial r + \rho\bar{F}_r \tag{12}$$

where the complex number external force (such as gravity) is written as

$$\bar{F}_r = F_r e^{j\theta_{Fr}} = - \partial\bar{W}/\partial\bar{r} \tag{13}$$

and where $\bar{v}$ = complex number velocity, $\bar{t}$ = complex number time and $\bar{r}$ = complex number radial coordinate. The complex number velocity and space and time coordinates are written as[28]

$$\bar{r} = re^{j\theta_r} \tag{14}$$

$$\bar{t} = te^{j\theta_t} \tag{15}$$

$$\bar{v} = ve^{j\theta_v} \tag{16}$$

The measured values of the space and time coordinates and the particle velocity are given by[28]

$$r_m = r \cos \theta_r \tag{17}$$

$$t_m = t \cos \theta_t \tag{18}$$

$$v_m = v \cos \theta_v \tag{19}$$

For matter in equilibrium equation (12) becomes

$$\rho d\bar{v}/d\bar{t} = - \cos \beta_{r,r}\ \partial\bar{P}/\partial r + \rho\bar{F}_r = 0 \tag{20}$$

where[28]

$$d\bar{v}/d\bar{t} = [(dv/dt)^2 + (vd\theta_v/dt)^2]^{1/2} \cos \beta_{t,t}\ e^{j\Phi_v} \tag{21}$$

$$\cos \beta_{r,r}\ \partial\bar{P}/\partial r = [(\partial P/\partial r)^2 + (P\partial\theta_P/\partial r)^2]^{1/2} \cos \beta_{r,r}\ e^{j\Phi_P} \tag{22}$$

$$\Phi_v = \theta_v + \beta_{v,t} - \theta_t - \beta_{t,t} \tag{23}$$

$$= \theta_r + \beta_{r,r} + \beta_{v,t} - 2(\theta_t + \beta_{t,t})$$

$$\Phi_P = \theta_P + \beta_{P,r} \tag{24}$$

$$\tan \beta_{v,t} = \frac{vd\theta_v/dt}{dv/dt} \tag{25}$$

$$\tan \beta_{t,t} = t\partial\theta_t/\partial t \tag{26}$$

$$\tan \beta_{P,r} = \frac{P\partial\theta_P/\partial r}{\partial P/\partial r} \tag{27}$$

$$\tan \beta_{r,r} = r\partial\theta_r/\partial r \tag{28}$$

To obtain the second relation in equation (23) the following relationship is used[28]

$$\theta_v = \theta_r + \beta_{r,r} - \theta_t - \beta_{t,t} \tag{29}$$

Combining equations (21) through (29) shows that the equilibrium condition given by equation (20) is equivalent to

$$\Phi_v = 0 \quad (30)$$

$$dv/dt = 0 \quad (31)$$

$$d\theta_v/dt = 0 \quad (32)$$

and[28]

$$\Phi_P = \theta_{Fr} \quad (33A)$$

$$[(\partial P/\partial r)^2 + (P\partial\theta_P/\partial r)^2]^{1/2} \cos \beta_{r,r} = \rho F_r \quad (33B)$$

From equations (23) and (30) it follows that for equilibrium

$$0 = \theta_v + \beta_{v,t} - \theta_t - \beta_{t,t} \quad (34)$$

$$= \theta_r + \beta_{r,r} + \beta_{v,t} - 2(\theta_t + \beta_{t,t})$$

Neglecting the β's in equation (34) gives the following approximation

$$\theta_v \sim \theta_t \sim \theta_r/2 \quad (35)$$

The relationship between θ_r and θ_P is obtained from equations (33A) and (33B). For gravity, equations (33A) and (33B) yield a set of coupled differential equations for P, θ_P and θ_r.[28] An approximate solution of equation (33A) gives the following equation for matter in a gravity field[28]

$$\theta_r \sim -\theta_P \quad (36)$$

Then equations (35) and (36) give for a gravitating system

$$\theta_v \sim \theta_t \sim -\theta_P/2 \quad (37)$$

For a general system one has

$$\theta_r \sim -\sigma\theta_P \quad (38)$$

where σ = index that describes the state equation for matter. Equations (30) through (33) give the general conditions of equilibrium. For photons $\theta_t = \theta_r$, and the light speed has a zero internal phase angle.

From equation (36) and the relation $\bar{V} \sim 4/3\pi\bar{r}^3$ it follows that the phase angle for the volume is given by

$$\theta_V \sim 3\theta_r \sim -3\theta_P \quad (40)$$

for a gravitating system. For a uniform system the volume is given by

$$\bar{V} = e^{j\theta_V} 4/3\pi r^3 = e^{j\theta_V} 4/3\pi r_a^3 \quad (41)$$

so that

$$V = V^a \quad (42)$$

The renormalized and unrenormalized scalar coordinates are parameters related by[28]

$$r = r^a \quad (43)$$

$$\psi = \psi^a \quad (44)$$

$$\phi = \phi^a \quad (45)$$

where r = magnitude of radial coordinate, ψ = magnitude of the complex number zenith angle and ϕ = magnitude of the complex number azimuthal angle. Thus V and V^a are simply equivalent parameters in the trace equations (1) and (2).

The determination of the broken symmetry phase angle of the temperature θ_T is determined from the energy equipartition theorem which can be written for a complex number particle velocity as

$$\bar{\varepsilon} = \langle 1/2m\bar{v}^2 \rangle = k\bar{T} \quad (46)$$

where $\bar{\varepsilon}$ = complex number average kinetic energy per particle and where m = particle mass and k = Boltzmann constant. The real and imaginary parts of equation (46) can be written as

$$\varepsilon_R = m/2 \int_0^\infty v^2 \cos(2\theta_v) g(v) dv = kT \cos \theta_T \quad (47)$$

$$\varepsilon_I = m/2 \int_0^\infty v^2 \sin(2\theta_v) g(v) dv = kT \sin \theta_T \quad (48)$$

where $g(v)$ = renormalized molecular velocity distribution function. Then

$$\tan \theta_T = \varepsilon_I/\varepsilon_R \quad (49)$$

$$kT = (\varepsilon_R^2 + \varepsilon_I^2)^{1/2} \quad (50)$$

which are the equations of θ_T and T. For a gravitating system θ_v is given by equation (37), and for this case θ_v is independent of the molecular speeds. Therefore for this case it follows from equation (37) and equations (46) through (48) that

$$\theta_T = 2\theta_v \sim -\theta_P \quad (51)$$

Because the magnitude of the temperature T must appear on both sides of the basic trace equation (1) and (2) it follows that

$$T = T^a \tag{52}$$

where

$$kT^a = m/2 \int_0^\infty v_a^2 g^a(v^a)dv^a \tag{53}$$

In fact equation (52) implies the validity of equation (51) and the relation $g = g^a$. Therefore T and T^a simply play the roles of equivalent parameters in the trace equations (1) and (2). The measured temperature is obtained from equations (11), (51) and (52) to be

$$T_m = T \cos \theta_T = T \cos \theta_P = T^a \cos \theta_P \tag{54}$$

In view of the complex number values of the volume and temperature it might be thought that the trace equation (2) should be written in the following completely asymmetric form

$$\bar{U} + \bar{T}\left(\frac{d\bar{U}}{d\bar{T}}\right)_{\bar{P}\bar{V}} - 3\bar{V}\,\frac{d}{d\bar{V}}(\bar{P}\bar{V})_{\bar{U}} = U^a + T^a\left(\frac{dU^a}{dT^a}\right)_{p^a v^a} \tag{55}$$

but this is not correct as can be seen by applying equation (55) to the real classical gases. The experimental fact that the first term of the virial expansion (the ideal gas) and the second virial coefficient must be unaffected by equation (55) requires that the temperature and volume terms that appear as complex numbers in the left hand side of equation (55) must in fact actually appear as real numbers equal to the magnitudes of the complex number volume and temperature as shown correctly in equations (2), (42) and (52).[27] The trace equation corresponds to a uniform pressure and energy density system so that equations (42) and (52) are implicitly assumed in equations (1) and (2).

For systems with nonuniform pressure fields, the determination of the internal phase angles of the coordinates generally involves the solution of coupled differential equations.[28] Thus for gravitating stars or planets the determination of θ_r involves the solution of the following two equations[28]

$$\cos \beta_{r,r}\,\frac{1}{r^2}\,\frac{\partial}{\partial r}\left\{\frac{r^2}{\rho}\,\frac{\partial P}{\partial r}\cos \beta_{r,r}\left[1 + P^2\left(\frac{\partial\theta_P/\partial r}{\partial P/\partial r}\right)^2\right]^{1/2}\right\} = -\,4\pi G\rho \tag{56}$$

$$\theta_P + \tan^{-1}\left(P\,\frac{\partial\theta_P/\partial r}{\partial P/\partial r}\right) = -\,2\theta_r + \pi \tag{57}$$

combined with the solution of the relativistic trace equation (2) which links

θ_P and P to density and temperature for gases, solids or Fermi liquids with internal phase. Equation (56) is the combined equation that arises from the following equilibrium equation[28]

$$\partial P/\partial r \cos \beta_{r,r} \left[1 + P^2 \left(\frac{\partial \theta_P/\partial r}{\partial P/\partial r} \right)^2 \right]^{1/2} = - GM\rho/r^2 \qquad (58)$$

and the relationship of mass and density (which will be treated in Section 3) given by[28]

$$\cos \beta_{r,r} \; \partial M/\partial r = 4\pi r^2 \rho \qquad (59)$$

When the internal phase angles are set to zero the equilibrium equation (58) reduces to the standard result[30]

$$\partial P/\partial r = - GM\rho/r^2 \qquad (60)$$

The small gradient approximation to equation (57) is

$$\theta_P + P \frac{\partial \theta_P/\partial r}{\partial P/\partial r} = - 2\theta_r \qquad (61)$$

which will be used in Section 5 for approximate solutions for θ_r .

Newton's gravitational law can be written for spatial coordinates with broken internal symmetry as[28]

$$\bar{g} = - GM/\bar{r}^2 \qquad (62)$$

where $\bar{g}$ = complex number acceleration of gravity. The measured acceleration of gravity is given by the real value of equation (62) as follows[28]

$$g_m = - GM/r^2 \cos(2\theta_r) \qquad (63)$$

Written in terms of the measured radial coordinate given by equation (17) gives[28]

$$g_m = - GM/r_m^2 \cos(2\theta_r) \cos^2 \theta_r \qquad (64)$$

These formulas are valid for spherical masses. The conventional value of the acceleration of gravity is expressed in terms of the measured radial distance as[28]

$$g_c = - GM/r_m^2 = - GM/r^2 \cos^{-2} \theta_r \qquad (65)$$

and therefore[28]

$$g_m - g_c = GM/r_m^2 [1 - \cos(2\theta_r) \cos^2 \theta_r] \tag{66}$$

$$\sim 3\theta_r^2 GM/r_m^2$$

The derivatives of g_m and g_c with respect to r are given by[28]

$$\partial g_m/\partial r = 2GM/r^3 [\cos(2\theta_r) + r\partial\theta_r/\partial r \sin(2\theta_r)] - 4\pi G\rho \cos(2\theta_r) \tag{67}$$

$$\partial g_c/\partial r = 2GM/(r^3 \cos^2 \theta_r)(1 - \tan \theta_r \, r\partial\theta_r/\partial r) - 4\pi G\rho/\cos^2 \theta_r \tag{68}$$

Then a parameter D can be defined given by[28]

$$D = \frac{\partial g_m/\partial r_m - \partial g_c/\partial r_m}{\partial g_c/\partial r_m} = \frac{\partial g_m/\partial r - \partial g_c/\partial r}{\partial g_c/\partial r} \tag{69}$$

$$= \frac{A + B}{C}$$

where

$$A = [\cos(2\theta_r) \cos^2 \theta_r - 1](1 - 2\pi\rho r^3/M) \tag{70}$$

$$B = r\partial\theta_r/\partial r [\sin(2\theta_r) \cos^2 \theta_r + \tan \theta_r] \tag{71}$$

$$C = 1 - \tan \theta_r \, r\partial\theta_r/\partial r - 2\pi\rho r^3/M \tag{72}$$

For small θ_r the parameter D can be written as[28]

$$D = - 3\theta_r^2(1 - \eta) \tag{73}$$

$$\eta = \frac{r/\theta_r \partial\theta_r/\partial r}{1 - 2\pi\rho r^3/M} \tag{74}$$

From equation (64) it follows that for spherical bodies the Newtonian law of gravitation in space with broken internal symmetry requires a coordinate dependent effective gravitational constant given by

$$G_r = G \cos(2\theta_r) \cos^2 \theta_r \tag{75}$$

For small values of θ_r equation (75) becomes

$$G_r = G(1 - 3\theta_r^2 + \cdots) \tag{76}$$

Therefore for space with broken internal symmetry $G_r < G$, and G represents the ideal case of gravitation in totally symmetric space ($\theta_r = 0$) . For a homogeneous spherical planet or star the internal phase angle of the radial coordinate will be a function of the radial coordinate magnitude $\theta_r = \theta_r(r)$ and is obtained as a solution to the coupled gravitational equilibrium (56) and (57). In general however $\theta_r = \theta_r(r,\psi,\phi)$ for an inhomogeneous body such as the earth, and therefore G_r will depend on latitude and longitude. Equations (64) and (75) are valid for both the interior and exterior of a spherical planet.

From equation (75) it follows that

$$\partial G_r/\partial r = - GE\partial\theta_r/\partial r \tag{77}$$

$$\partial G_r/\partial\psi = - GE\partial\theta_r/\partial\psi \tag{78}$$

$$\partial G_r/\partial\phi = - GE\partial\theta_r/\partial\phi \tag{79}$$

where

$$E = \sin(2\theta_r)(4\cos^2\theta_r - 1) \tag{80}$$

For small θ_r , $E \sim 6\theta_r$. In Section 5 it is shown that $\theta_r < 0$ and $\partial\theta_r/\partial r > 0$ for idealized planets so that $\partial G_r/\partial r > 0$. From equations (77) through (79) it follows that

$$r/G_r\partial G_r/\partial r = - 2Hr\partial\theta_r/\partial r \tag{81}$$

$$\psi/G_r\partial G_r/\partial\psi = - 2H\psi\partial\theta_r/\partial\psi \tag{82}$$

$$\phi/G_r\partial G_r/\partial\phi = - 2H\phi\partial\theta_r/\partial\phi$$

where

$$H = \tan(2\theta_r) + \tan\theta_r \tag{84}$$

For small θ_r , $H \sim 3\theta_r$.

It is the variation of the acceleration of gravity of the earth that is determined in gravity measurements, and it is important to have a measure of the difference between the rates of change of g_m and g_c with respect to radial distance. One such measure is given in equation (69). Another measure might consider the difference of the normalized rates of change of the acceleration of gravity as follows

$$D_2 = \frac{r/g_m \partial g_m/\partial r - r/g_c \partial g_c/\partial r}{r/g_c \partial g_c/\partial r} \tag{85}$$

$$= \frac{r_m/g_m \partial g_m/\partial r_m - r_m/g_c \partial g_c/\partial r_m}{r_m/g_c \partial g_c/\partial r_m}$$

From equation (63) it follows that

$$r/g_m \partial g_m/\partial r = -2[1 + \tan(2\theta_r) r\partial\theta_r/\partial r - 2\pi r^3 \rho/M] \tag{86}$$

while from equation (65) it follows that

$$r/g_c \partial g_c/\partial r = -2[1 - \tan\theta_r \; r\partial\theta_r/\partial r - 2\pi r^3 \rho/M] \tag{87}$$

Then

$$D_2 = \frac{Hr\partial\theta_r/\partial r}{1 - \tan\theta_r \; r\partial\theta_r/\partial r - 2\pi r^3 \rho/M} \tag{88}$$

where H is given by equation (84). For small values of θ_r

$$D_2 \sim + 3\theta_r^2 \eta \tag{89}$$

where η is given by equation (74).

3. MEASUREMENT AND GEOMETRY OF SPACE AND TIME. It has been assumed that the complex number space and time coordinates are Euclidian and that[28]

$$\bar{x}^2 + \bar{y}^2 = \bar{r}^2 \tag{90}$$

$$\sin^2 \bar{\phi} + \cos^2 \bar{\phi} = 1 \tag{91}$$

$$\tan \bar{\phi} = \bar{y}/\bar{x} \tag{92}$$

where

$$\bar{x} = xe^{j\theta_x} \tag{93}$$

$$\bar{y} = ye^{j\theta_y} \tag{94}$$

$$\bar{\phi} = \phi e^{j\theta_\phi} \tag{95}$$

$$\sin \bar{\phi} = S_\phi e^{j\theta_{s\phi}} \tag{96}$$

$$\cos\bar{\phi} = C_\phi e^{-j\theta_{c\phi}} \tag{97}$$

where $\bar{\phi}$ = complex number azimuthal angle, and where[28]

$$S_\phi = [\sin^2(\phi\cos\theta_\phi) + \sinh^2(\phi\sin\theta_\phi)]^{1/2} \tag{98}$$

$$C_\phi = [\cos^2(\phi\cos\theta_\phi) + \sinh^2(\phi\sin\theta_\phi)]^{1/2} \tag{99}$$

$$\tan\theta_{s\phi} = \cot(\phi\cos\theta_\phi)\tanh(\phi\sin\theta_\phi) \tag{100}$$

$$\tan\theta_{c\phi} = \tan(\phi\cos\theta_\phi)\tanh(\phi\sin\theta_\phi) \tag{101}$$

The component equations of equation (90) determine r and θ_r and are written as

$$x^2\cos(2\theta_x) + y^2\cos(2\theta_y) = r^2\cos(2\theta_r) \tag{102}$$

$$x^2\sin(2\theta_x) + y^2\sin(2\theta_y) = r^2\sin(2\theta_r) \tag{103}$$

while the component equations of equation (91) are

$$S_\phi^2\cos(2\theta_{s\phi}) + C_\phi^2\cos(2\theta_{c\phi}) = 1 \tag{104}$$

$$S_\phi^2\sin(2\theta_{s\phi}) - C_\phi^2\sin(2\theta_{c\phi}) = 0 \tag{105}$$

Equations (90) through (105) also give

$$S_\phi/C_\phi = y/x \tag{106}$$

$$\theta_{s\phi} + \theta_{c\phi} = \theta_y - \theta_x \tag{107}$$

$$S_\phi^2 = \sin(2\theta_{c\phi})/\sin[2(\theta_{c\phi} + \theta_{s\phi})] \tag{108}$$

$$C_\phi^2 = \sin(2\theta_{s\phi})/\sin[2(\theta_{c\phi} + \theta_{s\phi})] \tag{109}$$

The measured coordinates and angles are given by[28]

$$x_m = x\cos\theta_x \tag{110}$$

$$y_m = y\cos\theta_y \tag{111}$$

$$r_m = r\cos\theta_r \tag{112}$$

$$\phi_m = \phi\cos\theta_\phi \tag{113}$$

Substituting equations (110) through (112) into equations (102) and (103) shows that

$$x_m^2 + y_m^2 \neq r_m^2 \qquad (114)$$

which indicates that in a broken symmetry system the measured coordinates are non-Euclidian.

A. Length of Curves

The length of a curve in complex number space is given by

$$\bar{L} = Le^{j\theta_L} = \int[\bar{r}^2 + (d\bar{r}/d\bar{\phi})^2]^{1/2}\, d\bar{\phi} \qquad (115)$$

where

$$\frac{d\bar{r}}{d\bar{\phi}} = e^{j\Phi_{r\phi}} \frac{\sec \beta_{r,r}}{\sec \beta_{\phi,\phi}} \frac{dr}{d\phi} \qquad (116)$$

$$\Phi_{r\phi} = \theta_r + \beta_{r,r} - \theta_\phi - \beta_{\phi,\phi} \qquad (117)$$

$$\sec \beta_{r,r} = [1 + (r\partial\theta_r/\partial r)^2]^{1/2} \qquad (118)$$

$$\sec \beta_{\phi,\phi} = [1 + (\phi\partial\theta_\phi/\partial\phi)^2]^{1/2} \qquad (119)$$

The measured length is given by

$$L_m = L \cos \theta_L \qquad (120$$

For a circle with $\partial r/\partial\phi = 0$ the complex number length is

$$\bar{L} = r \int_0^{2\pi} e^{j(\theta_r+\theta_\phi+\beta_{\phi,\phi})} \sec \beta_{\phi,\phi}\, d\phi \qquad (121)$$

If in addition θ_ϕ and θ_r are independent of ϕ equation (121) becomes

$$\bar{L} = 2\pi re^{j(\theta_r+\theta_\phi)} \qquad (122)$$

$$L = 2\pi r \qquad (123)$$

$$\theta_L = \theta_r + \theta_\phi \qquad (124)$$

The measured circumference is

$$L_m = L \cos \theta_L = 2\pi r \cos(\theta_r + \theta_\phi) \qquad (125)$$

Note that if $\theta_\phi = 0$

$$L = 2\pi r \tag{126}$$

$$\theta_L = \theta_r \tag{127}$$

$$L_m = 2\pi r \cos\theta_r = 2\pi r_m \tag{128}$$

which is the result obtained in Reference 28.

B. Area of a Plane Curve.

The area enclosed by a plane curve in complex number space is

$$\bar{A} = \frac{1}{2}\int \bar{r}^2 d\bar{\phi} = \frac{1}{2}\int r^2 e^{j(2\theta_r+\theta_\phi)}(d\phi + j\phi d\theta_\phi) \tag{129}$$

$$= \frac{1}{2}\int r^2 e^{j(2\theta_r+\theta_\phi+\beta_{\phi,\phi})} \sec\beta_{\phi,\phi}\, d\phi$$

For a circle with r = constant

$$\bar{A} = \frac{r^2}{2}\int e^{j(2\theta_r+\theta_\phi+\beta_{\phi,\phi})} \sec\beta_{\phi,\phi}\, d\phi \tag{130}$$

If θ_r and θ_ϕ are constants

$$\bar{A} = Ae^{j\theta_A} = \pi r^2 e^{j(2\theta_r+\theta_\phi)} \tag{131}$$

and therefore

$$A = \pi r^2 \tag{132}$$

$$\theta_A = 2\theta_r + \theta_\phi \tag{133}$$

$$A_m = \pi r^2 \cos(2\theta_r + \theta_\phi) = \pi r_m^2 \cos(2\theta_r + \theta_\phi)/\cos^2\theta_r \tag{134}$$

where A_m = measured area. Finally, if $\theta_\phi = 0$

$$A = \pi r^2 \tag{135}$$

$$\theta_A = 2\theta_r \tag{136}$$

$$A_m = \pi r^2 \cos(2\theta_r) = \pi r_m^2 \cos(2\theta_r)/\cos^2\theta_r \tag{137}$$

which is the result obtained in Reference 28.

C. Area of Surface.

The area of a surface is given in complex number spherical coordinates as

$$\bar{S} = \int \bar{r}^2 \sin\bar{\psi}\, d\bar{\psi}\, d\bar{\phi} \tag{138}$$

$$= \int r^2 S_\psi e^{j(2\theta_r+\theta_{s\psi}+\theta_\psi+\theta_\phi)} (d\psi + j\psi d\theta_\psi)(d\phi + j\phi d\theta_\phi)$$

$$= \int r^2 S_\psi \sec\beta_{\psi,\psi} \sec\beta_{\phi,\phi}\, e^{j\Phi_S}\, d\psi\, d\phi$$

where $\bar{\psi}$ = complex number zenith angle given by

$$\bar{\psi} = \psi e^{j\theta_\psi} \tag{139}$$

and where

$$\sec\beta_{\psi,\psi} = [1 + (\psi\partial\theta_\psi/\partial\psi)^2]^{1/2} \tag{140}$$

$$\sec\beta_{\phi,\phi} = [1 + (\phi\partial\theta_\phi/\partial\phi)^2]^{1/2} \tag{141}$$

$$\Phi_S = 2\theta_r + \theta_{s\psi} + \theta_\psi + \beta_{\psi,\psi} + \theta_\phi + \beta_{\phi,\phi} \tag{142}$$

$$S_\psi = [\sin^2(\psi\cos\theta_\psi) + \sinh^2(\psi\sin\theta_\psi)]^{1/2} \tag{143}$$

$$\tan\theta_{s\psi} = \cot(\psi\cos\theta_\psi)\tanh(\psi\sin\theta_\psi) \tag{144}$$

If θ_ψ and θ_ϕ are constants then

$$\bar{S} = e^{j(\theta_{s\psi}+\theta_\psi+\theta_\phi)} \int r^2 S_\psi e^{j2\theta_r}\, d\psi\, d\phi \tag{145}$$

For r and θ_r independent of angles

$$\bar{S} = e^{j(2\theta_r+\theta_{s\psi}+\theta_\psi+\theta_\phi)}\, 2\pi r^2 \int S_\psi\, d\psi \tag{146}$$

Directly from equation (138) it follows that if r, θ_r, θ_ψ and θ_ϕ are independent of angles (a crude approximation)

$$\bar{S} \sim 2\pi r^2 e^{j(2\theta_r+\theta_\phi)}[1 - \cos(\pi e^{j\theta_\psi})] \tag{147}$$

It is easy to show that equations (146) and (147) are equivalent. If $\theta_\psi = 0$ it follows that

$$\bar{S} = 4\pi r^2 e^{j(2\theta_r+\theta_\phi)} \tag{148}$$

Finally, if $\theta_\phi = 0$

$$\bar{S} = 4\pi r^2 e^{j2\theta_r} \tag{149}$$

$$S = 4\pi r^2 \tag{150}$$

$$\theta_S = 2\theta_r \tag{151}$$

$$S_m = 4\pi r^2 \cos(2\theta_r) \tag{152}$$

$$= 4\pi r_m^2 \cos(2\theta_r)/\cos^2 \theta_r$$

From equations (62) and (147) it follows that Gauss's law for spatial coordinates with broken internal symmetry is given by

$$\int \bar{g} d\bar{S} \sim -2\pi GMe^{j\theta_\phi}[1 - \cos(\pi e^{j\theta_\psi})] \tag{153}$$

assuming θ_ϕ and θ_ψ are constants (which can only be a crude approximation).

D. Volume

The volume contained within a closed surface is written in complex number spherical polar coordinates as

$$\bar{V} = \int \bar{r}^2 \sin \bar{\psi}\, d\bar{\psi}\, d\bar{\phi}\, d\bar{r} \tag{154}$$

$$= \int r^2 S_\psi \sec \beta_{\psi,\psi} \sec \beta_{\phi,\phi} \sec \beta_{r,r}\, e^{j\Phi_V}\, d\psi\, d\phi\, dr$$

where

$$\Phi_V = 3\theta_r + \beta_{r,r} + \theta_{s\psi} + \theta_\psi + \beta_{\psi,\psi} + \theta_\phi + \beta_{\phi,\phi} \tag{155}$$

If θ_ψ and θ_ϕ are constants

$$\bar{V} = e^{j(\theta s\psi + \theta\psi + \theta\phi)} \int r^2 S_\psi \sec \beta_{r,r}\, e^{j(3\theta r + \beta r,r)}\, d\psi\, d\phi\, dr \tag{156}$$

For θ_ψ and $\theta_\phi = 0$

$$\bar{V} = \int r^2 \sin \psi \sec \beta_{r,r}\, e^{j(3\theta_r + \beta_{r,r})}\, d\psi\, d\phi\, dr \tag{157}$$

For r and θ_r independent of ψ and ϕ (a sphere)

$$\bar{V} = 4\pi \int r^2 \sec \beta_{r,r}\, e^{j(3\theta_r + \beta_{r,r})}\, dr \tag{158}$$

The component parts of equation (158) are

$$\dot{V} \cos \theta_V = 4\pi \int r^2 \sec \beta_{r,r} \cos(3\theta_r + \beta_{r,r}) dr \tag{159}$$

$$V \sin \theta_V = 4\pi \int r^2 \sec \beta_{r,r} \sin(3\theta_r + \beta_{r,r}) dr \tag{160}$$

which determines V and θ_V for a sphere in a gravitational field. For θ_r = constant

$$\bar{V} = e^{j3\theta_r} 4\pi/3r^3 = 4\pi/3\bar{r}^3 \tag{161}$$

$$V = 4\pi/3r^3 \tag{162}$$

$$\theta_V = 3\theta_r \tag{163}$$

$$V_m = 4\pi/3r^3 \cos(3\theta_r) = 4\pi/3r_m^3 \cos(3\theta_r)/\cos^3 \theta_r \tag{164}$$

E. Density

The rest mass of a body does not have an internal phase because it is invariant under the effects of the basic trace equation (2).[29] The instantaneous density is given by

$$\bar{\rho} = \rho e^{j\theta_\rho} = dM/d\bar{V} = \cos \beta_{V,V}\, dM/dV\, e^{-j(\theta_V + \beta_{V,V})} \tag{165}$$

where

$$\tan \beta_{V,V} = V d\theta_V/dV \tag{166}$$

Therefore

$$\rho = \cos \beta_{V,V}\, dM/dV \tag{167}$$

$$\theta_\rho = -\theta_V - \beta_{V,V} \tag{168}$$

Combining equations (165) and (154) gives the following results for the density

$$\rho = (r^2 S_\psi \sec \beta_{\psi,\psi} \sec \beta_{\phi,\phi} \sec \beta_{r,r})^{-1} \frac{dM}{d\psi\, d\phi\, dr} \tag{169}$$

$$\theta_\rho = -\Phi_V \tag{170}$$

where Φ_V is given by equation (155). For radial symmetry equations (158) and (165) give

$$\rho \sim \frac{\cos \beta_{r,r}}{4\pi r^2} \frac{\partial M}{\partial r} \tag{171}$$

$$\theta_\rho \sim - 3\theta_r - \beta_{r,r} \tag{172}$$

Equation (171) combined with the following stellar equilibrium equation for broken symmetry matter[28]

$$M = - \frac{r^2 \partial P/\partial r}{G\rho} \cos \beta_{r,r} \left[1 + \left(P \frac{\partial\theta_P/\partial r}{\partial P/\partial r}\right)^2\right]^{1/2} \tag{173}$$

$$= - \frac{r^2 \partial P/\partial r}{G\rho} \cos \beta_{r,r} \sec \beta_{P,r}$$

gives the combined stellar equilibrium (56) for ordinary stars. The angle $\beta_{P,r}$ that appears in equation (173) is defined by

$$\tan \beta_{P,r} = P \frac{\partial\theta_P/\partial r}{\partial P/\partial r} \tag{174}$$

The measured density is given by

$$\rho_m = \rho \cos \theta_\rho \tag{175}$$

For a relativistic interacting system having a complex number internal energy $\bar{U}$, the mass is given by $\bar{M} = \bar{U}/c^2$ and the instantaneous density is[27]

$$\bar{\rho}_r = c^{-2}\, d\bar{U}/|d\bar{V}| = d\bar{M}/|d\bar{V}| \tag{176}$$

Combining equations (158) and (176) gives

$$\rho_r \sim \frac{\cos \beta_{r,r}}{4\pi r^2} \left[(\partial M/\partial r)^2 + (M \partial\theta_M/\partial r)^2\right]^{1/2} \tag{177}$$

$$\theta_{\rho r} \sim \theta_U + \beta_{U,r} \tag{178}$$

where $\theta_M = \theta_U$, and $\beta_{U,r}$ is given by

$$\tan \beta_{U,r} = U \frac{\partial\theta_U}{\partial r} \Big/ \frac{\partial U}{\partial r} = M \frac{\partial\theta_M}{\partial r} \Big/ \frac{\partial M}{\partial r} \tag{179}$$

If special relativity is included the pressure adds to the internal energy density, and the inertial mass density becomes[31-33]

$$\bar{\rho}_I = \bar{\rho}_r + \bar{P}/c^2 \tag{180}$$

while the gravitational mass density is

$$\bar{\rho}_G = \bar{\rho}_r + 3\bar{P}/c^2 \tag{181}$$

Equations (180) and (181) can be simplified by combining them with equation (158). It should be pointed out that the complex number values for coordinates is also suggested by string theory.[34]

4. NEWTONIAN GRAVITY FOR NONSPHERICAL MASSES WITH BROKEN INTERNAL SYMMETRY.

A. Complex Number Gravitational Potential.

By analogy to the standard scalar form of the gravitational potential for a nonspherical body, the following expression for the complex number gravitational potential for a nonspherical body existing in space with broken internal symmetries is postulated[35-39]

$$\bar{V} = -GM/\bar{r}[1 - \sum_{n=2}^{\infty} \bar{I}_n (\bar{a}/\bar{r})^n \bar{P}_n(\cos\bar{\psi})] \tag{182}$$

where $\bar{V}$ = complex number potential, $\bar{a}$ = complex number equatorial radius, $\bar{r}$ = complex number radial coordinate of a point outside of the body, $\bar{I}_n$ = complex number coefficients, and $\bar{P}_n(\cos\bar{\psi})$ = complex number Legendre polynomials corresponding to the complex number zenith angle $\bar{\psi}$. The complex number quantities appearing in equation (182) can be written as

$$\bar{V} = Ve^{j\theta_V} \tag{183}$$

$$\bar{a} = ae^{j\theta_a} \tag{184}$$

$$\bar{I}_n = I_n e^{j\theta_{In}} \tag{185}$$

$$\bar{P}_n = P_n e^{j\theta_{Pn}} \tag{186}$$

where, for instance, P_n and θ_{Pn} = magnitude and phase angle of the complex number Legendre polynomials. The real and imaginary parts of equations (182) are given by

$$V\cos\theta_V = -\frac{GM}{r}[\cos\theta_r - I_2P_2(a/r)^2\cos(\theta_{I2} + \theta_{P2} + 2\theta_a - 3\theta_r) - \cdots] \tag{187}$$

$$V\sin\theta_V = \frac{GM}{r}[\sin\theta_r + I_2P_2(a/r)^2\sin(\theta_{I2} + \theta_{P2} + 2\theta_a - 3\theta_r) + \cdots] \tag{188}$$

Equations (187) and (188) can be used to determine V and θ_V. For instance

$$\tan\theta_V = -\frac{\sin\theta_r + I_2P_2(a/r)^2\sin(\theta_{I2} + \theta_{P2} + 2\theta_a - 3\theta_r) + \cdots}{\cos\theta_r - I_2P_2(a/r)^2\cos(\theta_{I2} + \theta_{P2} + 2\theta_a - 3\theta_r) - \cdots} \tag{189}$$

while squaring and adding equations (187) and (188) gives V^2 . In the limit $a/r \rightarrow 0$ equations (187) and (188) become

$$V_m = V_R = V \cos \theta_V = - GM/r \cos \theta_r = - GM/r_m \cos^2 \theta_r \qquad (190)$$

$$V_I = V \sin \theta_V = GM/r \sin \theta_r = GM/r_m \sin \theta_r \cos \theta_r \qquad (191)$$

which corresponds to a point mass whose complex number potential is given by

$$\bar{V} = - GM/\bar{r} \qquad (192)$$

The various terms in the gravitational potential will now be considered.

B. Complex Number Legendre Polynomials.

The appearance of $\bar{P}_n(\cos \bar{\psi})$ in equation (182) needs some explanation. Following the standard prescription for obtaining Legendre polynomials for scalar angles, the following generalizations to complex angles are given[40]

$$\bar{P}_o = 1 \qquad (193)$$

$$\bar{P}_1 = \cos \bar{\psi} \qquad (194)$$

$$\bar{P}_2 = 1/2(3 \cos^2 \bar{\psi} - 1) \qquad (195)$$

$$\bar{P}_3 = 1/2(5 \cos^3 \bar{\psi} - 3 \cos \bar{\psi}) \qquad (196)$$

$$\bar{P}_4 = 1/8(35 \cos^4 \bar{\psi} - 30 \cos^2 \bar{\psi} + 3) \qquad (197)$$

$$\bar{P}_5 = 1/8(63 \cos^5 \bar{\psi} - 70 \cos^3 \bar{\psi} + 15 \cos \bar{\psi}) \qquad (198)$$

$$\bar{P}_6 = 1/16(231 \cos^6 \bar{\psi} - 315 \cos^4 \bar{\psi} + 105 \cos^2 \bar{\psi} - 5) \qquad (199)$$

where[28]

$$\bar{\psi} = \psi e^{j\theta_\psi} \qquad (200)$$

$$\cos \bar{\psi} = C_\psi e^{-j\theta_{c\psi}} \qquad (201)$$

$$C_\psi = [\cos^2(\psi \cos \theta_\psi) + \sinh^2(\psi \sin \theta_\psi)]^{1/2} \qquad (202)$$

$$\tan \theta_{c\psi} = \tan(\psi \cos \theta_\psi) \tanh(\psi \sin \theta_\psi) \qquad (203)$$

From equations (186) and (194) it follows that

$$P_1 = C_\psi \qquad (204)$$

$$\theta_{P1} = - \theta_{c\psi} \qquad (205)$$

Equations (186) and (195) give

$$P_2 \cos \theta_{P2} = 1/2[3C_\psi^2 \cos(2\theta_{c\psi}) - 1] \tag{206}$$

$$P_2 \sin \theta_{P2} = - 3/2\, C_\psi^2 \sin(2\theta_{c\psi}) \tag{207}$$

which gives

$$P_2 = 1/2[9C_\psi^4 - 6C_\psi^2 \cos(2\theta_{c\psi}) + 1]^{1/2} \tag{208}$$

$$\tan \theta_{P2} = \frac{- 3C_\psi^2 \sin(2\theta_{c\psi})}{3C_\psi^2 \cos(2\theta_{c\psi}) - 1} \tag{209}$$

From equation (186) and (196) it follows that

$$P_3 \cos \theta_{P3} = 1/2[5C_\psi^3 \cos(3\theta_{c\psi}) - 3C_\psi \cos \theta_{c\psi}] \tag{210}$$

$$P_3 \sin \theta_{P3} = 1/2[-5C_\psi^3 \sin(3\theta_{c\psi}) + 3C_\psi \sin \theta_{c\psi}] \tag{211}$$

which gives

$$P_3 = 1/2[25C_\psi^6 - 30C_\psi^4 \cos(2\theta_{c\psi}) + 9C_\psi^2]^{1/2} \tag{212}$$

$$\tan \theta_{P3} = \frac{- 5C_\psi^3 \sin(3\theta_{c\psi}) + 3C_\psi \sin \theta_{c\psi}}{5C_\psi^3 \cos(3\theta_{c\psi}) - 3C_\psi \cos \theta_{c\psi}} \tag{213}$$

and so on for P_n and θ_{Pn} .

The complex number Legendre polynomials can also be written in terms of the complimentary angle $\bar{\chi}$ which is defined by

$$\sin \bar{\chi} = \cos \bar{\psi} \tag{214}$$

$$\cos \bar{\chi} = \sin \bar{\psi} \tag{215}$$

where $\bar{\chi}$ = complex number latitude and[28]

$$\sin \bar{\chi} = S_\chi e^{j\theta_{s\chi}} \tag{216}$$

$$\cos \bar{\chi} = C_\chi e^{-j\theta_{c\chi}} \tag{217}$$

and where

$$S_\chi = [\sin^2(\chi \cos \theta_\chi) + \sinh^2(\chi \sin \theta_\chi)]^{1/2} \quad (218)$$

$$C_\chi = [\cos^2(\chi \cos \theta_\chi) + \sinh^2(\chi \sin \theta_\chi)]^{1/2} \quad (219)$$

$$\tan \theta_{s\chi} = \cot(\chi \cos \theta_\chi) \tanh(\chi \sin \theta_\chi) \quad (220)$$

$$\tan \theta_{c\chi} = \tan(\chi \cos \theta_\chi) \tanh(\chi \sin \theta_\chi) \quad (221)$$

The defining relations given by equations (214) and (215) can be written as

$$S_\chi = C_\psi \quad (222)$$

$$C_\chi = S_\psi \quad (223)$$

$$\theta_{s\chi} = -\theta_{c\psi} \quad (224)$$

$$\theta_{s\psi} = -\theta_{c\chi} \quad (225)$$

which relate χ and θ_χ to ψ and θ_ψ .

Combining equation (214) with equations (193) through (199) gives

$$\bar{P}_o = 1 \quad (226)$$

$$\bar{P}_1 = \sin \bar{\chi} \quad (227)$$

$$\bar{P}_2 = 1/2(3 \sin^2 \bar{\chi} - 1) \quad (228)$$

$$\bar{P}_3 = 1/2(5 \sin^3 \bar{\chi} - 3 \sin \bar{\chi}) \quad (229)$$

$$\bar{P}_4 = 1/8(35 \sin^4 \bar{\chi} - 30 \sin^2 \bar{\chi} + 3) \quad (230)$$

$$\bar{P}_5 = 1/8(63 \sin^5 \bar{\chi} - 70 \sin^3 \bar{\chi} + 15 \sin \bar{\chi}) \quad (231)$$

$$\bar{P}_6 = 1/16(231 \sin^6 \bar{\chi} - 315 \sin^4 \bar{\chi} + 105 \sin^2 \bar{\chi} - 5) \quad (232)$$

Note also that

$$P_1 = S_\chi \quad (233)$$

$$\theta_{P1} = \theta_{s\chi} \quad (234)$$

$$P_2 = 1/2[9S_\chi^4 - 6S_\chi^2 \cos(2\theta_{s\chi}) + 1]^{1/2} \quad (235)$$

$$\tan \theta_{P2} = \frac{3S_\chi^2 \sin(2\theta_{s\chi})}{3S_\chi^2 \cos(2\theta_{s\chi}) - 1} \quad (236)$$

$$P_3 = 1/2[25S_\chi^6 - 30S_\chi^4 \cos(2\theta_{s\chi}) + 9S_\chi^2]^{1/2} \tag{237}$$

$$\tan \theta_{P3} = \frac{5S_\chi^3 \sin(3\theta_{s\chi}) - 3S_\chi \sin \theta_{s\chi}}{5S_\chi^3 \cos(3\theta_{s\chi}) - 3S_\chi \cos \theta_{s\chi}} \tag{238}$$

Consider now the special case of the equator and the north pole. At the equator equations (193) through (238) give

$$\chi = 0 \qquad \psi = \pi/2 \qquad \theta_\psi(\pi/2) = 0 \tag{239}$$

$$\theta_{c\chi}(0) = 0 \qquad \theta_{s\chi}(0) = \theta_\chi(0) \qquad \theta_{s\psi}(\pi/2) = 0 \tag{240}$$

$$\theta_{c\psi}(\pi/2) = - \theta_\chi(0) \tag{241}$$

$$S_\chi(0) = 0 \qquad C_\chi(0) = 1 \tag{242}$$

$$S_\psi(\pi/2) = 1 \qquad C_\psi(\pi/2) = 0 \tag{243}$$

$$\begin{array}{lll} P_o = 1 & \theta_{Po} = 0 & \bar{P}_o = 1 \\ P_1 = 0 & \theta_{P1} = \theta_\chi(0) & \bar{P}_1 = 0 \\ P_2 = 1/2 & \theta_{P2} = \pi & \bar{P}_2 = -1/2 \\ P_3 = 0 & \theta_{P3} = \theta_\chi(0) & \bar{P}_3 = 0 \\ P_4 = 3/8 & \theta_{P4} = 0 & \bar{P}_4 = 3/8 \end{array} \tag{244}$$

where $\theta_\chi(0)$ = value of θ_χ at the equator.

At the north pole the following relationships are obtained from equations (193) through (238)

$$\chi = \pi/2 \qquad \psi = 0 \qquad \theta_\chi(\pi/2) = 0 \tag{245}$$

$$\theta_{c\psi}(0) = 0 \qquad \theta_{s\psi}(0) = \theta_\psi(0) \qquad \theta_{s\chi}(\pi/2) = 0 \tag{246}$$

$$\theta_{c\chi}(\pi/2) = - \theta_\psi(0) \tag{247}$$

$$S_\psi(0) = 0 \qquad C_\psi(0) = 1 \tag{248}$$

$$S_\chi(\pi/2) = 1 \qquad C_\chi(\pi/2) = 0 \tag{249}$$

$$
\begin{array}{lll}
P_o = 1 & \theta_{Po} = 0 & \bar{P}_o = 1 \\
P_1 = 1 & \theta_{P1} = 0 & \bar{P}_1 = 1 \\
P_2 = 1 & \theta_{P2} = 0 & \bar{P}_2 = 1 \\
P_3 = 1 & \theta_{P3} = 0 & \bar{P}_3 = 1 \\
P_4 = 1 & \theta_{P4} = 0 & \bar{P}_4 = 1
\end{array} \tag{250}
$$

where $\theta_\psi(0)$ = value of θ_ψ at the north pole.

Equations (214) and (215) are valid for complimentary angles. Combining these conditions with equations (218) through (221) shows that if $\chi = 0$ then $\psi = \pi/2$ and if $\psi = 0$ then $\chi = \pi/2$ so that

$$\theta_\psi(\pi/2) = 0 \tag{251}$$

$$\theta_\chi(\pi/2) = 0 \tag{252}$$

as indicated in equations (239) and (245). This shows that for complementary angles

$$\bar{\psi} + \bar{\chi} = \pi/2 \tag{253}$$

and the right angle $\pi/2$ is not associated with an internal phase angle. The component parts of equation (253) are given by

$$\psi \cos \theta_\psi + \chi \cos \theta_\chi = \pi/2 \tag{254}$$

$$\psi \sin \theta_\psi + \chi \sin \theta_\chi = 0 \tag{255}$$

Equation (254) states that

$$\psi_m + \chi_m = \pi/2 \tag{256}$$

which states that the sum of the measured complementary angles is equal to $\pi/2$.

The angle $\pi/2$ apparently is the only angle whose internal phase angle is zero, all other angles exhibit an internal phase. Consider an angle $\bar{\psi}$ which is composed of two component parts $\bar{\psi}_1$ and $\bar{\psi}_2$ so that

$$\bar{\psi} = \bar{\psi}_1 + \bar{\psi}_2 \tag{257}$$

$$\psi \cos \theta_\psi = \psi_1 \cos \theta_{\psi 1} + \psi_2 \cos \theta_{\psi 2} \tag{258}$$

$$\psi \sin \theta_\psi = \psi_1 \sin \theta_{\psi 1} + \psi_2 \sin \theta_{\psi 2} \tag{259}$$

Equation (258) states that

$$\psi_m = \psi_{1m} + \psi_{2m} \tag{260}$$

which agrees with reality in that the measured total angle is the sum of each measured part. For the special case of $\pi/2$, equations (254) and (255) give

$$\psi = \pi/2 \sin \theta_\chi \,/(\cos \theta_\psi \sin \theta_\chi - \sin \theta_\psi \cos \theta_\chi) \tag{261}$$

$$\chi = - \pi/2 \sin \theta_\psi \,/(\cos \theta_\psi \sin \theta_\chi - \sin \theta_\psi \cos \theta_\chi) \tag{262}$$

Alternatively equations (254) and (255) can be written as

$$\cos \theta_\psi = (1 - \chi^2/\psi^2 \sin^2 \theta_\chi)^{1/2} \tag{263}$$

$$\psi(1 - \chi^2/\psi^2 \sin^2 \theta_\chi)^{1/2} + \chi \cos \theta_\chi = \pi/2 \tag{264}$$

Figure 1 shows the variation of θ_ψ and S_ψ .

C. Complex Number Gravity Potential Coefficients $\bar{I}_n$.

The dimensionless complex number coefficients $\bar{I}_n$ that appear in equation (182) describe the distribution of mass within a planet. Equation (182) shows that $\bar{I}_o = 1$ and $\bar{I}_1 = 0$ because the origin of coordinates can be located at the center of mass of the planet. For practical calculations only the second order coefficient $\bar{I}_2$ is retained. The value of $\bar{I}_2$ is obtained as an obvious generalization of the standard scalar form for this coefficient as follows[35]

$$\bar{I}_2 = I_2 e^{j\theta I2} = (\bar{C} - \bar{A})/(M\bar{a}^2) \tag{265}$$

where $\bar{C}$ = complex number moment of inertia about the polar axis, and $\bar{A}$ = complex number moment of inertia about one of the transverse axes. If the z axis is taken to be the polar axis[35]

$$\bar{C} = Ce^{j\theta C} = \int(\bar{x}^2 + \bar{y}^2)dM \tag{266}$$

$$\bar{A} = Ae^{j\theta A} = \int(\bar{x}^2 + \bar{z}^2)dM = \int(\bar{y}^2 + \bar{z}^2)dM \tag{267}$$

The real and imaginary parts of equation (265) are

$$I_2 \cos \theta_{I2} = 1/(Ma^2)[C \cos(\theta_C - 2\theta_a) - A \cos(\theta_A - 2\theta_a)] \tag{268}$$

$$I_2 \sin \theta_{I2} = 1/(Ma^2)[C \sin(\theta_C - 2\theta_a) - A \sin(\theta_A - 2\theta_a)] \tag{269}$$

From equations (172) and (173) it follows that

$$I_2^2 = 1/(M^2a^4)[C^2 + A^2 - 2AC\cos(\theta_C - \theta_A)] \qquad (270)$$

$$\tan\theta_{I2} = \frac{C\sin(\theta_C - 2\theta_a) - A\sin(\theta_A - 2\theta_a)}{C\cos(\theta_C - 2\theta_a) - A\cos(\theta_A - 2\theta_a)} \qquad (271)$$

The measured values of I_2, C and A are given by

$$I_{2m} = I_2 \cos\theta_{I2} \qquad (272)$$

$$C_m = C\cos\theta_C \qquad (273)$$

$$A_m = A\cos\theta_A \qquad (274)$$

D. Rotational Effects of a Gravitating Planet for Space and Time with Broken Internal Symmetries.

For a rotating planet (or star) the total potential also includes a rotational term, and is written as an obvious generalization to the standard scalar form as follows[35-39]

$$\bar{U} = -GM/\bar{r} + GM/\bar{r}\bar{I}_2\bar{P}_2(\bar{a}/\bar{r})^2 - 1/2\,\bar{r}^2\bar{\omega}^2\cos^2\bar{\chi} \qquad (275)$$

where $\bar{U}$ = total complex number potential and $\bar{\omega}$ = complex number angular speed. The complex number angular speed has already been considered in mechanical problems.[28] Combining equation (275) with equations (216), (217) and (228) gives

$$\bar{U} = -GM/\bar{r} + GM/\bar{r}\bar{I}_2(\bar{a}/\bar{r})^2\,1/2(3S_\chi^2e^{2j\theta_{s\chi}} - 1) - 1/2\,\bar{r}^2\bar{\omega}^2C_\chi^2e^{-2j\theta_{c\chi}} \qquad (276)$$

The total potential can be evaluated at the equator $\chi = 0$ and at the north pole $\chi = \pi/2$ as follows

$$\bar{U}_o = -GM/\bar{a} - GM\bar{I}_2/(2\bar{a}) - 1/2\,\bar{a}^2\bar{\omega}^2 \qquad \text{equator} \qquad (277)$$

$$\bar{U}_o = -GM/\bar{c} + GM\bar{a}^2\bar{I}_2/\bar{c}^3 \qquad \text{north pole} \qquad (278)$$

where $S_\chi(0) = 0$ was used to obtain equation (277), and $S_\chi(\pi/2) = 1$ and $\theta_{s\chi}(\pi/2) = 0$ were used to obtain equation (278), and where $\bar{c} = ce^{j\theta_c}$ = complex number radius at the poles. The potentials in equations (277) and (278) must have the same value for the geoid so that the complex number flattening is given by the following generalization of the standard scalar result[35]

$$\bar{f} = fe^{j\theta_f} = (\bar{a} - \bar{c})/\bar{a} = 3/2\bar{I}_2 + \bar{\omega}^2\bar{a}^3/(2GM) \qquad (279)$$

where it is assumed that $\bar{a} \sim \bar{c}$ to obtain this equation. From equation (279) it follows that

$$f \cos \theta_f = 3/2\ I_2 \cos \theta_{I2} + \omega^2 a^3/(2GM) \cos(2\theta_\omega + 3\theta_a) \qquad (280)$$

$$f \sin \theta_f = 3/2\ I_2 \sin \theta_{I2} + \omega^2 a^3/(2GM) \sin(2\theta_\omega + 3\theta_a) \qquad (281)$$

These equations determine f and θ_f . The measured value of the flattening is $f_m = f \cos \theta_f$.

The real and imaginary parts of equation (275) are

$$U \cos \theta_U = - GM/r \cos \theta_r + GMI_2P_2a^2/r^3 \cos(\theta_{I2} + \theta_{P2} + 2\theta_a - 3\theta_r) \qquad (282)$$

$$- 1/2\ r^2\omega^2C_\chi^2 \cos(2\theta_r + 2\theta_\omega - 2\theta_{c\chi})$$

$$U \sin \theta_U = GM/r \sin \theta_r + GMI_2P_2a^2/r^3 \sin(\theta_{I2} + \theta_{P2} + 2\theta_a - 3\theta_r) \qquad (283)$$

$$- 1/2\ r^2\omega^2C_\chi^2 \sin(2\theta_r + 2\theta_\omega - 2\theta_{c\chi})$$

Equivalently equation (276) can be used to write

$$U \cos \theta_U = - GM/r \cos \theta_r + 3/2\ GMI_2a^2S_\chi^2/r^3 \cos(\theta_{I2}+2\theta_{s\chi}+2\theta_a-3\theta_r) \qquad (284)$$

$$- 1/2\ GMI_2a^2/r^3 \cos(\theta_{I2}+2\theta_a-3\theta_r) - 1/2\ r^2\omega^2C_\chi^2 \cos(2\theta_r+2\theta_\omega-2\theta_{c\chi})$$

$$U \sin \theta_U = GM/r \sin \theta_r + 3/2\ GMI_2a^2S_\chi^2/r^3 \sin(\theta_{I2}+2\theta_{s\chi}+2\theta_a-3\theta_r) \qquad (285)$$

$$- 1/2\ GMI_2a^2/r^3 \sin(\theta_{I2}+2\theta_a-3\theta_r) - 1/2\ r^2\omega^2C_\chi^2 \sin(2\theta_r+2\theta_\omega-2\theta_{c\chi})$$

E. Acceleration of Gravity

The acceleration of gravity of a rotating planet is obtained from equation (275) by

$$\bar{g} = ge^{j\theta_g} = - \partial\bar{U}/\partial\bar{r} \qquad (286)$$

$$= - GM/\bar{r}^2 + 3GM/\bar{r}^2(\bar{a}/\bar{r})^2\bar{I}_2\bar{P}_2 + \bar{r}\bar{\omega}^2 \cos^2 \bar{\chi}$$

The real and imaginary parts of equation (286) are

$$g \cos \theta_g = - GM/r^2 \cos(2\theta_r) + 3GMa^2/r^4I_2P_2 \cos(\theta_{I2}+\theta_{P2}+2\theta_a-4\theta_r) \qquad (287)$$

$$+ r\omega^2C_\chi^2 \cos(\theta_r+2\theta_\omega-2\theta_{c\chi})$$

$$g \sin \theta_g = GM/r^2 \sin(2\theta_r) + 3GMa^2/r^4 I_2 P_2 \sin(\theta_{I2}+\theta_{P2}+2\theta_a-4\theta_r) \tag{288}$$

$$+ r\omega^2 C_\chi^2 \sin(\theta_r+2\theta_\omega-2\theta_{c\chi})$$

Equation (286) can also be written as

$$\bar{g} = - GM/\bar{r}^2 + 3GM/\bar{r}^2(\bar{a}/\bar{r})^2 \bar{I}_2\ 1/2(3S_\chi^2 e^{2j\theta s\chi} - 1) + \bar{r}\bar{\omega}^2 C_\chi^2 e^{-2j\theta c\chi} \tag{289}$$

and therefore

$$g \cos \theta_g = - GM/r^2 \cos(2\theta_r) + 9/2\ GMa^2/r^4 I_2 S_\chi^2 \cos(\theta_{I2}+2\theta_{s\chi}+2\theta_a-4\theta_r) \tag{290}$$

$$- 3/2\ GMa^2/r^4 I_2 \cos(\theta_{I2}+2\theta_a-4\theta_r) + r\omega^2 C_\chi^2 \cos(\theta_r+2\theta_\omega-2\theta_{c\chi})$$

$$g \sin \theta_g = GM/r^2 \sin(2\theta_r) + 9/2\ GMa^2/r^4 I_2 S_\chi^2 \sin(\theta_{I2}+2\theta_{s\chi}+2\theta_a-4\theta_r) \tag{291}$$

$$- 3/2\ GMa^2/r^4 I_2 \sin(\theta_{I2}+2\theta_a-4\theta_r) + r\omega^2 C_\chi^2 \sin(\theta_r+2\theta_\omega-2\theta_{c\chi})$$

The measured acceleration of gravity is given by $g \cos \theta_g$. Equations (287) and (288) or (290) and (291) can be used to determine g and θ_g. These equations can be written in terms of measured quantities by making the substitutions

$$r = r_m/\cos \theta_r \tag{292}$$

$$a = a_m/\cos \theta_a \tag{293}$$

$$c = c_m/\cos \theta_c \tag{294}$$

$$\chi = \chi_m/\cos \theta_\chi \tag{295}$$

$$\omega = \omega_m/\cos \theta_\omega \tag{296}$$

$$I_2 = I_{2m}/\cos \theta_{I2} \tag{297}$$

$$f = f_m/\cos \theta_f \tag{298}$$

where r_m = measured radial coordinate, a_m = measured equatorial radius, c_m = measured polar radius, χ_m = measured latitude, ω_m = measured rotational speed, I_{2m} = measured mass distribution coefficients and f_m = measured flattening. Substituting equations (292) through (298) into equation (287) gives

$$g_m = - GM/r_m^2 \cos(2\theta_r) \cos^2 \theta_r \tag{299}$$

$$+ (3GMa_m^2 I_{2m} P_2 \cos^4 \theta_r)/(r_m^4 \cos^2 \theta_a \cos \theta_{I2}) \cos(\theta_{I2}+\theta_{P2}+2\theta_a-4\theta_r)$$

$$+ (r_m \omega_m^2 C_\chi^2)/(\cos \theta_r \cos^2 \theta_\omega) \cos(\theta_r+2\theta_\omega-2\theta_{c\chi})$$

Equivalently, substituting equations (292) through (298) into equation (290) gives the measured acceleration of gravity as

$$g_m = - GM/r_m^2 \cos(2\theta_r) \cos^2 \theta_r \tag{300}$$

$$+ (9GMa_m^2 I_{2m} S_\chi^2 \cos^4 \theta_r)/(2r_m^4 \cos^2 \theta_a \cos \theta_{I2}) \cos(\theta_{I2}+2\theta_{s\chi}+2\theta_a-4\theta_r)$$

$$- (3GMa_m^2 I_{2m} \cos^4 \theta_r)/(2r_m^4 \cos^2 \theta_a \cos \theta_{I2}) \cos(\theta_{I2}+2\theta_a-4\theta_r)$$

$$+ (r_m \omega_m^2 C_\chi^2)/(\cos \theta_r \cos^2 \theta_\omega) \cos(\theta_r+2\theta_\omega-2\theta_\chi)$$

From equations (218) and (219) it follows that S_χ and C_χ can be expressed in terms of the measured latitude by

$$S_\chi = [\sin^2 \chi_m + \sinh^2(\chi_m \tan \theta_\chi)]^{1/2} \tag{301}$$

$$C_\chi = [\cos^2 \chi_m + \sinh^2(\chi_m \tan \theta_\chi)]^{1/2} \tag{302}$$

For small θ_χ it follows that

$$S_\chi \sim \sin \chi_m \tag{303}$$

$$C_\chi \sim \cos \chi_m \tag{304}$$

From equations (220) and (221) it follows that

$$\tan \theta_{s\chi} = \cot \chi_m \tanh(\chi_m \tan \theta_\chi) \tag{305}$$

$$\tan \theta_{c\chi} = \tan \chi_m \tanh(\chi_m \tan \theta_\chi) \tag{306}$$

F. Acceleration of Gravity on the Geoid

The approximate shape of the geoid for a planet with broken internal symmetries is written as a simple generalization of the standard scalar expression as follows[35]

$$\bar{r} = \bar{a}(1 - \bar{f} \sin^2 \bar{\chi}) \tag{307}$$

From equation (307) it follows that

$$\bar{r}^{-2} = \bar{a}^{-2}(1 + 2\bar{f} \sin^2 \bar{\chi} + \cdots) \tag{308}$$

Using equation (308) in the first term of equation (286) gives for the acceleration on the geoid

$$\bar{g} = - GM/\bar{a}^2(1 + 2\bar{f} \sin^2 \bar{\chi}) + 3GM/\bar{a}^2\bar{J}_2\bar{P}_2 + \bar{a}\bar{\omega}^2 \cos^2 \bar{\chi} \tag{309}$$

The real and imaginary parts of equation (309) are written as

$$g \cos \theta_g = - GM/a^2 \cos(2\theta_a) - 2GMfS_\chi^2/a^2 \cos(\theta_f+2\theta_{s\chi}-2\theta_a) \quad (310)$$

$$+ 3GMI_2P_2/a^2 \cos(\theta_{I2}+\theta_{P2}-2\theta_a) + a\omega^2C_\chi^2 \cos(\theta_a+2\theta_\omega-2\theta_{c\chi})$$

$$= - GM/a^2 \cos(2\theta_a) - 2GMfS_\chi^2/a^2 \cos(\theta_f+2\theta_{s\chi}-2\theta_a)$$

$$+ 9/2GMI_2S_\chi^2/a^2 \cos(\theta_{I2}+2\theta_{s\chi}-2\theta_a) - 3/2GMI_2/a^2 \cos(\theta_{I2}-2\theta_a)$$

$$+ a\omega^2C_\chi^2 \cos(\theta_a+2\theta_\omega-2\theta_{c\chi})$$

$$g \sin \theta_g = GM/a^2 \sin(2\theta_a) - 2GMfS_\chi^2/a^2 \sin(\theta_f+2\theta_{s\chi}-2\theta_a) \quad (311)$$

$$+ 3GMI_2P_2/a^2 \sin(\theta_{I2}+\theta_{P2}-2\theta_a) + a\omega^2C_\chi^2 \sin(\theta_a+2\theta_\omega-2\theta_{c\chi})$$

$$= GM/a^2 \sin(2\theta_a) - 2GMfS_\chi^2/a^2 \sin(\theta_f+2\theta_{s\chi}-2\theta_a)$$

$$+ 9/2GMI_2S_\chi^2/a^2 \sin(\theta_{I2}+2\theta_{s\chi}-2\theta_a) - 3/2GMI_2/a^2 \sin(\theta_{I2}-2\theta_a)$$

$$+ a\omega^2C_\chi^2 \sin(\theta_a+2\theta_\omega-2\theta_{c\chi})$$

Equations (310) and (311) can be written in terms of measured quantities by using equations (292) through (298). The measured acceleration on the geoid is given by equation (310).

The acceleration of gravity at the equator is obtained by using equation (309) with equations (239) through (244) that describe the equator, with the result that

$$\bar{g}_e = - GM/\bar{a}^2 - 3/2GM\bar{I}_2/\bar{a}^2 + \bar{a}\bar{\omega}^2 \quad (312)$$

The real and imaginary parts of equation (312) give

$$g_e \cos \theta_{ge} = - GM/a^2 \cos(2\theta_a) - 3/2GMI_2/a^2 \cos(\theta_{I2}-2\theta_a) + a\omega^2 \cos(\theta_a+2\theta_\omega) \quad (313)$$

$$g_e \sin \theta_{ge} = GM/a^2 \sin(2\theta_a) - 3/2GMI_2/a^2 \sin(\theta_{I2}-2\theta_a) + a\omega^2 \sin(\theta_a+2\theta_\omega) \quad (314)$$

from which g_e and θ_{ge} can be determined. Combining equations (279), (309) and (312) and neglecting higher order terms gives

$$\bar{g} = \bar{g}_e[1 + (5/2\bar{m} - \bar{f})\sin^2 \bar{\chi}] \quad (315)$$

where

$$\bar{m} = me^{j\theta_m} = \bar{\omega}^2\bar{a}^3/(GM) \qquad (316)$$

and where $\bar{m}$ is related to the complex number flattening $\bar{f}$ by equation (279) which can be rewritten as

$$\bar{f} = 3/2\bar{I}_2 + \bar{m}/2 \qquad (317)$$

Equation (315) is the complex number generalization of Clairut's equation.[35] Equation (315) can be written as

$$g \cos \theta_g = g_e \cos \theta_{ge} + 5/2 g_e m S_\chi^2 \cos(\theta_{ge}+\theta_m+2\theta_{s\chi}) - g_e f S_\chi^2 \cos(\theta_{ge}+\theta_f+2\theta_{s\chi}) \qquad (318)$$

$$g \sin \theta_g = g_e \sin \theta_{ge} + 5/2 g_e m S_\chi^2 \sin(\theta_{ge}+\theta_m+2\theta_{s\chi}) - g_e f S_\chi^2 \sin(\theta_{ge}+\theta_f+2\theta_{s\chi}) \qquad (319)$$

G. Apparent Non-Newtonian Effects

The measured acceleration of gravity given by equation (299) has to be compared to the conventionally calculated acceleration of gravity in order to estimate the magnitude of the discrepancy. The conventionally calculated acceleration of gravity is just the scalar form of equation (286) in which the measured distances and angles appear as follows[35]

$$\begin{aligned} g^c &= - GM/r_m^2 + 3GMa_m^2/r_m^4 I_{2c} P_{2c} + r_m \omega_m^2 \cos^2 \chi_m \qquad (320) \\ &= - GM/r_m^2 + 9/2GMa_m^2/r_m^4 I_{2c} \sin^2 \chi_m \\ &\quad - 3/2GMa_m^2/r_m^4 I_{2c} + r_m \omega_m^2 \cos^2 \chi_m \end{aligned}$$

where the conventionally calculated second order Legendre polynomial is written as[29]

$$P_{2c} = 1/2(3 \sin^2 \chi_m - 1) \qquad (321)$$

The conventionally calculated mass distribution coefficient I_{2c} is similar to equation (265) in that[35]

$$I_{2c} = (C_c - A_c)/(Ma_m^2) \qquad (322)$$

where the conventional moments of inertia are given by[35]

$$C_c = \int (x_m^2 + y_m^2)dM \qquad (323)$$

$$A_c = \int (x_m^2 + z_m^2)dM = \int (y_m^2 + z_m^2)dM \qquad (324)$$

Thus the conventional calculations are done using the measured coordinates of geodesy r_m , x_m , y_m , z_m and χ_m .

Comparing equation (320) with equation (299) gives

$$g_m - g_c = GM/r_m^2[1 - \cos(2\theta_r)\cos^2\theta_r] \tag{325}$$

$$+ \frac{3MGa_m^2}{r_m^4}\left[\frac{\cos^4\theta_r}{\cos^2\theta_a} I_2P_2 \cos(\theta_{I2}+\theta_{P2}+2\theta_a-4\theta_r) - I_{2c}P_{2c}\right]$$

$$+ r_m\omega_m^2\left[\frac{C_\chi^2}{\cos\theta_r\cos^2\theta_\omega}\cos(\theta_r+2\theta_\omega-2\theta_{c\chi}) - \cos^2\chi_m\right]$$

where P_2 and θ_{P2} are expressed in terms of the measured latitude χ_m by using equations (235), (236), (301), (302) and (306). In a similar fashion combining equation (300) and (320) gives

$$g_m - g_c = GM/r_m^2[1 - \cos(2\theta_r)\cos^2\theta_r] \tag{326}$$

$$+ \frac{9GMa_m^2}{2r_m^4}\left[\frac{\cos^4\theta_r}{\cos^2\theta_a} I_2S_\chi^2 \cos(\theta_{I2}+2\theta_{s\chi}+2\theta_a-4\theta_r) - I_{2c}\sin^2\chi_m\right]$$

$$- \frac{3GMa_m^2}{2r_m^4}\left[\frac{\cos^4\theta_r}{\cos^2\theta_a} I_2 \cos(\theta_{I2}+2\theta_a-4\theta_r) - I_{2c}\right]$$

$$+ r_m\omega_m^2\left[\frac{C_\chi^2}{\cos\theta_r\cos^2\theta_\omega}\cos(\theta_r+2\theta_\omega-2\theta_{c\chi}) - \cos^2\chi_m\right]$$

The first term in equations (325) or (326) gives the dominant effect of the broken symmetry of space on the discrepancy between the measured and conventionally calculated values of the Newtonian acceleration of gravity. From these two equations the parameters D given by equation (69) and D_2 given by equation (85) can be calculated.

5. MINE SHAFT, BOREHOLE, TOWER AND EÖTVÖS EXPERIMENTS. This section considers the apparent deviations from Newtonian gravity that have recently been reported in the literature.[1-25] These discrepancies have been found in laboratory Eötvös experiments where the validity of Newton's gravitation law is examined over short ranges for deviations from the inverse square law and to detect a possible dependence on the composition (baryon number) of the attracting masses.[6-15] Deviations from the inverse square law have also been found in the measurement of the acceleration of gravity over vertical distances of hundreds of meters in mine shaft, borehole and tower experiments.[1-5,24,25] An analysis of these apparent discrepancies is given in this section which is based on the broken symmetry of space that is induced by a pressure field.[28] A spherical earth assumption is made for the calculations done here so that the acceleration of gravity and the effective radial gravitational constant are given by equations (64) and (75) respectively in terms of the internal phase angle θ_r of the radial coordinates.

A. **Small Argument Approximation to the Equilibrium Equation for the Internal Phase Angles of the Radial Coordinates.**

For the case where θ_P varies slowly with radial distance, the following approximations can be written for equation (57)

$$\theta_P + P\frac{\partial\theta_P/\partial r}{\partial P/\partial r} = -2\theta_r \tag{327}$$

The solution of equation (327) determines $\theta_r(r)$ in terms of $P(r)$ and $\theta_P(r)$ and hence by equations (64) and (75) the acceleration of gravity and the effective gravitational constant are also obtained. As a simple example consider the case

$$P = P(0)e^{-\alpha r} \tag{328}$$

$$\theta_P = \theta_P(0)e^{-\beta r} \tag{329}$$

where $P(0)$ and $\theta_P(0)$ = pressure and its internal phase angle at the center of the earth. Combining equations (327) through (329) gives

$$\theta_r = -1/2(\alpha+\beta)/\alpha\theta_P(0)e^{-\beta r} \tag{330}$$

$$= -1/2(\alpha+\beta)/\alpha\theta_P$$

For the center of the earth $r = 0$ and equation (330) gives

$$\theta_r(0) = -1/2(\alpha+\beta)/\alpha\theta_P(0) \tag{331}$$

At the earth's surface $r = R$ and

$$\theta_r(R) = -1/2(\alpha+\beta)/\alpha\theta_P(0)e^{-\beta R} \tag{332}$$

$$= -1/2(\alpha+\beta)/\alpha\theta_P(R)$$

where $\theta_P(R)$ = internal phase angle of pressure at the earth's surface given by

$$\theta_P(R) = \theta_P(0)e^{-\beta R} \tag{333}$$

Note also that the pressure at the earth's surface is given by

$$P(R) = P(0)e^{-\alpha R} \tag{334}$$

Equations (333) and (334) can be used to evaluate α and β .

For the case of a linear variation of the pressure and its internal phase angle of the form

$$P = P(0) - \alpha r \tag{335}$$

$$\theta_P = \theta_P(0) - \beta r \tag{336}$$

it follows from equation (327) that

$$\theta_r = -1/2[\theta_p(0) + \beta/\alpha P(0)] + \beta r \quad (337)$$

The values of α and β can be obtained by evaluating equations (335) and (336) at the earth's surface as follows

$$P(R) = P(0) - \alpha R \quad (338)$$

$$\theta_p(R) = \theta_p(0) - \beta R \quad (339)$$

At the center of the earth equation (337) gives

$$\theta_r(0) = -1/2[\theta_p(0) + \beta/\alpha P(0)] \quad (340)$$

while for the surface of the earth

$$\theta_r(R) = -1/2[\theta_p(0) + \beta/\alpha P(0)] + \beta R \quad (341)$$

Equations (330) and (337) show that in general $\theta_r < 0$ within a gravitating body.

B. Theory of the Apparent Non-Newtonian Behaviour of Gravity in Mine Shaft, Borehole and Tower Experiments.

Measurements of the variation of the acceleration of gravity up the heights of a tower and down the depths of a mine shaft or borehole have indicated discrepancies with the inverse square law of Newtonian gravity. A possible explanation of these discrepancies has been given by assuming the validity of Newtonian gravitation in matter with broken internal symmetries.[28] The result is that the acceleration of gravity for a spherical earth is given by equation (64). In order to apply this equation to an analysis of mine shaft, borehole and tower gravity measurements it is first necessary to calculate the internal phase angle θ_r from equation (327). Let the coordinates measured up a tower from the earth's surface be designated by h, so that the distance from the center of the earth to a point on the tower is given by

$$r = R + h \quad (342)$$

where R = magnitude of the earth's radius at the base of the tower. Equation (342) applies to a mine shaft or borehole if $h < 0$. Combining equations (327) and (342) gives

$$\theta_p + P\frac{\partial\theta_p/\partial h}{\partial P/\partial h} = -2\theta_r \quad (343)$$

as the equation for determining θ_r.

The magnitude of the atmospheric pressure at points on a tower, mine shaft or borehole can be written in its simplest form by the following linear equation

$$P = P(R) - \rho(R)g(R)h = \rho(R)g(R)(h^a - h) \quad (344)$$

where the equivalent height of the atmosphere is given by

$$h^a = P(R)/[\rho(R)g(R)] \quad (345)$$

and where $P(R)$, $\rho(R)$ and $g(R)$ = magnitudes of the pressure, air density and acceleration of gravity respectively at the earth's surface. The measured values of these quantities are given by $P_m(R) = P(R)\cos\theta_P(R)$, $\rho_m(R) = \rho(R)\cos\theta_\rho(R)$ and $g_m(R) = g(R)\cos\theta_g(R)$ respectively. The measured pressure is given by $P_m = P\cos\theta_P$. The internal phase angle of the pressure will be written in a form similar to equation (344) as follows

$$\theta_P = \theta_P(R) - \eta h \tag{346}$$

Equations (344) and (346) are the simplest equations that can be chosen to describe the variation with height (or depth) of the atmospheric pressure and its internal phase angle. Strictly speaking P , θ_P and θ_r should be determined simultaneously from equations (56) and (57) and the renormalized state equation which is given by a solution of the complex number relativistic trace equation (2). Such a simultaneous solution is difficult to obtain. Equations (344) and (346) represent a crude solution to equations (2), (56) and (57). These assumed solutions will now be used to obtain θ_r from equation (343). Combining equations (343), (344) and (346) gives

$$\theta_r = -1/2\theta_P(R) - 1/2\eta(h^a - 2h) \tag{347}$$

$$= -1/2[\theta_P(R) + \eta h^a] + \eta h$$

At the earth's surface

$$\theta_r(R) = -1/2[\theta_P(R) + \eta h^a] \tag{348}$$

Consider now the case where the pressure and its internal phase vary according to the following exponential forms

$$P = P(R)e^{-\delta h} \tag{349}$$

$$\theta_P = \theta_P(R)e^{-\kappa h} \tag{350}$$

Combining equations (343), (349) and (350) gives

$$\theta_r = -1/2(\delta + \kappa)/\delta\theta_P(R)e^{-\kappa h} \tag{351}$$

and the value at the earth's surface is

$$\theta_r(R) = -1/2(\delta + \kappa)/\delta\theta_P(R) \tag{352}$$

The values of θ_r determine the apparent deviation of the acceleration of gravity from Newton's law of gravity as is shown in equations (64) and (75). Figures 2 and 3 show sketches of the variation of P and θ_P for the solid earth, ocean and for the atmosphere in an air-filled mine shaft or borehole or adjacent to a tower. The expected variation of θ_r in the solid earth, ocean and atmosphere is shown in Figure 4, while Figure 5 shows the corresponding variation of G_r as given by equation (75). Figure 5 shows that local measurements of the acceleration of gravity will yield values of G_r which are less than the value of G. The value of G is associated with the complete symmetry of time and space, and as such it cannot be measured directly. The result $G_r < G$ is due to the effects of the complex number atmospheric pressure in the case of mine shaft, borehole and tower experiments, and to the complex number water pressure for measurements of G_r carried out in the depths of the ocean. For measurements of the variation of G_r in a mine shaft, borehole or up a tower the characteristic range for the variation of G_r should be about 7 km because the atmospheric pressure decrease with height has a characteristic attenuation distance of about 7 km.[43]

Equation (75) and Figure 5 also show that were it possible to measure the variation of the acceleration of gravity with depth in solid rock the values for G_r would be less than those measured in an air-filled mine shaft or in the ocean. This is because P , θ_P and $|\theta_r|$ are larger in rock than in the ocean or in an air-filled mine shaft at a corresponding depth. Measurements of G_r in the ocean should yield weaker gravity (smaller G_r) than corresponding measurements in an air-filled mine shaft or borehole because values of $|\theta_r|$ in the ocean are larger than their corresponding values in an air-filled mine shaft or borehole at the same depth (see equation (75) and Figure 4).

C. Internal Phase Theory of the Eötvös Experiment and its Relationship to Mine Shaft, Borehole and Tower Experiments.

This part of the paper describes a theoretical analysis of the Eötvös experiment in terms of Newtonian gravity and the broken symmetry internal phase angles of the relevant coordinates of the experiment. The Eötvös experiment has been thoroughly described in the literature and only the briefest review is given in this paragraph.[6-15] This experiment measures the horizontal force of gravity between two spheres of material that are suspended in close proximity to each other. Conventional Newtonian theory predicts the measured gravity force to be dependent on the inverse square of the separation distance and on the product of the masses of the two spheres, but recent experiments suggest the possibility of composition dependent effects and deviations from the inverse square law.[6-15]

Consider now the Eötvös experiment from the perspective of the internal phase theory of coordinates. The two spheres can be oriented in any direction between the north-south direction (whose separation is described by a decrement of the zenith angle) or in the east-west direction (whose separation is then described by a decrement of the azimuthal angle). For the north-south orientation the complex number distance between the two spheres is written as

$$d\bar{\ell}_\psi = d\ell_\psi e^{j\theta_{\ell\psi}} = \bar{r}d\bar{\psi} = \bar{r}e^{j\theta_\psi}(d\psi + j\psi d\theta_\psi) \qquad (353)$$

which gives

$$d\ell_\psi = r \sec \beta_{\psi,\psi}\, d\psi \qquad (354)$$

$$\theta_{\ell\psi} = \theta_r + \theta_\psi + \beta_{\psi,\psi} \qquad (355)$$

where $\bar{r} = \bar{R} + \bar{h}$ = complex number distance of the two spheres from the center of the earth, $\bar{R}$ = complex number earth's radius at the position of the two spheres, $\bar{h}$ = complex number distance above (or below) the earth's surface at which the

Eötvös experiment is conducted, $d\bar{\psi}$ = complex number zenith angle separation of the two spheres situated in the north-south direction (longitudinal plane) and $\beta_{\psi,\psi}$ is given by equation (140). The measured distance between the two spheres situated in the north-south orientation is given by

$$d\ell_{\psi m} = d\ell_{\psi} \cos \theta_{\ell\psi} \tag{356}$$

For the east-west orientation the complex number distance between the two spheres is

$$d\bar{\ell}_{\phi} = d\ell_{\phi} e^{j\theta_{\ell\phi}} = \bar{r} \sin \bar{\psi} \, d\bar{\phi} \tag{357}$$

where $d\bar{\phi}$ = complex number azimuthal angle separation of the two spheres that are situated in a plane of latitude $\bar{\chi} = \pi/2 - \bar{\psi}$, and therefore

$$d\ell_{\phi} = rS_{\psi} \sec \beta_{\phi,\phi} \, d\phi \tag{358}$$

$$\theta_{\ell\phi} = \theta_r + \theta_{s\psi} + \theta_{\phi} + \beta_{\phi,\phi} \tag{359}$$

where S_{ψ} and $\theta_{s\psi}$ are given by equations (143) and (144) respectively, and where $\beta_{\phi,\phi}$ is given by equation (141). The measured distance between the two spheres situated in the east-west direction is given by

$$d\ell_{\phi m} = d\ell_{\phi} \cos \theta_{\ell\phi} \tag{360}$$

The gravitational force between the two spheres situated in the north-south direction is

$$\bar{F}_{\psi} = - Gm^2/(d\bar{\ell}_{\psi})^2 = - Gm^2/(d\ell_{\psi})^2 e^{-2j\theta_{\ell\psi}} \tag{361}$$

where m = mass of one sphere. The measured gravitational force between the two spheres in the north-south direction is

$$\begin{aligned} F_{\psi m} &= - Gm^2/(d\ell_{\psi})^2 \cos(2\theta_{\ell\psi}) \\ &= - Gm^2/(d\ell_{\psi m})^2 \cos(2\theta_{\ell\psi}) \cos^2 \theta_{\ell\psi} \end{aligned} \tag{362}$$

The conventional calculation of the Newtonian gravitational force between the two Eötvös spheres is given by

$$F_{\psi c} = - Gm^2/(d\ell_{\psi m})^2 \tag{363}$$

Therefore the difference between the measured and conventionally predicted forces in the north-south direction is

$$\begin{aligned} \Delta F_{\psi} &= F_{\psi m} - F_{\psi c} \\ &= GM^2/(d\ell_{\psi m})^2 [1 - \cos^2 \theta_{\ell\psi} \cos(2\theta_{\ell\psi})] \\ &\sim 3\theta_{\ell\psi}^2 Gm^2/(d\ell_{\psi m})^2 \end{aligned} \tag{364}$$

In a similar fashion the complex number gravitational force and the measured force between the two spheres situated in the east-west orientation are given respectively by

$$\bar{F}_{\phi} = - Gm^2/(d\bar{\ell}_{\phi})^2 = - Gm^2/(d\ell_{\phi})^2 e^{-2j\theta_{\ell\phi}} \tag{365}$$

$$F_{\phi m} = - Gm^2/(d\ell_{\phi})^2 \cos(2\theta_{\ell\phi}) \tag{366}$$

$$= - Gm^2/(d\ell_{\phi m})^2 \cos(2\theta_{\ell\phi}) \cos^2 \theta_{\ell\phi}$$

The conventionally calculated gravitational force is given by

$$F_{\phi c} = - Gm^2(d\ell_{\phi m})^2 \tag{367}$$

and the difference between equations (366) and (367) is

$$\Delta F_{\phi} = F_{\phi m} - F_{\phi c} \tag{368}$$

$$= Gm^2/(d\ell_{\phi m})^2[1 - \cos^2 \theta_{\ell\phi} \cos(2\theta_{\ell\phi})]$$

$$\sim 3\theta^2_{\ell\phi} Gm^2/(d\ell_{\phi m})^2$$

A measurement of the discrepancies between the measured and predicted values of the gravity force for the north-south and east-west orientations of the Eötvös experiment will give values of $\theta_{\ell\psi}$ and $\theta_{\ell\phi}$.

From equations (75), (362) and (366) it follows that there are three effective gravitational constants each associated with a direction (r , ψ or ϕ) of measurement of the gravitational force, so that

$$G_r = G \cos(2\theta_r) \cos^2 \theta_r \sim G(1 - 3\theta_r^2 + \cdots) \tag{369}$$

$$G_{\psi} = G \cos(2\theta_{\ell\psi}) \cos^2 \theta_{\ell\psi} \sim G(1 - 3\theta^2_{\ell\psi} + \cdots) \tag{370}$$

$$G_{\phi} = G \cos(2\theta_{\ell\phi}) \cos^2 \theta_{\ell\phi} \sim G(1 - 3\theta^2_{\ell\phi} + \cdots) \tag{371}$$

Due to the internal phase structure of the coordinates the effective Newtonian gravitational constant has three distinct values along the three orthogonal directions at a point on the earth's surface. Because $|\theta_{\psi}|$ and $|\theta_{\phi}|$ are expected to be smaller than $|\theta_r|$ it follows from equations (355) and (359) that for the same height (or depth)

$$\left.\begin{array}{l} |\theta_{\ell\phi}| < |\theta_{\ell\psi}| < |\theta_r| \\ G_{\phi} > G_{\psi} > G_r \end{array}\right\} \theta_r < 0 , \theta_{\phi} > 0 , \theta_{\psi} > 0 \tag{372}$$

$$|\theta_{\ell\phi}| > |\theta_{\ell\psi}| > |\theta_r| \quad G_\phi < G_\psi < G_r \quad \Big\} \quad \theta_r < 0 \ , \ \theta_\phi < 0 \ , \ \theta_\psi < 0 \qquad (373)$$

where for both cases $\theta_{\ell\phi} < 0$ and $\theta_{\ell\psi} < 0$. According to this theory, depending on the signs of θ_ϕ and θ_ψ the values of G_ψ and G_ϕ measured by the Eötvös experiment can be greater or less than the value of G_r determined by gravimeter measurements in a mine shaft, borehole or tower experiment. References 1 through 3 suggest that equation (373) are the correct conditions while references 5 and 25 suggest that equation (372) gives the correct relationship between G_ϕ , G_ψ and G_r . The experimental results are not yet clear enough to decide between $\theta_\phi > 0$ and $\theta_\psi > 0$ or $\theta_\phi < 0$ and $\theta_\psi < 0$.

On account of equations (369) through (371) it follows that G_ψ and G_ϕ measured by an Eötvös experiment should have a similar variation with depth (or height) as does G_r . This is shown in Figure 5. The validity of equations (370) and (371) can possibly be tested by conducting Eötvös experiments in the depths of the ocean, down mine shafts, or up a tower in order to see if G_ψ and G_ϕ vary in the same sense as G_r . For a tower or air-filled mine shaft Eötvös experiment the characteristic length over which G_ψ and G_ϕ change should be about 7 km because this is the characteristic variation distance of the atmospheric pressure in the vertical direction.[43] The characteristic distance for the decrease of G_ψ and G_ϕ with depth in the ocean (or solid earth if such experiments were possible) should be much larger than 7 km because the pressure changes in these cases are over hundreds and thousands of kilometers.[35-40,43-45] Another possible test would be to perform the Eötvös experiment in a pressure chamber and measure the pressure dependence of G_ψ and G_ϕ in order to verify that G_ψ and G_ϕ are decreasing functions of the ambient pressure as suggested by equations (370) and (371). In any case, equations (369) through (371) show that the local measurements of gravity do not directly determine the Newtonian gravitational constant G . Approximate values of G can be determined directly from satellite or solar system measurements where the effects of ambient pressure are negligible, but in this case the values θ_r^v , θ_ψ^v and θ_ϕ^v of the broken symmetry vacuum must be taken into consideration. Thus even for measurements in the vacuum G cannot be directly measured.

Consider the variation of the gravitational constant G_r given by equation (369) from which it follows that

$$G_r(R-h) - G_r(R) \sim 3G[\theta_r^2(R) - \theta_r^2(R-h)] \qquad (374)$$

$$G_r(R+h) - G_r(R) \sim 3G[\theta_r^2(R) - \theta_r^2(R+h)] \qquad (375)$$

A Taylor series expansion of θ_r gives

$$\theta_r(R-h) = \theta_r(R) - h\partial\theta_r/\partial h + \cdots \qquad (376)$$

$$\theta_r(R+h) = \theta_r(R) + h\partial\theta_r/\partial h + \cdots \qquad (377)$$

Combining equations (374) through (377) gives for small h

$$[G_r(R-h) - G_r(R)]/G \sim - s_r h \tag{378}$$

$$[G_r(R+h) - G_r(R)]/G \sim s_r h \tag{379}$$

where

$$s_r = 6|\theta_r(R)| \; \partial\theta_r/\partial r|_R > 0 \tag{380}$$

remembering that $\theta_r < 0$. Combining equations (378) and (379) gives

$$G_r(R-h) < G_r(R) < G_r(R+h) \tag{381}$$

as shown in Figure 5. Note that $G_r(R)$ is <u>not</u> the value of the gravitational constant that is measured at the earth's surface by the Eötvös experiment. The value of the gravitational constant measured by the Eötvös experiment is given by $G_\psi(R)$ or $G_\phi(R)$.

The Eötvös experiment can be done at various depths and heights. From equations (370) and (371) it follows that

$$G_\psi(R\pm h) - G_\psi(R) \sim 3G[\theta^2_{\ell\psi}(R) - \theta^2_{\ell\psi}(R\pm h)] \tag{382}$$

$$G_\phi(R\pm h) - G_\phi(R) \sim 3G[\theta^2_{\ell\phi}(R) - \theta^2_{\ell\phi}(R\pm h)] \tag{383}$$

A Taylor series is used to obtain

$$\theta_{\ell\psi}(R\pm h) = \theta_{\ell\psi}(R) \pm h\partial\theta_{\ell\psi}/\partial r|_R + \cdots \tag{384}$$

$$\theta_{\ell\phi}(R\pm h) = \theta_{\ell\phi}(R) \pm h\partial\theta_{\ell\phi}/\partial r|_R + \cdots \tag{385}$$

Combining equations (382) through (385) gives for small h

$$[G_\psi(R\pm h) - G_\psi(R)]/G \sim \pm s_\psi h \tag{386}$$

$$[G_\phi(R\pm h) - G_\phi(R)]/G \sim \pm s_\phi h \tag{387}$$

where

$$s_\psi = 6|\theta_{\ell\psi}(R)| \; \partial\theta_{\ell\psi}/\partial r|_R > 0 \tag{388}$$

$$s_\phi = 6|\theta_{\ell\phi}(R)| \; \partial\theta_{\ell\phi}/\partial r|_R > 0 \tag{389}$$

because $\theta_{\ell\psi} < 0$ and $\theta_{\ell\phi} < 0$. Therefore

$$G_\psi(R-h) < G_\psi(R) < G_\psi(R+h) \tag{390}$$

$$G_\phi(R-h) < G_\phi(R) < G_\phi(R+h) \tag{391}$$

Thus G_ψ and G_ϕ are increasing functions of height.

The Eötvös experiments done at the same height but in the east-west and north-south directions give the following difference obtained from equations

(370) and (371)

$$(G_\psi - G_\phi)/G \sim 3(\theta_{\ell\phi}^2 - \theta_{\ell\psi}^2) \tag{392}$$
$$= 3[(\theta_r + \theta_{s\psi} + \theta_\phi + \beta_{\phi,\phi})^2 - (\theta_r + \theta_\psi + \beta_{\psi,\psi})^2]$$
$$\sim 6\theta_r(\theta_{s\psi} + \theta_\phi + \beta_{\phi,\phi} - \theta_\psi - \beta_{\psi,\psi})$$
$$\sim 6\theta_r\theta_\phi$$

Therefore

$$G_\phi > G_\psi \qquad \theta_\phi > 0 \tag{393}$$
$$G_\phi < G_\psi \qquad \theta_\phi < 0 \tag{394}$$

The measurement of the east-west/north-south asymmetry will give the sign (and approximate magnitude if θ_r is known) of the internal phase angle of the angular coordinates.

In general the Eötvös experiment is done at the earth's surface and determines $G_\psi(R)$ and $G_\phi(R)$, while the vertical gravity measurements using gravimeters are done down mine shafts and boreholes or up towers and determine $G_r(R\pm h)$. Consider a comparison of $G_r(R\pm h)$ and $G_\psi(R)$ which can be obtained using equations (355), (369) and (370)

$$[G_r(R\pm h) - G_\psi(R)]/G \sim 3[\theta_{\ell\psi}^2(R) - \theta_r^2(R\pm h)] \tag{395}$$
$$= 3\{[\theta_r(R) + \theta_\psi(R) + \beta_{\psi,\psi}(R)]^2 - \theta_r^2(R\pm h)\}$$

Combining equations (376), (377) and (395) gives

$$[G_r(R\pm h) - G_\psi(R)]/G \sim \pm s_r h + 6\theta_r(R)[\theta_\psi(R) + \beta_{\psi,\psi}(R)] + 3[\theta_\psi(R) + \beta_{\psi,\psi}(R)]^2$$
$$\sim \pm s_r h + 6\theta_r(R)[\theta_\psi(R) + \beta_{\psi,\psi}(R)] \tag{396}$$

From equation (396) it follows that

$$[G_r(R) - G_\psi(R)]/G \sim 6\theta_r(R)[\theta_\psi(R) + \beta_{\psi,\psi}(R)] + 3[\theta_\psi(R) + \beta_{\psi,\psi}(R)]^2 \tag{397}$$
$$\sim 6\theta_r(R)[\theta_\psi(R) + \beta_{\psi,\psi}(R)]$$

It follows from equations (381), (396) and (397) that

$$G_r(R-h) < G_r(R) < G_r(R+h) < G_\psi(R) < G \qquad \theta_\psi(R) > 0 \tag{398}$$
$$G_\psi(R) < G_r(R-h) < G_r(R) < G_r(R+h) < G \qquad \theta_\psi(R) < 0 \tag{399}$$

The inequalities in equations (398) and (399) hold only for small h.

Now consider the case of the east-west oriented Eötvös experiment. Combining equations (359), (369) and (371) gives

$$[G_r(R\pm h) - G_\phi(R)]/G \sim 3[\theta^2_{\ell\phi}(R) - \theta^2_r(R\pm h)] \tag{400}$$

$$= 3\{[\theta_r(R) + \theta_{s\psi}(R) + \theta_\phi(R) + \beta_{\phi,\phi}(R)]^2 - \theta^2_r(R\pm h)\}$$

Combining equations (376), (377) and (400) gives

$$[G_r(R\pm h) - G_\phi(R)]/G \sim \pm s_r h + 6\theta_r(R)[\theta_{s\psi}(R) + \theta_\phi(R) + \beta_{\phi,\phi}(R)] \tag{401}$$

$$+ 3[\theta_{s\psi}(R) + \theta_\phi(R) + \beta_{\phi,\phi}(R)]^2$$

$$\sim \pm s_r h + 6\theta_r(R)[\theta_{s\psi}(R) + \theta_\phi(R) + \beta_{\phi,\phi}(R)]$$

From equation (401) it follows that

$$[G_r(R) - G_\phi(R)]/G \sim 6\theta_r(R)[\theta_{s\psi}(R) + \theta_\phi(R) + \beta_{\phi,\phi}(R)] \tag{402}$$

$$+ 3[\theta_{s\psi}(R) + \theta_\phi(R) + \beta_{\phi,\phi}(R)]^2$$

$$\sim 6\theta_r(R)[\theta_{s\psi}(R) + \theta_\phi(R) + \beta_{\phi,\phi}(R)]$$

From equations (381), (401) and (402) it follows that

$$G_r(R-h) < G_r(R) < G_r(R+h) < G_\phi(R) < G \quad \theta_\phi(R) > 0\ ,\ \theta_\psi(R) > 0 \tag{403}$$

$$G_\phi(R) < G_r(R-h) < G_r(R) < G_r(R+h) < G \quad \theta_\phi(R) < 0\ ,\ \theta_\psi(R) < 0 \tag{404}$$

The inequalities in equations (403) and (404) hold only for small h.

Inequalities (398) and (403) are supported by the experimental data in References 5, 25 and 49 and suggest that mine shaft, borehole and tower determinations of the gravitational constant will be less than the gravitational constant determined by an Eötvös experiment performed at the earth's surface. On the other hand, the inequalities (399) and (404) are supported by the experimental data given in References 1 through 3 and indicate that the gravitational constant determined from a mine shaft, borehole or tower experiment will be larger than the value of the gravitational constant obtained by an Eötvös experiment conducted at the earth's surface. Only one set of data can be correct. When the correct set of experimental data is finally determined the proper signs and approximate magnitudes of θ_ϕ and θ_ψ will be fixed. Neither mine shaft, borehole, tower or Eötvös experiments directly measure the constant G .

If it is possible to conduct an Eötvös experiment at various depths in a mine shaft or at different heights up a tower, it becomes important to compare G_r with G_ψ and G_ϕ at the same depth or height. From equations (369) through (371) it follows that

$$(G_r - G_\psi)/G \sim 3(\theta^2_{\ell\psi} - \theta^2_r) \tag{405}$$
$$= 3[(\theta_r + \theta_\psi + \beta_{\psi,\psi})^2 - \theta^2_r]$$
$$= 3[2\theta_r(\theta_\psi + \beta_{\psi,\psi}) + (\theta_\psi + \beta_{\psi,\psi})^2]$$
$$\sim 6\theta_r(\theta_\psi + \beta_{\psi,\psi})$$
$$\sim 12\theta_r\theta_\psi$$

In a similar fashion

$$(G_r - G_\phi)/G \sim 3(\theta^2_{\ell\phi} - \theta^2_r) \tag{406}$$
$$= 3[(\theta_r + \theta_{s\psi} + \theta_\phi + \beta_{\phi,\phi})^2 - \theta^2_r]$$
$$= 3[2\theta_r(\theta_{s\psi} + \theta_\phi + \beta_{\phi,\phi}) + (\theta_{s\psi} + \theta_\phi + \beta_{\phi,\phi})^2]$$
$$\sim 6\theta_r(\theta_{s\psi} + \theta_\phi + \beta_{\phi,\phi})$$
$$\sim 18\theta_r\theta_\phi$$

The inequalities (372) and (373) can also be deduced from equations (405) and (406).

The variation of G_r , G_ϕ and G_ψ with depth in a mine shaft and borehole or with height up a tower is due to Newtonian gravity in broken symmetry space combined with the variation of the broken symmetry atmospheric pressure. The variation of the broken symmetry atmospheric pressure with radial distance induces a variation of θ_r , θ_ϕ and θ_ψ with radial distance (Figure 4) and this determines the non-Newtonian variation of G_r , G_ϕ and G_ψ according to equations (369) through (371). This apparent non-Newtonian behaviour of gravity has been interpreted as being due to the existence of graviscalar (spin 0) and graviphoton (spin 1) component forces of gravity (the "fifth" and "sixth" forces).[1-4,19-25] The hypothetical graviscalar is an attractive force while the hypothetical graviphoton mediates a repulsive force, and both are described by finite range Yukawa terms that are added to the ordinary Newtonian potential. But in fact these hypothetical forces are not required to describe the experimental results. The apparent non-Newtonian behaviour is due to ordinary Newtonian gravity in matter and space whose pressure and coordinate fields exhibit broken internal symmetries. The relative magnitudes of $G_r(R\pm h)$ and $G_\psi(R)$ or $G_\phi(R)$ are not related to new gravitation forces but rather to the broken symmetry of pressure and spatial coordinates.

6. NUMERICAL VALUES OF THE INTERNAL PHASE ANGLES. This section determines numerical values of θ_r , θ_ϕ , θ_ψ and θp within the atmosphere in the vicinity of the earth's surface. Two methods are used. The discrepancy between the measured and predicted values of the gravitational red shift of γ-rays in the Pound-Rebka-Snider experiment.[32,33,46-48] The second method is based on the measurement of the apparent departure of the force of gravity from Newtonian behaviour in mine shaft, borehole and tower experiments.[1-3,5,25,49]

A. Measurement of the Gravitational Red Shift

The experiments of Pound, Rebka and Snider measured the gravitational red shift of a γ-ray falling in the earth's gravitational field. The conventional expression for the red shift in frequency is given by[32,33]

$$z_c = (\Delta\nu/\nu)_c = [V(r_{2m}) - V(r_{1m})]/c^2 \tag{407}$$

where the conventionally calculated gravitational potential is written as[28]

$$V(r_m) = - GM/r_m \tag{408}$$

and therefore

$$z_c = GM/c^2(1/r_{1m} - 1/r_{2m}) \tag{409}$$

In this paper the theory of coordinates with internal phase requires a complex number red shift given by

$$\bar{z} = ze^{j\theta_z} = \Delta\bar{\nu}/\bar{\nu} = [\bar{V}(\bar{r}_2) - \bar{V}(\bar{r}_1)]/c^2 \tag{410}$$

$$= GM/c^2(1/\bar{r}_1 - 1/\bar{r}_2)$$

The measured gravitational red shift is given by the real part of equation (410)

$$z_m = z_R = z\cos\theta_z = GM/c^2(1/r_1 \cos\theta_{r1} - 1/r_2 \cos\theta_{r2}) \tag{411}$$

while the imaginary part of equation (430) is

$$z_I = z\sin\theta_z = - GM/c^2(1/r_1 \sin\theta_{r1} - 1/r_2 \sin\theta_{r2}) \tag{412}$$

where

$$\bar{r}_1 = r_1 e^{j\theta_{r1}} \tag{413}$$

$$\bar{r}_2 = r_2 e^{j\theta_{r2}} \tag{414}$$

Because $r_{1m} = r_1 \cos\theta_{r1}$ and $r_{2m} = r_2 \cos\theta_{r2}$ it follows from equation (411) that

$$z_m = GM/c^2(1/r_{1m} \cos^2\theta_{r1} - 1/r_{2m} \cos^2\theta_{r2}) \tag{415}$$

The difference between the measured and conventionally calculated gravitational red shift is obtained from equations (409) and (415) to be

$$z_c - z_m = GM/c^2(1/r_{1m} \sin^2\theta_{r1} - 1/r_{2m} \sin^2\theta_{r2}) \tag{416}$$

$$\sim GM/c^2(\theta_{r1}^2/r_{1m} - \theta_{r2}^2/r_{2m})$$

$$\sim \theta_r^2 z_c$$

Therefore approximately

$$\theta_r^2 \sim (z_c - z_m)/z_c \sim 0.01 \tag{417}$$

where according to References 47 and 48 the fractional difference between the calculated and measured gravitational red shift is 1%. From equation (417) it follows that

$$\theta_r \sim - 0.1 \text{ rad} = - 5.7° \tag{418}$$

In this way a value of θ_r is obtained from the measurement of the gravitational red shift. This laboratory value of θ_r is probably more accurate than the corresponding values of θ_r that may possibly be obtained from measurements of the apparent non-Newtonian variation of gravity in mine shafts, boreholes and towers. The value of θ_r given in equation (418) will be used to obtain values θ_ϕ and θ_ψ in section B.

B. Analysis of the Apparent Non-Newtonian Gravity Measurements.

Conflicting experimental data have been presented for the values of the gravitational constant derived from measurements of the variation of the force of gravity with distance in mine shafts, boreholes and towers. According to References 1 through 3, the values of the gravitational constant derived from mine shaft gravity variations are larger than those derived from Eötvös experiments conducted at the earth's surface. On the other hand, References 25 and 49 indicates that borehole measurements in the ice of a glacier produce values of the gravitational constant that are smaller than the values of the gravitational constant derived from Eötvös experiments performed at the surface of the earth. In addition, Reference 5 indicates that the gravitational constant derived from gravity measurements on a tower is smaller than that measured by the Eötvös experiments at the earth's surface. The experimental results given in References 1 through 3 are in conflict with the experimental results of References 5, 25 and 49. Therefore the numerical calculations in this section are done for each situation. According to the theory of Newtonian gravity in matter and vacuum with broken internal symmetries the discrepancies between the measured and conventionally predicted values of the force of gravity in mine shaft, borehole, tower and Eötvös experiments are related to the values of θ_r , θ_ψ and θ_ϕ .

Two cases will be examined in this section according to the relative magnitudes of the internal phase angles $|\theta_r|$, $|\theta_\psi|$ and $|\theta_\phi|$.

Case 1: $|\theta_r(R)| >> |\theta_\psi(R)|$ and $|\theta_r(R)| >> |\theta_\phi(R)|$

Combining equations (392), (396) and (401) gives for small h/R

$$[G_\psi(R) - G_\phi(R)]/G \sim 6xy \tag{419}$$

$$[G_r(R \pm h) - G_\psi(R)]/G \sim 12xy \tag{420}$$

$$[G_r(R \pm h) - G_\phi(R)]/G \sim 18xy \tag{421}$$

where

$$x = \theta_r(R) < 0 \tag{422}$$

$$y = \theta_\psi(R) \sim \beta_{\psi,\psi}(R) \sim \theta_{s\psi}(R) \sim \theta_\phi(R) \sim \beta_{\phi,\phi}(R) \tag{423}$$

so that $|x| >> |y|$. Because $x < 0$ it is the sign of y that determines the signs of the expressions in equations (419) through (421). The expressions in equations (420) and (421) can be either positive or negative, so that they can describe either the experimental results of References 1 through 3 (which requires $y < 0$) or the experimental results of References 5, 25 and 49 (which requires $y > 0$). Because the value of $x = \theta_r(R)$ is known from equation (418) the determination of any one of the three differences in equations (419) through (421) would immediately determine the value for y.

Consider first the determination of the gravitational constant from mine shafts which is about 0.6% larger than the value obtained from Eötvös experiments performed at the earth's surface.[1-3] Using the average of equations (420) and (421) (because of the unspecified orientation of the Eötvös experiments) yields

$$15xy = 15(-0.1)y = 0.006 \tag{424}$$

where $x = -0.1$ was obtained from equation (418). Equation (424) gives

$$\begin{aligned} x &= \theta_r(R) = -\ 0.1 \text{ rad} = -\ 5.7° \\ y &= \theta_\psi(R) = \theta_\phi(R) = -\ 0.004 \text{ rad} = -\ 0.23° \end{aligned} \tag{425}$$

From equation (419) the north-south/east-west asymmetry of the Eötvös experiment is given by

$$[G_\psi(R) - G_\phi(R)]/G \sim +\ 0.0024 \tag{426}$$

Borehole data from a Greenland glacier shows that the derived gravitational constant is about 2.8% smaller than the value of the gravitational constant derived from Eötvös experiments done at the earth's surface.[25,49] Also, measurements of the variation of the gravity force up a tower gives results for the gravitational constant that are 2.0% smaller than that obtained from Eötvös experiments at the surface of the earth.[5] Therefore using the average of the results of References 5 and 49 with the average of equations (420) and (421) gives

$$15xy = 15(-0.1)y = -\ 0.024 \tag{427}$$

where again $x = -0.1$ was obtained from equation (418). Then equation (427) yields

$$\begin{aligned} x &= \theta_r(R) = -\ 0.1 \text{ rad} = -\ 5.7° \\ y &= \theta_\psi(R) = \theta_\phi(R) = +\ 0.016 \text{ rad} = +\ 0.92° \end{aligned} \tag{428}$$

Equation (419) gives the north-south/east-west asymmetry of the Eötvös experiment to be

$$[G_\psi(R) - G_\phi(R)]/G \sim -\ 0.0096 \tag{429}$$

An independent determination of the north-south/east-west asymmetry of the Eötvös experiment would immediately determine the signs and values of θ_ψ (and θ_ϕ).

Case 2: $\theta_r(R) = \theta_\psi(R) = \theta_\phi(R) = \theta_{s\psi}(R) = \beta_{\psi,\psi}(R) = \beta_{\phi,\phi}(R)$

Combining equations (392), (395) and (400) gives for small h/R

$$[G_\psi(R) - G_\phi(R)]/G \sim 3[(4w)^2 - (3w)^2] = 21w^2 \quad (430)$$

$$[G_r(R\pm h) - G_\psi(R)]/G \sim 3[(3w)^2 - w^2] = 24w^2 \quad (431)$$

$$[G_r(R\pm h) - G_\phi(R)]/G \sim 3[(4w)^2 - w^2] = 45w^2 \quad (432)$$

where $w = \theta_r(R) = \theta_\psi(R) = \theta_\phi(R) = \theta_{s\psi}(R) = \beta_{\psi,\psi}(R) = \beta_{\phi,\phi}(R) < 0$ (433)

Thus $w < 0$ because $\theta_r(R) < 0$. The expressions in equations (431) and (432) are always positive and therefore Case 2 agrees only with the experimental data given in References 1 through 3. Using the average of equations (431) and (432) (because of the unspecified orientation of the Eötvös experiments) and the 0.006 positive fractional difference between the values of the gravitational constant measured in a mine shaft and by an Eötvös experiment performed at the surface of the earth given by References 1 through 3 yields

$$34.5w^2 = 0.006$$
$$w = \theta_r(R) = \theta_\psi(R) = \theta_\phi(R) = -0.76° \quad (434)$$

where w is given by equation (433). Equation (430) gives the north-south/east-west asymmetry of the Eötvös experiment as

$$[G_\psi(R) - G_\phi(R)]/G \sim 0.0037 \quad (435)$$

The predicted value $\theta_r(R) = -0.76°$ is much less in magnitude than the value $\theta_r(R) = -5.7°$ predicted by the Pound-Rebka-Snider experiment. Therefore Case 2 as represented in the assumption given in equation (433) may not be physically realistic.

7. CONCLUSION. Newtonian gravity in space and time with broken internal symmetries produces an apparent non-Newtonian behaviour of the acceleration of gravity, and the gravitational constant varies with the radial distance from the center of a planet. This is due to the fact that the pressure and coordinates in matter (and vacuum) exhibit broken symmetries that are represented by internal phase angles which vary with radial distance. The measured apparent non-Newtonian gravity effects are therefore due to the variation of the atmospheric pressure in mine shafts and boreholes and on towers, and this introduces an apparent 7 km finite range force component. New forces in addition to Newtonian gravitation are not required to explain the experimental observations. The values of the internal phase angles of the coordinates can be obtained from the Pound-Rebka-Snider gravitational red shift experiment, the measurements of the apparent non-Newtonian gravity field, and the Eötvös experiments. The internal phase angles of space and time will influence the basic calculations of astrophysics and geophysics.

ACKNOWLEDGEMENT

The author wishes to thank Elizabeth K. Klein for typing this paper.

REFERENCES

1. Stacey, F. D., Tuck, G. J., Moore, G. I., Holding, S. C., Goodwin, B. D. and Zhou, R., "Geophysics and the Law of Gravity," Rev. Mod. Phys., Vol. 59, No. 1, Jan 1987, p. 157.

2. Moore, G. I., Stacey, F. D., Tuck, G. J., Goodwin, B. D., Linthorne, N. P., Barton, M. A., Reid, D. M. and Agnew, G. D., "Determination of the Gravitational Constant at an Effective Mass Separation of 22m," Phys. Rev. D, Vol. 38, No. 4, 15 Aug. 1988, p. 1023.

3. Stacey, F. D., Tuck, G. J. and Moore, G. I., "Quantum Gravity: Observational Constraints on a Pair of Yukawa Terms," Phys. Rev. D, Vol. 36, No. 8, 15 Oct. 1987, p. 2374.

4. Bartlett, D. F. and Tew, W. L., "Comment on Quantum Gravity: Observational Constraints on a Pair of Yukawa Terms," Phys. Rev. D, Vol. 38, No. 12, 15 Dec. 1988, p. 3843.

5. Eckhardt, D. H., Jekeli, C., Lazarewicz, A. R., Romaides, A. J. and Sands, R. W., "Tower Gravity Experiment: Evidence for Non-Newtonian Gravity," Phys. Rev. Lett., Vol. 60, No. 25, 20 June 1988, p. 2567.

6. Fischbach, E., Sudarsky, D., Szafer, A., Talmadge, C. and Aronson, S. H., "Reanalysis of the Eötvös Experiment," Phys. Rev. Lett., Vol. 56, No. 1, 6 Jan. 1986, p. 3.

7. Thieberger, P., Phys. Rev. Lett., Vol. 58, 1987, p. 1066.

8. Stubbs, C. W., Adelberger, E. G., Raab, F. J., Gundlach, J. H., Heckel, B. R., McMurry, K. D., Swanson, H. E. and Watanabe, R., Phys. Rev. Lett., Vol. 58, 1987, p. 1070.

9. Adelberger, E. G., Stubbs, C. W., Rogers, W. F., Raab, F. J., Heckel, B. R., Gundlach, J. H., Swanson, H. E. and Watanabe, R., Phys. Rev. Lett., Vol. 59, 1987, p. 849.

10. Niebauer, T. M., McHugh, M. P. and Faller, J. E., Phys. Rev. Lett., Vol. 59, 1987, p. 609.

11. Boynton, P. E., Crosby, D., Ekstrom, P. and Szumilo, A., Phys. Rev. Lett., Vol. 59, 1987, p. 1385.

12. Fitch, V. L., Isaila, M. V. and Palmer, M. A., "Limits on the Existence of a Material-Dependent Intermediate-Range Force," Phys. Rev. Lett., Vol. 60, No. 18, 2 May 1988, p. 1801.

13. Speake, C. C. and Quinn, T. J., "Search for a Short-Range Isospin-Coupling Component of the Fifth Force with Use of a Beam Balance," Phys. Rev. Lett., Vol. 61, No. 12, 19 Sep. 1988, p. 1340.

14. Fischbach, E., Kloor, H. T., Talmadge, C., Aronson, S. H. and Gillies, G. T., "Possibility of Shielding the Fifth Force," Phys. Rev. Lett., Vol. 60, 4 Jan. 1988, p. 74.

15. Cowsik, R., Krishnan, N., Tandon, S. N. and Unnikrishnan, C. S., "Limit on the Strength of Intermediate-Range Forces Coupling to Isospin," Phys. Rev. Lett., Vol. 61, 7 Nov. 1988, p. 2179.

16. Damour, T., Gibbons, G. W. and Taylor, J. H., "Limits on the Variability of G Using Binary-Pulsar Data," Phys. Rev. Lett., Vol. 61, No. 10, 5 Sep. 1988, p. 1151.

17. Talmadge, C., Berthias, J. P., Hellings, R. W. and Standish, E. M., "Model-Independent Constraints on Possible Modifications of Newtonian Gravity," Phys. Rev. Lett., Vol. 61, No. 10, 5 Sep. 1988, p. 1159.

18. Burgess, C. P. and Cloutier, J., "Astrophysical Evidence for a Weak New Force," Phys. Rev. D, Vol. 38, No. 10, 15 Nov. 1988, p. 2944.

19. Goldman, T., Hughes, R. J. and Nieto, M. M., Phys. Lett. B, Vol. 171, 1986, p. 217.

20. Nieto, M. M., Goldman, T. and Hughes, R. J., "Phenomenological Aspects of New Gravitational Forces. IV. New Terrestial Experiments," Phys. Rev. D, Vol. 38, No. 10, 15 Nov. 1988, p. 2937.

21. Ander, M. E., Goldman, T., Hughes, R. J. and Nieto, M. M., "Possible Resolution of the Brookhaven and Washington Eötvös Experiments," Phys. Rev. Lett., Vol. 60, No. 13, 28 March 1988, p. 1225.

22. Niebauer, T. M., Faller, J. E. and Bender, P. L., "Comment on Possible Resolution of the Brookhaven and Washington Eötvös Experiments," Phys. Rev. Lett., Vol. 61, 7 Nov. 1988, p. 2272.

23. Stubbs, C. W., Adelberger, E. G. and Gregory, E. C., "Constraints of Proposed Spin-0 and Spin-1 Partners of the Graviton," Phys. Rev. Lett., Vol. 61, 21 Nov. 1988, p. 2409.

24. Schwarzschild, B., "From Mine Shafts to Cliff's - The 'Fifth Force' Remains Elusive," Physics Today, July 1988, p. 21.

25. Pool, R., "Was Newton Wrong?," Science, Vol. 241, 12 Aug. 1988, p. 789.

26. Weiss, R. A., "Scale Invariant Equations for Relativistic Waves," Fourth Army Conference on Applied Mathematics and Computing, Cornell University, Ithaca, NY, ARO 87-1, 27-30 May 1986, p. 307.

27. Weiss, R. A., "Thermodynamic Gauge Theory of Solids and Quantum Liquids with Internal Phase," Fifth Army Conference on Applied Mathematics and Computing, West Point, New York, ARO 88-1, June 15-18 1987, p. 649.

28. Weiss, R. A., "The Broken Symmetry of Space and Time in Bulk Matter and the Vacuum," Sixth Army Conference on Applied Mathematics and Computing, Boulder, Colorado, ARO 89-1, 31 May-3 June 1988, p. 317.

29. Weiss, R. A., Relativistic Thermodynamics, Vols. 1 and 2, Exposition Press, New York, 1976.

30. Chandrasekhar, S., "On Stars, Their Evolution and Their Stability," Revs. Mod. Phys., Vol. 56, April 1984, p. 137.

31. Tolman, R. C., Relativity Thermodynamics and Cosmology, Oxford, Clarendon Press, London, 1934.

32. Misner, C. W., Thorne, K. S. and Wheeler, J. A., Gravitation, W. H. Freeman and Company, San Francisco, 1973.

33. Weinberg, S., Gravitation and Cosmology, John Wiley, New York, 1972.

34. Witten, E., "Space-Time and Topological Orbifolds," Phys. Rev. Lett., Vol. 61, No. 6, 8 Aug. 1988, p. 670.

35. Stacey, F. D., Physics of the Earth, John Wiley, New York, 1977.

36. Garland, G. D., The Earth's Shape and Gravity, Pergamon Press, New York, 1965.

37. Jeffreys, H., The Earth, Cambridge University Press, New York, 1962.

38. Spencer Jones, H., "Dimensions and Rotation," article in The Earth as a Planet, edited by Kuiper, G. P., University of Chicago Press, Chicago, 1954.

39. Officer, C. B., Introduction to Theoretical Geophysics, Springer-Verlag, New York, 1974.

40. King-Hele, D. G., "The Earth's Gravitational Potential, Deduced from the Orbits of Artificial Satellites," article in The Earth Today, edited by Jeffreys, H., Interscience, New York, 1961.

41. Murphy, G. M., Ordinary Differential Equations and Their Solutions, Van Nostrand, New York, 1960.

42. Petit Bois, G., Tables of Indefinite Integrals, Dover, New York, 1961.

43. Humphreys, W. J., Physics of the Air, McGraw-Hill, New York, 1929.

44. Gutenberg, B., Physics of the Earth's Interior, Academic Press, New York, 1959.

45. Bullard, E., "The Interior of the Earth," Article in The Earth as a Planet, edited by Kuiper, G., University of Chicago Press, Chicago, 1954.

46. Pound, R. V. and Rebka, G. A., "Apparent Weight of Photons," Phys. Rev. Lett., Vol. 4, 1960, p. 337.

47. Pound, R. V. and Snider, J. L., "Effect of Gravity on Nuclear Resonance," Phys. Rev. Lett., Vol. 13, 1964, p. 539.

48. Pound, R. V. and Snider, J. L., "Effect of Gravity on Gamma Radiation," Phys. Rev. B, Vol. 140, 1965, p. 788.

49. Ander, M. E., Zumberge, M. A., Lautzenhiser, T., Parker, R. L., Aiken, C.L., Gorman, M. R., Nieto, M. M., Cooper, A. P., Ferguson, J. F., Fisher, E., McMechan, G. A., Sasagawa, G., Stevenson, J. M., Backus, G., Chave, A. D., Greer, J., Hammer, P., Hansen, B. L., Hildebrand, J. A., Kelty, J. R., Sidles, C. and Wirtz, J., "Test of Newton's Inverse-Square Law in the Greenland Ice Cap," Phys. Rev. Lett., Vol. 62, No. 9, 27 Feb. 1989.

ERRATA: Reference 28

equation (384) $-\cos\beta_{r,r}\,(\partial\bar{P}/\partial r + dt/dr\,\partial\bar{P}/\partial t) - \rho\,\partial\bar{W}/\partial\bar{r}$

equation (385) $-\cos\beta_{\phi,\phi}\,(1/r\,\partial\bar{P}/\partial\phi + 1/r\,dt/d\phi\,\partial\bar{P}/\partial t) - \rho/\bar{r}\,\partial\bar{W}/\partial\bar{\phi}$

equation (389) $-\cos\beta_{x,x}\,(\partial\bar{P}/\partial x + dt/dx\,\partial\bar{P}/\partial t) - \rho\,\partial\bar{W}/\partial\bar{x}$

equation (393) $\cos\beta_{r,r}\,\partial\bar{P}/\partial r = D_P e^{j\Phi_P}$

equation (396) $\Phi_P = \theta_P + \beta_{P,r}$

equation (408) $\theta_P + \beta_{P,r} = \theta_W + \beta_{W,r} - \theta_r - \beta_{r,r} + \pi$

equation (425) $\theta_P + \beta_{P,r} = -\,2\theta_r + \pi$

equation (426) $\theta_P + \tan^{-1}\left(\dot{P}\,\dfrac{\partial\theta_P/\partial r}{\partial P/\partial r}\right) = -\,2\theta_r + \pi$

equation (428) $\theta_P + \beta'_{P,r} = -\,2\theta_r$

equation (431) $\cos\beta_{r,r}\,\partial\bar{P}/\partial r = -\,GM\rho/\bar{r}^2$

equation (451) $\cos\beta_{r,r}\,\partial\rho/\partial r = -\,G\rho^2 M/(\bar{K}_S\bar{r}^2) = -\,G\rho M/(\bar{v}_S^2\bar{r}^2)$

equation (456) $\theta_{vS} = -\,\theta_r$

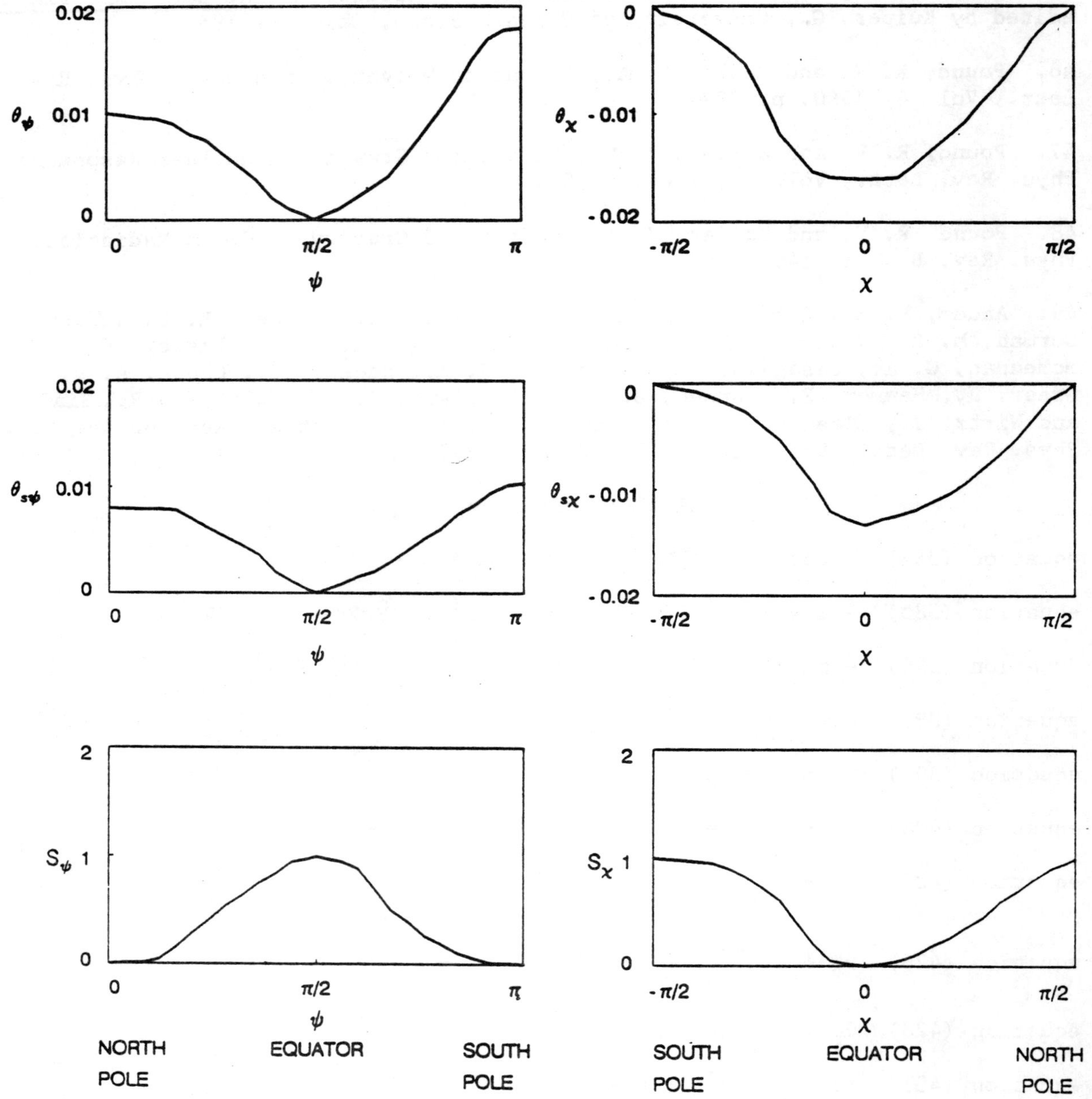

Figure 1. Sketch showing the dependence of θ_ψ, $\theta_{s\psi}$ and S_ψ on zenith angle ψ, and the dependence of θ_χ, $\theta_{s\chi}$ and S_χ on the latitude χ. The experimental situation is not yet clear and it may be that $\theta_\psi < 0$ and $\theta_\chi > 0$.

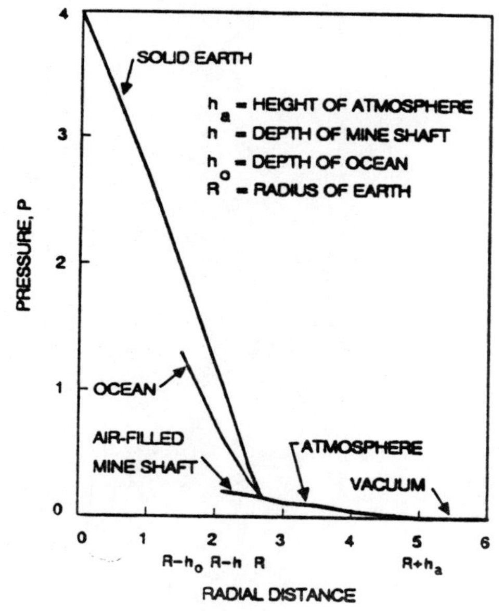

Figure 2. Sketch showing variation of pressure with radial distance for the atmosphere, ocean and solid earth (not to scale).

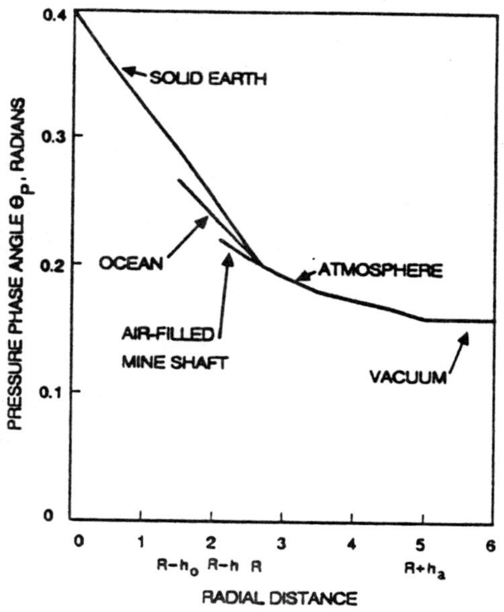

Figure 3. Sketch showing variation of the internal phase angle of the pressure with radial distance for the atmosphere, ocean and solid earth (not to scale).

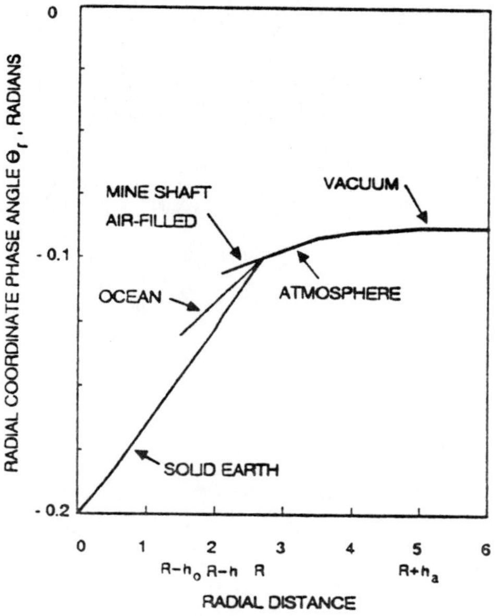

Figure 4. Sketch showing variation of the internal phase angle of the radial coordinate with radial distance for the atmosphere, ocean and solid earth (not to scale).

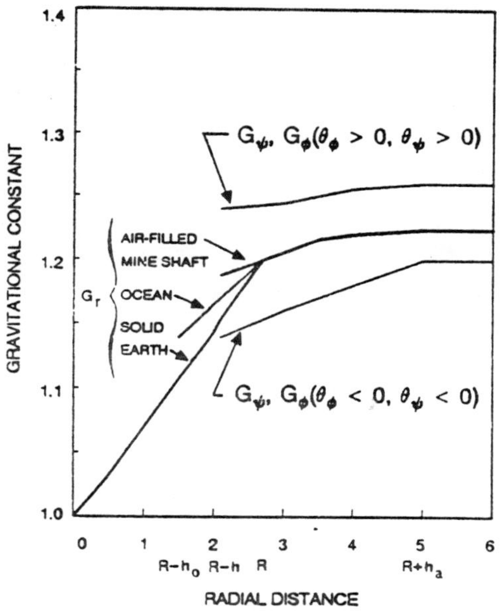

Figure 5. Sketch showing the variation of G_r, G_ψ and G_ϕ with radial distance. Two possible cases for G_ψ and G_ϕ are shown (not to scale).

K

THE INTERNAL PHASE STRUCTURE OF ATOMS

Richard A. Weiss
U. S. Army Engineer Waterways Experiment Station
Vicksburg, Mississippi 39180

ABSTRACT. The three dimensional Schrödinger equation for hydrogen-like atoms under pressure is solved and the spectra and eigenfunctions are calculated using the fact that in a pressure field the coordinates have internal phase angles. Because the coordinates have broken internal symmetries the energy eigenvalues are complex numbers whose real parts yield the measurable quantities that can be experimentally tested by examining the spectra of one-electron atoms under pressure. The magnetic, azimuthal and principal quantum numbers must be represented as complex numbers for hydrogen-like atoms under pressure. It is found that under pressure hydrogen-like atoms will exhibit a pressure dependent fine structure in which the energy levels of the valence electron depend on the magnetic quantum number as well as on the principal quantum number. The pressure dependence of the spectra of hydrogen-like atoms is determined. This research will have applications to stellar atmospheres and to gases at high pressures associated with conventional and nuclear explosions.

1. INTRODUCTION. The early development of atomic and nuclear physics made minimal use of gauge field theory because the only gauge field known was electromagnetism.[1] The importance of gauge fields was only fully realized in the past twenty-five years from a search for a unifying principle behind the four fundamental interactions.[2] Gauge theories are now important in many scientific and mathematics disciplines.[3,4] Recently a gauge theory of relativistic thermodynamics has been developed which suggests that the pressure in a bulk matter system has a broken internal symmetry and must be represented by a complex number.[5,6] A set of renormalization group equations has been developed which gives the recipe for calculating the magnitude and internal phase angle of the pressure as a function of temperature and density for an interacting bulk matter system.[5,6] This can be applied to gases, liquids or solids.

If the pressure has a broken internal symmetry then Euler's equations of motion suggest that space and time coordinates must also have broken internal symmetries and be treated as complex numbers.[7] Arguments from string theory also predict a complex number coordinate representation.[8] The complex number values of the space and time coordinates imply that the basic wave equations of classical and quantum physics must also contain these broken internal coordinate symmetries. For instance the Schrödinger and Dirac equations must be written as complex number coordinate equations whose eigenvalues and eigenfunctions have broken internal symmetries.[9] Therefore, indirectly, a gauge theory of relativistic thermodynamics predicts microscopic effects which affect the basic calculations of atomic physics and the structure of atoms. In fact, atoms located in a pressure field should exhibit an internal phase structure which depends on the magnitude of the ambient pressure.

The Bohr atom under zero external forces does not exhibit an internal phase structure. This can be seen by writing the complex number generalization of the Bohr atom equations as follows[10,11]

$$\mu\bar{r}^2\bar{\omega} = n\hbar \qquad (1)$$

$$\mu\bar{\omega}^2\bar{r} = Ze^2/\bar{r}^2 \qquad (2)$$

where the complex number radius and frequency are given by[7]

$$\bar{r} = re^{j\theta_r} \qquad (3)$$

$$\bar{\omega} = \omega e^{j\theta_\omega} \qquad (4)$$

and where μ = reduced mass of the electron, n = integer, $\hbar = h/(2\pi)$, e = electron charge and Z = atomic number. Equations (1) and (2) can be written as

$$\mu r^2\omega = n\hbar \qquad (5)$$

$$\mu\omega^2 r = Ze^2/r^2 \qquad (6)$$

$$2\theta_r + \theta_\omega = 0 \qquad (7)$$

$$3\theta_r + 2\theta_\omega = 0 \qquad (8)$$

The solution of equations (5) and (6) yield the standard expressions for the radius and total energy of the Bohr hydrogen atom

$$r = n^2\hbar^2/(\mu Ze^2) \qquad (9)$$

$$E_n = -\mu Z^2e^4/(2\hbar^2n^2) \qquad (10)$$

while equations (7) and (8) give

$$\theta_r = 0 \qquad (11)$$

$$\theta_\omega = 0 \qquad (12)$$

Thus the internal phase structure vanishes for an isolated Bohr atom, and the Bohr radius and energy levels are real numbers. Because all of the quantities on the right hand side of the energy equation (10) are constants it is not possible for the energy levels to be pressure dependent.

Atoms subjected to pressure or other external forces have an internal phase structure. This can be seen in a simple fashion by noting that for the case of a Bohr electron under the influence of an external radial force equation (2) can be written as

$$\mu\bar{\omega}^2\bar{r} = Ze^2/\bar{r}^2 + \bar{F} \qquad (13)$$

where $\bar{F} = Fe^{j\theta_F}$ = complex number external force acting on the electron. The external force can be transmitted to the electron by electrical forces from adjacent atoms and is utlimately related to the complex number pressure which can be written as

$$\bar{P} = Pe^{j\theta_P} \tag{14}$$

where θ_P = internal phase angle of the pressure. A more sophisticated approach to the problem of the Bohr atom in a pressure field is to solve Schrödinger's equation for this case. When this is done it is found that the presence of a broken symmetry pressure requires that the principal quantum number that appears in the Bohr quantization equation (1) must be a complex number so that

$$\mu\bar{r}^2\bar{\omega} = \bar{\eta}\hbar \tag{15}$$

$$\bar{\eta} = \eta e^{j\theta_\eta} \tag{16}$$

where $\bar{\eta}$ = complex number principal quantum number which is related to the broken symmetry of the azimuthal angle and ultimately to the complex number pressure (Section 5). For a hydrogen-like atom under pressure the quantization condition in equation (15) yields

$$\mu r^2\omega = \eta\hbar \tag{17}$$

$$2\theta_r + \theta_\omega = \theta_\eta \tag{18}$$

while the force balance equation (13) gives

$$\mu\omega^2 r \cos(2\theta_\omega + \theta_r) = Ze^2/r^2 \cos(2\theta_r) + F \cos \theta_F \tag{19}$$

$$\mu\omega^2 r \sin(2\theta_\omega + \theta_r) = - Ze^2/r^2 \sin(2\theta_r) + F \sin \theta_F \tag{20}$$

Equations (17) through (20) are four scalar equations which can be solved for r, θ_r, ω and θ_ω in terms of η, θ_η, F and θ_F or ultimately in terms of P and θ_P. For $F \neq 0$ the solution of the Schrödinger equation becomes extremely difficult and therefore this is not considered in this paper. Instead it is assumed that $\bar{F} \sim 0$ so that θ_r and θ_ω are obtained from equations (8) and (18) to be

$$\theta_r \sim 2\theta_\eta \qquad \theta_\omega \sim - 3\theta_\eta \tag{20A}$$

and the corresponding Schrödinger equation can also be solved.

This paper considers the solution of Schrödinger's equation for hydrogen-like atoms in a three dimensional space that exhibits broken internal symmetry. Section 2 gives the general form of Schrödinger's equation for a spherically symmetric potential with broken internal symmetries. Section 3 considers the azimuthal angle equation and introduces the complex magnetic quantum number. Section 4 treats the zenith angle equation which introduces the complex azimuthal quantum number, while Section 5 solves the radial equation which has a complex value of the principal quantum number. Section 6 presents the complex number wave functions for hydrogen-like atoms under pressure and Section 7 determines the complex number energy eigenvalues. Only bound states are considered in this paper.

2. SCHRÖDINGER'S EQUATION FOR HYDROGEN WITH BROKEN INTERNAL SYMMETRIES.

The three dimensional Schrödinger equation in spherical polar coordinates for an electron in a central force field in an atom whose space and time coordinates have broken internal symmetries is written as a generalization of the standard form of this equation as follows[12-14]

$$\frac{\partial^2\bar{\Psi}}{\partial\bar{r}^2} + \frac{2}{\bar{r}}\frac{\partial\bar{\Psi}}{\partial\bar{r}} + \frac{1}{\bar{r}^2 \sin\bar{\psi}}\frac{\partial}{\partial\bar{\psi}}\left(\sin\bar{\psi}\,\frac{\partial\bar{\Psi}}{\partial\bar{\psi}}\right) + \frac{1}{\bar{r}^2 \sin^2\bar{\psi}}\frac{\partial^2\bar{\Psi}}{\partial\bar{\phi}^2} + \frac{8\pi^2\mu}{h^2}[\bar{E} - \bar{V}(\bar{r})]\bar{\Psi} = 0 \quad (21)$$

where the complex number spherical polar coordinates are written as

$$\bar{r} = re^{j\theta_r} \quad (22)$$

$$\bar{\psi} = \psi e^{j\theta_\psi} \quad (23)$$

$$\bar{\phi} = \phi e^{j\theta_\phi} \quad (24)$$

where $\bar{\Psi}$ = complex number wave function, $\bar{r}$ = complex number radial coordinate, $\bar{\psi}$ = complex number zenith angle and $\bar{\phi}$ = complex number azimuthal angle. The broken symmetry of the coordinates are described by the internal phase angles θ_r, θ_ψ and θ_ϕ which are pressure dependent.[7] The measured values of the coordinates are given by the real parts of the complex number coordinates given in equations (22) through (24) and are written as[7]

$$r_m = r\cos\theta_r \quad (25)$$

$$\psi_m = \psi\cos\theta_\psi \quad (26)$$

$$\phi_m = \phi\cos\theta_\phi \quad (27)$$

The complex number potential is written as

$$\bar{V} = Ve^{j\theta_V} \quad (28)$$

which for the Coulomb potential becomes

$$\bar{V} = -Ze^2/\bar{r} \quad (29)$$

$$V = -Ze^2/r \quad (30)$$

$$\theta_V = -\theta_r \quad (31)$$

If the potential were directly measurable the measured value would be given by[7]

$$V_m = V\cos\theta_V \quad (32)$$

$$= -Ze^2/r\cos\theta_r$$

Note the effect of the external pressure is assumed only to make the coordinates complex numbers and not to change the basic form of the Coulomb potential. The complex number total energy is written as

$$\bar{E} = Ee^{j\theta_E} \tag{33}$$

while its measured value is given by

$$E_m = E \cos \theta_E \tag{34}$$

Both $\bar{E}$ and E_m are determined in Section 7. The complex number wave function is written as

$$\bar{\Psi} = \Psi e^{j\theta_\Psi} \tag{35}$$

and will be determined in Section 5. The measured wave function is given by

$$\Psi_m = \Psi \cos \theta_\Psi \tag{36}$$

The complex number Schrödinger equation (21) can be separated into three component equations by following the standard recipe of writing the wave function as a product of three independent functions as follows[12-14]

$$\bar{\Psi} = \bar{R}(\bar{r})\bar{W}(\bar{\psi})\bar{\Phi}(\bar{\phi}) \tag{37}$$

where

$$\bar{R} = Re^{j\theta_R} = \text{radial wave function} \tag{38}$$

$$\bar{W} = We^{j\theta_W} = \text{zenith angle wave function} \tag{39}$$

$$\bar{\Phi} = \Phi e^{j\theta_\Phi} = \text{azimuthal angle wave function} \tag{40}$$

Combining equations (37) through (40) gives

$$\Psi = RW\Phi \tag{41}$$

$$\theta_\Psi = \theta_R + \theta_W + \theta_\Phi \tag{42}$$

Placing equation (37) into equation (21) yields the following generalizations of the standard azimuthal angle equation, zenith angle equation and radial equation respectively[12-14]

$$d^2\bar{\Phi}/d\bar{\phi}^2 + \bar{M}^2\bar{\Phi} = 0 \tag{43}$$

$$1/\sin \bar{\psi}\ d/d\bar{\psi}(\sin \bar{\psi}\ d\bar{W}/d\bar{\psi}) + (\bar{\beta} - \bar{M}^2/\sin^2 \bar{\psi})\bar{W} = 0 \tag{44}$$

$$\bar{r}^2\, d^2\bar{R}/d\bar{r}^2 + 2\bar{r}\, d\bar{R}/d\bar{r} + (\bar{k}^2\bar{r}^2 - \bar{\beta})\bar{R} = 0 \tag{45}$$

where

$$\bar{k}^2 = 2\mu/\hbar^2(\bar{E} - \bar{V}) \tag{46}$$

$$\bar{M} = Me^{j\theta_M} \tag{47}$$

$$\bar{\beta} = \beta e^{j\theta_\beta} \tag{48}$$

where $\bar{M}$ = complex magnetic quantum number which will be determined in Section 3, and $\bar{\beta}$ = complex number constant that will be determined in Section 4. The solution of the complex number Schrödinger equation for a hydrogen-like atom requires first the solution of equations (43) through (45) and the determination of the six functions $R(r)$, $\theta_R(r)$, $W(\psi)$, $\theta_W(\psi)$, $\Phi(\phi)$ and $\theta_\Phi(\phi)$, and secondly the determination of the values of E and θ_E which give the complex number energy eigenvalues.

3. THE AZIMUTHAL ANGLE EQUATION FOR A HYDROGEN-LIKE ATOM WITH BROKEN INTERNAL SYMMETRY. This section determines the solution of the complex number azimuthal equation and gives the magnitude and internal phase angle of the complex magnetic quantum number $\bar{M}$. The formal solution of equation (43) is written as[12-14]

$$\bar{\Phi} = \bar{A}e^{i\overline{M\phi}} + \bar{B}e^{-i\overline{M\phi}} \tag{49}$$

It will now be shown that $\overline{M\phi}$ must be a real number if $\bar{\Phi}$ is to be symmetrical (unchanged) under a 2π change of the value of ϕ_m. In other words, because ϕ_m given by equation (27) is the measured azimuthal angle, the wave function in equation (49) must be unchanged under $\phi_m \rightarrow \phi_m + 2\pi$, and this implies the reality of $\overline{M\phi}$. The reality condition for $\overline{M\phi}$ can be written as

$$\overline{M\phi} = M\phi \tag{50}$$

$$\theta_M + \theta_\phi = 0 \tag{51}$$

where equations (24) and (47) were used. In order to verify this conclusion the complex numbers $\bar{M}$ and $\bar{\phi}$ are written in terms of their real and imaginary parts as follows

$$\bar{M} = M_R + jM_I = M(\cos\theta_M + j\sin\theta_M) \tag{52}$$

$$\bar{\phi} = \phi_R + j\phi_I = \phi(\cos\theta_\phi + j\sin\theta_\phi) \tag{53}$$

Using equations (52) and (53) allow $\overline{M\phi}$ to be written as

$$\overline{M\phi} = M_R\phi_R - M_I\phi_I + j(M_I\phi_R + M_R\phi_I) \tag{54}$$

If the imaginary part of equation (54) is zero, then

$$\phi_I = -M_I\phi_R/M_R \tag{55}$$

and substituting equation (55) into equation (54) gives

$$\overline{M\phi} = (M_R^2 + M_I^2)\phi_R/M_R = M^2\phi_R/M_R = M^2\phi_m/M_R \tag{56}$$

which shows that if $\overline{M\phi}$ is real it is also linear in ϕ_m the measured value of ϕ given by equation (27).

The linearity in ϕ_m shown by equation (56) allows the possibility of having the azimuthal wave function given in equation (49) unchanged under $\phi_m \rightarrow \phi_m + 2\pi$ by requiring

$$M^2/M_R = m \qquad (57)$$

where m = standard magnetic quantum number which is a positive or negative integer. Combining equations (51), (52) and (57) gives

$$\begin{aligned} M &= m \cos \theta_M \\ &= m \cos \theta_\phi \end{aligned} \qquad (58)$$

as the condition for symmetry under $\phi_m \to \phi_m + 2\pi$. The exponent terms in equation (49) can be written as

$$\overline{M}\overline{\phi} = M\phi = m\phi \cos \theta_\phi = m\phi_R = m\phi_m \qquad (59)$$

and therefore

$$\overline{\Phi} = \overline{A}e^{i|m|\phi_m} + \overline{B}e^{-i|m|\phi_m} \qquad (60)$$

The magnitude and internal phase angle of the complex magnetic quantum number $\overline{M}$ are given by equations (58) and (51) respectively. The real and imaginary parts of $\overline{M}$ are given by

$$M_R = m \cos^2 \theta_\phi \qquad (61)$$

$$M_I = - m \sin \theta_\phi \cos \theta_\phi \qquad (62)$$

$$\overline{M} = m \cos \theta_\phi \, e^{-j\theta_\phi} \qquad (63)$$

The interesting thing about equation (58) is that M is not an integer but reduces to an integer for the symmetrical case of $\theta_\phi = 0$.

4. <u>THE ZENITH ANGLE EQUATION FOR A HYDROGEN-LIKE ATOM WITH BROKEN INTERNAL SYMMETRY.</u> This section solves the complex number zenith angle equation (44) and determines the complex number parameter $\overline{\beta}$ that appears in this equation. This equation will be solved by a simple generalization of the standard technique used to solve the corresponding scalar form of this equation.[12-14] Define the complex number $\overline{\xi}$ by

$$\overline{\xi} = \cos \overline{\psi} = \xi e^{j\theta_\xi} = C_\psi e^{-j\theta_{c\psi}} \qquad (64)$$

so that[7]

$$\xi = C_\psi \qquad (65)$$

$$\theta_\xi = - \theta_{c\psi} \qquad (66)$$

$$C_\psi = [\cos^2(\psi \cos \theta_\psi) + \sinh^2(\psi \sin \theta_\psi)]^{1/2} \qquad (67)$$

$$\tan \theta_{c\psi} = \tan(\psi \cos \theta_\psi) \tanh(\psi \sin \theta_\psi) \qquad (68)$$

Then equation (44) can be written as

$$d/d\overline{\xi}[(1 - \overline{\xi}^2)d\overline{W}/d\overline{\xi}] + [\overline{\beta} - \overline{M}^2/(1 - \overline{\xi}^2)]\overline{W} = 0 \qquad (69)$$

where $\bar{M}$ is given by equation (63).

The solution of equation (69) follows from the standard procedure for the solution of the corresponding scalar equation by writing[12-14]

$$\bar{W} = (1 - \bar{\xi}^2)^{\bar{M}'/2}\bar{G}(\bar{\xi}) \tag{70}$$

where

$$\bar{M}' = |m| \cos\theta_\phi \, e^{-j\theta_\phi} \qquad M' = |m| \cos\theta_\phi \tag{70A}$$

$$M'_R = |m| \cos^2\theta_\phi \qquad M'_I = -|m| \sin\theta_\phi \cos\theta_\phi \tag{70B}$$

which gives[12-14]

$$(1 - \bar{\xi}^2)d^2\bar{G}/d\bar{\xi}^2 - 2(\bar{M}' + 1)\bar{\xi}\, d\bar{G}/d\bar{\xi} + [\bar{\beta} - \bar{M}'(\bar{M}' + 1)]\bar{G} = 0 \tag{71}$$

The solution of equation (71) is obtained by a power series expansion and gives the following generalization of the standard results for real numbers[12-14]

$$\bar{a}_{\sigma+2}/\bar{a}_\sigma = [(\sigma + \bar{M}')(\sigma + \bar{M}' + 1) - \bar{\beta}]/[(\sigma + 1)(\sigma + 2)] \tag{72}$$

where $\bar{a}_\sigma$ and $\bar{a}_{\sigma+2}$ = coefficients in a power series expansion of $\bar{G}$, and σ = integer. Equation (72) shows that the only way the power series breaks off at term $\bar{\xi}^\nu$ is to have

$$\bar{\beta} = \bar{\mathcal{L}}(\bar{\mathcal{L}} + 1) \tag{73}$$

where

$$\bar{\mathcal{L}} = \bar{M}' + \nu \tag{74}$$

where ν = integer. Equations (73) and (74) are recognized as complex number generalizations of the standard scalar results.[12-14] The important thing about equations (72) through (74) is that $\bar{\beta}$, $\bar{M}'$ and $\bar{\mathcal{L}}$ are complex numbers and not integers as in the standard case. The only integer requirement is that $\bar{\mathcal{L}}$ and $\bar{M}'$ differ by an integer as shown in equation (74). $\bar{\mathcal{L}}$ is a complex number generalization of the standard integer azimuthal quantum number ℓ .

The solution to equation (69) can then be written using equations (70), (72) and (73)

$$\begin{aligned}\bar{W} = {} & (1 - \bar{\xi}^2)^{\bar{M}'/2} c_0\{1 + 1/2[\bar{M}'(\bar{M}' + 1) - \bar{\mathcal{L}}(\bar{\mathcal{L}} + 1)]\bar{\xi}^2 \\ & + 1/24[\bar{M}'(\bar{M}' + 1) - \bar{\mathcal{L}}(\bar{\mathcal{L}} + 1)][(\bar{M}' + 2)(\bar{M}' + 3) - \bar{\mathcal{L}}(\bar{\mathcal{L}} + 1)]\bar{\xi}^4 + \cdots\} \\ & + (1 - \bar{\xi}^2)^{\bar{M}'/2} c_1\{\bar{\xi} + 1/2[(\bar{M}' + 1)(\bar{M}' + 2) - \bar{\mathcal{L}}(\bar{\mathcal{L}} + 1)]\bar{\xi}^3 \\ & + 1/120[(\bar{M}' + 1)(\bar{M}' + 2) - \bar{\mathcal{L}}(\bar{\mathcal{L}} + 1)][(\bar{M}' + 3)(\bar{M}' + 4) - \bar{\mathcal{L}}(\bar{\mathcal{L}} + 1)]\bar{\xi}^5 + \cdots\}\end{aligned} \tag{75}$$

which clearly breaks off when equation (74) is satisfied for a series of integers $\nu = 0, 1, 2, \cdots$. The solutions given in equation (75) are simple generalizations of the standard associated Legendre functions, so that formally

$$\bar{W} = \bar{P}_{\bar{\mathcal{L}}}^{\bar{M}'}(\bar{\xi}) \tag{76}$$

Consider now the specific solutions corresponding to $\nu = \ell - |m| = 0, 1, 2, \cdots$.

For $\nu = 0$, $\ell = |m|$, $\bar{\mathcal{L}} = \bar{M}'$

$$\bar{P}_{\bar{M}'}^{\bar{M}'}(\bar{\xi}) = (1 - \bar{\xi}^2)^{\bar{M}'/2} = \sin^{\bar{M}'} \bar{\psi} \tag{77}$$

For $\nu = 1$, $\ell = |m| + 1$, $\bar{\mathcal{L}} = \bar{M}' + 1$

$$\bar{P}_{\bar{M}'+1}^{\bar{M}'}(\bar{\xi}) = (1 - \bar{\xi}^2)^{\bar{M}'/2} \bar{\xi} = \sin^{\bar{M}'} \bar{\psi} \cos \bar{\psi} \tag{78}$$

For $\nu = 2$, $\ell = |m| + 2$, $\bar{\mathcal{L}} = \bar{M}' + 2$

$$\bar{P}_{\bar{M}'+2}^{\bar{M}'}(\bar{\xi}) = (1 - \bar{\xi}^2)^{\bar{M}'/2}[(2\bar{M}' + 3)\bar{\xi}^2 - 1] \tag{79}$$

$$= \sin^{\bar{M}'} \bar{\psi} [(2\bar{M}' + 3)\cos^2 \bar{\psi} - 1]$$

For $\nu = 3$, $\ell = |m| + 3$, $\bar{\mathcal{L}} = \bar{M}' + 3$

$$\bar{P}_{\bar{M}'+3}^{\bar{M}'}(\bar{\xi}) = (1 - \bar{\xi}^2)^{\bar{M}'/2}[(2\bar{M}' + 5)\bar{\xi}^3 - 3\bar{\xi}] \tag{80}$$

$$= \sin^{\bar{M}'} \bar{\psi} [(2\bar{M}' + 5)\cos^3 \bar{\psi} - 3 \cos \bar{\psi}]$$

The value of $\bar{M}'$ that appears in equations (75) through (80) is given by equation (70A). In this way the solutions of the azimuthal equation for a hydrogen-like atom with broken internal symmetry are obtained. The solution given in equations (75) through (80) are given in Section 6 for specific atomic shells.

It is clear that the integer ν in equation (74) must be given by $\nu = \ell - |m|$ because equation (74) is also valid for the case of zero internal phase, and therefore

$$\bar{\mathcal{L}} = \bar{M}' + \ell - |m| \tag{81}$$

$$= |m| \cos \theta_\phi \, e^{-j\theta_\phi} + \ell - |m|$$

$$= \ell - |m| \sin^2 \theta_\phi - j|m| \sin \theta_\phi \cos \theta_\phi$$

where $\bar{M}'$ is given by equation (70A).

The complex azimuthal angular momentum quantum number $\bar{\mathcal{L}}$ can be written as

$$\bar{\mathcal{L}} = \mathcal{L}e^{j\theta_{\mathcal{L}}} \tag{82}$$

Combining equations (74), (81) and (82) gives

$$\mathcal{L} \cos \theta_{\mathcal{L}} = M' \cos \theta_{M'} + \ell - |m| \tag{83}$$

$$\mathcal{L} \sin \theta_{\mathcal{L}} = M' \sin \theta_{M'} \tag{84}$$

where M' and $\theta_{M'}$ are given by equations (70A) and (51) respectively. Combining equations (51), (58), (83) and (84) gives

$$\tan \theta_{\mathcal{L}} = (M' \sin \theta_{M'})/(M' \cos \theta_{M'} + \ell - |m|) \tag{85}$$

$$= -(|m| \sin \theta_\phi \cos \theta_\phi)/(\ell - |m| \sin^2 \theta_\phi)$$

$$\theta_{\mathcal{L}} \sim - |m|\theta_\phi/\ell \tag{85A}$$

$$\mathcal{L}^2 = \ell^2[1 - |m|/\ell(2 - |m|/\ell)\sin^2 \theta_\phi] \tag{86}$$

$$= \ell^2 - |m|(2\ell - |m|)\sin^2 \theta_\phi$$

Note that if $\theta_\phi = 0$ then $\theta_{\mathcal{L}} = 0$ and $\mathcal{L} = \ell$. Also if $m = 0$ then $\theta_{\mathcal{L}} = 0$ and $\mathcal{L} = \ell$. From equation (86) it follows that in general $\mathcal{L} \leqslant \ell$. The interesting point is that equation (86) shows that $\mathcal{L}$ is not an integer and depends on the values of $|m|$ as follows:

$\nu = 0$, $\ell = |m|$

$$\mathcal{L} = \ell \cos \theta_\phi = |m| \cos \theta_\phi \tag{87}$$

$\nu = 1$, $\ell = |m| + 1$

$$\mathcal{L}^2 = (|m| + 1)^2[1 - |m|(|m| + 2)/(|m| + 1)^2 \sin^2 \theta_\phi] \tag{88}$$

$$= 1 + |m|(|m| + 2)\cos^2 \theta_\phi$$

$$= \ell^2 - (\ell^2 - 1)\sin^2 \theta_\phi$$

$\nu = 2$, $\ell = |m| + 2$

$$\mathcal{L}^2 = (|m| + 2)^2[1 - |m|(|m| + 4)/(|m| + 2)^2 \sin^2 \theta_\phi] \tag{89}$$

$$= 4 + |m|(|m| + 4)\cos^2 \theta_\phi$$

$$= \ell^2 - (\ell^2 - 4) \sin^2 \theta_\phi$$

$\nu = 3$, $\ell = |m| + 3$

$$\mathcal{L}^2 = (|m| + 3)^2[1 - |m|(|m| + 6)/(|m| + 3)^2 \sin^2 \theta_\phi] \tag{90}$$

$$= 9 + |m|(|m| + 6)\cos^2 \theta_\phi$$

$$= \ell^2 - (\ell^2 - 9)\sin^2 \theta_\phi$$

5. THE RADIAL EQUATION FOR HYDROGEN-LIKE ATOMS WITH BROKEN INTERNAL SYMMETRIES. In this section the complex number radial equation (45) is solved and the complex principal quantum number is introduced. Combining equations (45)

and (73) gives

$$\bar{r}^2 d^2\bar{R}/d\bar{r}^2 + 2\bar{r}d\bar{R}/d\bar{r} + [\bar{k}^2\bar{r}^2 - \bar{\mathcal{L}}(\bar{\mathcal{L}} + 1)]\bar{R} = 0 \tag{91}$$

where $\bar{\mathcal{L}}$ is given by equations (74) or (81) and $\bar{k}$ is given by

$$\bar{k}^2 = 2\mu/\hbar^2(\bar{E} + Ze^2/\bar{r}) \tag{92}$$

The solution to equation (91) can be found by a generalization of the standard method developed for the real number form of the radial equation.[12-14] A change of dependent and independent variables is made by writing

$$\bar{\rho} = 2\bar{\alpha}\bar{r} = \rho e^{j\theta_\rho} \tag{93}$$

$$\bar{R} = \bar{\rho}^{\bar{\mathcal{L}}} e^{-\bar{\rho}/2}\bar{L}(\bar{\rho}) = Re^{j\theta_R} \tag{94}$$

where

$$\bar{\alpha}^2 = -2\mu\bar{E}/\hbar^2 = \alpha^2 e^{2j\theta_\alpha} \tag{95}$$

Substituting equations (93) through (95) into equation (91) gives

$$\bar{\rho}d^2\bar{L}/d\bar{\rho}^2 + [2(\bar{\mathcal{L}} + 1) - \bar{\rho}]d\bar{L}/d\bar{\rho} + (\bar{\eta} - \bar{\mathcal{L}} - 1)\bar{L} = 0 \tag{96}$$

where

$$\bar{\eta} = \eta e^{j\theta_\eta} = \mu Ze^2/(\bar{\alpha}\hbar^2) = Z/(a_o\bar{\alpha}) \tag{97}$$

$$a_o = \hbar^2/(\mu e^2)$$

where $\bar{\eta}$ = complex principal quantum number and a_o = Bohr radius. From equation (95) it follows that

$$\alpha^2 = -2\mu E/\hbar^2 \tag{99}$$

$$2\theta_\alpha = \theta_E \tag{100}$$

while from equation (97) it follows that

$$\eta = \mu Ze^2/(\alpha\hbar^2) = Z/(a_o\alpha) \tag{101}$$

$$\theta_\eta = -\theta_\alpha = -\theta_E/2 \tag{102}$$

Finally from equations (93) and (97) it follows that

$$\bar{\rho} = 2Z\bar{r}/(a_o\bar{\eta}) \tag{103}$$

$$\rho = 2Zr/(a_o\eta) \tag{104}$$

$$\theta_\rho = \theta_r + \theta_\alpha = \theta_r - \theta_\eta = \theta_r + \theta_E/2 \tag{105}$$

Thus θ_ρ introduces the radial coordinate internal phase angle θ_r. The values of η and θ_η will be determined later in this section.

The solution of equation (96) can be obtained by a power series whose terms have the form $a_{\nu'}\rho^{\nu'}$ where $a_{\nu'}$ are determined by the following generalization of the standard recursion formula[12]

$$(\bar{\eta} - \bar{\mathcal{L}} - 1 - \nu')\bar{a}_{\nu'} + [2(\nu' + 1)(\bar{\mathcal{L}} + 1) + \nu'(\nu' + 1)]\bar{a}_{\nu'+1} = 0 \qquad (106)$$

where ν' = integer. The condition that $a_N\rho^N$ be the last non-zero term is that $\bar{a}_{N+1} = 0$ so that the break off condition is

$$\bar{\eta} = \bar{\mathcal{L}} + 1 + N \qquad (107)$$

where N = integer. Equation (107) is the complex number generalization of the standard scalar result.[12] For a hydrogen-like atom with broken internal symmetries the principal quantum number is a complex number related to $\bar{\mathcal{L}}$ by equation (107). Equations (106) and (107) show that a break off solution to equation (96) is possible for complex principal and azimuthal quantum numbers provided that $\bar{\eta} - \bar{\mathcal{L}}$ = integer. Because equation (107) is also valid for zero internal phase angles it follows that

$$n = \ell + 1 + N \qquad (108)$$

where n and ℓ = standard integer principal and azimuthal quantum numbers respectively.

Combining equations (107) and (108) gives

$$\bar{\eta} = \bar{\mathcal{L}} + n - \ell \qquad (109)$$

where $\bar{\mathcal{L}}$ is given by equations (74) or (81). The real and imaginary parts of equation (109) give

$$\eta \cos\theta_\eta = \mathcal{L}\cos\theta_{\mathcal{L}} + n - \ell \qquad (110)$$

$$\eta \sin\theta_\eta = \mathcal{L}\sin\theta_{\mathcal{L}} \qquad (111)$$

Combining equations (110) and (111) gives

$$\tan\theta_\eta = (\mathcal{L}\sin\theta_{\mathcal{L}})/(\mathcal{L}\cos\theta_{\mathcal{L}} + n - \ell) \qquad (112)$$

$$\eta^2 = \mathcal{L}^2 + 2\mathcal{L}(n - \ell)\cos\theta_{\mathcal{L}} + (n - \ell)^2 \qquad (113)$$

where $\theta_{\mathcal{L}}$ and $\mathcal{L}$ are given by equations (85) and (86) respectively. Alternatively, equation (109) can be rewritten using equation (81) with the result

$$\bar{\eta} = \bar{M}' + n - |m| \qquad (114)$$

$$= |m| \cos\theta_\phi \, e^{-j\theta_\phi} + n - |m|$$

where $\bar{M}'$ is given by equation (70A). Taking the real and imaginary parts of equation (114) gives

$$\eta \cos\theta_\eta = M' \cos\theta_{M'} + n - |m| \tag{115}$$
$$= |m| \cos^2\theta_\phi + n - |m|$$
$$= n - |m| \sin^2\theta_\phi$$

$$\eta \sin\theta_\eta = M' \sin\theta_{M'} \tag{116}$$
$$= - |m| \sin\theta_\phi \cos\theta_\phi$$

Combining equations (115) and (116) gives

$$\tan\theta_\eta = - (|m| \sin\theta_\phi \cos\theta_\phi)/(n - |m| \sin^2\theta_\phi) \tag{117}$$

$$\eta^2 = n^2[1 - |m|/n(2 - |m|/n)\sin^2\theta_\phi] \tag{118}$$
$$= n^2 - |m|(2n - |m|)\sin^2\theta_\phi$$

Equations (117) and (118) give the internal phase angle and magnitude respectively of the complex principal quantum number for a hydrogen-like atom with broken internal symmetries. The values of θ_η and η depend on both n and $|m|$. For $m = 0$ it follows that $\theta_\eta = 0$ and $\eta = n$. For the symmetric case with $\theta_\phi = 0$ it follows that $\theta_\eta = 0$ and $\eta = n$. From equation (118) it follows that for small θ_ϕ

$$\eta \sim n[1 - 1/2|m|/n(2 - |m|/n)\sin^2\theta_\phi] \tag{119}$$

Equation (118) shows that $\eta \leqslant n$. Also, from equation (117) it follows that for small θ_ϕ

$$\theta_\eta \sim - |m|/n\theta_\phi \tag{120}$$

while equations (102) and (105) show that

$$\theta_E \sim 2|m|/n\theta_\phi \tag{121}$$

$$\theta_\rho \sim \theta_r + |m|/n\theta_\phi \tag{122}$$

Equation (96) is a complex number generalization of the standard differential equation satisfied by associated Laguerre polynomials.[12] In fact the solution of equation (96) is a complex number associated Laguerre polynomial of degree $\bar{\eta} - \bar{\mathcal{L}} - 1 = N$ and order $2\bar{\mathcal{L}} + 1$. Thus the degree is a real number (integer) but the order is a complex number. The solution of equation (96) can be written formally as[12]

$$\bar{L} = \bar{L}^{2\bar{\mathcal{L}}+1}_{\bar{\eta}+\bar{\mathcal{L}}}(\bar{\rho}) \tag{123}$$

The argument $\bar{\rho}$ is a complex number given by equation (93). The series solution to equation (96) is obtained from the recursion relations in equation (106), so that

$$\bar{L}_{\bar{n}+\bar{\mathcal{L}}}^{2\bar{\mathcal{L}}+1}(\bar{\rho}) = 1 - \frac{N}{2(\bar{\mathcal{L}}+1)}\bar{\rho} + \frac{N(N-1)}{2(\bar{\mathcal{L}}+1)[4(\bar{\mathcal{L}}+1)+2]}\bar{\rho}^2 \tag{124}$$

$$- \frac{N(N-1)(N-2)}{2(\bar{\mathcal{L}}+1)[4(\bar{\mathcal{L}}+1)+2][6(\bar{\mathcal{L}}+1)+6]}\bar{\rho}^3 + \cdots$$

out to $\bar{\rho}^N$. Because the order of the polynomial $2\bar{\mathcal{L}}+1$ is a complex number it does not make sense to derive the complex number associated Laguerre polynomials in terms of derivatives of the Laguerre polynomials.[12] For the case of complex number order, equation (96) must be considered as the fundamental defining relation, and equation (124) is the basic solution. Also, the complex number associated Laguerre polynomials cannot be derived from a generating function as this requires the order to be an integer.[12] Finally, the complex number associated Laguerre polynomials can always be written as

$$\bar{L}_{\bar{n}+\bar{\mathcal{L}}}^{2\bar{\mathcal{L}}+1}(\bar{\rho}) = L_{\bar{n}+\bar{\mathcal{L}}}^{2\bar{\mathcal{L}}+1}(\rho,\theta_\rho,\mathcal{L},\theta_{\mathcal{L}})e^{j\theta_L} \tag{125}$$

where θ_L = internal phase angle of the complex associated Laguerre polynomial.

The first few complex number associated Laguerre polynomials are obtained from equation (124) to be:

$N = 0$, $\bar{n} = \bar{\mathcal{L}} + 1$, $n = \ell + 1$

$$\bar{L}_{2\bar{\mathcal{L}}+1}^{2\bar{\mathcal{L}}+1} = 1 \tag{126}$$

$N = 1$, $\bar{n} = \bar{\mathcal{L}} + 2$, $n = \ell + 2$

$$\bar{L}_{2\bar{\mathcal{L}}+2}^{2\bar{\mathcal{L}}+1} = 1 - \frac{\bar{\rho}}{2(\bar{\mathcal{L}}+1)} = \frac{2(\bar{\mathcal{L}}+1) - \bar{\rho}}{2(\bar{\mathcal{L}}+1)} \tag{127}$$

$N = 2$, $\bar{n} = \bar{\mathcal{L}} + 3$, $n = \ell + 3$

$$\bar{L}_{2\bar{\mathcal{L}}+3}^{2\bar{\mathcal{L}}+1} = 1 - \frac{2\bar{\rho}}{2(\bar{\mathcal{L}}+1)} + \frac{2\bar{\rho}^2}{2(\bar{\mathcal{L}}+1)[4(\bar{\mathcal{L}}+1)+2]} \tag{128}$$

$$= \frac{2(\bar{\mathcal{L}}+1)(2\bar{\mathcal{L}}+3) - 2(2\bar{\mathcal{L}}+3)\bar{\rho} + \bar{\rho}^2}{2(\bar{\mathcal{L}}+1)(2\bar{\mathcal{L}}+3)}$$

$N = 3$, $\bar{\eta} = \bar{\mathcal{L}} + 4$, $n = \ell + 4$

$$L_{2\bar{\mathcal{L}}+4}^{-2\bar{\mathcal{L}}+1} = 1 - \frac{3\bar{\rho}}{2(\bar{\mathcal{L}} + 1)} + \frac{6\bar{\rho}^2}{2(\bar{\mathcal{L}} + 1)[4(\bar{\mathcal{L}} + 1) + 2]} \tag{129}$$

$$- \frac{6\bar{\rho}^3}{2(\bar{\mathcal{L}} + 1)[4(\bar{\mathcal{L}} + 1) + 2][6(\bar{\mathcal{L}} + 1) + 6]}$$

$$= \frac{4(\bar{\mathcal{L}} + 1)(2\bar{\mathcal{L}} + 3)(\bar{\mathcal{L}} + 2) - 6(2\bar{\mathcal{L}} + 3)(\bar{\mathcal{L}} + 2)\bar{\rho} + 6(\bar{\mathcal{L}} + 2)\bar{\rho}^2 - \bar{\rho}^3}{4(\bar{\mathcal{L}} + 1)(2\bar{\mathcal{L}} + 3)(\bar{\mathcal{L}} + 2)}$$

The solution of the radial equation for hydrogen-like atoms with broken internal symmetries is then obtained from equations (94) and (123) to be

$$\bar{R} = \bar{C}\bar{\rho}^{\bar{\mathcal{L}}} e^{-\bar{\rho}/2} L_{\bar{\eta}+\bar{\mathcal{L}}}^{-2\bar{\mathcal{L}}+1}(\bar{\rho}) \tag{130}$$

which is a complex number generalization of the standard result.[12] The term $\bar{\rho}^{\bar{\mathcal{L}}}$ can be written explicitly as follows

$$\bar{\rho}^{\bar{\mathcal{L}}} = e^{\bar{\mathcal{L}}\, \ell n\, \bar{\rho}} \tag{131}$$

where

$$\ell n\, \bar{\rho} = \ell n\, \rho + j\theta_\rho \tag{132}$$

Using equation (81) for $\bar{\mathcal{L}}$ gives

$$\bar{\rho}^{\bar{\mathcal{L}}} = e^{A+jB} \tag{133}$$

where

$$A = (\ell - |m| \sin^2 \theta_\phi)\ell n\, \rho + |m|\theta_\rho \sin \theta_\phi \cos \theta_\phi \tag{134}$$

$$B = - |m| \sin \theta_\phi \cos \theta_\phi\, \ell n\, \rho + \theta_\rho(\ell - |m| \sin^2 \theta_\phi) \tag{135}$$

Note also that

$$e^{-\bar{\rho}/2} = e^{-\rho/2(\cos \theta_\rho + j \sin \theta_\rho)} \tag{136}$$

The quantities ρ and θ_ρ that appear in equations (134) through (136) depend on η and θ_η through equations (104) and (105), and in turn η and θ_η depend on n, $|m|$ and θ_ϕ through equations (118) and (117) respectively. Combining equations (125), (130), (133) and (136) gives

$$\bar{R} = \bar{C}e^{(A-\rho/2 \cos \theta_\rho)} e^{j(B-\rho/2 \sin \theta_\rho+\theta_L)}\, L_{\bar{\eta}+\bar{\mathcal{L}}}^{2\bar{\mathcal{L}}+1}(\rho,\theta_\rho,\mathcal{L},\theta_{\mathcal{L}}) \tag{137}$$

The interesting thing about the complex number broken symmetry radial solution in equation (137) is that it depends on the magnetic quantum number $|m|$ an well as on the principal quantum number n . In fact equations (104) and (119) show that

$$\rho = 2Zr/(a_o n) \tag{138}$$

$$\sim 2Zr/(a_o n)[1 + 1/2|m|/n(2 - |m|/n)\sin^2 \theta_\phi]$$

where the approximation is for small θ_ϕ . Also from equation (122) θ_ρ introduces both θ_r and θ_ϕ . Therefore for $m \neq 0$ both θ_r and θ_ϕ appear in the solution of the complex number radial equation. As the simplest case consider $n = 1$, $\ell = 0$ and $m = 0$. Then $A = 0$, $B = 0$ and $\eta = 1$ and

$$\rho = 2Zr/a_o \tag{139}$$

$$\theta_\rho = \theta_r \tag{140}$$

$$\bar{R} = C_o e^{-Z\bar{r}/a_o} \tag{141}$$

$$R_m = C_o e^{-Zr/a_o \cos \theta_r} \cos(Zr/a_o \sin \theta_r) \tag{142}$$

The wave functions for hydrogen-like atoms with broken internal symmetries are listed by atomic shells in Section 6.

6. WAVE FUNCTIONS FOR A HYDROGEN-LIKE ATOM WITH BROKEN INTERNAL SYMMETRIES.

This section consists of a table of wave functions for the atomic shells of hydrogen-like atoms with broken internal symmetries. The broken internal symmetry is due basically to the broken internal symmetries of the coordinates which are described by θ_r , θ_ψ and θ_ϕ as discussed in Sections 3 through 5. The complex magnetic quantum number given by equation (70A) appears frequently in the atomic wave functions and will be written as

$$\bar{M}' = |m|\bar{y} \qquad M' = |m|y \tag{143}$$

where

$$\bar{y} = \cos\theta_\phi\, e^{-j\theta_\phi} = \cos^2\theta_\phi - j\cos\theta_\phi \sin\theta_\phi \tag{144}$$

$$y = \cos\theta_\phi \tag{145}$$

The following is a table of complete wave functions for the K , L , M and N shells. This table is a generalization of the standard scalar results.[12]

K shell: n = 1

$\ell = 0$, $m = 0$, $\bar{M}' = 0$, $M' = 0$, $\bar{\mathcal{L}} = 0$, $\mathcal{L} = 0$, $\bar{n} = 1$, $\eta = 1$,
$2\bar{\mathcal{L}} + 1 = 1$, $\bar{n} + \bar{\mathcal{L}} = 1$, $\bar{\rho} = 2Z\bar{r}/a_o$, $N = 0$

$$\bar{\Psi}_{1s_o} = \bar{C}_{1s_o} e^{-\bar{\rho}/2} \qquad (146)$$

L shell: n = 2

$\ell = 0$, $m = 0$, $\bar{M}' = 0$, $M' = 0$, $\bar{\mathcal{L}} = 0$, $\mathcal{L} = 0$, $\bar{n} = 2$, $\eta = 2$,
$2\bar{\mathcal{L}} + 1 = 1$, $\bar{n} + \bar{\mathcal{L}} = 2$, $\bar{\rho} = 2Z\bar{r}/(2a_o)$, $N = 1$

$$\bar{\Psi}_{2s_o} = \bar{C}_{2s_o} 1/2(2 - \bar{\rho})e^{-\bar{\rho}/2} \qquad (147)$$

$\ell = 1$, $m = 0$, $\bar{M}' = 0$, $M' = 0$, $\bar{\mathcal{L}} = 1$, $\mathcal{L} = 1$, $\bar{n} = 2$, $\eta = 2$,
$2\bar{\mathcal{L}} + 1 = 3$, $\bar{n} + \bar{\mathcal{L}} = 3$, $\bar{\rho} = 2Z\bar{r}/(2a_o)$, $N = 0$

$$\bar{\Psi}_{2p_o} = \bar{C}_{2p_o} \bar{\rho}\, e^{-\bar{\rho}/2} \cos \bar{\psi} \qquad (148)$$

$\ell = 1$, $m = 1$, $\bar{M}' = \bar{y}$, $M' = y$, $\bar{\mathcal{L}} = \bar{y}$, $\mathcal{L} = y$, $\bar{n} = \bar{y} + 1$,
$\eta = 2(1 - 3/4 \sin^2 \theta_\phi)^{1/2}$, $2\bar{\mathcal{L}} + 1 = 2\bar{y} + 1$, $\bar{n} + \bar{\mathcal{L}} = 2\bar{y} + 1$,
$\bar{\rho} = 2Z\bar{r}/[(\bar{y} + 1)a_o]$, $N = 0$

$$\bar{\Psi}_{2p_1} = \bar{C}_{2p_1} e^{-\bar{\rho}/2} \bar{\rho}^{\bar{y}} \sin^{\bar{y}} \bar{\psi} \cos(y\phi) \qquad (149)$$

$\ell = 1$, $m = -1$, **same as for equation (149)**

$$\bar{\Psi}_{2p_{-1}} = \bar{C}_{2p_{-1}} e^{-\bar{\rho}/2} \bar{\rho}^{\bar{y}} \sin^{\bar{y}} \bar{\psi} \sin(y\phi) \qquad (150)$$

M shell: n = 3

$\ell = 0$, $m = 0$, $\bar{M}' = 0$, $M' = 0$, $\bar{\mathcal{L}} = 0$, $\mathcal{L} = 0$, $\bar{n} = 3$, $n = 3$,
$2\bar{\mathcal{L}} + 1 = 1$, $\bar{n} + \bar{\mathcal{L}} = 3$, $\bar{\rho} = 2Z\bar{r}/(3a_o)$, $N = 2$

$$\bar{\Psi}_{3s_o} = \bar{C}_{3s_o} e^{-\bar{\rho}/2}\ 1/6(6 - 6\bar{\rho} + \bar{\rho}^2) \tag{151}$$

$\ell = 1$, $m = 0$, $\bar{M}' = 0$, $M' = 0$, $\bar{\mathcal{L}} = 1$, $\mathcal{L} = 1$, $\bar{n} = 3$, $n = 3$,
$2\bar{\mathcal{L}} + 1 = 3$, $\bar{n} + \bar{\mathcal{L}} = 4$, $\bar{\rho} = 2Z\bar{r}/(3a_o)$, $N = 1$

$$\bar{\Psi}_{3p_o} = \bar{C}_{3p_o} e^{-\bar{\rho}/2}\ \bar{\rho}/4(4 - \bar{\rho})\cos\bar{\psi} \tag{152}$$

$\ell = 1$, $m = 1$, $\bar{M}' = \bar{y}$, $M' = y$, $\bar{\mathcal{L}} = \bar{y}$, $\mathcal{L} = y$, $\bar{n} = \bar{y} + 2$,
$n = 3(1 - 5/9 \sin^2 \theta_\phi)^{1/2}$, $2\bar{\mathcal{L}} + 1 = 2\bar{y} + 1$, $\bar{n} + \bar{\mathcal{L}} = 2\bar{y} + 2$,
$\bar{\rho} = 2Z\bar{r}/[(\bar{y} + 2)a_o]$, $N = 1$

$$\bar{\Psi}_{3p_1} = \bar{C}_{3p_1} e^{-\bar{\rho}/2}\ \bar{\rho}^{\bar{y}}\ \frac{[2(\bar{y} + 1) - \bar{\rho}]}{2(\bar{y} + 1)} \sin^{\bar{y}} \bar{\psi} \cos(y\phi) \tag{153}$$

$\ell = 1$, $m = -1$, same as for equation (153)

$$\bar{\Psi}_{3p_{-1}} = \bar{C}_{3p_{-1}} e^{-\bar{\rho}/2}\ \bar{\rho}^{\bar{y}}\ \frac{[2(\bar{y} + 1) - \bar{\rho}]}{2(\bar{y} + 1)} \sin^{\bar{y}} \bar{\psi} \sin(y\phi) \tag{154}$$

$\ell = 2$, $m = 0$, $\bar{M}' = 0$, $M' = 0$, $\bar{\mathcal{L}} = 2$, $\mathcal{L} = 2$, $\bar{n} = 3$, $n = 3$,
$2\bar{\mathcal{L}} + 1 = 5$, $\bar{n} + \bar{\mathcal{L}} = 5$, $\bar{\rho} = 2Z\bar{r}/(3a_o)$, $N = 0$

$$\bar{\Psi}_{3d_o} = \bar{C}_{3d_o} e^{-\bar{\rho}/2}\ \bar{\rho}^2(3 \cos^2 \bar{\psi} - 1) \tag{155}$$

$\ell = 2$, $m = 1$, $\bar{M}' = \bar{y}$, $M' = y$, $\bar{\mathcal{L}} = \bar{y} + 1$, $\mathcal{L} = 2(1 - 3/4 \sin^2 \theta_\phi)^{1/2}$,
$\bar{n} = \bar{y} + 2$, $n = 3(1 - 5/9 \sin^2 \theta_\phi)^{1/2}$, $2\bar{\mathcal{L}} + 1 = 2\bar{y} + 3$, $\bar{n} + \bar{\mathcal{L}} = 2\bar{y} + 3$,
$\bar{\rho} = 2Z\bar{r}/[(\bar{y} + 2)a_o]$, $N = 0$

$$\bar{\Psi}_{3d_1} = \bar{C}_{3d_1} e^{-\bar{\rho}/2} \bar{\rho}^{\bar{y}+1} \sin^{\bar{y}} \bar{\psi} \cos \bar{\psi} \cos(y\phi) \quad (156)$$

$\ell = 2$, $m = -1$, same as for equation (156)

$$\bar{\Psi}_{3d_{-1}} = \bar{C}_{3d_{-1}} e^{-\bar{\rho}/2} \bar{\rho}^{\bar{y}+1} \sin^{\bar{y}} \bar{\psi} \cos \bar{\psi} \sin(y\phi) \quad (157)$$

$\ell = 2$, $m = 2$, $\bar{M}' = 2\bar{y}$, $M' = 2y$, $\bar{\mathcal{L}} = 2\bar{y}$, $\mathcal{L} = 2y$, $\bar{n} = 2\bar{y} + 1$,
$n = 3(1 - 8/9 \sin^2 \theta_\phi)^{1/2}$, $2\bar{\mathcal{L}} + 1 = 4\bar{y} + 1$, $\bar{n} + \bar{\mathcal{L}} = 4\bar{y} + 1$,
$\bar{\rho} = 2Z\bar{r}/[(2\bar{y} + 1)a_o]$, $N = 0$

$$\bar{\Psi}_{3d_2} = \bar{C}_{3d_2} e^{-\bar{\rho}/2} \bar{\rho}^{2\bar{y}} \sin^{2\bar{y}} \bar{\psi} \cos(2y\phi) \quad (158)$$

$\ell = 2$, $m = -2$, same as for equation (158)

$$\bar{\Psi}_{3d_{-2}} = \bar{C}_{3d_{-2}} e^{-\bar{\rho}/2} \bar{\rho}^{2\bar{y}} \sin^{2\bar{y}} \bar{\psi} \sin(2y\phi) \quad (159)$$

N shell: $n = 4$

$\ell = 0$, $m = 0$, $\bar{M}' = 0$, $M' = 0$, $\bar{\mathcal{L}} = 0$, $\mathcal{L} = 0$, $\bar{n} = 4$, $n = 4$,
$2\bar{\mathcal{L}} + 1 = 1$, $\bar{n} + \bar{\mathcal{L}} = 4$, $\bar{\rho} = 2Z\bar{r}/(4a_o)$, $N = 3$

$$\bar{\Psi}_{4s_o} = \bar{C}_{4s_o} e^{-\bar{\rho}/2} 1/24(24 - 36\bar{\rho} + 12\bar{\rho}^2 - \bar{\rho}^3) \quad (160)$$

$\ell = 1$, $m = 0$, $\bar{M}' = 0$, $M' = 0$, $\bar{\mathcal{L}} = 1$, $\mathcal{L} = 1$, $\bar{\eta} = 4$, $\eta = 4$,
$2\bar{\mathcal{L}} + 1 = 3$, $\bar{\eta} + \bar{\mathcal{L}} = 5$, $\bar{\rho} = 2Z\bar{r}/(4a_o)$, $N = 2$

$$\bar{\Psi}_{4p_o} = \bar{C}_{4p_o} e^{-\bar{\rho}/2} \bar{\rho}/20(20 - 10\bar{\rho} + \bar{\rho}^2)\cos\bar{\psi} \qquad (161)$$

$\ell = 1$, $m = 1$, $\bar{M}' = \bar{y}$, $M' = y$, $\bar{\mathcal{L}} = \bar{y}$, $\mathcal{L} = y$, $\bar{\eta} = \bar{y} + 3$,
$\eta = 4(1 - 7/16 \sin^2 \theta_\phi)^{1/2}$, $2\bar{\mathcal{L}} + 1 = 2\bar{y} + 1$, $\bar{\eta} + \bar{\mathcal{L}} = 2\bar{y} + 3$,
$\bar{\rho} = 2Z\bar{r}/[(\bar{y} + 3)a_o]$, $N = 2$

$$\bar{\Psi}_{4p_1} = \bar{C}_{4p_1} e^{-\bar{\rho}/2} \bar{\rho}^{\bar{y}} \frac{[2(\bar{y} + 1)(2\bar{y} + 3) - 2(2\bar{y} + 3)\bar{\rho} + \bar{\rho}^2]}{2(\bar{y} + 1)(2\bar{y} + 3)} \sin^{\bar{y}}\bar{\psi}\cos(y\phi) \qquad (162)$$

$\ell = 1$, $m = -1$, same as for equation (162)

$$\bar{\Psi}_{4p_{-1}} = \bar{C}_{4p_{-1}} e^{-\bar{\rho}/2} \bar{\rho}^{\bar{y}} \frac{[2(\bar{y} + 1)(2\bar{y} + 3) - 2(2\bar{y} + 3)\bar{\rho} + \bar{\rho}^2]}{2(\bar{y} + 1)(2\bar{y} + 3)} \sin^{\bar{y}}\bar{\psi}\sin(y\phi) \qquad (163)$$

$\ell = 2$, $m = 0$, $\bar{M}' = 0$, $M' = 0$, $\bar{\mathcal{L}} = 2$, $\mathcal{L} = 2$, $\bar{\eta} = 4$, $\eta = 4$,
$2\bar{\mathcal{L}} + 1 = 5$, $\bar{\eta} + \bar{\mathcal{L}} = 6$, $\bar{\rho} = 2Z\bar{r}/(4a_o)$, $N = 1$

$$\bar{\Psi}_{4d_o} = \bar{C}_{4d_o} e^{-\bar{\rho}/2} \bar{\rho}^2\ 1/6(6 - \bar{\rho})(3\cos^2\bar{\psi} - 1) \qquad (164)$$

$\ell = 2$, $m = 1$, $\bar{M}' = \bar{y}$, $M' = y$, $\bar{\mathcal{L}} = \bar{y} + 1$, $\mathcal{L} = 2(1 - 3/4 \sin^2 \theta_\phi)^{1/2}$,
$\bar{\eta} = \bar{y} + 3$, $\eta = 4(1 - 7/16 \sin^2 \theta_\phi)^{1/2}$, $2\bar{\mathcal{L}} + 1 = 2\bar{y} + 3$, $\bar{\eta} + \bar{\mathcal{L}} = 2\bar{y} + 4$,
$\bar{\rho} = 2Z\bar{r}/[(\bar{y} + 3)a_o]$, $N = 1$

$$\bar{\Psi}_{4d_1} = \bar{C}_{4d_1} e^{-\bar{\rho}/2} \bar{\rho}^{\bar{y}+1} \frac{[2(\bar{y} + 2) - \bar{\rho}]}{2(\bar{y} + 2)} \sin^{\bar{y}}\bar{\psi}\cos\bar{\psi}\cos(y\phi) \qquad (165)$$

$\ell = 2$, $m = -1$, same as for equation (165)

$$\bar{\Psi}_{4d_{-1}} = \bar{C}_{4d_{-1}} e^{-\bar{\rho}/2} \bar{\rho}^{\bar{y}+1} \frac{[2(\bar{y} + 2) - \bar{\rho}]}{2(\bar{y} + 2)} \sin^{\bar{y}} \bar{\psi} \cos \bar{\psi} \sin(y\phi) \qquad (166)$$

$\ell = 2$, $m = 2$, $\bar{M}' = 2\bar{y}$, $M' = 2y$, $\bar{\mathcal{L}} = 2\bar{y}$, $\mathcal{L} = 2y$, $\bar{n} = 2\bar{y} + 2$,
$n = 4(1 - 3/4 \sin^2 \theta_\phi)^{1/2}$, $2\bar{\mathcal{L}} + 1 = 4\bar{y} + 1$, $\bar{n} + \bar{\mathcal{L}} = 4\bar{y} + 2$,
$\bar{\rho} = 2Z\bar{r}/[(2\bar{y} + 2)a_o]$, $N = 1$

$$\bar{\Psi}_{4d_2} = \bar{C}_{4d_2} e^{-\bar{\rho}/2} \bar{\rho}^{2\bar{y}} \frac{[2(2\bar{y} + 1) - \bar{\rho}]}{2(2\bar{y} + 1)} \sin^{2\bar{y}} \bar{\psi} \cos(2y\phi) \qquad (167)$$

$\ell = 2$, $m = -2$, same as for equation (167)

$$\bar{\Psi}_{4d_{-2}} = \bar{C}_{4d_{-2}} e^{-\bar{\rho}/2} \bar{\rho}^{2\bar{y}} \frac{[2(2\bar{y} + 1) - \bar{\rho}]}{2(2\bar{y} + 1)} \sin^{2\bar{y}} \bar{\psi} \sin(2y\phi) \qquad (168)$$

$\ell = 3$, $m = 0$, $\bar{M}' = 0$, $M' = 0$, $\bar{\mathcal{L}} = 3$, $\mathcal{L} = 3$, $\bar{n} = 4$, $n = 4$,
$2\bar{\mathcal{L}} + 1 = 7$, $\bar{n} + \bar{\mathcal{L}} = 7$, $\bar{\rho} = 2Z\bar{r}/(4a_o)$, $N = 0$

$$\bar{\Psi}_{4f_o} = \bar{C}_{4f_o} e^{-\bar{\rho}/2} \bar{\rho}^3 (5 \cos^3 \bar{\psi} - 3 \cos \bar{\psi}) \qquad (169)$$

$\ell = 3$, $m = 1$, $\bar{M}' = \bar{y}$, $M' = y$, $\bar{\mathcal{L}} = \bar{y} + 2$, $\mathcal{L} = 3(1 - 5/9 \sin^2 \theta_\phi)^{1/2}$,
$\bar{n} = \bar{y} + 3$, $n = 4(1 - 7/16 \sin^2 \theta_\phi)^{1/2}$, $2\bar{\mathcal{L}} + 1 = 2\bar{y} + 5$, $\bar{n} + \bar{\mathcal{L}} = 2\bar{y} + 5$,
$\bar{\rho} = 2Z\bar{r}/[(\bar{y} + 3)a_o]$, $N = 0$

$$\bar{\Psi}_{4f_1} = \bar{C}_{4f_1} e^{-\bar{\rho}/2} \bar{\rho}^{\bar{y}+2} \sin^{\bar{y}} \bar{\psi} [(2\bar{y} + 3)\cos^2 \bar{\psi} - 1]\cos(y\phi) \qquad (170)$$

$\ell = 3$, $m = -1$, same as for equation (170)

$$\bar{\Psi}_{4f_{-1}} = \bar{C}_{4f_{-1}} e^{-\bar{\rho}/2} \bar{\rho}^{\bar{y}+2} \sin^{\bar{y}} \bar{\psi} \, [(2\bar{y} + 3)\cos^2 \bar{\psi} - 1]\sin(y\phi) \qquad (171)$$

$\ell = 3$, $m = 2$, $\bar{M}' = 2\bar{y}$, $M' = 2y$, $\bar{\mathcal{L}} = 2\bar{y} + 1$, $\mathcal{L} = 3(1 - 8/9 \sin^2 \theta_\phi)^{1/2}$,
$\bar{\eta} = 2\bar{y} + 2$, $\eta = 4(1 - 3/4 \sin^2 \theta_\phi)^{1/2}$, $2\bar{\mathcal{L}} + 1 = 4\bar{y} + 3$, $\bar{\eta} + \bar{\mathcal{L}} = 4\bar{y} + 3$,
$\bar{\rho} = 2Z\bar{r}/[(2\bar{y} + 2)a_o]$, $N = 0$

$$\bar{\Psi}_{4f_2} = \bar{C}_{4f_2} e^{-\bar{\rho}/2} \bar{\rho}^{2\bar{y}+1} \sin^{2\bar{y}} \bar{\psi} \cos \bar{\psi} \cos(2y\phi) \qquad (172)$$

$\ell = 3$, $m = -2$, same as for equation (172)

$$\bar{\Psi}_{4f_{-2}} = \bar{C}_{4f_{-2}} e^{-\bar{\rho}/2} \bar{\rho}^{2\bar{y}+1} \sin^{2\bar{y}} \bar{\psi} \cos \bar{\psi} \sin(2y\phi) \qquad (173)$$

$\ell = 3$, $m = 3$, $\bar{M}' = 3\bar{y}$, $M' = 3y$, $\bar{\mathcal{L}} = 3\bar{y}$, $\mathcal{L} = 3y$, $\bar{\eta} = 3\bar{y} + 1$,
$\eta = 4(1 - 15/16 \sin^2 \theta_\phi)^{1/2}$, $2\bar{\mathcal{L}} + 1 = 6\bar{y} + 1$, $\bar{\eta} + \bar{\mathcal{L}} = 6\bar{y} + 1$,
$\bar{\rho} = 2Z\bar{r}/[(3\bar{y} + 1)a_o]$, $N = 0$

$$\bar{\Psi}_{4f_3} = \bar{C}_{4f_3} e^{-\bar{\rho}/2} \bar{\rho}^{3\bar{y}} \sin^{3\bar{y}} \bar{\psi} \cos(3y\phi) \qquad (174)$$

$\ell = 3$, $m = -3$, same as for equation (174)

$$\bar{\Psi}_{4f_{-3}} = \bar{C}_{4f_{-3}} e^{-\bar{\rho}/2} \bar{\rho}^{3\bar{y}} \sin^{3\bar{y}} \bar{\psi} \sin(3y\phi) \qquad (175)$$

The reader should be aware that the value of $\bar{\rho}$ that appears in the above table depends on $\bar{\eta}$ and is thus dependent on n and $|m|$.

7. ENERGY LEVELS AND RADII OF HYDROGEN-LIKE ATOMS WITH BROKEN INTERNAL SYMMETRY. This section calculates the measurable values of the energy levels and atomic radii of hydrogen-like atoms with broken internal symmetries. The complex number energy levels are obtained from equations (95) and (97) to be

$$\bar{E}_\eta = E_\eta e^{j\theta_{E\eta}} = -\hbar^2/(2\mu)\bar{\alpha}^2 = -\mu Z^2 e^4/(2\hbar^2\bar{\eta}^2) \tag{176}$$

where $\bar{\eta}$ is given by equations (109) or (114). From equation (176) it follows that

$$E_\eta = -\mu Z^2 e^4/(2\hbar^2\eta^2) \tag{177}$$
$$\sim -\mu Z^2 e^4/(2\hbar^2 n^2)[1 + |m|/n(2 - |m|/n)\sin^2\theta_\phi]$$

$$\theta_{E\eta} = -2\theta_\eta \tag{178}$$

where η and θ_η are given by equations (118) and (117) respectively. The measured energy levels are given by[7]

$$E_{\eta m} = E_\eta \cos\theta_{E\eta} \tag{179}$$
$$= -\mu Z^2 e^4/(2\hbar^2\eta^2)\cos(2\theta_\eta)$$

From equation (116) it follows that

$$\sin\theta_\eta = -|m|/\eta \cos\theta_\phi \sin\theta_\phi \tag{180}$$

$$\cos(2\theta_\eta) = 1 - 2\sin^2\theta_\eta \tag{181}$$
$$= 1 - 2(|m|/\eta)^2\cos^2\theta_\phi \sin^2\theta_\phi$$
$$\sim 1 - 2(|m|/n)^2\cos^2\theta_\phi \sin^2\theta_\phi$$

Combining equations (177), (179) and (181) gives

$$E_{\eta m} \sim -\mu Z^2 e^4/(2\hbar^2 n^2)(1 + F\sin^2\theta_\phi) \tag{182}$$

where

$$F = |m|/n[2 - |m|/n(1 + 2\cos^2\theta_\phi)] \tag{183}$$
$$\sim |m|/n(2 - 3|m|/n)$$

The values of F are given approximately for small θ_ϕ as follows

$$F = 0 \quad m = 0 \text{ or } |m|/n = 2/3 \qquad (184)$$

$$F > 0 \quad |m|/n < 2/3 \qquad (185)$$

$$F < 0 \quad |m|/n > 2/3 \qquad (186)$$

Equations (177), (179) and (182) reduce to the standard Bohr result in equation (10) for the case $\theta_\phi = 0$.

For the K and L shells $|m|/n < 1/2$ so that $F \geqslant 0$. For the M and P shells, etc, it is possible to have the situation $|m|/n = 2/3$ and $F = 0$. For the N , O , P , ··· shells it is possible to have $|m|/n > 2/3$ and $F = < 0$. In any case it is clear from equations (182) and (183) that the measured energy eigenvalues of hydrogen-like atoms with broken internal symmetry should depend on $|m|$ as well as on the principal quantum number n . In addition, the asymmetry factor in equation (182) is $F \sin^2 \theta_\phi$ where θ_ϕ depends on the pressure acting on the system of atoms. Thus the broken symmetry of the azimuthal angle destroys the degeneracy associated with the energy eigenvalues given by equation (10). The energy eigenvalues do not depend explicitly on ℓ and therefore some remaining degeneracy still exists.

A transition from a state n' , m' to the state n , m is according to equation (182) associated with the energy difference

$$\Delta E_{\eta m} = \mu Z^2 e^4/(2\hbar^2)[1/n^2 - 1/n'^2 + (F/n^2 - F'/n'^2)\sin^2 \theta_\phi] \qquad (187)$$

where

$$F/n^2 = |m|/n^3[2 - |m|/n(1 + 2\cos^2 \theta_\phi)] \qquad (188)$$

$$F'/n'^2 = |m'|/n'^3[2 - |m'|/n'(1 + 2\cos^2 \theta_\phi)] \qquad (189)$$

$$F/n^2 - F'/n'^2 = 2(|m|/n^3 - |m'|/n'^3) - (|m|^2/n^4 - |m'|^2/n'^4)(1 + 2\cos^2 \theta_\phi) \qquad (190)$$

$$\sim 2(|m|/n^3 - |m'|/n'^3) - 3(|m|^2/n^4 - |m'|^2/n'^4)$$

It is the internal phase angle θ_ϕ of the azimuthal angle that introduces the magnetic quantum number into the energy eigenvalues given in equation (177) and in the formula for the transition energy given in equation (187).

The pressure variation of the energy eigenvalues will now be calculated. From equation (179) it follows that

$$\partial E_{\eta m}/\partial P = \mu Z^2 e^4/(\hbar^2 n^2)[\cos(2\theta_\eta)1/\eta\partial\eta/\partial P + \sin(2\theta_\eta)\partial\theta_\eta/\partial P] \qquad (191)$$

The derivatives on the right hand side of equation (191) are easily evaluated. The value of $\partial\eta/\partial P$ is obtained from equation (118) to be

$$1/\eta\partial\eta/\partial P = -\kappa_1 \partial\theta_\phi/\partial P \tag{192}$$

where

$$\kappa_1 = \frac{|m|/n(2 - |m|/n)\sin\theta_\phi \cos\theta_\phi}{1 - |m|/n(2 - |m|/n)\sin^2\theta_\phi} \tag{193}$$

$$\sim |m|/n(2 - |m|/n)\sin\theta_\phi \cos\theta_\phi$$

The value of $\partial\theta_\eta/\partial P$ can be obtained from equation (117) with the result

$$\partial\theta_\eta/\partial P = -\kappa_2 \partial\theta_\phi/\partial P \tag{194}$$

where

$$\kappa_2 = \frac{n|m|\cos(2\theta_\phi) + |m|^2 \sin^2\theta_\phi}{n^2 - |m|(2n - |m|)\sin^2\theta_\phi} \tag{195}$$

$$\sim |m|/n$$

Placing equations (192) and (194) into equation (191) gives

$$\partial E_{\eta m}/\partial P = -\mu Z^2 e^4/(\hbar^2 n^2)[\kappa_1 \cos(2\theta_\eta) + \kappa_2 \sin(2\theta_\eta)]\partial\theta_\phi/\partial P \tag{196}$$

where from equations (115) and (116)

$$\cos(2\theta_\eta) \sim 1 - 2(|m|/n)^2 \cos^2\theta_\phi \sin^2\theta_\phi \tag{197}$$

$$\sim 1$$

$$\sin(2\theta_\eta) = -2|m|/\eta^2(n - |m|\sin^2\theta_\phi)\sin\theta_\phi \cos\theta_\phi \tag{198}$$

$$\sim -2|m|/n \sin\theta_\phi \cos\theta_\phi$$

$$\theta_\eta \sim -|m|/n\ \theta_\phi \tag{199}$$

for small θ_ϕ . Therefore combining equations (193) and (195) through (198) gives

$$\partial E_{\eta m}/\partial P \sim -\mu Z^2 e^4/(\hbar^2 n^2)|m|/n(2 - 3|m|/n)\sin\theta_\phi \cos\theta_\phi\ \partial\theta_\phi/\partial P \tag{200}$$

The rate of change of the energy eigenstates with pressure can be positive or negative according to the value of $|m|/n$. In particular if $\partial\theta_\phi/\partial P > 0$ equation (200) gives

$$\left.\begin{array}{ll}\partial E_{nm}/\partial P \leqslant 0 & |m|/n \leqslant 2/3 \\ \partial E_{nm}/\partial P > 0 & 2/3 < |m|/n < 1\end{array}\right\} \qquad \begin{array}{r}(201)\\(202)\end{array}$$

For $|m|/n < 2/3$ an increase in pressure will lead to a greater binding energy per electron while for $|m|/n > 2/3$ an increase in pressure will produce less binding energy for the valence electron. These conclusions also follow directly from equations (182) and (183).

Finally, the simplest generalization of equation (9) for the radii of the Bohr orbits is given by

$$\bar{r}_n = \bar{n}^2\hbar^2/(\mu Z e^2) \tag{203}$$

and therefore combining equations (97), (118) and (203) gives

$$r_n = n^2\hbar^2/(\mu Z e^2) = n^2\hbar^2/(\mu Z e^2)[1 - |m|/n(2 - |m|/n)\sin^2\theta_\phi] \tag{204}$$

$$\theta_{rn} = 2\theta_n \sim -2|m|/n\theta_\phi \tag{205}$$

where θ_n is given by equations (117) and (120). The radii of the Bohr orbits of an atom with broken internal symmetry are pressure dependent through $\theta_\phi(P)$. The measured Bohr radii are given by

$$r_{nm} = r_n \cos\theta_{rn} \tag{206}$$

$$\sim n^2\hbar^2/(\mu Z e^2)(1 - G\sin^2\theta_\phi)$$

where

$$G = |m|/n[2 + |m|/n\cos(2\theta_\phi)] \tag{207}$$

$$\sim |m|/n(2 + |m|/n)$$

where $G \geqslant 0$. From equations (206) and (207) it follows that the measured Bohr radii are pressure dependent and

$$\partial r_{nm}/\partial P \sim -2n^2\hbar^2/(\mu Z e^2)|m|/n(2 + |m|/n)\sin\theta_\phi\cos\theta_\phi\,\partial\theta_\phi/\partial P \tag{208}$$

Therefore if $\partial\theta_\phi/\partial P > 0$ it follows that $\partial r_{nm}/\partial P < 0$. The analysis leading to equations (203) through (208) is simplistic because higher order terms associated with the azimuthal angular momentum $\underline{L}$ need to be inserted into equation (203).[12] The results for the energy levels and atomic radii may possibly be tested using circular atoms for which $|m| = \ell = n - 1$.[15]

8. CONCLUSION. The pressure dependence of the energy levels and radii of hydrogen-like atoms can be determined by taking into account the broken symmetries of the coordinates of the electron and the nucleus. The broken symmetries of the azimuthal and zenith angles are due essentially to a broken symmetry pressure field. But the vacuum state also exhibits this broken symmetry.[16] The broken internal symmetries of the coordinates requires the magnetic, azimuthal and principal quantum numbers to be complex numbers which are associated with internal phase angles. The internal phase angles of the three quantum numbers are expected to be pressure dependent. Schrödinger's equation for a hydrogen-like atom can be solved with complex quantum numbers, and break off solutions to the azimuthal angle, zenith angle and radial equations can be obtained.

The broken symmetry of the azimuthal angle may be associated with a vector boson which can be called the "muthon". The muthon may have important physical effects in systems having large scale broken azimuthal symmetry such as perhaps in the layered copper oxide structures of high temperature superconductors where it may serve as the intermediary particle for the formation of Cooper pairs of electrons or holes. Alternatively, the muthon could also play a role as the component of dark matter in galaxies where large scale broken azimuthal symmetry may occur due to the broken internal symmetry of the gravitational field.

ACKNOWLEDGEMENT

The author wishes to thank Elizabeth K. Klein for typing this paper.

REFERENCES

1. Itzykson, C. and Zuber, J. B., Quantum Field Theory, McGraw-Hill, New York, 1980.

2. Becher, P., Böhm, M. and Joos, H., Gauge Theories of Strong and Electroweak Interactions, John Wiley, New York, 1984.

3. Ryder, L. H., Quantum Field Theory, Cambridge University Press, 1985.

4. DeWit, B. and Smith, J., Field Theory in Particle Physics, North-Holland, New York, 1986.

5. Weiss, R. A., "Scale Invariant Equations for Relativistic Waves," Fourth Army Conference on Applied Mathematics and Computing, Cornell University, Ithaca, NY, ARO 87-1, 27-30 May 1986, p. 307.

6. Weiss, R. A., "Thermodynamic Gauge Theory of Solids and Quantum Liquids with Internal Phase," Fifth Army Conference on Applied Mathematics and Computing, West Point, New York, ARO 88-1, 15-18 June 1987, p. 649.

7. Weiss, R. A., "The Broken Symmetry of Space and Time in Bulk Matter and the Vacuum," Sixth Army Conference on Applied Mathematics and Computing, Boulder, Colorado, ARO 89-1, 31 May-3 June 1988, p. 317.

8. Witten, E., "Space-Time and Topological Orbifolds," Phys. Rev. Lett., Vol. 61, No. 6, 8 Aug 1988, p. 670.

9. Weiss, R. A., "Gauge Theory of Atomic Processes," Sixth Army Conference on Applied Mathematics and Computing, Boulder, Colorado, ARO 89-1, 31 May-3 June 1988, p. 223.

10. Richtmyer, F. K., Kennard, E. H. and Lauritsen, T., Introduction to Modern Physics, McGraw-Hill, New York, 1955.

11. Leighton, R. B., Principles of Modern Physics, McGraw-Hill, New York, 1959.

12. Pauling, L. and Wilson, E. B., Introduction to Quantum Mechanics, McGraw-Hill, New York, 1935.

13. Merzbacher, E., Quantum Mechanics, John Wiley, New York, 1961.

14. Powell, J. L. and Crasemann, B., Quantum Mechanics, Addison-Wesley, Reading, Massachusetts, 1961.

15. Hare, J., Gross, M. and Goy, P., "Circular Atoms Prepared by a New Method of Crossed Electric and Magnetic Fields," Phys. Rev. Lett., Vol. 61, No. 17, 24 Oct. 1988, p. 1938.

16. Weiss, R. A., "Maxwell's Equations with Broken Internal Symmetries," Sixth Army Conference on Applied Mathematics and Computing, Boulder, Colorado, ARO 89-1, 31 May-3 June 1988, p.271.

L

WAVE PROPAGATION IN ASYMMETRIC MEDIA

Richard A. Weiss
U. S. Army Engineer Waterways Experiment Station
Vicksburg, Mississippi 39180

ABSTRACT. The coordinates of space and time have broken internal symmetries for a region of spacetime located in a pressure field and perhaps even for the vacuum. Geometrical angles themselves have internal phase angles. A wave propagating in matter or the vacuum with broken internal symmetries will exhibit internal phase angles in its amplitude and dispersion characteristics. Cylindrical and spherical wave propagation in asymmetric matter is treated and the solution of the wave equation with broken internal symmetries is obtained. The observed periodicities of waves in measured time and measured space requires the propagation constants to be complex numbers but the phase must be a real number. A pressure field is associated with a broken internal symmetry, and therefore waves propagating in matter under pressure are expected to exhibit broken symmetry effects in the propagation parameters. Applications to acoustic and seismic waves are suggested.

1. INTRODUCTION. Matter and radiation exists within the continuum of spacetime, and it has been suggested that spacetime imprints measurable effects on the properties of bulk matter and radiation. These effects have been calculated by the development of a gauge theory of relativistic thermodynamics.[1] The effects of spacetime structure on matter and radiation occur in two ways, the first is by the effects of the Grüneisen parameter and bulk modulus which enter the relativistic trace equation as a requirement of gauge invariance.[1] The second way the metric of spacetime affects the state equation of matter and radiation is by requiring the thermodynamic functions such as pressure and internal energy to exhibit broken internal symmetries.[2] At the same time the coordinates of points located within matter, radiation or the vacuum also have broken internal symmetries and the internal phase angles of the coordinates must be determined simultaneously with the internal phase angles of the thermodynamic functions.[3]

All physical phenomena occuring within matter, radiation or the vacuum are affected by the broken symmetries of space and time. Electromagnetic and mechanical waves are expected to exhibit the effects of broken spacetime symmetry in both the wave amplitude and dispersion equation. The wave amplitude, wavelength and frequency are characterized by internal phase angles and therefore must be represented as complex numbers. The speed of sound and electromagnetic waves in matter must be represented as complex numbers, but the light speed in vacuum is a real number.

The broken symmetry of the pressure in matter or vacuum is derived from a relativistic trace equation.[2] In bulk matter or vacuum the space and time coordinates are complex numbers and are written as follows[3]

$$\bar{t} = te^{j\theta_t} \tag{1}$$

for the time, while the cartesian space coordinates are

$$\bar{x} = xe^{j\theta_x} \qquad \bar{y} = ye^{j\theta_y} \qquad \bar{z} = ze^{j\theta_z} \tag{2}$$

the cylindrical coordinates are

$$\bar{r} = re^{j\theta_r} \qquad \bar{\phi} = \phi e^{j\theta_\phi} \qquad \bar{z} = ze^{j\theta_\psi} \tag{3}$$

and for spherical coordinates

$$\bar{r} = re^{j\theta_r} \qquad \bar{\phi} = \phi e^{j\theta_\phi} \qquad \bar{\psi} = \psi e^{j\theta_\psi} \tag{4}$$

The measured time coordinate is just the real part of the complex number time[3]

$$t_m = t \cos \theta_t \tag{5}$$

while the measured space coordinates are given by

$$x_m = x \cos \theta_x \qquad y_m = y \cos \theta_y \qquad z_m = z \cos \theta_z \tag{6}$$

$$r_m = r \cos \theta_r \qquad \phi_m = \phi \cos \theta_\phi \qquad \psi_m = \psi \cos \theta_\psi \tag{7}$$

where t_m = measured time and x_m , y_m , z_m , r_m , ϕ_m and ψ_m = measured space coordinates.

This paper considers the solution of a complex number wave equation whose space and time coordinates have broken internal symmetries. Section 2 considers the time dependence of periodic waves in broken symmetry matter where the periodicity occurs in the real part (the measured part) of the complex number time coordinate. Section 3 considers cylindrical waves with broken internal symmetry and it is shown that the azimuthal angle equation has a complex number separation constant. The real part of the complex number azimuthal angle has the $0 \rightarrow 2\pi$ symmetry. The remaining coordinate equations also have complex number separation constants which are determined by the requirement that the wave periodicity occurs in the real parts of the complex number radial and z coordinates. Section 4 considers spherical wave propagation in asymmetric matter, and develops the equations describing the conditions of periodicity in the real parts of the radial, azimuthal and zenith angle coordinates.

2. PERIODIC VIBRATIONS IN SPACETIME WITH BROKEN INTERNAL SYMMETRIES. This section determines the relationship between the measured period and frequency from the experimental observation that waves and vibrations are periodic in measured space and time coordinates. The equation describing the time dependence of a periodic phenomena in space and time with broken internal symmetries is written as a generalization of the standard scalar equation[4-7]

$$d^2\bar{\mathfrak{T}}/d\bar{t}^2 + \bar{\omega}^2\bar{\mathfrak{T}} = 0 \tag{8}$$

whose solution is

$$\bar{\mathfrak{T}} = \bar{\mathfrak{T}}_o e^{i\bar{\omega}\bar{t}} \tag{9}$$

where $\bar{t}$ is given by equation (1) and where

$$\bar{\omega} = \omega e^{j\theta_\omega} = 2\pi/\bar{T} = 2\pi/Te^{-j\theta_T} \tag{10}$$

and therefore

$$\omega = 2\pi f = 2\pi/T \qquad \theta_\omega = -\theta_T \tag{11}$$

where $\bar{\omega}$ = complex number angular frequency whose magnitude and phase are ω and θ_ω respectively, and $\bar{T}$ = complex number period whose magnitude and phase are T and θ_T respectively.

The requirement that equation (9) represents a periodic wave in measured time implies that $\bar{\omega}\bar{t}$ is a real number. This can be seen by first writing $\bar{\omega}$ and $\bar{t}$ as complex numbers as follows

$$\bar{\omega} = \omega_R + j\omega_I = \omega(\cos\theta_\omega + j\sin\theta_\omega) \tag{12}$$

$$\bar{t} = t_R + jt_I = t(\cos\theta_t + j\sin\theta_t) \tag{13}$$

then

$$\bar{\omega}\bar{t} = \omega_R t_R - \omega_I t_I + j(\omega_I t_R + \omega_R t_I) \tag{14}$$

The reality condition gives

$$t_I = -\omega_I t_R/\omega_R \tag{15}$$

and therefore

$$\bar{\omega}\bar{t} = t_R\omega^2/\omega_R = t_m\omega^2/\omega_m = \omega t \tag{16}$$

where $\omega_m = \omega_R$ = measured angular speed. Therefore the reality of the phase $\bar{\omega}\bar{t}$ requires that the phase be linear in both the measured time t_m and the time magnitude t . The fact that the phase is linear in the measured time agrees with the experimental fact that vibrating systems are periodic in the measured time coordinate.

The measured period $T_m = T_R = T\cos\theta_T$ is obtained from equations (12) and (16) to be

$$T_R\omega^2/\omega_R = T_R\omega/\cos\theta_\omega = \omega T = 2\pi \tag{17}$$

and therefore

$$T_R = 2\pi/\omega \cos\theta_\omega = 2\pi/\omega_R \cos^2\theta_\omega = 1/f_R \cos^2\theta_\omega \qquad (18)$$

where $f_R = \omega_R/(2\pi)$ = real part of the frequency. Therefore the relationship between the measured period and the measured angular frequency is

$$T_m = 2\pi/\omega_m \cos^2\theta_\omega = 2\pi/\omega \cos\theta_\omega = 1/f \cos\theta_\omega \qquad (19)$$

or for the measured frequency and the measured period

$$f_m = 1/T_m \cos^2\theta_\omega = 1/T_m \cos^2\theta_T \qquad (20)$$

Note that $f_m \neq 1/T_m$. From equations (10), (11) and (16) it follows that

$$\theta_\omega = \theta_{\bar{f}} = -\theta_T = -\theta_t \qquad (21)$$

Periodicity requires the internal phase of the frequency to adjust itself so as to satisfy equation (21). Combining equations (11), (16), (19) and (21) gives

$$T = 1/f \qquad (22)$$

$$t/f = t_m/f_m \qquad (23)$$

$$tf = t/T = t_m/T_m \qquad (24)$$

Thus the phase in equation (16) can be written as

$$\bar{\omega}\bar{t} = \omega t = 2\pi t_m/T_m = \omega_m t_m/\cos^2\theta_\omega \qquad (25)$$

and

$$f_R/T_R = f_m/T_m = f^2 \qquad (26)$$

$$t_m^2/(f_m T_m) = t^2 \qquad (27)$$

The phase $\bar{\omega}\bar{t}$ has a period T when expressed in terms of t , and a period T_m when expressed in terms of the measured time t_m . The general solution of equation (8) is

$$\bar{\mathfrak{C}} = \bar{A}e^{i2\pi t_m/T_m} + \bar{B}e^{-i2\pi t_m/T_m} \qquad (28)$$

$$= \bar{A}e^{i2\pi t/T} + \bar{B}e^{-i2\pi t/T}$$

The conclusions for broken symmetry space and time that f_m and T_m are related by equation (20) and that $f_m \neq 1/T_m$ may possibly be experimentally verified if f_m and T_m can be independently measured for the same periodic phenomenon. Finally, the conclusions of this section are based on the observed fact that periodic physical systems have definite periods in measured time.

3. CYLINDRICAL WAVES IN ASYMMETRIC MATTER. The wave equation for cylindrical waves in matter with broken internal symmetry is written as a generalization of the standard scalar wave equation as follows[4-7]

$$\partial^2\bar{u}/\partial\bar{r}^2 + 1/\bar{r}\,\partial\bar{u}/\partial\bar{r} + 1/\bar{r}^2\,\partial^2\bar{u}/\partial\bar{\phi}^2 + \partial^2\bar{u}/\partial\bar{z}^2 = 1/\bar{v}^2\,\partial^2\bar{u}/\partial\bar{t}^2 \quad (29)$$

where $\bar{u}$ = wave amplitude with internal phase, and $\bar{v}$ = complex number phase velocity. Equation (29) can be solved by the standard technique of separation of variables[4-7]

$$\bar{u} = \bar{Z}(\bar{z})\bar{\Phi}(\bar{\phi})\bar{R}(\bar{r})\bar{\mathfrak{T}}(\bar{t}) \quad (30)$$

which gives the following simple complex number generalization of the standard equations for cylindrical waves[4-7]

$$d^2\bar{\Phi}/d\bar{\phi}^2 + \bar{M}^2\bar{\Phi} = 0 \quad (31)$$

$$d^2\bar{Z}/d\bar{z}^2 + \bar{k}_z^2\bar{Z} = 0 \quad (32)$$

$$\bar{r}^2\,d^2\bar{R}/d\bar{r}^2 + \bar{r}\,d\bar{R}/d\bar{r} + (\bar{k}_r^2\bar{r}^2 - \bar{M}^2)\bar{R} = 0 \quad (33)$$

$$d^2\bar{\mathfrak{T}}/d\bar{t}^2 + \bar{\omega}^2\bar{\mathfrak{T}} = 0 \quad (34)$$

where $\bar{M}$ = constant, and $\bar{k}_z$ and $\bar{k}_r$ are constants that are related by[4-7]

$$\bar{k}_z^2 = \bar{k}^2 - \bar{k}_r^2 \quad (35)$$

where $\bar{k}$ is defined by

$$\bar{k} = \bar{\omega}/\bar{v} \quad (36)$$

Using

$$\bar{v} = ve^{j\theta_v} \qquad \bar{k} = ke^{j\theta_k} \quad (37)$$

gives

$$k = \omega/v \qquad \theta_k = \theta_\omega - \theta_v \quad (38)$$

which determines k and θ_k . Writing $\bar{k}_z$ and $\bar{k}_r$ as

$$\bar{k}_z = k_z e^{j\theta kz} = k_{zR} + jk_{zI} \quad (39)$$

$$\bar{k}_r = k_r e^{j\theta kr} = k_{rR} + jk_{rI} \quad (40)$$

where $k_{zR} = k_z \cos\theta_{kz}$, $k_{zI} = k_z \sin\theta_{kz}$, $k_{rR} = k_r \cos\theta_{kr}$ and $k_{rI} = k_r \sin\theta_{kr}$, allows equation (35) to be written as

$$k_z^2 \cos(2\theta_{kz}) = k^2 \cos(2\theta_k) - k_r^2 \cos(2\theta_{kr}) \tag{41}$$

$$k_z^2 \sin(2\theta_{kz}) = k^2 \sin(2\theta_k) - k_r^2 \sin(2\theta_{kr}) \tag{42}$$

Corresponding to the phase velocity given by equation (36) the complex number group velocity is given by

$$\bar{v}_g = v_g e^{j\theta_{vg}} = d\bar{\omega}/d\bar{k} \tag{43}$$

Therefore

$$v_g = \cos\beta_{k,k} \left[(d\omega/dk)^2 + (\omega\, d\theta_\omega/dk)^2\right]^{1/2} \tag{44}$$

$$\theta_{vg} = \theta_\omega + \beta_{\omega,\omega} - \theta_k - \beta_{k,k} = \theta_v + \beta_{\omega,\omega} - \beta_{k,k} \tag{45}$$

$$\tan\beta_{\omega,\omega} = \frac{\omega\, d\theta_\omega/dk}{d\omega/dk} = \omega\, d\theta_\omega/d\omega \tag{46}$$

$$\tan\beta_{k,k} = k\, d\theta_k/dk \tag{47}$$

The solution to equations (31) through (33) will now be considered.

A. Solution of $\bar{\Phi}$ Equation.

Consider now the solution of equation (31) and the determination of the complex number constant $\bar{M}$. The solution of equation (31) can be written as

$$\bar{\Phi} = \bar{A}e^{i\bar{M}\bar{\phi}} + \bar{B}e^{-i\bar{M}\bar{\phi}} \tag{48}$$

It will now be shown that $\bar{M}\bar{\phi}$ must be a real number if Φ is to be a periodic function of the measured azimuthal angle $\phi_m = \phi \cos\theta_\phi$. Writing the complex number $\bar{M}$ as

$$\bar{M} = Me^{j\theta_M} = M_R + jM_I \tag{49}$$

allows the phase $\bar{M}\bar{\phi}$ in equation (48) to be written as

$$\bar{M}\bar{\phi} = M_R\phi_R - M_I\phi_I + j(M_I\phi_R + M_R\phi_I) \tag{50}$$

The reality of $\bar{M}\bar{\phi}$ gives

$$M_I\phi_R + M_R\phi_I = 0 \tag{51}$$

and

$$\bar{M}\bar{\phi} = M\phi = \phi_R M^2/M_R = \phi_m M^2/M_R \tag{52}$$

For a periodicity of the form $\bar{\Phi}(\phi_R) = \bar{\Phi}(\phi_R + 2\pi)$ it is required that

$$M^2/M_R = m \tag{53}$$

where m = positive integer. From equation (49) it follows that

$$M_R = M \cos \theta_M = M \cos \theta_\phi \tag{54}$$

because $\theta_M = -\theta_\phi$ from equation (52). Combining equations (53) and (54) gives

$$M = m \cos \theta_\phi \tag{55A}$$

$$M_R = m \cos^2 \theta_\phi \tag{55B}$$

$$M_I = - m \cos \theta_\phi \sin \theta_\phi \tag{55C}$$

as the condition for the function $\bar{\Phi}$ to be periodic in ϕ_R with period 2π. Therefore M is not an integer and equations (48) and (52) show that the wave amplitude is not periodic in the variable ϕ. Equation (45) can be written as

$$\bar{\Phi} = \bar{A}e^{im\phi_R} + \bar{B}e^{-im\phi_R} \tag{56}$$

which is periodic in ϕ_R. Note that $M\phi = m\phi_R$, and that $\bar{M}$ is a complex number in equation (31). Traditionally equation (31) accepts only integer values of the separation constant, but for waves in asymmetric space and time the separation constant is the complex number $\bar{M}$. The reality condition on the phase $\bar{M}\bar{\phi}$ is

$$\theta_M = - \theta_\phi \tag{57}$$

which is the equation for evaluating the phase angle θ_M. The internal phase angle of the magnetic quantum number must adjust itself in such a way that equation (57) is valid for periodic waves.

B. Solution of the $\bar{Z}$ Equation.

Equation (32) has the following formal solution

$$\bar{Z} = \bar{C}e^{i\bar{k}_z\bar{z}} + \bar{D}e^{-i\bar{k}_z\bar{z}} \tag{58}$$

The exponent term in equation (58) can be written as

$$\bar{k}_z\bar{z} = k_{zR}z_R - k_{zI}z_I + j(k_{zI}z_R + k_{zR}z_I) \tag{59}$$

and the reality requirement for the exponent term in internal space gives

$$\bar{k}_z\bar{z} = k_z z = z_R k_z^2/k_{zR} = z_R k_z/\cos \theta_{kz} \tag{60}$$

and which also gives $\theta_{kz} = - \theta_z$. If the waves propagate in the z direction with a measured spatial wavelength $L_{zm} = L_{zR} = L_z \cos\theta_z$ then equation (60) gives

$$L_{zR} k_z / \cos\theta_{kz} = 2\pi \tag{61}$$

for periodicity with wavelength L_{zR}. Therefore

$$k_z = 2\pi/L_{zR} \cos\theta_{kz} = 2\pi/L_z \tag{62}$$

$$k_{zR} = 2\pi/L_{zR} \cos^2\theta_{kz} = 2\pi/L_z \cos\theta_{kz} = k_z \cos\theta_{kz} \tag{63}$$

The reality condition on $\bar{k}_z \bar{z}$ shows that

$$\theta_{kz} = -\theta_z = -\theta_{Lz} \tag{64}$$

Equation (63) shows that $k_{zR} \neq 2\pi/L_{zR}$.

The solution given in equation (58) can be rewritten as

$$\bar{Z} = \bar{C}e^{ik_z z} + \bar{D}e^{-ik_z z} \tag{65}$$

$$= \bar{C}e^{i2\pi z_R/L_{zR}} + \bar{D}e^{-i2\pi z_R/L_{zR}}$$

The internal phase angle of the wave number must adjust itself to the local broken symmetry of spacetime such that equation (64) is satisfied for periodic waves. Equations (58) or (65) are the general solutions for plane waves in the z direction and equation (64) holds for $-\infty < z < \infty$. It is possible that $\bar{k}_z$ is an imaginary number in real space so that $\bar{k}_z = i\bar{\kappa}_z$, then the solution to equation (32) is attenuating in nature and given by

$$\bar{Z} = \bar{C}e^{\bar{\kappa}_z \bar{z}} + \bar{D}e^{-\bar{\kappa}_z \bar{z}} \tag{66}$$

and apparently $\bar{\kappa}_z \bar{z}$ need not be a real number in internal space because there is no periodicity requirement in the z direction for this case.

C. Solution of the $\bar{R}$ Equation.

The radial equation (33) is similar to the standard radial equation of vibration theory except that $\bar{R}$, $\bar{r}$, $\bar{k}_r$ and $\bar{M}$ are complex numbers. The formal solution to equation (33) can be written as a generalization of the standard result for scalar coordinates as follows[4,7]

$$\bar{R} = \bar{A}J_{\bar{M}}(\bar{k}_r \bar{r}) + \bar{B}N_{\bar{M}}(\bar{k}_r \bar{r}) \tag{67}$$

which represent standing waves, where $J_{\bar{M}}$ = complex number Bessel function and $N_{\bar{M}}$ = Neumann function of complex order $\bar{M}$. The progressive wave solutions to equation (33) are[4,7]

$$\bar{R} = \bar{A}\bar{H}^{(1)}_{\bar{M}}(\bar{k}_r \bar{r}) + \bar{B}\bar{H}^{(2)}_{\bar{M}}(\bar{k}_r \bar{r}) \tag{68}$$

where $\bar{H}^{(1)}_{\bar{M}}$ and $\bar{H}^{(2)}_{\bar{M}}$ = complex number Hankel functions of the first and second kind of order $\bar{M}$. The asymptotic values of the Hankel functions are given by the following generalization of the standard results[4,7]

$$\bar{H}^{(1)}_{\bar{M}}(\bar{k}_r\bar{r}) \sim [2/(\pi\bar{k}_r\bar{r})]^{1/2}\ e^{i(\bar{k}_r\bar{r}-\bar{M}\pi/2-\pi/4)} \tag{69}$$

$$\bar{H}^{(2)}_{\bar{M}}(\bar{k}_r\bar{r}) \sim [2/(\pi\bar{k}_r\bar{r})]^{1/2}\ e^{-i(\bar{k}_r\bar{r}-\bar{M}\pi/2-\pi/4)} \tag{70}$$

If equations (69) and (70) represent in-going and out-going waves which have a periodicity in the measured radial coordinate $r_R = r\cos\theta_r$ it follows that the phase $\bar{k}_r\bar{r}$ must be a real number in the far field, with $r \to \infty$, so that

$$\bar{k}_r\bar{r} = k_r r = r_R k_r^2/k_{rR} = r_R k_r/\cos\theta_{kr} \tag{71}$$

$$\theta_{kr} = -\theta_r(r = \infty) \tag{72}$$

where $k_{rR} = k_r \cos\theta_{kr}$.

Let the waves in the far field propagate in the r direction with a measured spatial wavelength $L_{rR} = L_r\cos\theta_r$ where L_r = intrinsic spatial wavelength in the r direction. Then from equation (71)

$$L_{rR}k_r/\cos\theta_{kr} = 2\pi \tag{73}$$

$$k_r = 2\pi/L_{rR}\cos\theta_{kr} = 2\pi/L_r \tag{74}$$

$$k_{rR} = 2\pi/L_{rR}\cos^2\theta_{kr} = 2\pi/L_r\cos\theta_{kr} = k_r\cos\theta_{kr} \tag{75}$$

so that $k_{rR} \neq 2\pi/L_{rR}$. Equations (71) through (75) hold only in the far field because only in this region is the concept of the wavelengths L_r and L_{rR} defined. The presence of periodic waves requires the broken symmetry of the wavelength to adjust itself so as to satisfy equation (72) at large distances from the source of the waves. In the far field of asymmetric waves equations (69) and (70) can be rewritten as

$$\bar{H}^{(1)}_{\bar{M}}(k_r r) \sim [2/(\pi k_r r)]^{1/2}\ e^{ik_r r}\ e^{-i\pi/2(\bar{M}+\frac{1}{2})} \tag{76}$$

$$\bar{H}^{(2)}_{\bar{M}}(k_r r) \sim [2/(\pi k_r r)]^{1/2}\ e^{-ik_r r}\ e^{i\pi/2(\bar{M}+\frac{1}{2})} \tag{77}$$

These solutions represent progressive waves.

As a special case consider the solution of standing waves in a vibrating membrane located in space and time with broken internal symmetries. The solution of the wave equation for this case is an obvious generalization of the standard results for scalar quantities[6]

$$\bar{u} = \bar{C}J_{\bar{M}}(\bar{k}_r\bar{r})\cos(\bar{M}\bar{\phi} + \bar{\alpha})\ e^{i\bar{\omega}\bar{t}} \tag{78}$$

$$= \bar{C}J_{\bar{M}}(\bar{k}_r\bar{r})\cos(m\phi_m + \bar{\alpha})\ e^{i\omega_m t_m/\cos^2\theta_\omega} \tag{79}$$

where for a membrane $k_r = k$. The small argument expansion of the Bessel function of order $\bar{M}$ is given by the following generalization of the standard scalar results

$$\bar{J}_{\bar{M}}(\bar{k}_r\bar{r}) = (\bar{k}_r\bar{r})^{\bar{M}}\{1 - \bar{k}_r^2\bar{r}^2/[4(\bar{M} + 1)] + \bar{k}_r^4\bar{r}^4/[32(\bar{M} + 1)(\bar{M} + 2)] - \cdots\} \tag{80}$$

Using equations (49) and (80) gives

$$\bar{J}_{\bar{M}} = (k_r r)^x e^{-y+jw}\{1 - \bar{k}_r^2\bar{r}^2/[4(\bar{M} + 1)] + \bar{k}_r^4\bar{r}^4/[32(\bar{M} + 1)(\bar{M} + 2)] - \cdots\} \tag{81}$$

where

$$x = M \cos \theta_M = m \cos^2 \theta_M \tag{82}$$

$$y = M(\theta_{kr} + \theta_r)\sin \theta_M \tag{83}$$

$$= m(\theta_{kr} + \theta_r)\sin \theta_M \cos \theta_M$$

$$w = m(\theta_{kr} + \theta_r)\cos^2 \theta_M + m\ \ell n(k_r r) \cos \theta_M \sin \theta_M \tag{84}$$

The case $\bar{M} = 0$ gives

$$\bar{J}_o = 1 - 1/4\bar{k}_r^2\bar{r}^2 + 1/64\bar{k}_r^4\bar{r}^4 - \cdots \tag{85}$$

In equation (78) and (79) $\bar{k}_r\bar{r}$ is not a real number because the vibrations are not periodic in the radial direction.

4. SPHERICAL WAVES IN ASYMMETRIC MATTER. A simple generalization of the standard wave equation for spherical waves gives the following equation that describes spherical waves in space and time with broken internal symmetries[4-7]

$$\partial^2\bar{u}/\partial\bar{r}^2 + 2/\bar{r}\,\partial\bar{u}/\partial\bar{r} + 1/(\bar{r}^2 \sin \bar{\psi})\,\partial/\partial\bar{\psi}(\sin \bar{\psi}\,\partial\bar{u}/\partial\bar{\psi}) \tag{86}$$

$$+ 1/(\bar{r}^2 \sin^2 \bar{\psi})\partial^2\bar{u}/\partial\bar{\phi}^2 = 1/\bar{v}^2\,\partial^2\bar{u}/\partial\bar{t}^2$$

Separating the complex number wave amplitude as

$$\bar{u} = \bar{R}(\bar{r})\bar{W}(\bar{\psi})\bar{\Phi}(\bar{\phi})\bar{\mathfrak{C}}(\bar{t}) \tag{87}$$

gives

$$(1 - \bar{\mu}^2)d^2\bar{W}/d\bar{\mu}^2 - 2\bar{\mu}\,d\bar{W}/d\bar{\mu} + [\bar{\mathcal{L}}(\bar{\mathcal{L}} + 1) - \bar{M}^2/(1 - \bar{\mu}^2)]\bar{W} = 0 \tag{88}$$

$$\bar{r}^2\,d^2\bar{R}/d\bar{r}^2 + 2\bar{r}\,d\bar{R}/d\bar{r} + [\bar{k}^2\bar{r}^2 - \bar{\mathcal{L}}(\bar{\mathcal{L}} + 1)]\bar{R} = 0 \tag{89}$$

$$d^2\bar{\Phi}/d\bar{\phi}^2 + \bar{M}^2\bar{\Phi} = 0 \tag{90}$$

$$d^2\bar{\mathfrak{C}}/d\bar{t}^2 + \bar{\omega}^2\bar{\mathfrak{C}} = 0 \tag{91}$$

where $\bar{\mu} = \cos \bar{\psi}$ and $\bar{k} = \bar{\omega}/\bar{v}$.

A. Solution of the $\bar{W}$ Equation.

The solution of equation (88) can be obtained as a complex number generalization of the associated Legendre polynomials.[7] The complex number associated Legendre polynomials can be obtained from equation (88) by writing[7]

$$\bar{W} = (1 - \bar{\mu}^2)^{\bar{M}/2}\bar{f}(\bar{\mathcal{L}},\bar{M},\bar{\mu}) \tag{92}$$

where

$$\bar{f} = \sum_{s=o}^{\infty} \bar{c}_s\bar{\mu}^s \tag{93}$$

by direct substitution one finds[4-7]

$$\bar{c}_2 = 1/2[\bar{M}(\bar{M} + 1) - \bar{\mathcal{L}}(\bar{\mathcal{L}} + 1)]\bar{c}_o \tag{94}$$

$$\bar{c}_3 = 1/6[(\bar{M} + 1)(\bar{M} + 2) - \bar{\mathcal{L}}(\bar{\mathcal{L}} + 1)]\bar{c}_1 \tag{95}$$

$$\bar{c}_4 = 1/12[(\bar{M} + 2)(\bar{M} + 3) - \bar{\mathcal{L}}(\bar{\mathcal{L}} + 1)]\bar{c}_2 \tag{96}$$

$$\bar{c}_5 = 1/20[(\bar{M} + 3)(\bar{M} + 4) - \bar{\mathcal{L}}(\bar{\mathcal{L}} + 1)]\bar{c}_3 \tag{97}$$

The following is the complex number generalization of the standard scalar results[4-7]

$$\bar{c}_{\nu+2}/\bar{c}_{\nu} = [(\nu + \bar{M})(\nu + \bar{M} + 1) - \bar{\mathcal{L}}(\bar{\mathcal{L}} + 1)]/[(\nu + 1)(\nu + 2)] \tag{98}$$

where ν = integer. Break off polynomial solutions can exist even when $\bar{M}$ and $\bar{\mathcal{L}}$ are complex numbers provided that they are related by

$$\bar{\mathcal{L}} = \bar{M} + \nu \tag{99}$$

where the integer ν must have the value

$$\nu = \ell - m \tag{100}$$

where ℓ and m = integer separation constants. Combining equations (99) and (100) gives

$$\bar{\mathcal{L}} = \bar{M} + \ell - m \tag{101}$$

Equation (99) reduces to equation (100) for symmetric spacetime.

From equation (101) it follows that

$$\mathcal{L} \cos \theta_{\mathcal{L}} = M \cos \theta_M + \ell - m \tag{102}$$

$$\mathcal{L} \sin \theta_{\mathcal{L}} = M \sin \theta_M \tag{103}$$

From equations (102) and (103) it follows that

$$\tan \theta_{\mathcal{L}} = (M \sin \theta_M)/(M \cos \theta_M + \ell - m) \tag{104}$$

$$\mathcal{L} = \ell[1 - m/\ell(2 - m/\ell)\sin^2 \theta_M]^{1/2} \tag{105}$$

where from the analysis of the Φ equation given in Section 3 it follows that

$$M = m \cos \theta_M \tag{106}$$

$$\theta_M = - \theta_\phi \tag{107}$$

The complex number associated Legendre polynomial solutions can then be written as

$$\bar{W} = (1 - \bar{\mu}^2)^{\bar{M}/2} \qquad \bar{\mathcal{L}} = \bar{M} \tag{108}$$

$$\bar{W} = (1 - \bar{\mu}^2)^{\bar{M}/2}\, \bar{\mu} \qquad \bar{\mathcal{L}} = \bar{M} + 1 \tag{109}$$

$$\bar{W} = (1 - \bar{\mu}^2)^{\bar{M}/2}[(2\bar{M} + 3)\bar{\mu}^2 - 1] \qquad \bar{\mathcal{L}} = \bar{M} + 2 \tag{110}$$

$$\bar{W} = (1 - \bar{\mu}^2)^{\bar{M}/2}[(2\bar{M} + 5)\bar{\mu}^3 - 3\bar{\mu}] \qquad \bar{\mathcal{L}} = \bar{M} + 3 \tag{111}$$

Note that formally $\bar{W}$ is given by the following associated Legendre polynomials

$$\bar{W} = \bar{P}_{\bar{\mathcal{L}}}^{\bar{M}}(\bar{\mu}) \tag{112}$$

B. Solution of the $\bar{R}$ Equation

The solution of the complex number radial equation (89) can be obtained by formal analogy to the solution of the real number version of equation (89) and the result for standing waves in asymmetric matter is[4-7]

$$\bar{R} = \bar{A}\bar{j}_{\bar{\mathcal{L}}}(\bar{k}\bar{r}) + \bar{B}\bar{n}_{\bar{\mathcal{L}}}(\bar{k}\bar{r}) \tag{113}$$

while for progressive waves is asymmetric matter[4-7]

$$\bar{R} = \bar{A}\bar{h}_{\bar{\mathcal{L}}}^{(1)}(\bar{k}\bar{r}) + \bar{B}\bar{h}_{\bar{\mathcal{L}}}^{(2)}(\bar{k}\bar{r}) \tag{114}$$

where $\bar{j}_{\bar{\mathcal{L}}}(\bar{k}\bar{r})$ = complex number spherical Bessel function of order $\bar{\mathcal{L}}$, $\bar{n}_{\bar{\mathcal{L}}}(\bar{k}\bar{r})$ = complex number spherical Neumann function of order $\bar{\mathcal{L}}$, $\bar{h}_{\bar{\mathcal{L}}}^{(1)}(\bar{k}\bar{r})$ = complex number spherical Hankel function of first kind of order $\bar{\mathcal{L}}$ and $\bar{h}_{\bar{\mathcal{L}}}^{(2)}(\bar{k}\bar{r})$ = complex number spherical Hankel function of the second kind of order $\bar{\mathcal{L}}$. These functions are defined as generalizations of the corresponding real valued functions as follows[4-7]

$$\bar{j}_{\bar{L}}(\bar{k}\bar{r}) = [\pi/(2\bar{k}\bar{r})]^{1/2}\ \bar{J}_{\bar{L}+\frac{1}{2}}(\bar{k}\bar{r}) \tag{115}$$

$$\bar{n}_{\bar{L}}(\bar{k}\bar{r}) = [\pi/(2\bar{k}\bar{r})]^{1/2}\ \bar{N}_{\bar{L}+\frac{1}{2}}(\bar{k}\bar{r}) \tag{116}$$

$$\bar{h}^{(1)}_{\bar{L}}(\bar{k}\bar{r}) = [\pi/(2\bar{k}\bar{r})]^{1/2}[\bar{J}_{\bar{L}+\frac{1}{2}}(\bar{k}\bar{r}) + i\bar{N}_{\bar{L}+\frac{1}{2}}(\bar{k}\bar{r})] \tag{117}$$

$$\bar{h}^{(2)}_{\bar{L}}(\bar{k}\bar{r}) = [\pi/(2\bar{k}\bar{r})]^{1/2}[\bar{J}_{\bar{L}+\frac{1}{2}}(\bar{k}\bar{r}) - i\bar{N}_{\bar{L}+\frac{1}{2}}(\bar{k}\bar{r})] \tag{118}$$

The asymptotic expansions derived from equations (115) through (118) are[4-7]

$$\bar{j}_{\bar{L}}(\bar{k}\bar{r}) \to 1/(\bar{k}\bar{r})\sin(\bar{k}\bar{r} - \bar{L}\pi/2) \tag{119}$$

$$\bar{n}_{\bar{L}}(\bar{k}\bar{r}) \to 1/(\bar{k}\bar{r})\cos(\bar{k}\bar{r} - \bar{L}\pi/2) \tag{120}$$

$$\bar{h}^{(1)}_{\bar{L}}(\bar{k}\bar{r}) \to 1/(\bar{k}\bar{r})e^{i[\bar{k}\bar{r}-(\bar{L}+1)\pi/2]} \tag{121}$$

$$\bar{h}^{(2)}_{\bar{L}}(\bar{k}\bar{r}) \to 1/(\bar{k}\bar{r})e^{-i[\bar{k}\bar{r}-(\bar{L}+1)\pi/2]} \tag{122}$$

In order for equations (119) through (122) to describe periodic waves in the far field the following conditions must hold

$$\left.\begin{array}{l} \bar{k}\bar{r} = kr \\ \theta_k + \theta_r = 0 \end{array}\right\} \quad r \to \infty \tag{123}$$

Thus for instance the replacement $\bar{k}\bar{r} = kr$ can be made in the right hand sides of equations (119) through (122). Therefore the internal phase angle of spherical waves in the far field must adjust itself to the local broken symmetry or space such that

$$\theta_k = -\theta_r(r = \infty) \tag{124}$$

5. CONCLUSION. Waves propagating in matter or spacetime with broken symmetries will have complex number separation constants $\bar{M}$ and $\bar{L}$. For spherical waves the separation constants must be related by $\bar{L} - \bar{M}$ = integer. The observed periodicities of the waves in measured time and measured space requires complex number separation constants, wave numbers, frequencies and coordinates. However the quantities $\bar{\omega}\bar{t}$, $\bar{M}\phi$ and $\bar{k}\bar{r}$ must be real numbers for periodic waves. Applications to seismic and acoustic waves are possible because the earth's gravity induces a broken symmetry in the coordinates.

ACKNOWLEDGEMENT

The author wishes to thank Elizabeth K. Klein for typing this paper.

REFERENCES

1. Weiss, R. A., Relativistic Thermodynamics, Vols. 1 and 2, Exposition Press, New York, 1976.

2. Weiss, R. A., "Thermodynamic Gauge Theory of Solids and Quantum Liquids with Internal Phase," Fifth Army Conference on Applied Mathematics and Computing, West Point, New York, ARO 88-1, June 15-18, 1987, p. 649.

3. Weiss, R. A., "The Broken Symmetry of Space and Time in Bulk Matter and the Vacuum," Sixth Army Conference on Applied Mathematics and Computing, Univ. of Colorado, Boulder, ARO 89-1, 31 May-3 June, 1988, p. 317.

4. Morse, P. M. and Feshbach, H., Methods of Theoretical Physics, Vols. 1 and 2, McGraw-Hill, New York, 1953.

5. Morse, P. M., Vibration and Sound, McGraw-Hill, New York, 1948.

6. Lindsay, R. B., Mechanical Radiation, McGraw-Hill, New York, 1960.

7. Skudrzyk, E., The Foundations of Acoustics, Springer-Verlag, New York, 1971.

M

ULTRAFAST THERMODYNAMIC PROCESSES

Richard A. Weiss
U. S. Army Engineer Waterways Experiment Station
Vicksburg, Mississippi 39180

ABSTRACT. The conventional thermodynamic description of a rapid reversible process assumes that the process is adiabatic and no heat is exchanged between the thermodynamic system and the environment so that the entropy of the system remains constant. This paper suggests the possibility of processes that occur so fast that the magnitudes of both the entropy and internal energy of the system remain constant. For such a system there is an exchange of heat with the environment in the form of a change of the internal phases of the thermodynamic system. The thermodynamic equations for internal phase changing processes are developed, and a general procedure is developed for relating the temperature and density for a system undergoing an ultrafast process. The magnitude and internal phase angle of the pressure associated with an ultrafast thermodynamic process are calculated. The rapid processes that occur in supernovae may possibly be described by these calculations. Applications to the early stages of chemical reactions are suggested.

1. INTRODUCTION. Processes that occur very fast appear in both astrophysical and laboratory situations. For example, rapid nuclear processes occur in stars before and during supernova explosions.[1-5] These include electron capture by protons and the rapid capture of neutrons by atomic nuclei. In addition there are the processes associated with the core bounce and the subsequent generation of shock waves. Finally, associated with stellar core collapse is the generation of neutrinos which interact with the stellar atmosphere and often produce pressures that are sufficient to blow off the atmosphere.[1-5] These processes occur on very short time scales and the question of the adequacy of the adiabatic assumption of ordinary thermodynamics arises because the adiabatic process requires the internal energy to change and this may occur on a slower time scale than the short time scale of the physical process itself.

The description of the interaction of gravity waves with matter, as in the case of a laboratory gravitational wave detector, needs to account for the rapid distortion of atoms and molecules due to the rapid change of the curvature of spacetime.[6,7] A description of these ultrafast gravity wave interactions requires a description of a state equation for matter which includes parameters that determine the effects of gravity waves on the atomic structure of matter. Such a state equation has been developed for the real gases.[8] Again the question arises as to whether the adiabatic assumption is a valid description of the interaction of gravity waves with matter or whether something more sophisticated is required to describe this extremely fast process.

Rapid processes also occur in more conventional laboratory experiments. Consider the actual processes that occur during chemical reactions such as chemical bond breaking and formation which may occur on the femtosecond time scale.[9-11] Another example of an ultrafast process that may require reinterpretation is the case of subpicosecond laser pulses interacting with matter.[12-14] A better understanding of the state equations of matter and of ultrafast thermodynamic processes are needed to describe these physical processes.

A theory of relativistic thermodynamic state equations has been developed in order to account for a difficulty with the state equation of matter at high densities, namely the fact that the state equation is not nearly as stiff as is predicted by conventional calculations.[15] The four dimensional Minkowski space-time of special relativity was introduced through the development of a relativistic trace equation, and specific solutions of this equation were developed for solids, quantum liquids and the real classical gases.[15,16] In order to have the Lie group $e^{\pm j\phi}$ (and $e^{\pm\phi}$) as the gauge group of relativistic thermodynamics, the concept of thermodynamic variables with internal phase angles was introduced.[17,18]

The trace equation for completely symmetric matter is given by[15]

$$U + T(dU/dT)_{PV} - 3V\, d/dV(PV)_{U} = U^{a} + T(dU^{a}/dT)_{P^{a}V} \qquad (1)$$

where U and P = relativistic internal energy and pressure respectively, U^a and P^a = unrenormalized energy and pressure respectively, and T and V = temperature and volume respectively. The trace equation for matter whose thermodynamic functions have broken symmetries is given by

$$\bar{U} + T(d\bar{U}/dT)_{\bar{P}V} - 3V\, d/dV(\bar{P}V)_{\bar{U}} = U^{a} + T(dU^{a}/dT)_{P^{a}V} \qquad (2)$$

where $\bar{U}$ and $\bar{P}$ = complex number representations of the renormalized internal energy and pressure respectively. Equation (2) can be further simplified by using the following form of the Gibbs-Helmholtz-Maxwell equation[18]

$$\partial\bar{U}/\partial V = T(\partial\bar{P}/\partial T)_{V} - \bar{P} \qquad (3)$$

The complex numbers U and P that appear in equations (2) and (3) are written as[18]

$$\bar{U} = Ue^{j\theta_U} \qquad (4)$$

$$\bar{P} = Pe^{j\theta_P} \qquad (5)$$

where U , P , θ_U and θ_P are obtained from a solution of equations (2) and (3). In a similar fashion the complex number entropy is written as[18]

$$\bar{S} = Se^{j\theta_S} \qquad (6)$$

As an illustrative example of the use of equations (1) and (2) they can be applied to real gases.

The pressure of an ordinary real gas is written as[19]

$$P^{a} = nR^{a}T(1 + B^{a}n + C^{a}n^{2} + \cdots) \qquad (7)$$

where n = reciprocal volume, and R^a , B^a and C^a = ordinary gas constant, second virial coefficient and third virial coefficient respectively. The corresponding pressure for a symmetric relativistic real gas is written as[8,15]

$$P = nRT(1 + Bn + Cn^2 + \cdots) \tag{8}$$

where R, B and C = corresponding relativistic gas constant, second virial coefficient and third virial coefficient respectively which are given by[8,15]

$$R = R^a \tag{9}$$

$$B = B^a \tag{10}$$

$$C = C^a - 3B_a^2 \, \ell n \, \psi^a \tag{11}$$

where

$$\psi^a = \frac{T}{T_R} \left| \frac{B^a(T)}{B^a(T_R)} \right|^{2/3} \tag{12}$$

where T_R = species dependent relativity temperature constant. For the case of a relativistic real gas with broken internal symmetry the pressure is written as[17]

$$\bar{P} = nRT(1 + Bn + \bar{C}n^2 + \cdots) \tag{13}$$

where $\bar{C}$ is obtained from a solution of equation (2) and is given by[17]

$$\bar{C} \sim C_a - 3B_a^2 \, \ell n \, \psi^a \, e^{j\theta_f} \tag{14}$$

where θ_f is given by the solution of a set of coupled differential equations.[17] Ultrafast thermodynamic processes, for which both the entropy and the magnitude of the internal energy are fixed, are only possible in systems like the real gases that have a parameter T_R which varies during the process. The parameter T_R depends on the species of atoms in the gas and therefore T_R changes for processes that alter the composition of the gas.

This paper considers thermodynamic processes that are sufficiently rapid to keep both the entropy and the magnitude of the internal energy constant or to keep the magnitudes of both the entropy and the internal energy fixed. Such processes change the internal phases of the entropy and internal energy, and the entropy and internal energy vectors essentially rotate (in internal space) but do not stretch. This is a special case of the general situation where the complex number thermodynamic functions rotate and stretch during thermodynamic processes.[18] A theory of ultrafast thermodynamic processes is developed and an expression for the pressure associated with these processes is derived.

2. ULTRAFAST THERMODYNAMIC PROCESSES.

This section considers the thermodynamic equations that describe ultrafast processes occuring in matter that has internal phase angles associated with the thermodynamic functions. The general thermodynamic equations of matter and radiation with internal phase have already been developed in the literature.[17,18] The expression for the first law of thermodynamics for matter with internal phase is written as[18]

$$d\bar{Q} = Td\bar{S} = d\bar{U} + \bar{P}dV + \sum_{\alpha} \bar{M}_{\alpha} d\alpha \tag{15}$$

where $\bar{Q}$ = complex number heat, $\bar{M}_{\alpha}$ = set of generalized complex number forces and α = set of generalized extensive variables.

For the special case where the change in entropy and internal energy are of the form of rotations it follows from equations (4) and (6) that[18]

$$d\bar{Q} = Td\bar{S} = jT\bar{S}d\theta_S \tag{16}$$

$$d\bar{U} = j\bar{U}d\theta_U \tag{17}$$

which combined with equation (15) gives the following result for an ultrafast process where S and U are both constant

$$jT\bar{S}d\theta_S = j\bar{U}d\theta_U + \bar{P}dV + \sum_{\alpha} \bar{M}_{\alpha} d\alpha \tag{18}$$

The generalized force $\bar{M}_{\alpha}$ can be written as

$$\bar{M}_{\alpha} = M_{\alpha} e^{j\theta_{M\alpha}} . \tag{19}$$

The real and imaginary parts of equation (18) can be written as

$$- TS \sin \theta_S \, d\theta_S = - U \sin \theta_U \, d\theta_U + P \cos \theta_P \, dV + \sum_{\alpha} M_{\alpha} \cos \theta_{M\alpha} \, d\alpha \tag{20}$$

$$+ TS \cos \theta_S \, d\theta_S = + U \cos \theta_U \, d\theta_U + P \sin \theta_P \, dV + \sum_{\alpha} M_{\alpha} \sin \theta_{M\alpha} \, d\alpha \tag{21}$$

For the special case of the relativistic real gas the generalized extensive variable is $\alpha = T_R$ and the generalized force is $\bar{M}_{\alpha} = \bar{S}_R$, where $\bar{S}_R$ is the complex number generalization of the scalar parameter S_R that appears in Reference 8. For the real gas equation (18) becomes

$$jT\bar{S}d\theta_S = j\bar{U}d\theta_U + \bar{P}dV + \bar{S}_R dT_R \tag{22}$$

where

$$\bar{S}_R = S_R e^{j\theta_{SR}} \tag{23}$$

is the following complex number generalization of the scalar result in Reference 8

$$\bar{S}_R = - 1/2NRTn^2 (\partial\bar{C}/\partial T_R)_T . \tag{24}$$

where

$$S_R = - 1/2NRTn^2 [(\partial C/\partial T_R)^2 + (C\partial\theta_C/\partial T_R)^2]^{1/2} \tag{25}$$

and

$$\tan \theta_{SR} = \frac{C\partial\theta_C/\partial T_R}{\partial C/\partial T_R} \tag{26}$$

and where N = number of moles. The real and imaginary parts of equation (22) can be written as

$$- TS \sin \theta_S \, d\theta_S = - U \sin \theta_U \, d\theta_U + P \cos \theta_P \, dV + S_R \cos \theta_{SR} \, dT_R \qquad (27)$$

$$+ TS \cos \theta_S \, d\theta_S = + U \cos \theta_U \, d\theta_U + P \sin \theta_P \, dV + S_R \sin \theta_{SR} \, dT_R \qquad (28)$$

which are the thermodynamic equations for an ultrafast process in a real gas with both S and U held constant. If the process is truly adiabatic with $\bar{S}$ = constant and dS = 0 and $d\theta_S = 0$ then the left hand sides of equations (27) and (28) must be set equal to zero. Note that in general $U = U(T,V,T_R)$, $\theta_U = \theta_U(T,V,T_R)$, $S = S(T,V,T_R)$ and $\theta_S = \theta_S(T,V,T_R)$.

In order to utilize equations (20) and (21) it is necessary to evaluate the differentials $d\theta_S$ and $d\theta_U$ for the case of an ultrafast process with both U and S held constant. These are obtained from the following total derivatives

$$(d\theta_S/d\alpha)_{U,S} = \partial\theta_S/\partial\alpha + \partial\theta_S/\partial V(dV/d\alpha)_{U,S} + \partial\theta_S/\partial T(dT/d\alpha)_{U,S} \qquad (29)$$

$$(d\theta_U/d\alpha)_{U,S} = \partial\theta_U/\partial\alpha + \partial\theta_U/\partial V(dV/d\alpha)_{U,S} + \partial\theta_U/\partial T(dT/d\alpha)_{U,S} \qquad (30)$$

where $(dV/d\alpha)_{U,S}$ and $(dT/d\alpha)_{U,S}$ are obtained from the following two conditions which state that S and U are constant

$$\partial S/\partial\alpha + \partial S/\partial V(dV/d\alpha)_{U,S} + \partial S/\partial T(dT/d\alpha)_{U,S} = 0 \qquad (31)$$

$$\partial U/\partial\alpha + \partial U/\partial V(dV/d\alpha)_{U,S} + \partial U/\partial T(dT/d\alpha)_{U,S} = 0 \qquad (32)$$

In general $S = S(\alpha,V,T)$, $\theta_S = \theta_S(\alpha,V,T)$, $U = U(\alpha,V,T)$ and $\theta_U = \theta_U(\alpha,V,T)$. From equations (31) and (32) it follows that

$$(dV/d\alpha)_{U,S} = \frac{(\partial U/\partial T)(\partial S/\partial\alpha) - (\partial S/\partial T)(\partial U/\partial\alpha)}{(\partial S/\partial T)(\partial U/\partial V) - (\partial U/\partial T)(\partial S/\partial V)} \qquad (33)$$

$$(dT/d\alpha)_{U,S} = \frac{(\partial S/\partial V)(\partial U/\partial\alpha) - (\partial U/\partial V)(\partial S/\partial\alpha)}{(\partial S/\partial T)(\partial U/\partial V) - (\partial U/\partial T)(\partial S/\partial V)} \qquad (34)$$

Eliminating $d\alpha$ from equations (33) and (34) gives

$$(dT/dV)_{U,S} = \frac{(\partial S/\partial V)(\partial U/\partial\alpha) - (\partial U/\partial V)(\partial S/\partial\alpha)}{(\partial U/\partial T)(\partial S/\partial\alpha) - (\partial S/\partial T)(\partial U/\partial\alpha)} \qquad (35)$$

Equation (35) relates T and V for the case of constant U and S . Only if $\partial U/\partial\alpha \neq 0$ and $\partial S/\partial\alpha \neq 0$ are T and V related. The derivative of the temperature with respect to the reciprocal volume at constant U and S is given by

$$n(dT/dn)_{U,S} = - V(dT/dV)_{U,S} \qquad (36)$$

If $\alpha = \alpha(V,T)$ is calculated from the condition $U(\alpha,T,V)$ = constant then the constant S condition can be written as

$$\partial S/\partial V + \partial S/\partial T(dT/dV)_{U,S} = 0 \tag{37}$$

$$\partial S/\partial T + \partial S/\partial V(dV/dT)_{U,S} = 0 \tag{38}$$

Similarly if $\alpha = \alpha(V,T)$ is eliminated by the condition $S(\alpha,V,T)$ = constant then the constant U condition can be written as

$$\partial U/\partial V + \partial U/\partial T(dT/dV)_{U,S} = 0 \tag{39}$$

$$\partial U/\partial T + \partial U/\partial V(dV/dT)_{U,S} = 0 \tag{40}$$

Neglecting the $d\alpha$ term in equations (20) and (21) gives

$$- TS \sin \theta_S \, d\theta_S \sim - U \sin \theta_U \, d\theta_U + P \cos \theta_P \, dV \tag{41}$$

$$+ TS \cos \theta_S \, d\theta_S \sim + U \cos \theta_U \, d\theta_U + P \sin \theta_P \, dV \tag{42}$$

Then

$$- P \cos \theta_P \sim TS \sin \theta_S \, (d\theta_S/dV)_{U,S} - U \sin \theta_U \, (d\theta_U/dV)_{U,S} \tag{43}$$

$$P \sin \theta_P \sim TS \cos \theta_S \, (d\theta_S/dV)_{U,S} - U \cos \theta_U \, (d\theta_U/dV)_{U,S} \tag{44}$$

Now assume that $\theta_S \sim \theta_U$ in the trigonometric terms

$$- P \cos \theta_P \sim \sin \theta_U \, [TS(d\theta_S/dV)_{U,S} - U(d\theta_U/dV)_{U,S}] \tag{45}$$

$$P \sin \theta_P \sim \cos \theta_U \, [TS(d\theta_S/dV)_{U,S} - U(d\theta_U/dV)_{U,S}] \tag{46}$$

From equations (45) and (46) it follows that

$$\tan \theta_P \sim - \tan \theta_U \tag{47}$$

$$\theta_P \sim \theta_U + \pi/2 \tag{48}$$

$$\cos \theta_P \sim - \sin \theta_U \tag{49}$$

$$\sin \theta_P \sim \cos \theta_U \tag{50}$$

Combining equations (45) through (50) gives

$$P \sim TS(d\theta_S/dV)_{U,S} - U(d\theta_U/dV)_{U,S} \tag{51}$$

where

$$(d\theta_S/dV)_{U,S} = \partial\theta_S/\partial V + \partial\theta_S/\partial T(dT/dV)_{U,S} + \partial\theta_S/\partial\alpha(d\alpha/dV)_{U,S} \quad (52)$$

$$(d\theta_U/dV)_{U,S} = \partial\theta_U/\partial V + \partial\theta_U/\partial T(dT/dV)_{U,S} + \partial\theta_U/\partial\alpha(d\alpha/dV)_{U,S} \quad (53)$$

where $(dT/dV)_{U,S}$ is given by equation (35) and $(d\alpha/dV)_{U,S}$ is given by equation (33). Equations (48) and (51) give the pressure associated with an ultrafast thermodynamic process that has both U and S held constant.

The magnitude of the pressure given by equation (51) can be written in terms of the reciprocal volume $n = 1/V$ as follows

$$P \sim Un^2(d\theta_U/dn)_{U,S} - TSn^2(d\theta_S/dn)_{U,S} \quad (54)$$

$$= En(d\theta_U/dn)_{U,S} - T\mathcal{S}n(d\theta_S/dn)_{U,S}$$

where the energy density E and the entropy density $\mathcal{S}$ are given by

$$E = U/V = nU \quad (55)$$

$$\mathcal{S} = S/V = nS \quad (56)$$

where U and S are constants in this paper. A further approximation for the pressure can be obtained by taking $\theta_S \sim \theta_U$ in equations (51) and (54) with the result

$$P \sim (TS - U)(d\theta_U/dV)_{U,S} \quad (57)$$

$$= (E - T\mathcal{S})n(d\theta_U/dn)_{U,S}$$

Equation (57) has a proper $T = 0$ limit.

Equation (54) is not an equation of state but rather gives the pressure for a thermodynamic process for which U and S are constants. Therefore $P = P(T)$ or $P = P(n)$ because V and T are related by equation (35). The total derivative dP/dn can be calculated from equation (54) as follows

$$(dP/dn)_{U,S} = U[2n(d\theta_U/dn)_{U,S} + n^2(d^2\theta_U/dn^2)_{U,S}] \quad (58)$$

$$- TS[2n(d\theta_S/dn)_{U,S} + n^2(d^2\theta_S/dn^2)_{U,S}]$$

$$- S(dT/dn)_{U,S}\, n^2(d\theta_U/dn)_{U,S}$$

Similarly

$$(dP/dT)_{U,S} = (dP/dn)_{U,S}(dn/dT)_{U,S} \quad (59)$$

where $(dn/dT)_{U,S}$ is given by equations (35) and (36).

Consider now the case of an ultrafast adiabatic process in which the entropy $\bar{S}$ and U remains fixed. For this case dS = 0 , $d\theta_S = 0$ and dU = 0 . Then equation (18) gives

$$j\bar{U}d\theta_U + \bar{P}dV + \sum_\alpha \bar{M}_\alpha d\alpha = 0 \tag{60}$$

Neglecting $d\alpha$ gives

$$\bar{P} = - U(d\theta_U/dV)_{U,\bar{S}}\ e^{j(\pi/2+\theta_U)} \tag{61}$$

$$P = - U(d\theta_U/dV)_{U,\bar{S}} \tag{62}$$

$$\theta_P = \pi/2 + \theta_U \tag{63}$$

For this case $\theta_U \neq \theta_S$ because θ_S = constant. The derivative in equation (62) is obtained from equations (33), (35) and (53). For this case two parameters α and β are required in equation (60) to evaluate the derivative in equation (62).

For the case of broken symmetry of space the first law of thermodynamics is written as

$$d\bar{Q} = d\bar{U} + \bar{P}|d\bar{V}| \tag{64}$$

where

$$|d\bar{V}| = \sec \beta_{V,V}\ dV \tag{65}$$

$$\tan \beta_{V,V} = V\partial\theta_V/\partial V \tag{66}$$

From equations (64) and (65) it follows that in order to obtain the basic equations of thermodynamics for broken symmetry space the substitution $\bar{P} \to \bar{P} \sec \beta_{V,V}$ is made in the basic thermodynamics equations such as those given in Reference 18. For instance, the trace equation (2) becomes

$$\bar{U} + T(d\bar{U}/dT)_{\bar{P}V \sec \beta_{V,V}} - 3Vd/dV(\bar{P}V \sec \beta_{V,V})_{\bar{U}} = U^a + T(dU^a/dT)_{p^aV} \tag{67}$$

while equation (3) becomes

$$\partial\bar{U}/\partial V = T\partial/\partial T(\bar{P} \sec \beta_{V,V}) - \bar{P} \sec \beta_{V,V} \tag{68}$$

$$= (T\partial\bar{P}/\partial T - \bar{P})\sec \beta_{V,V} + \bar{P}T\partial/\partial T(\sec \beta_{V,V})$$

For $\partial\theta_V/\partial T = 0$ equation (68) becomes

$$\cos \beta_{V,V}\ \partial\bar{U}/\partial V = T\partial\bar{P}/\partial T - \bar{P} \tag{69}$$

For T = 0 equation (68) or (69) gives

$$\bar{P} = -\cos\beta_{V,V}\ \partial\bar{U}/\partial V \qquad (70)$$

In general the broken symmetry of space lowers the calculated thermodynamic pressure. For example the broken internal symmetry of space requires equation (13) for the pressure of real gases with broken internal symmetry to be written as

$$\bar{P} = nRT\cos\beta_{V,V}\ (1 + Bn + \bar{C}n^2 + \cdots) \qquad (71)$$

For the case of an ultrafast thermodynamic process equation (64) becomes

$$jT\bar{S}d\theta_S = j\bar{U}d\theta_U + \bar{P}\sec\beta_{V,V}\ dV \qquad (72)$$

and equations (57) and (62) become respectively

$$P \sim \cos\beta_{V,V}\ (E - TS)n(d\theta_U/dn)_{U,S} \qquad (73)$$

$$P = -\cos\beta_{V,V}\ U(d\theta_U/dV)_{U,\bar{S}} \qquad (74)$$

For radial symmetry $\beta_{V,V} = \beta_{r,r}$ where r = radial coordinate. Because $\beta_{r,r}$ is related to the internal phase angle θ_r of the radial coordinate it follows that the laws of thermodynamics such as equations (67) through (74) depend on the internal phase structure of space. But the value of θ_r on a macroscopic scale depends on gravity. For instance for the earth's surface $\theta_r \sim -5.7°$. Therefore the calculations of thermodynamics must include the effects of gravity.

3. CONCLUSION. For systems, such as the real gas, with broken internal symmetries in the pressure, internal energy and entropy it is possible to have a thermodynamic process that occurs so fast as to keep the magnitudes of the entropy and internal energy constant. This is possible only for systems like the real gases which have a parameter (like the relativity temperature T_R) that changes during the process. Such a process involves a change of structure of the molecules or atoms of the system as in the case of chemical or nuclear reactions.

ACKNOWLEDGEMENT

The author wishes to thank Elizabeth K. Klein for typing this paper.

REFERENCES

1. Trimble, V., "Supernovae. Part I: The Events," Rev. Mod. Phys., Vol. 54, No. 4, Oct. 1982, p. 1183.

2. Fowler, W. A., "Experimental and Theoretical Astrophysics: The Quest for the Origin of the Elements," Rev. Mod. Phys., Vol. 56, No. 2, Apr. 1984, p. 149.

3. Woosley, S. E. and Phillips, M. M., "Supernova 1987A!," Science, Vol. 240, May 6, 1988, p. 750.

4. Bethe, H. A., "Nuclear Physics Needed for the Theory of Supernovae," Ann. Rev. Nucl. Part. Sci., Vol. 38, 1988, p. 1.

5. Trimble, V., "1987A: The Greatest Supernova Since Kepler," Rev. Mod. Phys., Vol. 60, No. 4, Oct. 1988, p. 859.

6. Thorne, K. S., "Gravitational Wave Research: Current Status and Future Prospects," Rev. Mod. Phys., Vol. 52, No. 2, Part 1, April 1980.

7. Press, W. H. and Thorne, K. S., "Gravitational Wave Astronomy," Ann. Rev. Astron. Astrophys., Vol. 10, 1972, p. 335.

8. Weiss, R. A., "Relativistic Wave Equations for Real Gases," Fourth Army Conference on Applied Mathematics and Computing, Cornell University, Ithaca, ARO 87-1, May 27-30, 1986, p. 341.

9. Zewail, A., "Laser Femtochemistry," Science, Vol. 242, 23 Dec. 1988, p. 1645.

10. Peters, K. S. and Snyder, G. J., "Time-Resolved Photoacoustic Calorimetry: Probing the Energetics and Dynamics of Fast Chemical and Biochemical Reactions," Science, Vol. 241, 26 Aug. 1988, p. 1053.

11. Davis, W. C., "The Detonation of Explosives," Scientific American, Vol. 256, No. 5, May 1987, p. 106.

12. Yablonovitch, E., "Energy Conservation in the Picosecond and Subpicosecond Photoelectric Effect," Phys. Rev. Lett., Vol. 60, No. 9, 29 Feb. 1988, p. 795.

13. Hicks, J. M., Urbach, L. E., Plummer, E. W. and Dai, H. L., "Can Pulsed Laser Excitation of Surfaces be Described by a Thermal Model?," Phys. Rev. Lett., Vol. 61, 28 Nov. 1988, p. 2588.

14. Murnane, M. M., Kapteyn, H. C. and Falcone, R. W., "High-Density Plasmas Produced by Ultrafast Laser Pulses," Phys. Rev. Lett., Vol. 62, 9 Jan. 1989, p. 155.

15. Weiss, R. A., <u>Relativistic Thermodynamics</u>, Vols. 1 and 2, Exposition Press, New York, 1976.

16. Weiss, R. A., "Relativistic Wave Equations for Solids and Low Temperature Quantum Systems," Third Army Conference on Applied Mathematics and Computing, Georgia Institute of Technology, ARO 86-1, May 13-16, 1985, p. 717.

17. Weiss, R. A., "Relativistic Thermodynamics of Real Gases with Broken Internal Symmetry," Sixth Army Conference on Applied Mathematics and Computing, Univ. of Colorado, Boulder, ARO 89-1, 31 May-3 June, 1988, p. 203.

18. Weiss, R. A., "Thermodynamic Gauge Theory of Solids and Quantum Liquids with Internal Phase," Fifth Army Conference on Applied Mathematics and Computing, West Point, New York ARO 88-1, June 15-18, 1987, p. 649.

19. Hirschfelder, J. O., Curtiss, C. F. and Bird, R. B., <u>Molecular Theory of Gases and Liquids</u>, John Wiley, New York, 1954.

INDEX

A

B

C

D

E

T

U

V

W

X-Y-Z